WEAR – MATERIALS, MECHANISMS AND PRACTICE

TIPS Tribology in Practice Series

Editors:
M.J. Neale, T.A. Polak and M. Priest

Guide to Wear Problems and Testing for Industry
M.J. Neale and M. Gee

Handbook of Surface Treatment and Coatings
M. Neale, T.A. Polak, and M. Priest (Eds)

Lubrication and Lubricant Selection – A Practical Guide, 3rd Edition
A.R. Lansdown

Rolling Contacts
T.A. Stolarski and S. Tobe

Total Tribology – Towards an integrated approach
I. Sherrington, B. Rowe and R. Wood (Eds)

Tribology – Lubrication, Friction and Wear
I.V. Kragelsky, V.V. Alisin, N.K. Myshkin and M.I. Petrokovets

Wear – Materials, Mechanisms and Practice
G. Stachowiak (Ed.)

WEAR – MATERIALS, MECHANISMS AND PRACTICE

Edited by

Gwidon W. Stachowiak

John Wiley & Sons, Ltd

Email (for orders and customer service enquiries): cs-books@wiley.co.uk
Visit our Home Page on www.wiley.com

Other Wiley Editorial Offices

John Wiley & Sons Inc., 111 River Street, Hoboken, NJ 07030, USA

Jossey-Bass, 989 Market Street, San Francisco, CA 94103-1741, USA

Wiley-VCH Verlag GmbH, Boschstr. 12, D-69469 Weinheim, Germany

John Wiley & Sons Australia Ltd, 42 McDougall Street, Milton, Queensland 4064, Australia

John Wiley & Sons (Asia) Pte Ltd, 2 Clementi Loop #02-01, Jin Xing Distripark, Singapore 129809

John Wiley & Sons Canada Ltd, 22 Worcester Road, Etobicoke, Ontario, Canada M9W 1L1

Wiley also publishes its books in a variety of electronic formats. Some content that appears in print may not be available in electronic books.

Library of Congress Cataloging in Publication Data

Wear – materials, mechanisms and practice / editor Gwidon W. Stachowiak.
p. cm.
Includes bibliographical references and index.
ISBN-13: 978-0-470-01628-2 (cloth : alk. paper)
ISBN-10: 0-470-01628-0 (cloth : alk. paper)
1. Mechanical wear. I. Stachowiak, G.W. (Gwidon W.)
TA418.4.W415 2005
620.1'1292—dc22 2005006833

British Library Cataloguing in Publication Data

A catalogue record for this book is available from the British Library

ISBN-13: 978-0-470-01628-2 (HB)
ISBN-10: 0-470-01628-0 (HB)

Typeset in 10/12pt Times by Integra Software Services Pvt. Ltd, Pondicherry, India
Printed and bound in Great Britain by Antony Rowe Ltd, Chippenham, Wiltshire
This book is printed on acid-free paper responsibly manufactured from sustainable forestry in which at least two trees are planted for each one used for paper production.

Contents

List of Contributors

Yves Berthier
LaMCoS UMR5514, INSA Lyon
Bat Jean D'Alembert, France

S.K. Biswas
Department of Mechanical Engineering
Indian Institute of Science
Bangalore, Karnataka, India

Brian J. Briscoe
Imperial College of Science and Technology and Medicine
Department of Chemical Engineering and Technology
London, UK

Christophe Donnet
Université Jean Monnet de Saint-Etienne
Laboratoire Traitement du Signal et Instrumentation, France

Ali Erdemir
Energy Technology Division
Argonne National Laboratory, Argonne, USA

Siegfried Fouvry
Laboratory of Tribology and Systems Dynamics
Ecole Centrale de Lyon, France

Klaus Friedrich
Universitat Kaiserslautern
Institut fur Verbundwerkstoffe GmbH, Germany

Richard S. Gates
National Institute of Standards and Technology
Gaithersburg, MD, USA

Andrew J. Gellman
Department of Chemical Engineering
Carnegie Mellon University
Pittsburg, PA, USA

Kenneth Holmberg
VTT Manufacturing Technology, Finland

Stephen Hsu
National Institute of Standards and Technology
Gaithersburg, MD, USA

Ian Hutchings
University of Cambridge
Institute for Manufacturing
Department of Engineering
Mill Lane, Cambridge, UK

S. Jahanmir
MiTi Heart Corporation
Gaithersburg, MD, USA

Philippe Kapsa
Laboratory of Tribology and Systems Dynamics
Ecole Centrale de Lyon, France

Koji Kato
Tribology Laboratory
School of Mechanical Engineering
Tohoku University, Sendai, Japan

Patrick Klein
Application Engineer Stationary Hydraulics
Busak+Shamban Deutschland GmbH Handwerkstr. 5–7
70565 Shettgart, Germany

Allan Matthews
Department of Engineering
Research Centre in Surface Engineering
University of Hull, UK

Ardian Morina
School of Mechanical Engineering
University of Leeds, UK

Ronald G. Munro
National Institute of Standards and Technology
Gaithersburg, MD, USA

Anne Neville
School of Mechanical Engineering
University of Leeds, UK

Dennis V. De Pellegrin
School of Mechanical Engineering
University of Western Australia
Crawley, Western Australia

Pawel Podsiadlo
School of Mechanical Engineering
University of Western Australia
Crawley, Western Australia

Brian J. Roylance
Tribology and Condition Monitoring Group
Department of Mechanical Engineering
University of Wales, Swansea, UK

Ming C. Shen
Zimmer, Austin, TX, USA

S.K. Sinha
Department of Mechanical Engineering
National University of Singapore, Singapore

Nicholas D. Spencer
Laboratory for Surface Science and Technology
Swiss Federal Institute of Technology
Zurich, Switzerland

Gwidon W. Stachowiak
School of Mechanical Engineering
University of Western Australia
Crawley, Western Australia

Grazyna B. Stachowiak
School of Mechanical Engineering
University of Western Australia
Crawley, Western Australia

Leo Vincent
Laboratory of Tribology and Systems Dynamics
Ecole Centrale de Lyon, France

Zhong Zhang
National Center for Nanoscience and Technology, China
No.2 1st North Road, Zhong-Guan-Cum
100080 Beijing, China

Series Editors' Foreword

Tribology is concerned with understanding the behaviour and performance of the components of machines and equipment, with surfaces that are subject to relative motion, either from other components or from loose materials. It therefore has a wide range of applications across many industries, and also in medicine in understanding the mechanism of operation of the joints between bones. The *Tribology in Practice Series* of books aims to make an understanding of tribology readily accessible and relevant to industry, so that it can be brought to bear on engineering problems.

This latest book in the series, *Wear – Mechanisms, Materials and Practice*, edited by Gwidon Stachowiak provides a comprehensive review of the current understanding of the wear of all kinds of matcrials, and how it can be controlled and reduced. The authors of the individual sections of the book are world experts in the various subject areas. They are therefore able to summarize the currently available knowledge and the ways in which it can be used to solve practical problems. The book will therefore provide a valuable reference work for engineers in industry, as well as being useful for research workers in the field by providing a summary of previous work.

As a Series, the *Tribology in Practice Series* is particularly concerned with design, failure investigation, and the application of tribological understanding to the products of various industries and to medicine. The scope of the series is as wide as the subject and applications of tribology. Wherever there is wear, rubbing, friction, or the need for lubrication, then there is scope for the introduction of practical, interpretative material. The Series Editors and the publishers would welcome suggestions and proposals for future titles in the Series.

M.J. Neale
Neale Consulting Engineers, UK
T.A. Polak
Neale Consulting Engineers, UK
M. Priest
University of Leeds, UK

Preface

Wear is the process occurring at the interfaces between interacting bodies and is usually hidden from investigators by the wearing components. However, this obstacle has been gradually overcome by scientists, revealing an intricate world of various wear modes and mechanisms. Since the early wear experiments our knowledge about wear has increased considerably and a significant progress in the description of wear mechanisms has been made. Over the past decades our views and understanding of wear have changed, including the classification of wear mechanisms. Concepts such as abrasion, adhesion and fatigue, originally used in the classification of wear mechanisms, are, now, insufficient. New materials and surface coatings wear in a specific manner. Complex reactions and transitions often take place on the wearing surfaces and our understanding of wear mechanisms occurring is critical to the effective utilization of these materials. Furthermore, if we understand how a material resists wear and friction, then it should be much easier to improve that material.

It is now clear that all known forms of friction and wear are controlled by thin films of material present between the interacting surfaces. It has been recognized since ancient times that supplying liquid or grease to a contact offers a lower friction and wear. If such a film is merely generated by wear of the bodies sliding in dry contact, the wear and friction are usually much higher. In general terms, this film formation controls wear to a large extent and is usually beneficial since it lowers friction and wear. However, there are also instances when film formation raises wear and friction.

In May 2000 Chris Taylor, the editor of the Journal of Engineering Tribology, had asked me to guest-edit a special issue of the *Journal of Engineering Tribology* on the topic of 'Wear/Lubricated Wear'. I found this to be a very good idea and agreed. The special issue was published almost two years later in 2002. The issue contained nine excellent papers covering a broad range of topics representing our state of knowledge on recent developments in the area of wear/lubricated wear. World-leading researchers in the area of wear and wear control, such as Koji Kato, Sanjay Biswas, Stephen Hsu, Ronald Munro, Ming Shen, Richard Gates, Andrew Gellman, Nic Spencer, Said Jahanmir, Ali Erdemir, Christophe Donnet, Brian Briscoe, Sujeet Sinha, Klaus Friedrich, Zhong Zhang, Patrick Klein and Gwidon Stachowiak, contributed papers on various topics to this issue. On completion of this work it became apparent that many researchers and engineers throughout the world would benefit from an expanded version in a book form. So I had presented the idea of publishing this special issue in a more expanded form to the Professional Engineering Publishers Ltd. The publisher was very supportive of the idea. In addition, the Editorial Board of the PEP book series made many valuable suggestions regarding the book content. As a result several additional experts, such as Ian Hutchings, Anne Neville, Ardian Morina, Kenneth Holmberg, Alan Matthews, Yves Berthier, Philippe Kapsa, Siegfried Fouvry, Leo Vincent and Brian Roylance, were invited to contribute chapters in specific areas of wear and wear control. Altogether, six additional chapters were invited and, as a result, a unique piece of work has emerged.

The resulting book represents the current state of art in the area of wear, wear mechanisms and materials. The chapters discuss the latest concepts in wear mechanism classification, wear of metals, wear of polymer and polymer composites, fretting wear, wear mapping of materials, friction and wear of diamond and diamond-like carbon films, wear of ceramics, concept of a third body in wear and friction problems and the tribology of engineered surfaces. Wear in boundary lubrication, effects of lubricant chemistry on wear, effects of surface chemistry in tribology, characterization and classification of particles and surfaces, and machine failure and its avoidance are also discussed. The strength of this book is in its current knowledge of topic and its frequent references to engineering practice. However, this book is not limited to presenting what is already known about wear. It also attempts to present the myriad of new emerging problems and the possible ways of solving them. It shows us that, although we already know a lot about wear, there are still some aspects of it to be yet uncovered and thoroughly investigated. It shows us that new ways and approaches to wear control are still being discovered and implemented in practice. The book also demonstrates what type of new problems we are most likely to be dealing with in the future.

I am very grateful to the authors for sharing with us their knowledge and for their hard work. In particular, I appreciate the time they dedicated to the meticulous preparation of their manuscripts. After all, it is not easy to put an extra task on top of the many other duties and commitments one already has. I am sure this book will provide a valuable reference for people with interest in wear and wear control.

Gwidon Stachowiak

1

The Challenge of Wear

I.M. Hutchings

Abstract

While accurate predictive models for wear rate are still an elusive goal, it is clear that significant recent progress has been made in our understanding of many aspects of wear mechanisms and that advances in materials, surface engineering and lubricants, as well as in design methods and condition monitoring, have led to major improvements in the efficiency, lifetime cost and performance of many engineering systems. There is still much potential for future development, and challenges in tribology, especially in the vital field of wear, remain.

1.1 Introduction

To understand the degradation processes known as wear, to predict the rate of wear and to reduce it still form some of the most problematic challenges facing the engineer. The understanding of wear often involves a detailed knowledge of mechanics, physics, chemistry and material science, while its quantitative prediction, even to within an order of magnitude, remains in many cases a distant goal. Although wear can often be reduced by lubrication, the extent of that reduction can almost never be predicted accurately. The following chapters focus on particular aspects of wear and review the current state of our knowledge on this vitally important topic.

1.2 Definitions and Development of Wear Studies

The widest definition of wear, which has been recognized for at least 50 years, includes the loss of material from a surface, transfer of material from one surface to another or movement of material within a single surface [1]. Although a narrower definition of wear has been proposed as 'progressive loss of substance from the operating surface of a body occurring as a result of relative motion at the surface' [2], the wide range of engineering applications of concern to

Wear – Materials, Mechanisms and Practice I.M. Hutchings

the tribologist is served better by a broader definition. A simple and useful statement is that wear is 'damage to a solid surface, generally involving progressive loss of material, due to relative motion between that surface and a contacting substance or substances' [3]. This includes (1) degradation by the displacement of material within the surface (leading to changes in surface topography without loss of material), as well as the more usual case of material removal; (2) the wear processes common in machines in which one surface slides or rolls against another, either with or without the presence of a deliberately applied lubricant; and (3) the more specialized types of wear which occur when the surface is abraded by hard particles moving across it, or is eroded by solid particles or liquid drops striking it or by the collapse of cavitation bubbles in a liquid. This definition, quite deliberately, tells us nothing about the mechanisms by which the degradation takes place. These may be purely mechanical, for example involving plastic deformation or brittle fracture, or they may involve significant chemical aspects, for example oxidation of a metal or hydration of a ceramic; in many practical cases, both chemical and mechanical processes play a role [4].

The study of tribology has a long history, extending for several centuries before the word itself was coined in 1965. Early studies of friction were performed by Leonardo da Vinci in the late sixteenth century, and the first quantitative understanding of fluid film lubrication originated with Beauchamp Tower in the late nineteenth century. Wear has entered the scientific arena rather more recently. The design and construction of early machines involved large clearances and rather slow speeds of operation, with the result that, provided gross adhesion or excessive friction could be avoided, changes in dimensions of sliding parts due to wear could often be tolerated with little adverse effect on performance. It was the development of the high-speed internal combustion engine in the early part of the twentieth century that provided the initial driving force for the study of wear which has grown in importance to the present day. Our understanding of wear mechanisms has developed most rapidly with the widespread use of electron microscopy and instrumental methods of microanalysis over the past 30 years. There are now many examples of advanced engineering products, some involving high-speed sliding or rolling and others small dimensions or hostile environments, whose development and successful use are possible only through the understanding and successful limitation of wear processes. These include gas turbine engines, artificial human joints, automotive engines and transmissions, tyres and brakes, hard disk drives for data storage and an increasing number of electromechanical devices for domestic and industrial use. Wear is, however, not always to be avoided: there are many manufacturing processes involving abrasive processing, for example, in which wear is used productively to form and shape surfaces [5].

1.3 Scope and Challenges

In some applications such as bearings, wear (and friction) is of primary concern, while for others the tribological performance of the system, although important, is not the main driver for its design. Thus, in modern engineering, we find increasing use of materials with more attractive combinations of density and mechanical properties than steel, or with benefits in cost, performance or formability, such as polymers, ceramics and various composite materials [6–8]. We also see a rapid increase in the use of surface engineering to provide a cost-effective combination of near-surface performance with desirable bulk properties in engineering components [9, 10]. These developments all pose particular challenges for the tribologist.

The origins of these challenges are many. The conditions to which a surface is exposed during wear are quite different from those involved in the measurement of conventional mechanical properties such as tensile strength, indentation hardness or fracture toughness. The dimensions of oxide or lubricant films, of surface height variation or of wear debris typically lie in the range from 10 nm to 10 μm [9]. In the absence of a thick lubricant film, surfaces make contact with each other at local high spots (asperities) which interact and induce high stresses (up to the yield point in some cases) over distances of the order of micrometres on a timescale of the order of microseconds. For typical speeds of relative motion, the strain rates at these microscopic sites of mechanical interaction can therefore be of the order of 10^4–$10^7 s^{-1}$. Not only are the timescales very short and strain rates high, but all the energy of frictional work is dissipated through the interactions of these contacts, often leading to high but transient local temperatures [11]. Even in a lubricated contact, the power density can be remarkably high: it has been estimated that in a thin elastohydrodynamic (EHL) oil film, the rate of viscous energy dissipation is of the order of 100 $TW\,m^{-3}$, equivalent to dissipating the entire electrical power output of USA in a volume of 5 l.

The difficulties involved in fully describing, and then in formulating models for, the behaviour of a wearing surface are not just associated with the extreme local conditions. The problem is much more complex than that, for at least three more reasons. First, the process of wear itself changes the composition and properties of the surface and near-surface regions; the material which separates two sliding surfaces can be treated as a distinct 'third-body' with its own evolutionary history and properties and these properties will often change during the lifetime of the system [12]. Second, the removal or displacement of material during wear leads to changes in surface topography. Third, the mechanisms by which wear occurs are often complex and can involve a mixture of mechanical and chemical processes: for example, in the unlubricated sliding of two steel surfaces, material may be removed by mechanical means after oxidation, while under conditions of boundary lubrication the source of wear is often the mechanical removal of the products of chemical reaction between the steel surface and the lubricant additives [13, 14]. Neither the mechanical nor the chemical interactions involved in sliding wear can yet be modelled accurately. The problem of fretting wear, in which contacting surfaces are exposed to small cyclic relative displacements, has some similarities to, but also many differences from, that of continuous sliding [15].

The case of abrasive wear is in principle slightly more tractable since chemical effects usually play a negligible role, but even here it would be necessary to model the deformation of the material to very large plastic strains, to incorporate realistic failure criteria (to account for ductile rupture or brittle fracture), to allow for changes in surface topography during wear, to account for the inhomogeneity of the material (associated with its microstructure, as it is initially and as it becomes modified during wear) at the length scale relevant to the unit interaction with an abrasive particle and to sum the individual effects of the interactions with perhaps many billions of abrasive particles.[1] The properties of the abrasive particles themselves would also have to be incorporated into a full model: these will include their bulk mechanical properties such as stiffness and strength, as well as a full description of their shape. The difficulties involved in describing the relevant aspects of particle shape alone have stimulated much research [16].

[1] One gram of abrasive particles 10 μm in diameter represents about 10^9 particles.

In view of the highly complex nature of wear processes and the difficulty of producing realistic models for them, it is not surprising that many discussions of sliding wear start with the simplest possible assumption of the relationship between wear rate and normal load:

$$Q = \frac{KW}{H} \tag{1}$$

where Q is the volume removed from the surface per unit sliding distance, W is the normal load applied to the surface by its counterbody and H is the indentation hardness of the wearing surface. K is a dimensionless quantity which is usually called the *wear coefficient* and which provides a valuable means of comparing the severity of different wear processes. If K, W and H remain constant during wear, then it is implicit in equation (1) that the volume of material lost from the surface is directly proportional to the relative sliding distance, or at constant sliding speed, to time. Equation (1) is usually called the *Archard wear equation*. Archard was perhaps the first to derive the relationship from a plausible physical model in 1953 [17], although Archard himself acknowledged the earlier work of Holm in 1946, and the empirical statement that wear is directly proportional to sliding distance and inversely proportional to normal load was made as early as 1927 by Preston in a study of the polishing of plate glass [18]. An equation identical to equation (1) was also stated by Taylor in 1948 without any indication that it was not already well known [19].[2]

For engineering applications, and especially for the wear of materials whose hardness cannot readily be defined (such as elastomers), the wear rate is commonly stated as $k = K/H = Q/W$. k is often called the *specific wear rate* and quoted in units of $mm^3\ N^{-1}\ m^{-1}$. For a material with a hardness H of 1 GPa (a soft steel, or a hard aluminium alloy, for example), the numerical value of k expressed in $mm^3\ N^{-1}\ m^{-1}$ is exactly the same as the value of K.

Figure 1.1 shows, very approximately, the range of values of K seen in various types of wear. Under unlubricated sliding conditions (so-called dry sliding), K can be as high as 10^{-2}, although it can also be as low as 10^{-6}. Often two distinct regimes of wear are distinguished, termed ‘severe’ and ‘mild’. Not only do these correspond to quite different wear rates (with K often above and below 10^{-4}, respectively), but they also involve significantly different mechanisms of material loss. In metals, ‘severe’ sliding wear is associated with relatively large particles of metallic debris, while in ‘mild’ wear the debris is finer and formed of oxide particles [13]. In the case of ceramics, the ‘severe’ wear regime is associated with brittle fracture, whereas ‘mild’ wear results from the removal of reacted (often hydrated) surface material.

When hard particles are present and the wear process involves abrasion (by sliding or rolling particles) or erosion (by the impact of particles), then the highest values of K occur;[3] the relatively high efficiency by which material is removed by abrasive or erosive wear explains why these processes can also be usefully employed in manufacturing [5].

The values of K which occur for unlubricated sliding, or for wear by hard particles, are generally intolerably high for practical engineering applications, and in most tribological designs lubrication is used to reduce the wear rate; the effect of lubrication in reducing wear

[2] Although the relevant paper was published in 1950, it had been presented at a conference in 1948.

[3] Equation (1) can be applied to abrasive wear as well as to sliding wear; an analogous equation can also be derived for erosion by solid particle impact, from which a value of K can be derived [20].

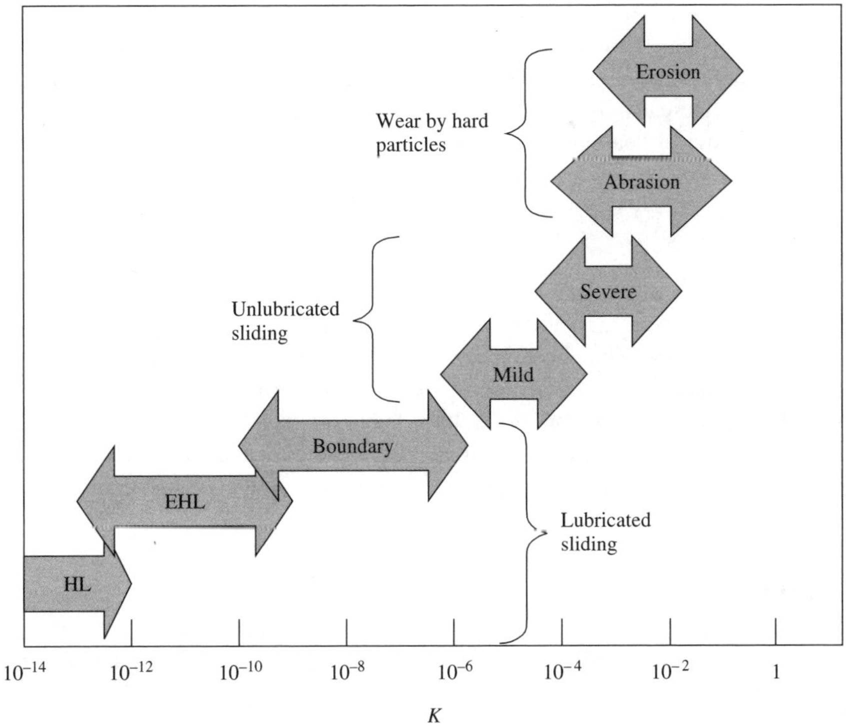

Figure 1.1 Schematic representation of the range of wear coefficient K exhibited under different conditions of wear. HL = hydrodynamic lubrication; EHL = elastohydrodynamic lubrication

is far more potent than its effect on friction, and the increase in life which results from the reduction in wear is generally much more important than the increase in efficiency from the lower frictional losses. As Figure 1.1 shows, even the least effective lubrication can reduce the wear rate by several orders of magnitude, and as the thickness of the lubricant film is increased in the progression from boundary to EHL and then to hydrodynamic lubrication, so the value of K falls rapidly. In the hydrodynamically lubricated components of a modern automotive engine, values of K as low as 10^{-19} are achieved [21].

There is a great deal of current interest in improving lubricants so as to achieve low wear rates with thinner films (associated with higher contact pressures). A good protective lubricant film requires the right combination of adhesion to the substrate, film formation and replenishment rate and shear strength [14]; allied with this is the need to find replacements for highly effective additives such as ZDDP (zinc dialkyl dithiophosphate) which contain elements which are detrimental both to the environment and to the long-term operation of automotive exhaust catalysts. Boron compounds are receiving much attention in this context, and there are also attractions in lubricants which can be transported to the sliding surfaces in the vapour phase, especially for very small-scale devices (e.g. MEMS) and for systems operating at high temperatures [14, 21, 22]. There is also active research into lubricants and lubricant additives which are effective for non-ferrous metals, ceramics and engineered surfaces [21].

As our understanding of wear processes deepens, it becomes increasingly important to be able to transfer that knowledge to engineers involved in both the design and operation of machines. Maps or wear regime diagrams provide powerful tools for the design process [4, 7, 23], while increasingly sophisticated methods have been developed to assess and monitor the tribological health of operating machinery and determine the appropriate levels of maintenance [24].

1.4 Conclusions

While accurate predictive models for wear rate are still an elusive goal, it is clear that significant recent progress has been made in our understanding of many aspects of wear mechanisms and that advances in materials, surface engineering and lubricants, as well as in design methods and condition monitoring, have led to major improvements in the efficiency, lifetime cost and performance of many engineering systems. There is still much potential for future development, and challenges in tribology, especially in the vital field of wear, remain.

References

1. Almen, J.O., in *Mechanical Wear* (ed J.T. Burwell), American Society for Metals, 1950, pp. 229–288.
2. Glossary of terms and definitions in the field of friction, wear and lubrication, Research Group on Wear of Engineering Materials, Organisation for Economic Co-operation and Development, 1969, reprinted in *Wear Control Handbook* (eds M.B. Peterson and W.O. Winer), American Society of Mechanical Engineers, 1980, pp. 1143–1303.
3. Standard terminology relating to wear and erosion, standard G-40-01, American Society for Testing and Materials, 2001.
4. Kato, K., 'Classification of Wear Mechanisms', in *Wear – Mechanisms, Materials and Practice* (ed G.W. Stachowiak), John Wiley & Sons, Ltd, Chichester, 2006, pp. 9–20.
5. Hutchings, I.M., 'Abrasion Processes in Wear and Manufacturing', *Proceedings of the Institution of Mechanical Engineers, Part J, Journal of Engineering Tribology*, **216**, 2002, 55–62.
6. Briscoe, B.J. and Sinha, S.K., 'Tribology of Polymeric Solids and Their Composites', in *Wear – Mechanisms, Materials and Practice* (ed G.W. Stachowiak), John Wiley & Sons, Ltd, Chichester, 2006, pp. 223–268.
7. Jahanmir, S., 'Wear of Ceramics: Wear Transitions and Tribochemical Reactions', in *Wear – Mechanisms, Materials and Practice* (ed G.W. Stachowiak), John Wiley & Sons, Ltd, Chichester, 2006, pp. 167–190.
8. Friedrich, K., Zhang, Z. and Klein, P., 'Wear of Polymer Composites', in *Wear – Mechanisms, Materials and Practice* (ed G.W. Stachowiak), John Wiley & Sons, Ltd, Chichester, 2006, pp. 269–290.
9. Holmberg, K. and Matthews, A., 'Tribology of Engineered Surfaces', in *Wear – Mechanisms, Materials and Practice* (ed G.W. Stachowiak), John Wiley & Sons, Ltd, Chichester, 2006, pp. 123–166.
10. Erdemir, A. and Donnet, C., 'Tribology of Diamond and Diamond-Like Carbon Films', in *Wear – Mechanisms, Materials and Practice* (ed G.W. Stachowiak), John Wiley & Sons, Ltd, Chichester, 2006, pp. 191–222.
11. Ashby, M.F., Abulawi, J. and Kong, H.S., 'Temperature Maps for Frictional Heating in Dry Sliding', *Tribology Transactions*, **34**, 1991, 577–587.
12. Berthier, Y., 'Third-Body Reality – Consequences and Use of the Third-Body Concept to Solve Friction and Wear Problems', in *Wear – Mechanisms, Materials and Practice* (ed G.W. Stachowiak), John Wiley & Sons, Ltd, Chichester, 2006, pp. 291–316.
13. Biswas, S.K., 'Wear of Metals: A Material Approach', in *Wear – Mechanisms, Materials and Practice* (ed G.W. Stachowiak), John Wiley & Sons, Ltd, Chichester, 2006, pp. 21–36.
14. Hsu, S.M., Munro, R.C., Shen, M.C. and Gates, R.S., 'Boundary Lubricated Wear', in *Wear – Mechanisms, Materials and Practice* (ed G.W. Stachowiak), John Wiley & Sons, Ltd, Chichester, 2006, pp. 37–70.
15. Kapsa, P., Fouvry, S. and Vincent, L., 'Basic Principles of Fretting', in *Wear – Mechanisms, Materials and Practice* (ed G.W. Stachowiak), John Wiley & Sons, Ltd, Chichester, 2006, pp. 317–338.

16. Stachowiak, G.W., Stachowiak, G.B., De Pellegrin, D.V. and Podsiadlo, P., 'Characterization and Classification of Abrasive Particles and Surfaces', in *Wear – Mechanisms, Materials and Practice* (ed G.W. Stachowiak), John Wiley & Sons, Ltd, Chichester, 2006, pp. 339–368.
17. Archard, J.F., 'Contact and Rubbing of Flat Surfaces', *Journal of Applied Physics*, **24**, 1953, 981–988.
18. Preston, F.W., 'The Theory and Design of Plate Glass Polishing Machines', *Journal of the Society of Glass Technologists*, **11**, 1927, 214–256.
19. Taylor, C.F., in *Mechanical Wear* (ed J.T. Burwell), American Society for Metals, 1950, pp. 1–7.
20. Hutchings, I.M. '*Tribology: Friction and Wear of Engineering Materials*', Butterworth Heinemann, Oxford, UK, 1992.
21. Neville, A. and Morina, A., 'Wear and Chemistry of Lubricants', in *Wear – Mechanisms, Materials and Practice* (ed G.W. Stachowiak), John Wiley & Sons, Ltd, Chichester, 2006, pp. 71–94.
22. Gellman, A.J. and Spencer, N.D., 'Surface Chemistry in Tribology', in *Wear – Mechanisms, Materials and Practice* (ed G.W. Stachowiak), John Wiley & Sons, Ltd, Chichester, 2006, pp. 95–122.
23. Hsu, S.M. and Shen, M.C., 'Wear Mapping of Materials', in *Wear – Mechanisms, Materials and Practice* (ed G.W. Stachowiak), John Wiley & Sons, Ltd, Chichester, 2006, pp. 369–424.
24. Roylance, B.J., 'Machine Failure and Its Avoidance – Tribology's Contribution to Effective Maintenance of Critical Machinery', in *Wear – Mechanisms, Materials and Practice* (ed G.W. Stachowiak), John Wiley & Sons, Ltd, Chichester, 2006, pp. 425–452.

2

Classification of Wear Mechanisms/Models

K. Kato

Abstract

Wear mechanisms may be briefly classified as mechanical, chemical and thermal wear whose wear modes are further classified into seven sub-classes. Some of them have wear models and mathematical expressions for wear rate, but many of them still do not have satisfactory wear models and wear equations for reliable predictions. This chapter reviews such current understanding on wear mechanisms and models.

2.1 Introduction

Wear is not a material property; however, it is a systems response [1, 2]. The wear rate of a material can vary from 10^{-3} to 10^{-10} mm^3/N m depending on contact conditions, such as the counterpart material, contact pressure, sliding velocity, contact shape, suspension stiffness, environment and the lubricant [3].

The wear rate changes through the repeated contact process under constant load and velocity. It is generally high in an initial unsteady state and relatively lower in the later steady one. Initial wear and steady wear are the terms used to describe wear rate changes resulting from wear due to repeated contact. In steel against steel contact, this change is caused by the change of the wear modes from adhesive wear to corrosive (oxidative) wear.

Running-in is also an expression used to describe the initial, higher wear rate in lubricated contacts. In the case of SiC to SiC contact in water, the wear rate changes from a high value of 10^{-6} mm^3/N m to a low value of 10^{-8} mm^3/N m. This is caused by the successive wear of surface asperities and the better conformity of smooth worn surfaces. In the early stage, wear is caused by brittle micro-fractures in the surface grains and in the later stage by tribochemical reaction [4–6].

Wear – Materials, Mechanisms and Practice Edited by G. Stachowiak

Lubrication reduces adhesive wear by reducing adhesion; however, it increases the abrasive wear by reducing the friction. Adhesive wear and the transfer layer reduce the subsequent adhesive wear when the friction between worn surface and transfer layer in air is low, for example with polytetrafluoroethylene [7]. Early corrosive wear reduces the subsequent corrosive wear when worn surfaces become smoother and conform better [8].

Wear is complicated in this way. It must be considered in terms of a multiparameter-sensitive phenomenon.

2.2 Classification of Wear Mechanisms and Wear Modes

2.2.1 Mechanical, Chemical and Thermal Wear

The classification of wear parameters, along with descriptive terms of the wear mechanisms, is shown in Table 2.1. 'Rolling wear', 'sliding wear', 'fretting wear' and 'impact wear' are the terms often used in practice and in literature. They are useful to describe the type of motion which results in wear; however, they do not describe possible wear mechanisms. 'Mechanical wear', 'chemical wear' and 'thermal wear' are terms used to describe briefly wear mechanisms occurring.

Table 2.1 Classification of wear parameters

Class	Parameter						
Friction type	Rolling	Rolling–sliding	Sliding	Fretting	Impact		
Contact shape	Sphere/ sphere	Cylinder/ cylinder	Flat/ flat	Sphere/ flat	Cylinder/ flat	Punch/ Flat	
Contact pressure level	Elastic	Elasto-plastic	Plastic				
Sliding speed or loading speed	Low	Medium	High				
Flash temperature	Low	Medium	High				
Mating contact material	Same	Harder	Softer	Compatible	Incompatible		
Environment	Vacuum	Gas	Liquid	Slurry			
Contact cycle	Low (single)	Medium	High				
Contact distance	Short	Medium	Long				
Phase of wear	Solid	Liquid	Gas	Atom	Ion		
Structure of wear particle	Original	Mechanically mixed	Tribochemically formed				
Freedom of wear particle	Free	Trapped	Embedded	Agglomerated			
Unit size of wear	mm scale	μm scale	nm scale				
Elemental physics and chemistry in wear	Physical adsorption, chemical adsorption, tribochemical activation and tribofilm formation, oxidation and delamination, oxidation and dissolution, oxidation and gas formation, phase transition, recrystallization, crack nucleation and propagation, adhesive transfer and retransfer						
Elemental system dynamics related to wear	Vertical vibration	Horizontal vibration	Self-excited vibration	Harmonic vibration	Stick-slip motion		
Dominant wear process	Fracture (ductile or brittle)	Plastic flow	Melt flow	Dissolution	Oxidation	Evaporation	
Wear mode	Abrasive	Adhesive	Flow	Fatigue	Corrosive	Melt	Diffusive
Wear type	Mechanical	Chemical	Thermal				

Mechanical wear describes wear governed mainly by the processes of deformation and fracturing. The deformation process has a substantial role in the overall wear of ductile materials and the fracturing process has a major role in the wear of brittle materials. Chemical wear describes wear governed mainly by the growth rate of a chemical reaction film. The growth rate of the film is accelerated mechanically by friction [9–11]. Therefore, chemical wear is called 'tribochemical wear'. Thermal wear describes wear governed mainly by local surface melting due to frictional heating [12]. Diffusive wear is also included in the term 'thermal wear', since it becomes noticeable only at high temperatures. The wear of brittle materials caused by fractures following thermal shocks may also be included in thermal wear [13].

These three descriptions of wear are necessary to characterize wear briefly; however, they are not sufficient to introduce wear models for wear rate predictions.

2.2.2 *Wear Modes: Abrasive, Adhesive, Flow and Fatigue Wear*

'Abrasive', 'adhesive', 'flow' and 'fatigue' wear are more descriptive expressions for mechanical wear, and their wear processes are schematically illustrated in Figure 2.1(a), (b), (c) and (d), respectively.

The abrasive wear of ductile materials is shown in Figure 2.1(a). Three-dimensional wear models of surface scratching by a hard asperity have been proposed and confirmed through quantitative agreements between experimental results and theoretical predictions [14, 15]. The wear volume, V, is given by the following expression:

$$V = \alpha\beta \frac{WL}{H_V} \tag{1}$$

where W is the load, L the sliding distance, α the shape factor of an asperity and β the degree of wear by abrasive asperity. Experimentally, α takes a value of about 0.1 and β varies between 0 and 1.0, depending on the value of the degree of penetration of an abrasive asperity, the shear strength at the contact interface and the mechanical properties of the wearing material. If the wear rate is given by a specific wear rate, w_s (= wear volume/load × sliding distance), or a wear coefficient, $K (= w_s H_V$, where H_V is the hardness), they are derived from equation (1) as follows:

$$w_s = \frac{\alpha\beta}{H_V} \tag{2}$$

$$K = \alpha\beta \tag{3}$$

These equations, however, are only valid for single-abrasive scratching. An abrasive wear mode map has been shown to be useful for abrasive wear rate prediction when multiple contacts are involved, where the degree of penetration and the contact shear strength are distributed among many asperity contacts in order to find statistical values [16].

As for the adhesive wear of ductile materials, schematically illustrated in Figure 2.1(b) [17], no predictive theories have been quantitatively confirmed by experiments. Assumptions have been made dealing with the unit volume of wear particles removed from the unit contact region [18, 19], but they do not agree well enough with experimental data to give a

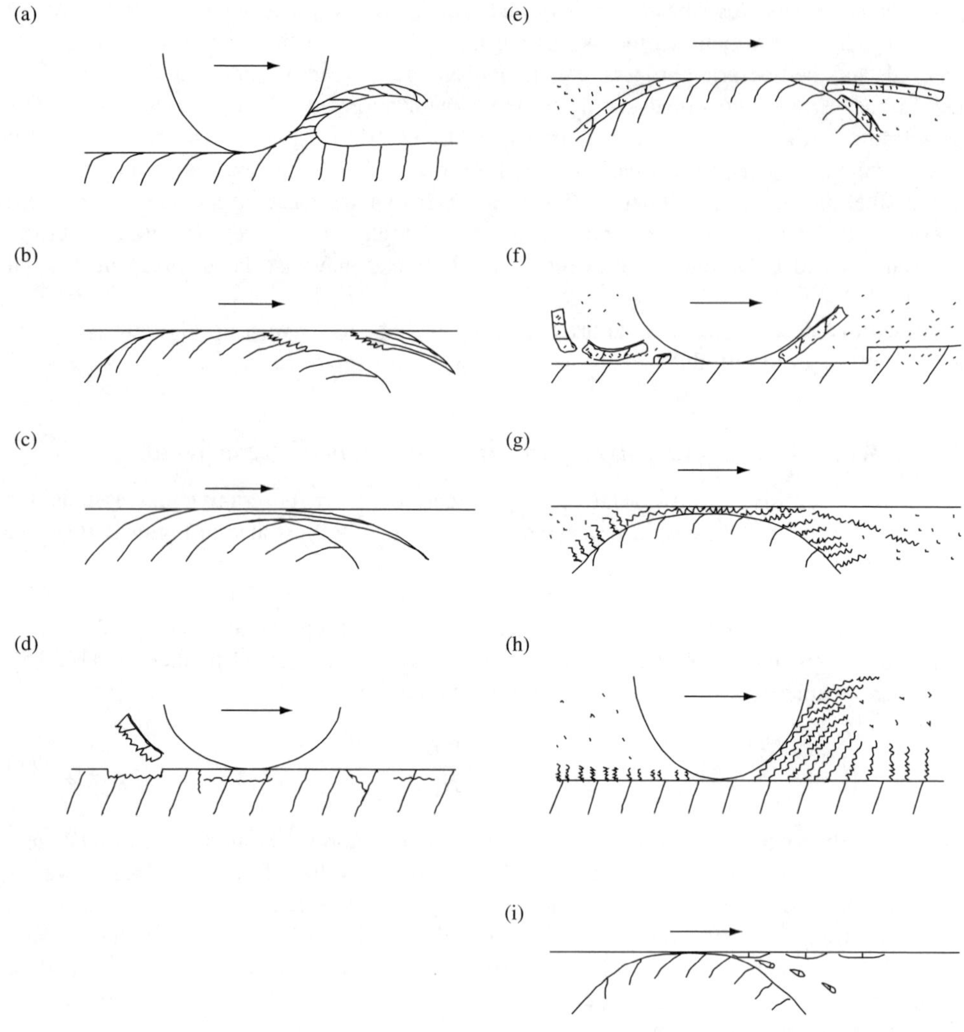

Figure 2.1 Schematic wear modes: (a) abrasive wear by microcutting of ductile bulk surface; (b) adhesive wear by adhesive shear and transfer; (c) flow wear by accumulated plastic shear flow; (d) fatigue wear by crack initiation and propagation; (e) corrosive wear by shear fracture of ductile tribofilm; (f) corrosive wear by delamination of brittle tribofilm; (g) corrosive wear by accumulated plastic shear flow of soft tribofilm; (h) corrosive wear by shaving of soft tribofilm; and (i) melt wear by local melting and transfer or scattering

basis for a quantitative theory. Wear equations for adhesive wear are given by the following expressions, which are similar to equation (1):

$$V = w_s WL \tag{4}$$

or

$$V = K\frac{WL}{H_V} \tag{5}$$

Experimental results for adhesive wear show that the wear volume increases almost linearly with the load and sliding distance. However, useful physical models were not found to explain the observed variation in values of w_s and K, where, for example, w_s varied from 10^{-2} to 10^{-10} mm^3/N m. Various statistical variables, such as material microstructure, surface roughness, flash temperatures, local contamination, adhesive transfers, free wear particles and microscale tribochemical reactions on the contact surfaces, are all related to the constants through the local values of the friction coefficient and the strength of contact materials. It must be noticed that the local friction coefficient is a function of the local shear strength at the contact interface and the local contact geometry.

As for flow wear, shown in Figure 2.1(c) [20, 21], experimental observations with steel are explained well by a theoretical model, called 'ratchetting', and the wear coefficient, K, is given as a function of the plasticity index, surface roughness and friction coefficient [22, 23]. Although the mechanism of flow wear is similar to that of low-cycle fatigue, crack initiation and propagation are not necessary to produce wear particles, and plastic flow is the major part in this wear mode. The question of whether the wear of a butter-like tribofilm covering a hard substrate could be treated as another form of flow wear still remains.

In the case of fatigue wear, predictions of high-cycle fatigue wear were made first. In this wear mode crack initiation and propagation, as shown in Figure 2.1(d), in a repeated contact stress cycle dominate. Stress conditions are assumed to be either elastic or elasto-plastic. Wear particle shape or unit wear volume is decided, therefore, by the path of crack propagation. The critical number N_f of rolling cycles for surface spalling by high-cycle fatigue in a steel ball bearing is experimentally given by the following equation:

$$N_f = bW^{-n} \tag{6}$$

where W is the load and b and n are experimental constants. The value of n is 3 for ball bearings [24]. Its basic premise is that spalling can be treated as a statistical fracture phenomenon following the theory of Weibull [25].

If the contact stress is at a level sufficiently high for plastic deformation to occur and repeated contact cycles are required to produce wear particles through crack initiation and propagation, a Coffin–Manson-type relation of the fatigue fracture can be used to model the low-cycle fatigue wear. The wear rate for this mode is theoretically introduced by creating a two-dimensional abrasive wear model of the plastic wave formation [26]. The wear coefficient, K, is given as follows:

$$K = \frac{9 \times 3^{1/2} r\mu}{C^D \gamma_t^{1-D}} \tag{7}$$

where r, μ and γ_t are all determined from the wave model as functions of the attack angle and the normalized shear strength of the contact interface, C is the monotonic effective shear strain and D an experimental constant used as the power in the low-cycle fatigue law.

Wear is supposed to occur when strain is accumulated to a critical value causing fracture. Equation (7) gives the predicted K value, ranging from 10^{-10} to 10^0 in relation to the changes of the friction coefficient and the asperity attack angle. This prediction seems to give a reasonable explanation for some experimental results [26].

The results obtained from both equations (6) and (7) are based on the experimental power law of fatigue fracture. The mechanism of fatigue can be analysed through linear fracture mechanics to a certain extent. Similarly, the fatigue wear mechanism may be understood by analysing the process of the crack initiation and propagation in forming a wear particle.

The crack propagation rate was theoretically calculated from a model of a subsurface crack, parallel to the surface, through linear fracture mechanics for elasto-plastic solids, and the following equation is proposed [27]:

$$\frac{\mathrm{d}a}{\mathrm{d}N} = c(\Delta K)^m \tag{8}$$

where a is the crack length, N the number of friction cycles, c and m are experimental constants and ΔK is the change in the stress intensity factor. The possible depth of the crack and the effective crack length are also calculated with linear elastic fracture mechanics [28].

Using a standard fatigue test of steel, it was confirmed that the crack (or void) nucleation period is much longer (more than 70% of the total life) than the period for the crack propagation to cause a failure. However, in cases of steel sliding in air, with a high friction coefficient ($\mu = 0.5$), the number of critical friction cycles needed for the generation of a void around a hard inclusion in the substrate was theoretically calculated as about 3 [29]. This means that further theoretical modelling of low-cycle fatigue wear could be advanced through the development of the crack propagation theory (equation (8)). On the other hand, it is well established that the initiation process of a crack and the nucleation process of a void are the rate-controlling processes in high-cycle fatigue wear [30].

2.2.3 *Corrosive Wear*

In cases of corrosive wear, thin films are assumed to form through a tribochemical reaction between contact surface materials and the surrounding media, such as air or a liquid lubricant. A hard tribofilm such as an iron oxide film on steel and a soft tribofilm such as a silica gel film on Si_3N_4 in water or a zinc dialkyl dithiophosphate (ZDDP) reaction film in oil are expected to be preferentially worn as shown in Figure 2.1(e) and (f) for the former and in Figure 2.1(g) and (h) for the latter. Hard, brittle iron oxide is presumed to delaminate by itself after reaching a critical thickness ξ. The wear coefficient K is expressed as follows:

$$K = \frac{\mathrm{d}A \exp(-Q/R_\mathrm{g}T)}{\xi^2 \rho^2 U} \tag{9}$$

where A is the Arrhenius constant, Q the activation energy, R_g the gas constant, T the absolute temperature, ρ the density of oxide, U the sliding velocity and d the distance along which the wearing contact was made [31]. It is generally assumed that the activation energy does not vary substantially between static and sliding conditions. With this assumption, the experimental wear results for the oxidation of steel during sliding gave a 10^3–10^{10} times larger value of Arrhenius constant in equation (9) than in static oxidation [32].

The same assumption was made for the critical oxide film thickness for its self-delamination. It was successfully used to model the oxidation wear of steel, which was used to construct the wear map of steel in a self-mated, unlubricated sliding [12].

As for a soft tribofilm on a relatively hard substrate, theoretical models for predicting wear rates have not yet been published, although experimental relationships have been observed between the wear rate and the frictional conditions, such as the load and the friction coefficient [33–35].

2.2.4 Melt and Diffusive Wear

The evidence that melt wear exists is obtained by observing spherical wear particles of unique surface morphology and also by observing a wear surface partially covered by droplets or a film on the smooth surface. This wear mode is not considered as a dominant steady wear mode in general tribo-elements. However, it is generated by unexpected contact conditions, such as hard inclusions at the contact interface or a sudden overloading due to vibration.

A wear model is introduced by assuming a melting zone caused by frictional heating. A wear map for this model allowing to predict wear rates has been proposed [12]. This model needs further experimental confirmations for quantitative prediction of wear rate.

From the viewpoint of the volume loss and the resultant shape change, diffusive wear on the atomic scale may not play an important role in practice, and a quantitative prediction of the wear volume may not be required. Even though the original material properties of a contact surface are degraded by losing significant chemical compositions, as a result of diffusion, the wear rate can still be increased by enhancing other wear modes, such as adhesive or abrasive wear. This occurs in the case of the wear of cutting tools under unlubricated conditions [36], where the temperature, T, is very high and the wear rate increases linearly as $\exp(-\Delta E/KT)$. However, models to predict the contribution of diffusive wear in such situations have not been found.

2.3 General Discussion of Wear Mechanisms and Their Models

2.3.1 Material Dependence

Traditional wear mechanisms have been classified into six wear modes: 'abrasive', 'adhesive', 'fatigue', 'corrosive', 'melt' and 'diffusive' wear. The expression 'flow wear' is newly introduced in Table 2.1 to give a more precise classification of the wear modes observed under repeated contact. Initiation of a crack and its propagation through stress cycles do not necessarily occur during mild wear; however, plastic flow and its accumulation through stress cycles are required to form a wear particle. These seven expressions have been proved to work mainly with metals. Each of these wear mode expressions gives a precise image of the wear process occurring in metals.

In the abrasive wear of metals, for example, there are three levels of wear: micro-cutting, wedge forming and ploughing [37]. In the repeated abrasive contacts, a long ribbon-like wear particle is generated through microcutting in the first cycle and wedge formation in the next cycle. The ploughing action with no wear particles comes last as a steady state of abrasive sliding. In the subsequent ploughing cycles, flow wear is observed [38].

On the other hand, SiC sliding against a diamond pin under similar abrasive contact conditions shows cutting type wear only after several sliding cycles. No wear particles are observed in the first sliding cycle [39]. If the load is large enough to cause an abrasive groove in a single scratch on ceramics, the wear volume, V, of one scratch groove is given by

$$V = \eta \frac{W^{9/8}}{K_c^{1/2} H^{5/8}} \left(\frac{E}{H}\right)^{4/5} L \tag{10}$$

where K_c is the fracture toughness, E the elastic modulus, H the hardness, L the sliding distance and η is a material-independent constant [40]. It is obvious that abrasive wear is now a function of the fracture toughness, K_c, which describes the wear of brittle fractures.

A similar material dependence of wear for the other wear modes of 'adhesive', 'flow', 'fatigue', 'corrosive', 'melt' and 'diffusive' wear should also be considered.

2.3.2 Wear Maps

In order to understand the relative relationships between different wear modes, the concept of 'wear map' is useful, where appropriate parameters have to be found to describe it.

The abrasive wear map of metals [37] describes the regimes of three wear modes, i.e. microcutting, wedge forming and ploughing, by introducing the degree of penetration D_p = indentation depth/contact radius and the normalized shear strength f = contact interface shear strength/wear material shear strength, as shown in Figure 2.2(a).

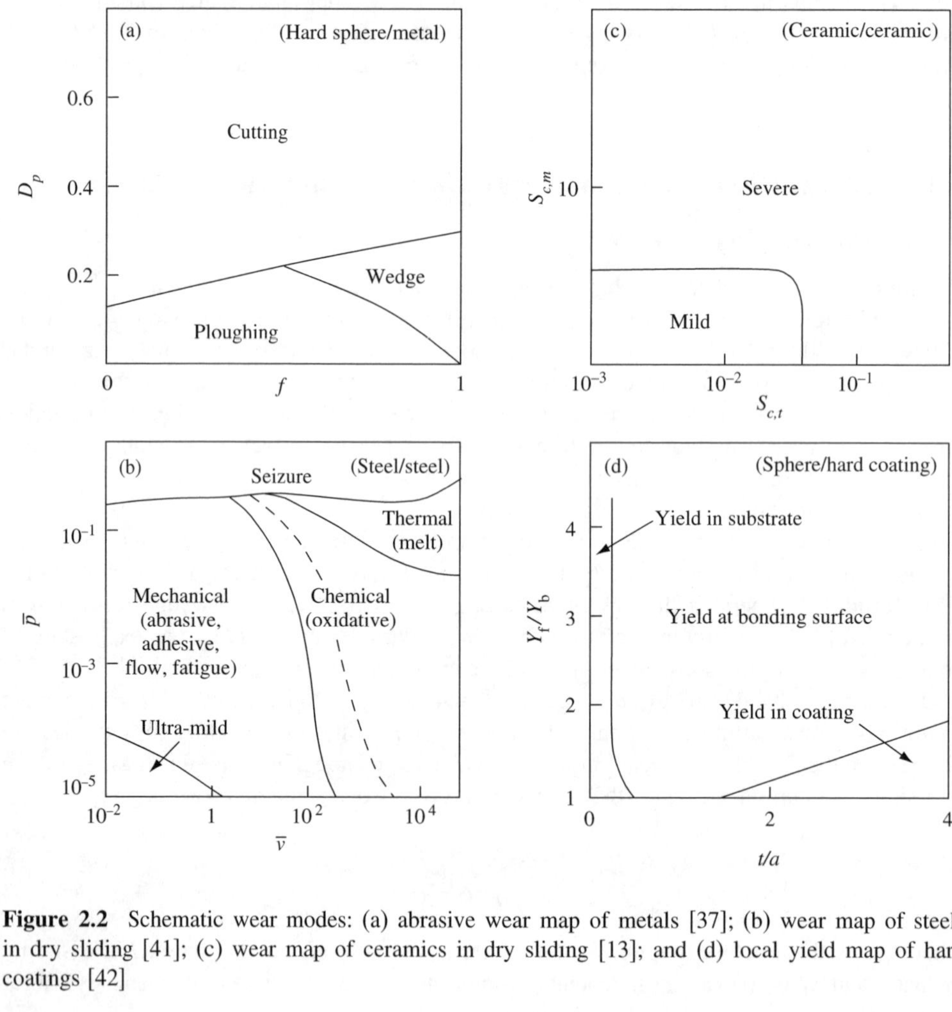

Figure 2.2 Schematic wear modes: (a) abrasive wear map of metals [37]; (b) wear map of steels in dry sliding [41]; (c) wear map of ceramics in dry sliding [13]; and (d) local yield map of hard coatings [42]

The wear map of steels in dry sliding [41] describes the regimes of wear mechanism, i.e. mechanical, chemical and thermal, by introducing the parameters of normalized pressure $\bar{p}$ = nominal contact pressure/hardness and normalized velocity $\bar{v}$ = velocity × contact radius/thermal diffusivity, as shown in Figure 2.2(b).

The wear map of ceramics in dry sliding [13] describes the regimes of mild and severe wear by introducing the parameters of mechanical severity of contact $S_{c,m}$ $(\mu, p_{max}, d, K_{IC})$ and thermal severity of contact $S_{c,t}$ $(\mu, r, \Delta T_s, V, H_V, K, \rho, c)$, as shown in Figure 2.2(c).

The local yield map of hard coatings [42] describes the regimes of coating delamination sites by introducing the parameters of normalized yield stress Y_f/Y_b = film yield stress/substrate yield stress and normalized coating thickness t/a = coating thickness/contact radius, as shown in Figure 2.2(d).

The parameters introduced in Figure 2.2(a–d) are indices which give the critical values for the transition of wear mode from one to another. They may differ substantially depending on the materials used, including coatings, as seen in equations (1), (9) and (10).

2.3.3 Wear Mode Transition

In the process of repeated contact, the microstructure and microgeometry of contact surfaces change as a result of wear. Rubbed surface work hardens and wear particles agglomerate with time and cover wear surfaces. Even in the case of contact of mirror surfaces of similar materials, the rubbing texture is always formed as a result of wear. This means that the abrasive contact is introduced after adhesive wear. It needs to be remembered that there is gradual transition in wear mode, at a microscopic contact point, from more severe initial contact to less severe in the following wear mode.

We may assume, for example, that in abrasive contact, the initial cutting mode transits to wedge forming and then to ploughing, as shown in Figure 2.2(a). In the case of dry sliding of steels, we may assume that the initial melt wear transits to mechanical wear and then to chemical (oxidative) wear, as shown in Figure 2.2(b). In the case of dry sliding of ceramics, we may expect the wear mode transition from severe to mild, as shown in Figure 2.2(c). If water is supplied to the contact interface of SiC, we can expect further wear mode transition from mechanical wear to tribochemical wear in the mild wear regime, as shown in Figure 2.2(c). Very low friction coefficient below 0.01 and low wear rate below 10^{-8} mm^3/N m will be generated after such wear mode transition of SiC in water in the repeated sliding contact [5].

In overall wear rate prediction after a long-time running, a series of transitions of the wear mode in the whole process have to be considered in this way and then the unit volume of wear in each wear mode has to be determined.

2.3.4 Erosion

Particle erosion, fluid erosion, cavitation erosion and spark erosion are the common types of erosive wear which are caused by the impact of solid particles, liquid droplets, bubbles or electrical sparks [3, 43]. The seven wear modes shown in Table 2.1 are thought to be working in various combinations for these four types of erosive wear. The three major categories of wear types, i.e. mechanical, chemical and thermal, can also be helpful in the understanding of these four common erosive wear mechanisms.

In the case of particle erosion, for example, fine hard particles of irregular shape impact and/or abrade the surface at high speed. Each particle acts as an abrasive of short time of contact and causes partial abrasive wear on the impacted surface. When the particle bounces, the top surface material adheres to the particle surface and is carried away by the mechanism of adhesive wear. Depending on the speed of impact, partial surface melting can be generated. The series of impact by the continuous flow of fine particles at the same portion of the surface generates fatigue wear by introducing cracks in the subsurface. If the temperature of the surroundings is high enough to cause extensive oxidation on the impacted surface, the wear mechanisms of abrasive, adhesive, fatigue and melt caused by the successive impact of particles are those of the oxide film on the impacted surface.

The erosion mechanism by fluid, cavitation or spark may be considered in a similar way. Because of this reason, 'erosion' is not included in the classification of wear modes or wear types in Table 2.1.

2.4 Conclusion

Three brief expressions of wear mechanisms (mechanical, chemical and thermal wear) and seven more detailed descriptions of wear modes (abrasive, adhesive, flow, fatigue, corrosive, melt and diffusive wear) are explained from the viewpoints of their classifications. Practically, any observed wear rate value is generated as a mixture of these different wear modes. Predictive models for 'abrasive', 'flow', 'fatigue' and 'corrosive' wear rates are briefly explained. The role of wear maps is also discussed.

Acknowledgements

The author would like to express his appreciation to Mr Boyko Stoimenov for his help in the preparation of the material for this chapter and to Mr Kurt Talke for his help in polishing the English.

This material has been based upon an article that first appeared in the *Journal of Engineering Tribology* – Proceedings Part J, 2002, Vol. 216, No. J6, ISSN 1350–6501, published by Professional Engineering Publishing. Permission is granted by the Institution of Mechanical Engineers.

References

1. Kato, K., 'Wear Mechanisms', in *New Direction in Tribology* (ed I. Hutchings), Mechanical Engineering Publications, London, 1997, pp. 39–56.
2. Bayer, R.G., '*Mechanical Wear Prediction and Prevention*', Dekker , New York, 1994, pp. 200–291.
3. Winner, W.O. (ed) '*Wear Control Handbook*', ASME, New York, 1980.
4. Fischer, T.E. and Tomizawa, H., 'Interaction of Tribochemistry and Microfracture in the Friction and Wear of Silicon Nitride', *Wear*, **105**, 1985, 29–45.
5. Cheng, M., Kato, K. and Adachi, K., 'Friction and Wear of Self-Mated SiC and Si3N4 Sliding in Water', Thirteenth International Conference on *Wear of Materials*, Vancouver, 2001, to be published in *Wear*.
6. Xu, J.G. and Kato, K., 'Formation of Tribochemical Layer of Ceramics Sliding in Water and Its Role for Low Friction', *Wear*, **245**, 2000, 67–75.
7. Bely, V.A., Sviridenok, A.I., Petrokovets, M.I. and Savkin, V.G., 'Friction and Wear in Polymer-Based Materials', Chapter 6, *Frictional Transfer*, Pergamon Press, Oxford, UK, 1982, pp. 195–212.

8. Stachowiak, G.W. and Batchelor, A.W., 'Engineering Tribology', Chapter 13, *Corrosive and Oxidative Wear*, Butterworth-Heinemann, Boston, USA, 2001, pp. 553–570.
9. Fischer, T.E. and Sexton, M.D., 'The Tribochemistry of Oxidative Wear', in *Physical Chemistry of the Solid State: Applications to Metals and Their Compounds* (ed P. Lacombe), Elsevier, Amsterdam, 1984, pp. 97–107.
10. Fischer, T.E. and Mullins, W.M., 'Chemical Aspects of Ceramic Tribology', *The Journal of Physical Chemistry*, **96**, 1992, 5690–5701.
11. Gates, R.S., Hsu, S.M. and Klaus, E.E. 'Tribochemical Mechanism of Alumina with Water', *Tribology Transactions, STLE*, **32**(3), 1989, 357–363.
12. Lim, S.C. and Ashby, M.F., 'Wear Mechanism Maps', *Acta Metallurgica*, **35**, 1987, 1–24.
13. Adachi, K., Kato, K. and Chen, N., 'Wear Map of Ceramics', *Wear*, **203–204**, 1997, 291–301.
14. Zum Gahr, K.H. '*Microstructure and Wear of Materials*', Tribology Series, Elsevier, Amsterdam, 1987, pp. 132–148.
15. Hokkirigawa, K. and Kato, K., 'Theoretical Estimation of Abrasive Wear Resistance Based on Microscopic Wear Mechanism', *Wear of Materials* (ed K.C. Ludema), ASME, New York, 1989, pp. 1–8.
16. Kato, K. and Hokkirigawa, K., 'Abrasive Wear Diagram', in *Proceedings of Eurotrib'85*, Vols 4–5, Elsevier, Amsterdam, 1985.
17. Kayaba, T. and Kato, K., 'The Adhesive Transfer of the Slip-Tongue and the Wedge', *ASLE Transactions*, **24**, 1981, 164–174.
18. Archard, J.F., 'Contact and Rubbing of Flat Surfaces', *Journal of Applied Physics*, **24**, 1953, 981–988.
19. Rabinowicz, E., '*Friction and Wear of Materials*', 2nd edn, Wiley-Interscience, New York, 1995.
20. Akagaki, T. and Kato, K., 'Plastic Flow Process of Surface Layer in Flow Wear Boundary Lubricated Conditions', *Wear*, **117**, 1987, 179–186.
21. Akagaki, T. and Kato, K., 'Simulation of Flow Wear in Boundary Lubrication Using a Vickers Indentation Method', *STLE Tribology Transactions*, **31**, 1988, 311–316.
22. Kapoor, A. and Johnson, K.L., 'Plastic Ratcheting as a Mechanism of Metallic Wear', *Proceedings of the Royal Society of London A*, **445**, 1994, 367–381.
23. Kapoor, A., Johnson, K.L. and Williams, J.A., 'A Model for the Mild Ratcheting Wear of Metals', *Wear*, **200**, 1996, 38–44.
24. Lundkey, G. and Palongrev, A., Dynamic Capacity of Rolling Bearings, *Ingeniorsveten-S kapsademicns*, no. 96, 1947.
25. Weibull, W., 'A Statistical Theory of the Strength of Materials', *Royal Swedish Academy of Engineering Science Proceedings*, **151**, 1930, 5–45.
26. Challen, J.M., Oxley, P.L.B. and Hockenhull, B.S., 'Prediction of Archard's Wear Coefficient for Metallic Sliding Friction Assuming a Low Cycle Fatigue Wear Mechanism', *Wear*, **111**, 1986, 275–288.
27. Fleming, J.R. and Suh, N.P., 'The Relationship Between Crack Propagation Rates and Wear Rates', *Wear*, **44**, 1977, 57–64.
28. Fleming, J.R. and Suh, N.P., 'Mechanics of Crack Propagation in Delamination Wear', *Wear*, **44**, 1977, 39–56.
29. Jahanmir, S. and Suh, N.P., 'Mechanics of Subsurface Void Nucleation in Delamination Wear', *Wear*, **44**, 1977, 17–38.
30. Yokobori, T., in *An Interdisciplinary Approach to Fracture and Strength of Solids* (ed J.D.C. Crisp), Wolters-Noordhoff, Groningen, 1968, pp. 191–192.
31. Quinn, T.F.J., 'The Effect of Hot-Spot Temperatures on the Unlubricated Wear of Steel', *ASLE Transactions*, **10**, 1967, 158–168.
32. Quinn, T.F.J., in *Proceedings of International Conference on Tribology – Friction, Wear and Lubrication*, Institution of Mechanical Engineer's Conference Series, Vol. 1987-5, Mechanical Engineering Publications, London, 1987, pp. 253–259.
33. Muratov, V.A., Luangvaranunt, T. and Fischer, T.E., 'The Tribochemistry and Silicon Nitride: Effects of Friction, Temperature and Sliding Velocity', *Tribology International*, **3**(10), 1998, 601–611.
34. Zhao, X.Z., Liu, J.J. and Fischer, T.E., 'Effects of Lubricant Rheology and Additive Chemistry in the Wear of Si_3N_4 Sliding on Steel', *Wear*, **223**, 1998, 37–43.
35. Kitaoka, S., Tsuji, T., Yamaguchi, Y. and Kashiwagi, K., 'Tribochemical Wear Theory of Non-Oxide Ceramics in High-Temperature and High-Pressure Water', *Wear*, **205**, 1997, 40–46.
36. Usui, E., Shirakashi, T. and Kitagawa, T., 'Analytical Prediction of Cutting Tool Wear', *Wear*, **100**, 1994, 129–151.
37. Hokkirigawa, K. and Kato, K., 'An Experimental and Theoretical Investigation of Ploughing, Cutting and Wedge Formation During Abrasive Wear', *Tribology International*, **21**, 1988, 151–157.

38. Kitsunai, H., Kato, K., Hokkirigawa, K. and Inoue, H., 'The Study of Microscopic Wear Mechanism by an SEM-Tribosystem: Transition of Microscopic Wear Mode During Repeated Sliding Friction', *Transactions of JSME*, **57**(535), 1991, 319–326.
39. Kato, K., In-situ SEM Observations of Wear Mode Transitions of SiC in Abrasive Sliding, by Experiment Study with SEM-Tribosystem, Tribology Laboratory of Tohoku University, Sendai, 1986 (unpublished).
40. Evans, A.G. and Marshall, D.B., 'Wear Mechanisms in Ceramics', in *Fundamentals of Friction and Wear of Materials* (ed D. Rigney), ASM, Metals Park, Ohio, 1980, pp. 439–452.
41. Lim, S.C. and Ashby, M.F., 'Wear Mechanism Maps', *Acta Metallurgica*, **35**, 1987, 1–24.
42. Diao, D.F., Kato, K. and Hayashi, K., 'The Local Yield Map of Hard Coatings Under Sliding Contact', in *Thin Films in Tribology* (ed D. Dowson *et al.*), Elsevier, Amsterdam, 1993, pp. 419–427.
43. Hutchings, I.M., '*Tribology*', Edward Arnold, London, 1992, pp. 182–186.

3

Wear of Metals: A Material Approach

S.K. Biswas

Abstract

Wear of metals in dry sliding is dictated by the material response to traction. This is demonstrated by considering the wear of aluminium and titanium alloys. In a regime of stable homogeneous deformation, the material approaching the surface from the bulk passes through microprocessing zones of flow, fracture, comminution and compaction to generate a protective tribofilm that retains the interaction in the mild wear regime. If the response leads to microstructural instabilities such as adiabatic shear bands, the near-surface zone consists of stacks of 500-nm layers situated parallel to the sliding direction. Microcracks are generated below the surface to propagate normally away from the surface through microvoids situated in the layers, until they reach a depth of 10–20 μm. Rectangular laminate debris consisting of a 20- to 40-layer stack is produced. The wear in this mode is severe.

3.1 Introduction

In the latter part of the twentieth century, there have been two major innovations to improve the wear resistance of engineering components in dry (continuous or intermittent) contact with a counterface. The first is the coating of the surface thinly with, generally, a hard ceramic and the second is the engineering of the surface itself by thermal and/or chemical treatments. Extensive work on the tribology of these new systems has brought out the fact that most of these innovations indeed lead to high wear resistance, but their reliability over long periods of use cannot always be taken for granted. The presence of defects and an interface that becomes a source of ready failure, coupled with the fact that expensive processing techniques need to be used to provide the protection, has limited their application to specific uses and no across-the-board panacea to the problem of wear appears to have been found. It is possibly

Wear – Materials, Mechanisms and Practice Edited by G. Stachowiak

for these reasons that there is renewed interest in the wear of metals and specifically in the mechanism of wear. The objective is to achieve a suitable wear-resistant microstructure by relatively cheap processing and alloying routes.

Another reason for the renewed focus on metal tribology is the advent of structurally sound light metals such as aluminium and titanium. High resistance to corrosion and the possibility of significant reduction of component weights have led to these metals being considered for high-performance applications where temperature, strain rates and stresses are maintained high to optimize efficiency. Aluminium is thus a serious candidate for replacing iron-based materials in internal combustion engines, while titanium- and nickel-based materials are used in aircraft engines. The tribology of these materials in these actual application conditions is receiving serious attention. The conventional continuum-based constitutive relations which have been used to model tribological behaviour of metals are no longer found adequate to explain tribology under these conditions as material response may no longer be homogeneous and isotropic. The information which is now sought to design materials for this new range of applications is related to dynamic evolutions of microstructure at the tribological interface and their influence on friction and wear. This has led to a search for alloying elements, heat treatment and processing routes which yield an optimum microstructure. For example, it is now established in industry that a precise combination of silicon, nickel and copper as alloying addition to aluminium cylinder liners cast at a critical cooling rate yields a microstructure which optimizes the friction coefficient at the top dead centre of a stroke. A mechanics of material-based approach to metal tribology thus provides a direction for future research.

As a function of imposed conditions such as normal load, velocity, environment and heating, the wear of metal may be mild or severe, one distinguished from the other by an order of difference in the wear rate or by an order or more of difference in the wear coefficient. Severe wear may lead to seizure and melting. An engineer is primarily interested in prolonging mild wear and thus preventing a transition to severe wear. The research work has thus concentrated on the mechanism of mild wear and the condition and changes that usher in severe wear, generally characterized by laminate-shaped metallic debris.

3.2 Mild Wear and Transition to Severe Wear

3.2.1 Mild Wear

The early notion [1] of mild wear was that it is primarily an oxidative phenomenon of growth and spalling of oxide at the asperity level. For iron-based metals, this may indeed be the protective mechanism, as has been proposed by Hanlon *et al.* [2] for bearing steels and by Rainforth *et al.* [3] for stainless steels. The strains in the near-surface levels in wear are high. For those microstructures with sufficient strain capacity and a high work-hardening index, the *in situ* changes may bring about protection [4, 5] by work hardening. Rigney [6], on the basis of some detailed TEM (transmission electron microscopy) studies, has shown how dislocation cell structures evolve in the near-surface region. Mechanically mixed layers (MMLs) have been found to protect copper [7], aluminium and its alloys [8–10], cadmium [11] and $MoSi_2$ [12]. In these wear tests, there is generally no oxygen found in the debris [9, 10, 13], though in some cases a little oxygen has been found in the MML itself [13].

On the basis of the present work on aluminium silicon alloy, a model has been proposed for the formation of an MML, as shown in Figure 3.1. As the material approaches the surface with wear, it flows and fractures. The fractured particles are comminuted to nanosized

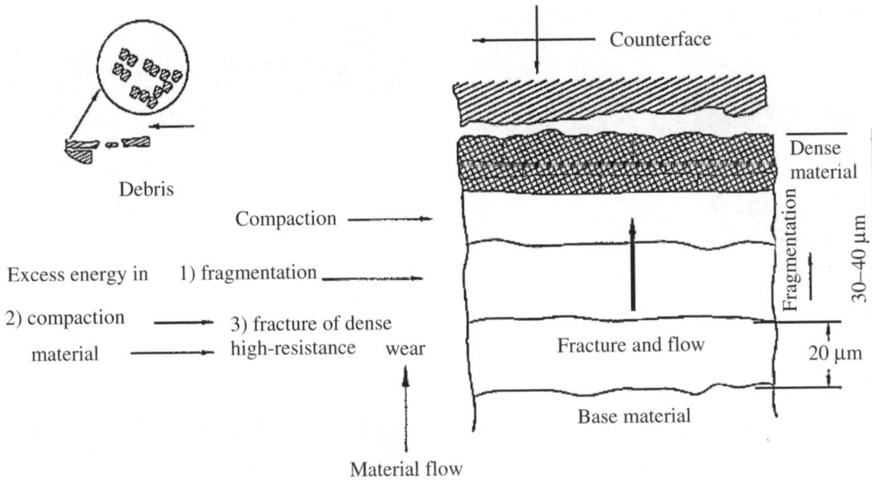

Figure 3.1 Schematic of a proposed model for the formation of the mechanically mixed layer (MML)

particles under high compressive and shear stresses in the 20–30 μm region below the surface. The comminuted material mixes with particles coming from the counterface (though not necessarily so), and under large hydrostatic compressive stresses which prevail in this region the mixture is compacted into smooth slabs. These slabs protect the surface and wear out by fracture. Figure 3.2 shows that this is a pressure- or stress-controlled process. Pressure has two effects mutually opposed to each other. Increasing pressure increases deformation and damage of the subsurface, which promotes wear. Increasing pressure, on the other hand, gives rise to greater compaction, which gives a denser MML and less wear. At 100 N, the compactive pressure is at the maximum, but this load also ushers in gross failure in the subsurface, which makes the MML unstable, thus moving the system into the severe wear region (Figure 3.2(e)). Figure 3.3 shows the corresponding wear characteristics.

It has also been found that it is possible to change the morphology of the MML by adding alloy elements to the bulk material. Addition of copper and magnesium to Al–Si alloy [8] and subsequent heat treatment spheroidized the second phase and gave rise to a very smooth (compared to that shown in Figure 3.2) MML. The hardness of the layer was found to be three to five times that of the bulk. This improved the attachment of the layer to the substrate [13] and the spheroidization reduced crack nucleation. The overall effect was to reduce the mild wear rate three times and to increase the transition load 1.5 times.

3.2.2 Transition to Severe Wear

Laminate debris is a characteristic of severe wear. A number of authors [9, 14–19] have noted crack nucleation at second-phase particles in a plastically flowing matrix, propagation of cracks parallel to the sliding direction and delamination leading to the formation of a debris, the debris thickness being related to the depth of the plastic zone. In an early work, Beesley and Eyre [20] found that the transition to severe wear was related to gross plastic flow in the subsurface that destabilizes the protective oxide layer. Wilson and Alpas [21], providing an excellent review of work in this area, observe that for a given material there is a critical

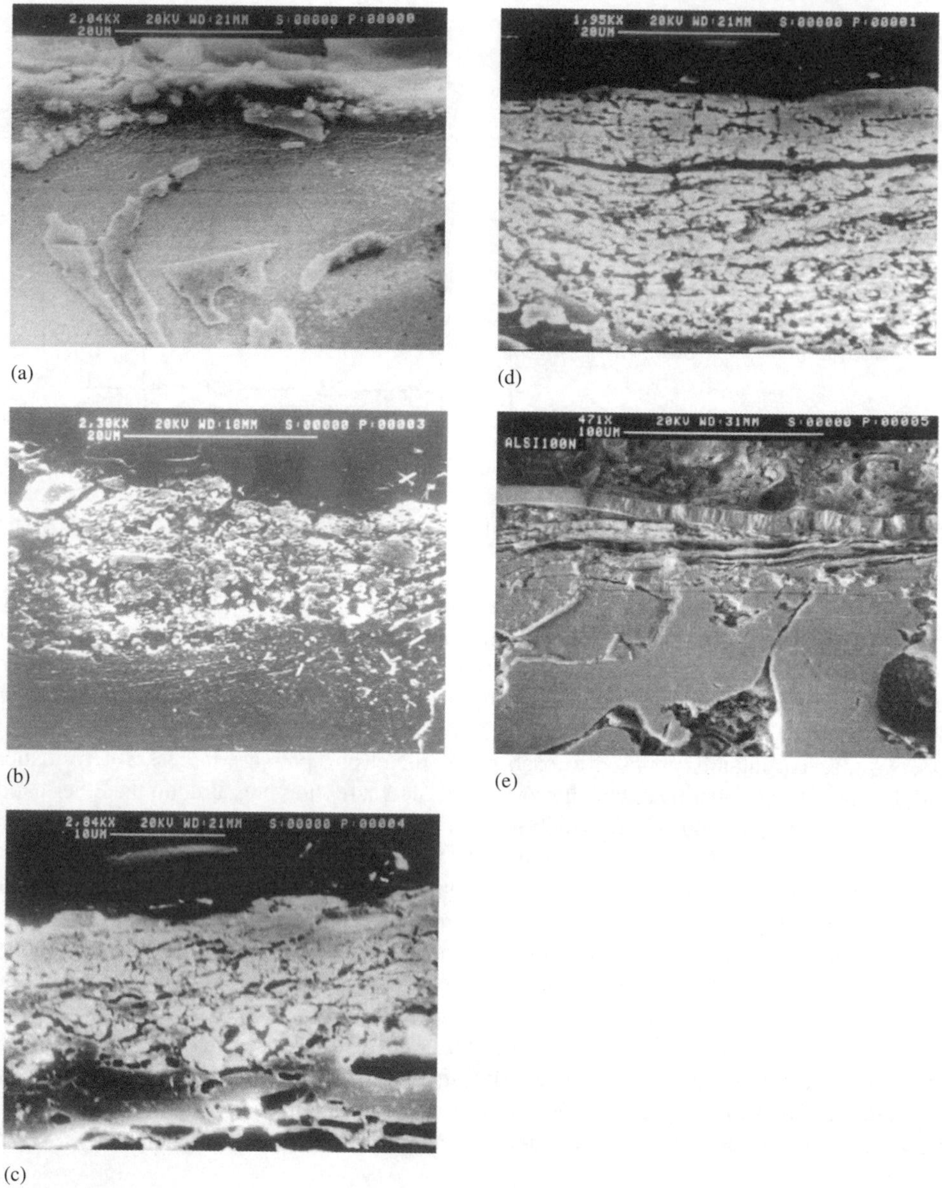

Figure 3.2 SEM micrographs of cross-sections taken parallel to the sliding direction for Al–7%Si alloy, speed 0.5 m/s and normal load: (a) 10 N; (b) 20 N; (c) 50 N; (d) 75 N; and (e) 100 N

flash temperature at which the mild-to-severe wear transition takes place. They observe that the load which marks the mild-to-severe wear transition decreases with increasing sliding velocity. They conclude that at low sliding velocities (or nominal strain rates), the material response is stable and isothermal, the debris being heavily strained equiaxed small particles. If the wear is mild, the MML consists of such particles compacted to a film. At higher

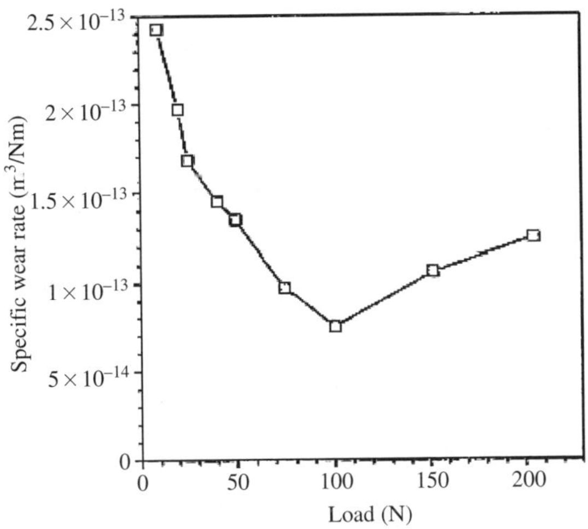

Figure 3.3 Specific wear rate varying with normal load at ambient condition for Al–7%Si alloy, speed 0.5 m/s

velocities, there is a critical strain for a given velocity which brings about strain localization and shear instabilities in an adiabatic environment. Such shear instabilities initiate high or severe wear and lead to laminate-shaped debris. This strain, which marks instability, decreases to a minimum and then increases with increasing nominal strain rate or sliding velocity. If such non-linear and non-monotonic behaviour of the critical strain (needed for shear instabilities) with strain rate is indeed true and it is assumed that the strain rate in a sliding subsurface increases from the bulk to the surface, it is possible for strain localization to be initiated at the surface or at some subsurface depth.

A heterogeneous or strain-localized microstructure in the near-surface zone is a feature of sliding wear of some metals. In their compression-cum-shear experiments on copper Dupont and Finnie [22] found adiabatic shear bands, the orientation of which was determined by the imposed traction. They opine that such bands are equivalent to elongated narrow subgrains found in wear experiments. Rigney [7] also found elongated (in the direction of sliding) subgrains with sharp cell walls in the near-surface regions, where the misorientation between the substructure elements is high.

Given this subsurface microstructure heterogeneity in sliding wear of metals, the question to be posed at this stage is how is debris generated. The evolved substructure implies local weaknesses and/or cracks in the cell walls. Such weaknesses have been shown [23, 24], depending on the material ductility, to undergo instability in mode II and to propagate, yielding laminate debris.

There thus appears to be two types of material response in metallic wear. It can be quasi-static isothermal or dynamic adiabatic. The former yields debris when fracture strain of a metal is exceeded or when there is unstable plastic flow as the shakedown limit is exceeded. The latter gives rise to microstructural instabilities and inhomogeneities. Considering that even in slow-speed experiments the strain rate near the interface can be high, it seems relevant to ask what the causes and parametric dependence of such instabilities are. For example, is such dependence monotonic with respect to strain rate and temperature? Further, in order to explain the comparative wear resistance of two metals, it may be important to delineate their proneness to an unstable material response.

In 1980, an important development took place in the field of material processing by large-scale deformation, which is discussed in [25–27]. These works demonstrate, in compression tests, the development of intrinsic instabilities such as adiabatic shear banding, twinning, wedge cracking and dynamic recrystallization. It was found that in specific regimes of strain rate and temperature, a metal is prone to dissipate a larger share of input power to evolve microstructural instabilities than it dissipates in homogeneous deformation and heating. These responses are unique to the initial microstructure of the metal. Such tendencies are reflected in the stress–strain characteristics of a metal as a peak stress followed by softening, serration and other unstable behaviour. Figure 3.4, for example, shows the stress–strain characteristic in compression of a hypereutectic aluminium alloy prone to an unstable strain rate response.

The importance of this work for tribology can be summarized as follows:

1. It provides a thermodynamics-based reasoning for the origin of localized deformation and instabilities. This is important for tribology as it occurs in the environment of large compressive stresses found in the subsurface of a sliding wear component.
2. It is related to initial microstructures and is not necessarily correlated with hardness. In tribology there are many instances of a lack of inverse correlation of wear with hardness [28–30].
3. The strain rate response is not monotonic in the temperature–strain rate space. This provides a clue for explaining some of the observed non-monotonic wear behaviours with load and velocity.

The wear behaviour of a number of metals, such as titanium [31], copper [32] and cadmium [11], has been investigated in the framework of their strain rate responses. First,

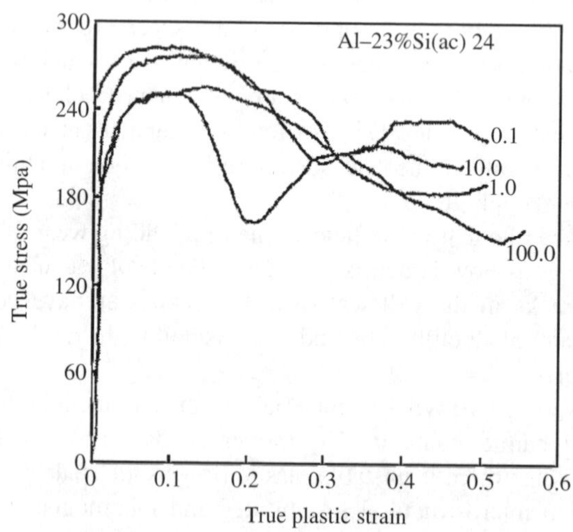

Figure 3.4 Stress–strain behaviour in room temperature compression at various strain rates (0.1–100/s) for Al–23%Si alloy

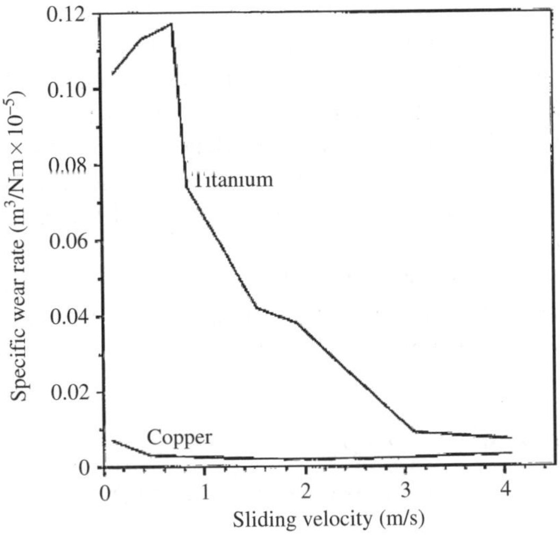

Figure 3.5 Sliding wear of OFHC copper and titanium pins against alumina in a pin-on-disc machine under the dry ambient condition

the strain rate response map of a metal in compression has been generated over a range of practical strain rates and temperatures. Then, the two estimated (surface layer) temperatures and strain rates obtained from the wear experiment have been superimposed on this map as a function of normal load and velocity. The 'compression' instability regimes were correlated with the variations in wear rate. Using this method, it was possible to explain the high wear resistance of oxygen-free high conductivity (OFHC) copper in comparison to that of a much harder metal, titanium (Figure 3.5). The presence of instability in the high strain rate–low-temperature regime also provided an explanation for the high wear rate of titanium at low sliding velocities.

It is not suggested here that this approach provides the only possible explanation for the wear of metals. For example, in some soft and ductile metals such as aluminium and copper, microstructural instabilities are rarely observed in practical ranges of temperature and strain rate. The wear takes place by homogeneous deformation in an isothermal mode. Material is removed by extrusion and formation of lips [32]. In this chapter, the wear of an engineering alloy Ti–6Al–4V used extensively in aerospace applications is investigated. The interesting feature of the wear of this alloy is that the wear with respect to normal load and velocity is non-monotonic and it is possible to demonstrate, with the help of a strain rate response map, how the wear characteristics over a monotonic change in velocity move from severe to mild regimes depending on the prevailing strain rate and temperature vis-á-vis those indicating 'compression' instabilities.

3.3 Strain Rate Estimates and Bulk Surface Temperature

In one series of experiments, Ti–6Al–4V (Ti64) pins of 9-mm diameter were machined out of hot-rolled Ti64 rods. The sliding tests were done on a pin-on-disc machine where a normally loaded Ti64 pin was slid against an alumina disc. The alumina disc (99.5%) made by hot isostatic processing was ground to a roughness of 0.2 μm CLA (central line average). The tests were carried out for a sliding distance of 1500 m in air at ambient temperature in a normal load range of 30–110 N and a sliding speed range of 1–11 m/s. Run-in was done at a sliding speed of 0.05 m/s and a normal load of 5 N for 30 min. The wear data are presented as an average of five experiments. The data were found to fall within 12% of the mean value. The worn surfaces were nickel coated and cross-sectioned using a low-speed diamond saw. The polished cross-sections were etched and viewed in the SEM (scanning electron microscope).

Figure 3.6 shows that at normal loads up to 70 N the wear rate decreases monotonically with increasing velocity. At higher loads, it is non-monotonic as a second peak happens at a velocity of 9 m/s, beyond which the wear decreases sharply to a level typical of lower loads.

3.3.1 Strain Rate Response Maps

The strain rate in the subsurface of a worn specimen is given by

$$\varepsilon\acute{Y} = \frac{d\varepsilon}{dt} = \left(\frac{\partial\varepsilon}{\partial x}\right)\left(\frac{dx}{dt}\right) + \left(\frac{\partial\varepsilon}{\partial y}\right)\left(\frac{dy}{dt}\right)$$

For steady-state sliding $(\partial\varepsilon/\partial y) = 0$, where y is the sliding direction. Thus

$$\varepsilon\acute{Y} = \left(\frac{\partial\varepsilon}{\partial x}\right)\left(\frac{dx}{dt}\right) = \sqrt{3}\left(\frac{d\varepsilon_x}{dx}\right)\left(\frac{dx}{dt}\right)$$

where $(d\varepsilon_x/dx)$ is the strain gradient with respect to the depth (x) from the surface to the bulk of the worn specimen and (dx/dt) is the wear rate, which can be taken as the rate at which the

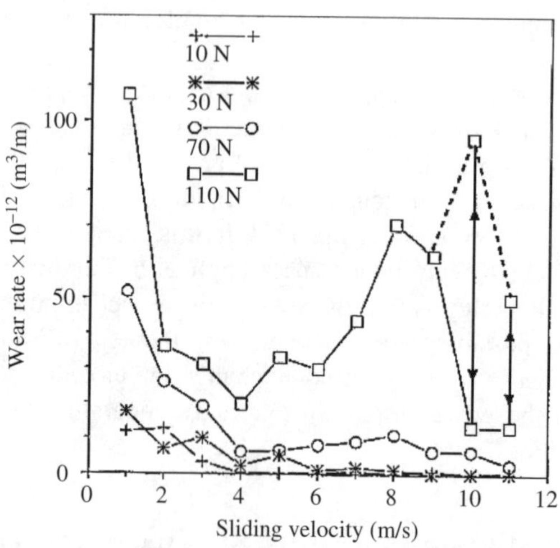

Figure 3.6 Wear rate versus sliding velocity of Ti–6Al–4V at constant normal loads, counterface alumina. For a 110-N load and 10 and 11 m/s sliding velocities arrows indicate periodically (sliding time) oscillating values. Standard deviation ($\pm\sigma$) is 12% of the mean shown in the figure

distance x of a point in the subsurface changes with time. For the present work, the gradient of strain obtained for copper by Alpas *et al.* [33] was used. This assumption is reasonable as the subsurface strain determined in the case of titanium [31], an hexagonal close packed (HCP) metal, using the deformation of grains in the subsurface [34] showed that the values were quite close to that determined by Alpas *et al.* [33]. Further, only an order of magnitude variation in the strain rate was found to affect interpretation of the present results.

Uniaxial compression tests were done on cylindrical billets of Ti–6A1–4V material in a strain rate range of 10^{-5}–10^{2}/s and a temperature range of 25–1000°C. The billets were compressed to 0.3 strain. The deformed billets were sectioned diametrically and their microstructure observed using an SEM. Figure 3.7 shows the zones in a strain rate–temperature space where adiabatic shear bands were observed in compression (for further details of these tests and strain rate response maps see [31, 32]). The steady-state temperatures and strain rates achieved in the wear tests (estimated as above) are plotted on the compression strain rate response maps. Figure 3.7 shows the strain rate–temperature coordinates of the wear experiments joined by straight lines.

The map (Figure 3.7) shows that at low velocities the subsurface microstructure is likely to contain adiabatic shear bands. If we assume that higher loads lead to more extensive propagation of cracks initiated by these localized shear bands, we would expect the wear rate in the low-velocity regime to rise with load as is in fact seen in Figure 3.6. Figure 3.7 shows that as the velocity increases, the strain rate–temperature vector emerges out of the unstable zone, and we may expect the subsurface deformation to become more homogeneous, yielding fracture debris that may aid in the formation of MML. This lowers the wear rate. The map in Figure 3.7 indicates the possibility of the strain rate–temperature vector entering a second zone of instability when the load and velocities are both high. In this regime, Figure 3.8 shows the friction to be also high, suggesting the additional possibility of high melt wear.

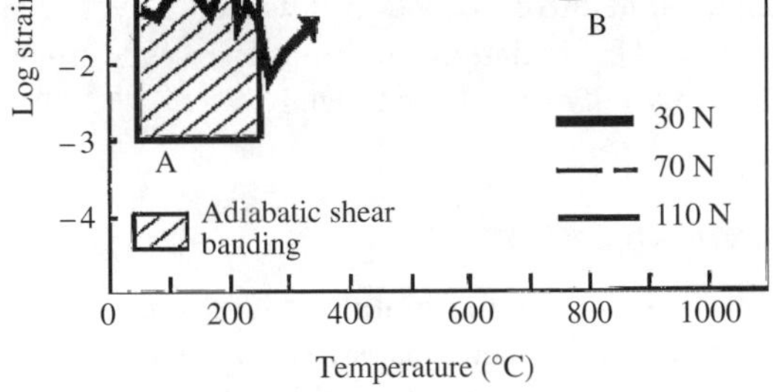

Figure 3.7 Constant normal load temperature–strain rate coordinates of wear experiments superposed on the strain rate response map of Ti–6Al–4V, in compression. The paths shown are obtained by joining wear experimental points by straight lines. The experimental points are strain rate–temperature coordinates, estimated from experimental data at a subsurface depth of 1 μm (1×10^{-6} m)

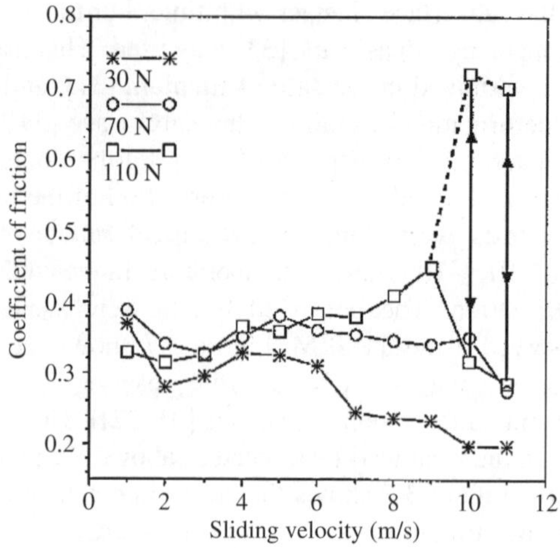

Figure 3.8 Coefficient of friction versus sliding velocity of Ti–6Al–4V alloy, in dry sliding at constant loads. At 110-N load and 10 and 11 m/s sliding velocities arrows indicate periodically (sliding time) oscillating values. Standard deviation ($\pm\sigma$) is 12% of the mean shown in the figure

3.3.2 Bulk Surface Temperature

Temperatures at different depths of 5 and 10 mm from the wearing surface were measured using chromel–alumel thermocouples; 1-mm holes were drilled horizontally at these two depths. The holes were drilled to reach the cylindrical axis. The thermocouples were set using silver paste. Following a step change in velocity, a time of 10 min was allowed before recording the temperature, although it was found that about 5 min was sufficient for the steady-state conditions to be established. Convective and radiative losses were considered in the estimation of the bulk surface temperatures. The convective heat transfer coefficient was calculated using the Churchill and Bernstein correlation for the forced convection in the cross-flow over cylinders [35]. The resultant velocity at any given axial location was obtained from the relations presented by Schlichting for flow over a rotating disc [36].

The average heat transfer coefficient was obtained by numerical averaging of the local heat transfer coefficients. An iterative method [37] was employed to obtain the temperature distribution along the pin. The radiation contribution was estimated from the Stefan–Boltzman law using the emissivity of polished metals.

3.3.3 The Phenomenological Argument

It has to be reiterated at this stage that the strain rate response map is for compression only. It may be justifiably argued that as these maps are sensitive to the stress state [38–40], a compression–shear map, more appropriate for a sliding wear test, may indeed be somewhat different from that shown in Figure 3.7. As such maps are not available at present, some qualitative observations will be made using the available 'compression map'.

In analysing the monotonic decrease in the abrasive wear of carbon steel, Nakajima and Mizutani [41] suggested that at low velocities the crystal lattice of surface layer material is unable to distort under load, giving rise to brittle failure. With increasing speed and rising temperature, a more homogeneous deformation mode takes over and the material fails by ductile failure. The present study leads to an extension of that argument to suggest that for Ti64 the power is dissipated in the low-temperature regime, giving rise to intense localized shear deformation. This nucleates microcracks. The important point to note here is that as long as the strain rate–temperature coordinate remains in the unstable regime of the map, the wear is high. As the nucleation sites and the microcracks become available in abundance due to adiabatic shear banding, the material is able to fracture easily.

The friction force trace at 110-N load and at velocities of more than 9 m/s showed spikes (Figure 3.8). Figure 3.9 shows melting on the specimen surface at 110-N load and 11 m/s. It may be surmised that when there is surface melting the coefficient of friction is much decreased and the strain rate–temperature coordinate in Figure 3.7 doubles back to the homogeneous deformation regime. This raises the friction and surface temperature for the interaction to re-enter regime B. The fluctuation repeats and this dual nature of the interaction is also reflected in the wear rate recorded at high load and high velocity (Figure 3.6).

3.3.4 Micrographic Observations

Figure 3.10(a) shows the subsurface of a specimen worn in the regime of low velocity (3 m/s) and high load (110 N). A highly layered structure (parallel to the sliding direction) exists in the top 10–20 μm region interspersed with cracks, which propagate fracturing through these layers, in a direction normal to the surface. Figure 3.10(b) shows a higher magnification view of the near-surface region made up of a stack of sheets/layers, each layer about 500 nm thick. Figure 3.10(c) shows another view of the cracks propagating normal to the surface. These cracks originate at the surface or very near the surface and propagate through the stack of sheets/layers until they reach a depth of 10–20 μm. It appears that at this stage the crack is deflected by 90° and travels along an interlayer interfacial weakness zone until it meets

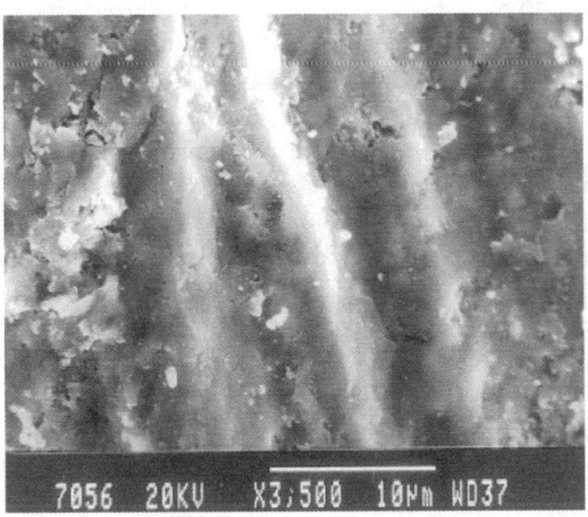

Figure 3.9 Micrograph of a worn surface of Ti–6Al–4V alloy showing melting, with load 110 N, speed 11 m/s

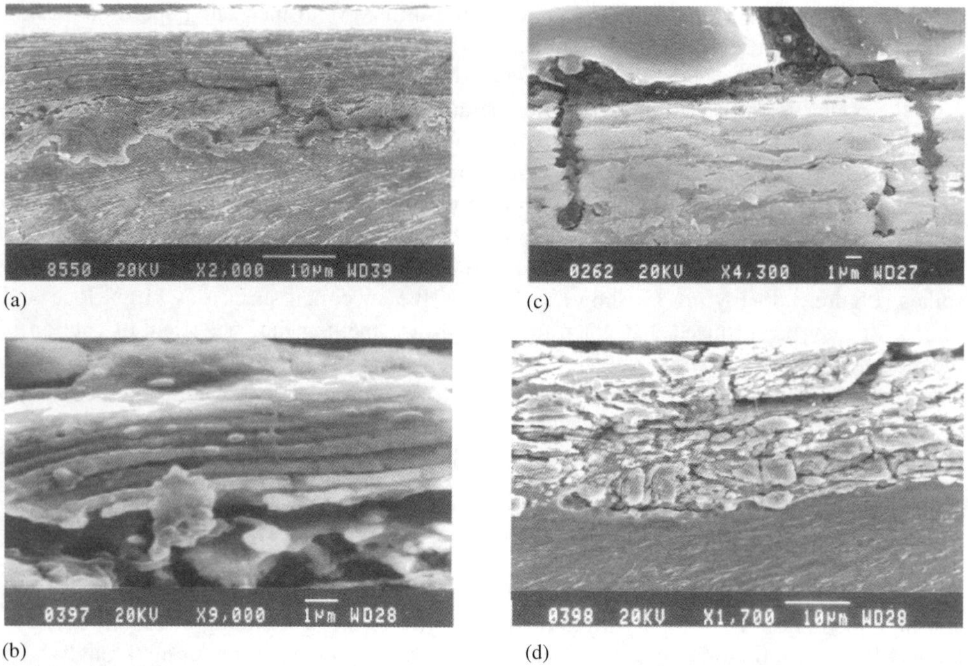

Figure 3.10 SEM micrographs of the worn subsurface, load 110 N, speed 3 m/s, cross-sections parallel to the sliding direction (Ti–6Al–4V): (a) the layered structure and debris formation; (b) details of the layered structure and interlayer cracking; (c) cracks propagating downwards normal to the worn surface; and (d) the generation of a debris

another normally propagating crack and a debris is generated (Figure 3.10(d)). If the normal load is now reduced to 30 N at this low velocity, the subsurface shows no sign of cracking (Figure 3.11(a)). On the other hand, it shows a smooth MML typical of mild wear. A model for the formation of the MML is shown in Figure 3.1. Compared to the morphology of the MML seen in the case of A1–Si alloy, the top slab of the MML here is always found to be continuous and smooth. Figure 3.11(b) and (c) shows the morphology of the MML recorded at a moderate velocity for a high load (5 m/s, 110 N) and Figure 3.11(d) at a very high velocity for a high load (11 m/s, 110 N). The Energy Dispersive Spectroscopy (EDS) of the MML showed (Figure 3.11(c)) some fine particles of alumina from the counterface finely dispersed in the tribofilm.

It is important to note that all the conditions that promote the MML give rise to low wear. Further, the temperature–strain rate coordinates of these conditions also fall outside the instability and inside the homogeneous deformation zones of Figure 3.7.

The subsurface morphology of samples worn at a high sliding velocity (9 m/s) and a high load (110 N) was found to be very similar to that found for the low-velocity, high-load sample. Figure 3.12 shows a tensile crack that originates just below the surface to propagate in a direction normal to the sliding surface.

Some problems have been faced in quantitative explanations. Figure 3.13 shows the constant velocity strain rate–temperature coordinates superposed on the strain rate response

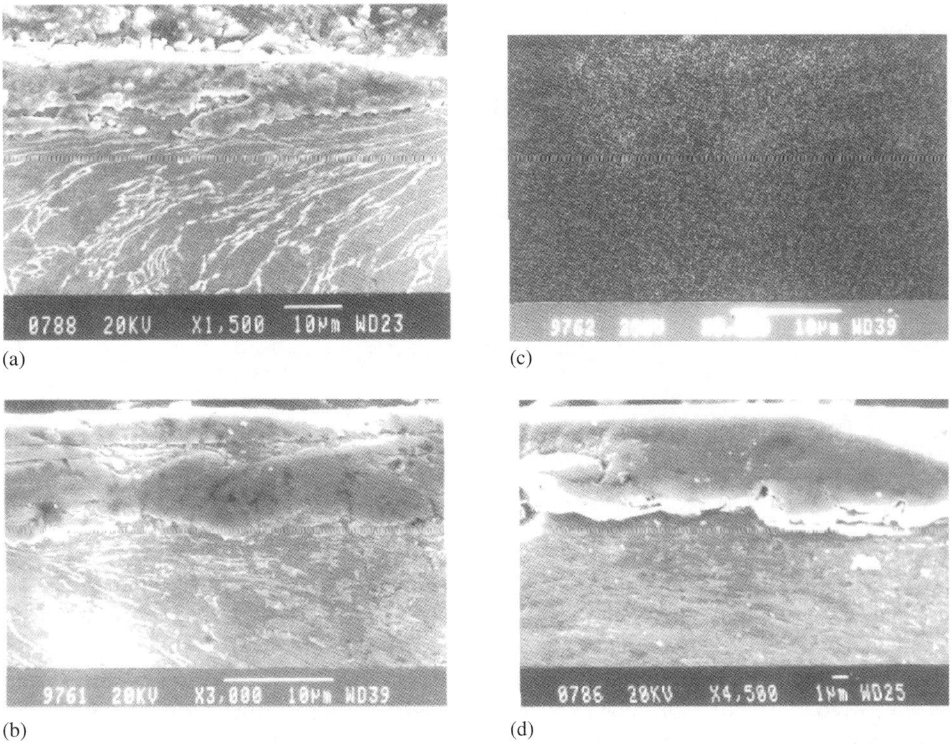

Figure 3.11 SEM micrographs of the worn subsurface, cross sections parallel to the sliding direction (Ti–6Al–4V): (a) normal load 30 N, speed 3 m/s, showing the mechanically mixed layer; (b) normal load 110 N, speed 5 m/s, showing the mechanically mixed layer; (c) EDS (aluminium) of (b); and (d) normal load 110 N, speed 11 m/s, showing the mechanically mixed layer

Figure 3.12 SEM micrograph of the subsurface, cross-section parallel to the sliding direction, load 110 N, speed 9 m/s, showing the microfracture (Ti–6Al–4V)

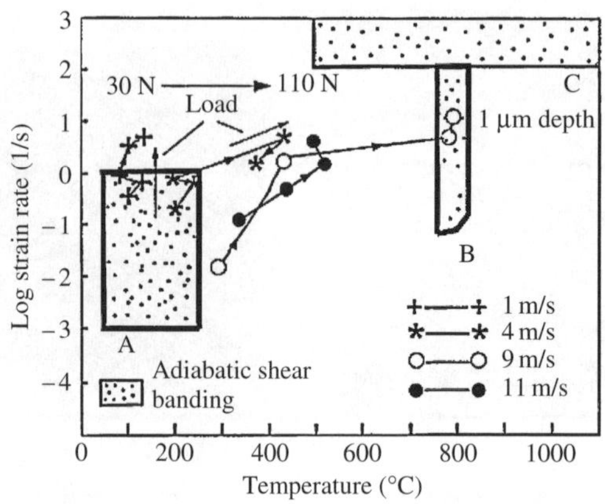

Figure 3.13 Constant sliding velocity temperature–strain rate coordinates of wear experiments superposed on the strain rate response map of Ti–6Al–4V, in compression

(compression) map of Ti64. The location of data presented for 1, 9 and 11 m/s matches reasonably well with the morphological (subsurface SEM micrograph) and wear data. High wear and low wear are seen at all loads for 1 and 11 m/s, respectively. For 9 m/s and low loads there is mild wear but at the same velocity high loads show severe wear and microstructural instability (zone B). For 4 m/s, however, the low-load data indicate microstructural instability and high wear, when in reality these experiments generate the MML and low wear. Such a discrepancy may indeed arise from the fact that the 'strain rate responses' are stress state related. The maps of relevance to tribology should therefore be compression–shear maps. Fabrication of such maps would help to provide a better understanding of the mechanism of wear in metals.

3.4 Summary

3.4.1 Homogeneous Deformation – Severe Wear

In this mode, the deformation varies continuously over the subsurface space, and continuum descriptions of elastic, plastic and fracture events are possible. In the absence of a protective film there is direct contact between the mating parts. The subsurface is likely to be in an isothermal state.

Some materials such as aluminium exhibit a homogeneous deformation mode over a wide range of temperature and strain rate. Other materials such as titanium and its alloys exhibit localized zones of intense deformation (instability) in some regimes of strain rate and temperature, while in others they deform homogeneously. When the deformation is homogeneous but there is large-scale flow in the subsurface, protective tribofilms are not allowed to settle and there is direct contact between the material and the counterface. The wear is severe and may happen by extrusion-aided lip formation [7]. Severe wear can also happen in a homogeneous deformation mode in sliding wear of a two-phase material where the cracks originating at the particle matrix interface can lead to laminate debris [9, 14, 15].

3.4.2 Homogeneous Deformation – Mild Wear

The deformation is homogeneous, as defined above, but a protective film at the interface prevents direct contact between the mating parts. Within the framework of homogeneous deformation a material may respond to traction in an altogether different way than in the case of severe wear in order to promote a mode of mild wear. The material from the bulk moving towards the surface, with wear, passes through different processing zones, flow, fracture, comminution and compaction, to yield a radically different near-surface microstructure. In the comminution stage, the fractured material is reduced to (100) nanometre-sized spherical particles. These particles mix in the final stage (next to the surface), with similar particles coming from the counterface as debris, and the mixture is compacted to yield a stable tribofilm at the surface. The film wears out by the intersection of cracks normal and parallel to the surface, generating small slabs of debris.

3.4.3 Inhomogeneous Deformation – Severe Wear

In this mode, zones of intense deformation are localized in the subsurface space where the thermal condition is near adiabatic. There is direct contact between the mating parts. When the strain rate–temperature combination promotes an inhomogeneous response in the form of localized instabilities, near-surface material is organized in stacks of extended layers. Microcracks are nucleated at points of instability and extend normal to the sliding direction. When such a crack meets an interlayer interface weakness at a distance of about 10–20 μm from the surface a debris is generated. Rosenfield [24] has shown that shear instabilities start just below the surface and can exist in a zone that spans a finite subsurface depth, this depth being a function of the normal load. It is likely that a weakness exists at the lower boundary of this zone, which facilitates the propagation of a crack in a direction parallel to the surface. No tribofilm is formed in such a situation and the wear is severe and high.

Acknowledgements

The author is grateful to a number of his students, Arvind Singh, Satish Vasu Kailas, Somi Reddy, B.N. Pramila Bai and Shivananda, for their contribution at various stages of this work and Mr H.S. Shamasunder for his technical support.

This material has been based upon an article that first appeared in the *Journal of Engineering Tribology* – Proceedings Part J, 2002, Vol. 216, No. J6, ISSN 1350–6501, published by Professional Engineering Publishing. Permission is granted by the Institution of Mechanical Engineers.

References

1. Quinn, T.F.J., '*Fundamentals of Tribology*', Vol. 117, MIT Press, MA, USA, 1995, p. 297.
2. Hanlon, D.N., Rainforth, W.M. and Sellars, C.M., *Wear*, **203–204**, 1997, 220.
3. Rainforth, W.M., Stevens, R. and Nutting, J., *Philosophical Magazine A*, **66**, 1992, 621.
4. Perrin, C. and Rainforth, W.M. *Wear*, **203–204**, 1997, 171.
5. Tabrett, C.P. and Sare, I.R., *Wear*, **203–204**, 1997, 206.
6. Rigney, D.A., *Annals of Material Science*, **18**, 1988, 141.
7. Rigney, D.A., in *Proceedings of ASM International Conference on Wear of Engineering Materials*, Materials Park, Ohio, 1998, p. 3.

8. Pramila Bai, B.N. and Biswas, S.K., *Acta Metallurgica*, **39**(5), 1991, 833.
9. Pramila Bai, B.N. and Biswas, S.K., *Transactions ASLE*, **29**, 1988, 116.
10. Antonio, R. and Borland, D.W., *Material Science Engineering*, **93**, 1987, 57.
11. Singh, R.A., Biswas, S.K. and Kailas, S.V., *Wear*, **225–229**, 1999, 770.
12. Shivananda, K., Ramasesha, S.K. and Biswas, S.K., in *Proceedings of International Tribology Conference*, Japanese Society of Tribologists, Nagasaki, 2000, p. 385.
13. Lepper, K., James, M., Chashechkina, J. and Rigney, D.A., *Wear*, **203–204**, 1997, 46.
14. Pramila Bai, B.N. and Biswas, S.K., *Lubrication Engineering*, **43**, 1987, 57.
15. Pramila Bai, B.N. and Biswas, S.K., *Journal of Material Science*, **19**, 1984, 3588.
16. Liu, Y., Asthana, R. and Rohatgi, P., *Journal of Material Science*, **26**, 1991, 99.
17. Zhang, J. and Alpas, A.T., *Material Science Engineering: A*, **160**, 1993, 25.
18. Wilson, S., *The Tribological Behaviour of Al Matrix Composites*, Ph.D. thesis, University of Capetown Libraries, 1993, microfilm.
19. Wilson, S. and Ball, A., '*Advances in Composites Tribology*', Vol. 8, Elsevier, Amsterdam, 1993, p. 311.
20. Beesley, C. and Eyre, T.S., *Tribology International*, **9**(2), 1976, 63.
21. Wilson, S. and Alpas, A.T., *Wear*, **225–229**, 1999, 440.
22. Dupont, F. and Finnie, I., *Wear*, **140**, 1990, 93.
23. Rosenfield, A.R., *Wear*, **72**, 1981, 97.
24. Rosenfield, A.R., *Wear*, **116**, 1987, 319.
25. Rishi, R., *Metal Transactions*, **12A**, 1981, 1089.
26. Gegel, H.L., Malas, J.C., Doraivelu, S.M. and Shande, V.A., *Metals Handbook*, **14**, 1988, 417.
27. Prasad, Y.V.R.K., *Indian Journal of Technology*, **28**, 1990, 435.
28. Rigney, D.A., *Wear*, **175**, 1994, 63.
29. Lancaster, J.K., *Wear*, **141**, 1990, 159.
30. Biswas, S.K., '*New Directions in Tribology*' (ed I.M. Hutchings), Mechanical Engineering Publications, London, 1997, p. 157.
31. Kailas, S.V. and Biswas, S.K., *Transactions of ASME, Journal of Tribology*, **119**, 1997, 31.
32. Kailas, S.V. and Biswas, S.K., *Transactions of ASME, Journal of Tribology*, **121**, 1999, 795.
33. Alpas, A.T., Hu, H. and Zhang, J., *Wear*, **162**, 1993, 188.
34. Dautzenberg, J.H. and Zaat, J.H., *Wear*, **126**, 1973, 105.
35. Churchill, S.W. and Bernstein, M., *Transactions of ASME, Journal of Heat Transfer*, **99**(2), 1977, 300.
36. Schlichting, H., '*Boundary-Layer Theory*', 7th edn, McGraw-Hill, New York, 1972.
37. Singh, R.A. and Biswas, S.K., *Tribology Letters*, **13**(3), 2002, 203.
38. Kailas, S.V., Prasad, Y.V.R.K. and Biswas, S.K., *Metallurgical and Materials Transactions*, **25A**, 1994, 2173.
39. Prasad, Y.V.R.K. and Seshacharyulu, T., *Materials Science Engineering*, **A243**, 1998, 82.
40. Lee, W.-S. and Lin, C.-P., *Materials Science Engineering*, **A241**, 1998, 48.
41. Nakajima, K. and Mizutani, Y., *Wear*, **13**, 1969, 283.

4

Boundary Lubricated Wear

S.M. Hsu, R.G. Munro, M.C. Shen and R.S. Gates

Abstract

This chapter reviews the fundamental nature of wear under lubricated conditions and presents a comprehensive view of our current understanding of the wear processes under boundary lubrication conditions. Wear under lubricated conditions can be classified into two main classes: well-lubricated systems and marginal lubricated systems. Past studies tend to focus on two phenomena: wear mechanism in the substrate beneath the surface and the chemical and physical mechanisms within the interfacial layer which includes the lubricant, lubricating film, and transfer films. In the wear literature, the focus is on understanding the wear mechanisms and how materials can be improved for wear resistance. Many of the wear studies are therefore either not lubricated or marginally lubricated. Wear data analyses also tend to focus on how wear progresses under "dry" wear conditions. In this chapter, we use dry wear as a baseline in assessing and comparing lubricated wear phenomena, wear measurement techniques, data interpretation, and the various assumptions behind normal wear interpretations. Lastly, the current modeling of lubricated wear is reviewed.

4.1 Introduction

The definition of wear, in a strict sense, is the removal of material from two surfaces under the mechanical action of the two surfaces rubbing together. This strict definition, while useful, does not take into account subsurface deformation, surface damage, or chemical corrosion. Therefore a broader definition of wear, the study of surface degradation and material loss, often is more appropriate. In both contexts, wear is a highly complex phenomenon because the very nature of the wear process is transient and depends upon the historical timeline of the process; i.e. wear is a cumulative process, and what happens at one time is a function of all of the events that occurred previously. Damaged surfaces do not wear the same as undamaged surfaces of the same material under the same test conditions, and the final amount

Wear – Materials, Mechanisms and Practice Edited by G. Stachowiak

of material loss may be much higher in the first case than in the second case. Material inhomogeneity, misalignment, and contamination also contribute to differences in wear behavior. Consequently, there is a certain unpredictability associated with any wear evaluation.

In the late 1970s, wear research increased significantly in response to a technological demand for new materials with longer lifetimes and higher performance levels in extreme environments. Many new materials were evaluated for wear resistance under dry and lubricated wear conditions, and the successful candidate materials went on to be tested in simulators and component testing under field conditions. Over the years, a methodology for assessing the wear characteristics of the materials has developed, and a large literature database has come into existence filled with such evaluations. The lubricant and additive industries, of course, have conducted wear tests to evaluate lubricants and additives since the 1940s. A different set of test methodologies has evolved for their purposes, and their data have been evaluated on the basis of a different set of assumptions. Their results have been published mostly in chemical journals.

The aim for material evaluation most often is to improve material performance. Wear methodology contributing to this effort tends to explore the material's wear characteristics with respect to increasing stress. As a result, high wear under severe conditions is often used to gain insights into the material's dominant wear mechanisms. In contrast, to evaluate how well lubricants and additives prevent wear, the effectiveness of a lubricating substance is ranked by how little wear it permits during a simulation cycle. For lubricants, the focal interest is found in the interfacial layer between the materials and not in the material's response to mechanical stresses. Instead, the material's response to chemistry is of greater importance. Given that the chemical research community and the materials research community do not often interact closely, it is understandable that two separate evaluation methodologies have emerged in the literature, often with cause for confusion. In this chapter, we attempt to resolve some of this confusion by contrasting the two evaluation methodologies and emphasizing their different purposes.

4.2 Lubricated Wear Classification

We can classify lubricated wear studies into two broad categories: well-lubricated wear systems and marginally lubricated wear systems. A well-lubricated wear system has very low wear (wear coefficient on the order of 10^{-8}–10^{-12}), and the wear process is controlled by the boundary lubricating films formed primarily from antiwear chemical additives. A marginally lubricated wear system typically involves *non-reactive* lubricants such as purified paraffinic oils, and the wear coefficient is on the order of 10^{-4}–10^{-6}.

Lubricated wear mechanism studies can be further classified into two nominally distinct categories: investigations of the wear mechanisms active in the substrate, and studies of the effectiveness of the chemical reaction films and the associated chemical degradation mechanisms of the lubricant in the interface between two surfaces.

4.3 Lubricated Wear Versus "Dry" Wear

A wear system consists of the contacting surfaces, the interfacial layer, and the operating environment under which the contact takes place. To understand the wear process, one needs to measure wear precisely within the context of the environment without interference from unintended parameters such as vibration, alignment, and contamination. Therefore, lubricated wear is the particular aspect of tribology that is concerned with both the measurement methodology and the mechanisms of surface degradation or protection by chemically active films.

It is useful and advantageous to use "dry wear" as a baseline to form a context within which lubricated wear can be most clearly delineated. Dry wear simply means that no intentional lubricant was used to achieve effective lubrication. As we now know from several decades of research, even dry wear tests may experience some form of lubrication arising from native oxides or hydroxides formed under oxidative and/or tribochemical conditions as a result of unintended reactions with moisture or other substances in the ambient environment. Such cases may need special interpretation. However, the majority of tests run under dry wear conditions tend to be dominated by the material properties of the two opposing surfaces. These latter cases form a reasonable baseline for highlighting differences that result from differing material properties.

The material wear community tends to use pin-on-disk as a primary tool for evaluation purposes. As such, the wear usually is relatively quite severe, but eventually the wear scar is limited by the pin radius. The amount of wear can be measured by profilometry. Many studies suggest a linear relationship between wear volume and time or the distance slid, as illustrated in Figure 4.1(a). The dimensionless wear coefficient, K, expresses this relationship as

$$K = W_v H / LD$$

where W_v is the wear volume, H the hardness, L the load, and D the distance slid. The assumptions necessary for this relation are (1) wear is proportional to load; (2) wear is proportional to the distance slid (or the amount of time during which sliding contact occurred); and (3) wear is inversely proportional to the hardness of the surface being worn away. In reality, these assumptions rarely hold except for very clean systems in tightly controlled environmental conditions. Even in dry sliding, wear-in effects are frequently significant. The inverse proportionality to the hardness of the surface being worn away holds only for systems where the difference in hardness between the two surfaces is large. Surface roughness is not a factor in this measure of wear because the wear is assumed to be severe enough that surface roughness difference is insignificant. In spite of these shortcomings, the wear coefficient is a very useful normalization parameter for comparing the wear behavior of a large range of materials. The widespread use of this parameter sometimes obscures the assumptions underlying this expression.

In well-lubricated wear, the wear is controlled by the effectiveness of the interfacial layer (boundary lubricating film) in alleviating the shear forces. If the film is effective, wear approaches a steady state as controlled by the removal of the film and the chemical processes that regenerate the film. Thus, the wear behavior exhibits a characteristic dependence on time approaching a steady state or a constant value, as illustrated in Figure 4.1(b). In this case, the dependence of wear on time is not linear, and the assumed equivalence to wear volume per distance slid is no longer valid. Note that in this case the wear volume may quickly reach a steady state (sometimes approaching an asymptotic value) once the effectiveness of the lubricating film is established. After that, the primary wear occurs in the film (removal of the film) and not the surface material. (If we define wear only in the context of the surface material, then "wear" of the film is "zero wear.") In this case, dividing the wear volume by

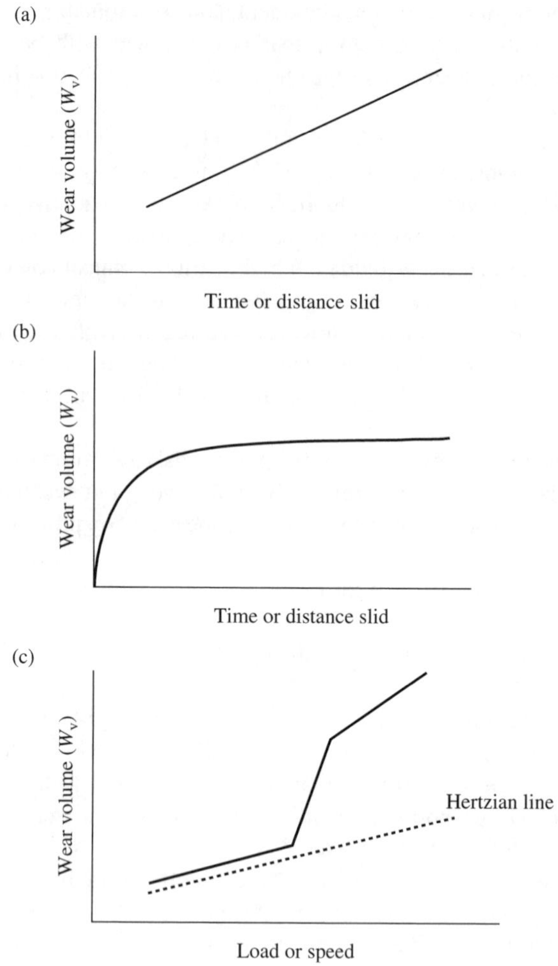

Figure 4.1 Comparison of idealized (a) linear and (b) nonlinear wear behaviors as a function of time, and (c) a schematic of the more typical wear behavior of materials as a function of load or sliding speed

the distance slid just makes the wear coefficient smaller and smaller with increasing time. This is one of the major points of confusion existing in the wear literature.

Clearly, the first issue to be resolved is what is being measured and what factors affect that measurement. In dry wear, the most obvious measure of wear is the amount of surface material removed as a result of the physical interactions between the two interacting surfaces. The mechanisms by which this volume is removed are dominated by the material properties of the interacting materials. In marginally lubricated systems (i.e. the lubricating films are not very effective, or the operating conditions far exceed the load limits of the chemical films), wear behavior can resemble the dry case. In some cases, transient nonlinear behavior prevents the system from ever reaching a steady state. This situation often occurs when material scientists add a "liquid lubricant" to their evaluation protocol without understanding the chemical reactivity between the surface and the lubricant. Often, the result is that an ineffective film is formed, and sometimes no film is formed. In well-lubricated wear, a very different situation occurs. After reaching steady state, the primary wear occurs in the lubricating film. Consequently, the effectiveness of the boundary lubricating film becomes the focus of attention, and what is measured, in a sense, is the wear of the interface. This distinction is critical to understanding boundary lubricated wear behavior.

When the focus is on the wear of the boundary lubricating film, there are additional factors that need to be considered, such as the temperature limit and load capacity of the film and the available reservoir of chemicals used to continually replenish the film. When any of these limitations is reached, the protective film no longer functions, and a wear transition (a sudden increase in wear rate) occurs as illustrated in Figure 4.1(c). This kind of transition can happen as a function of load or speed or temperature increase or time or a combination of these factors. Wear then goes through a transition from an effective lubrication regime (very mild wear, usually 10^{-8}–10^{-12} wear coefficient) to a different wear regime (usually more severe). Since the wear is controlled by chemistry, oftentimes, lubricant formulators add different mixes of additives so that when a transition occurs, the more severe contact conditions activate a different chemical film to form thus continuing to protect the system, albeit at a higher steady value. So in lubricated wear systems, it is not uncommon to find multiple transitions before reaching eventual seizure. Multiple transitions rarely happen in dry sliding cases.

The mechanisms driving the wear transitions often differ between the dry and the well-lubricated cases. In dry wear, wear transition usually can be attributed to (1) the increase in load that precipitates different wear mechanism such as fracture or delamination; (2) the introduction of third-body wear at the interface as a result of fatigue or substrate delamination; (3) the asperity flash temperature reaching a critical temperature that introduces melting or thermal shock or other additional stresses; or (4) the onset of a different dominant wear mechanism such as fatigue, fracture, grain pull-outs or plastic flow as load and speed or their combination reaches a critical value. In well-lubricated cases, the transition can be due to (1) depletion of the antiwear additive; (2) the lubricating film reaching its melting temperature; (3) the lubricating film reaching its load limit (adhesive strength); (4) the introduction of wear particles from delamination or grain pull-outs that destroy the film; or (5) the onset of other chemical processes such as corrosion or oxidation that form different chemical species which become dominant in the surface chemistry.

For well-lubricated cases, Beerbower [1] proposed a wear mode diagram, Figure 4.2, to illustrate the theoretical possibilities as a function of lubricant film thickness and wear. The diagram is an intelligent conjecture of what must have caused wear in a well-lubricated system. The reasoning is based on the assumption that lubrication is primarily controlled by the hydrodynamic and elastohydrodynamic fluid mechanics, and therefore if the film thickness is high, wear should be low. If the wear is high, then the wear must have been caused by other mechanisms such as chemical corrosion or fatigue. The wear values and the fluid film thickness ratio are only approximate but it does illustrate the various possible mechanisms in a well-lubricated system. In reality, wear is controlled by chemistry and the boundary lubricating film which is a function of reactivity and system operating severity. An alternative approach would be to plot the wear mode diagram for a given chemistry and material system as a function of load, speed, and duty cycle rather than wear and fluid film thickness. If the chemistry or the material system is changed, then a different diagram would result.

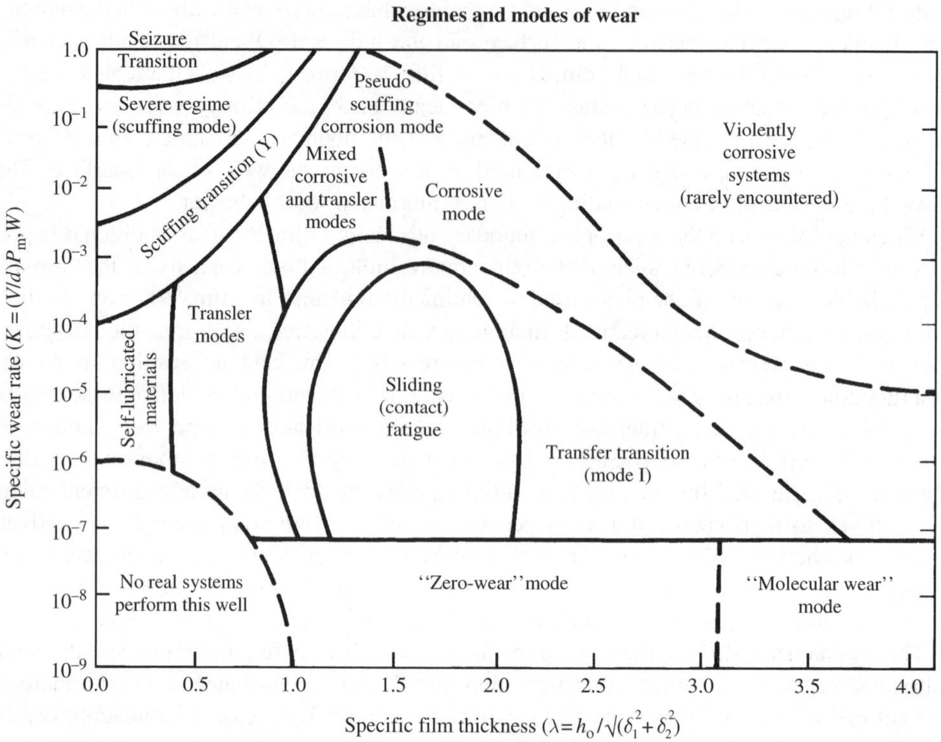

Figure 4.2 Lubricated wear mode diagram

The presence of wear transitions in a well-lubricated system makes the comparison of different materials and/or lubricant chemistry difficult. Since these transitions are nonlinear behaviors, unless the whole curve is shown, ordinal ranking of lubricants and materials at a particular speed or load can be misleading and sometimes result in a wrong conclusion.

So, dry wear measures material loss and lubricated wear measures the effectiveness of the interfacial layer. Because the intent of each measurement is different, the methodologies used for the two measurements are different.

4.4 Wear Measurement in Well-Lubricated Systems

The four-ball wear tester (FBWT) (Figure 4.3) is the predominant wear tester used by the oil industry to study lubricant chemistry. It has been used widely to study the lubricating properties of oils and to study chemical interactions at wearing contacts [2–7]. For these applications, the FBWT offers several advantages. Low-cost, high-precision wear samples in the form of bearing balls are readily available; the contact geometry itself ensures alignment (self-aligned); the rotating axis can be made with high-precision bearings and can be maintained with high trueness in rotation; and the vibrational frequency of the machine can be balanced easily. Further, the fact that there are three separate and distinct worn samples per test allows some degree of statistical averaging to account for material inhomogeneity.

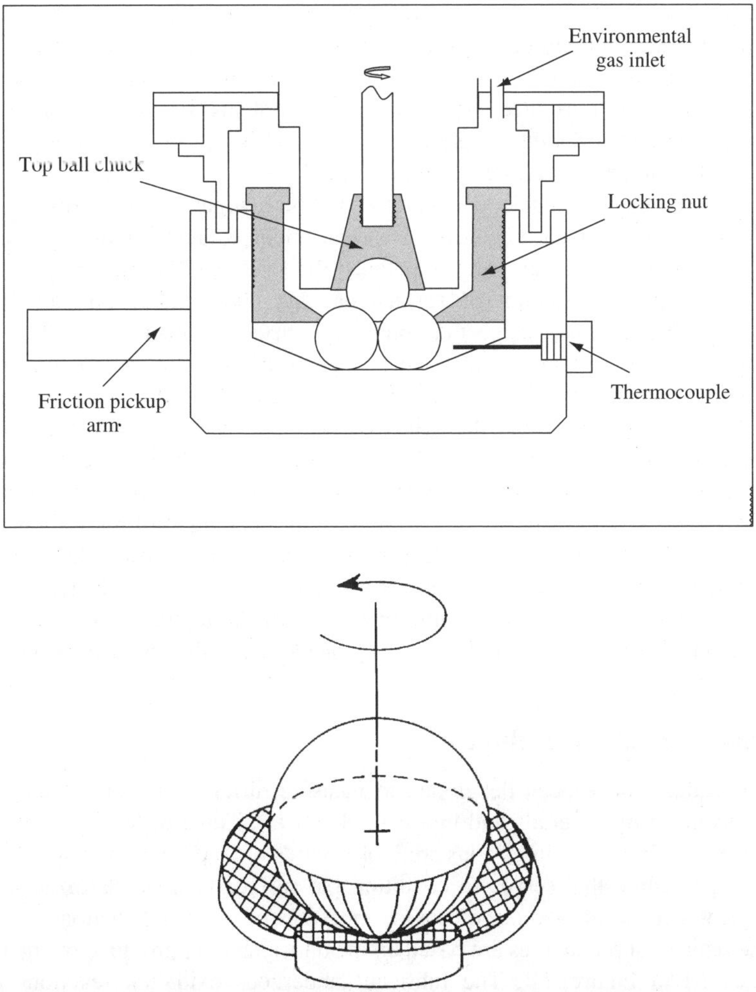

Figure 4.3 Schematic of a four-ball wear tester and its ball on three flats modification

It also provides multiple samples for sophisticated surface analysis techniques, some of which are destructive in nature.

A key issue in the selection of the FBWT as the primary measurement device for lubricant effectiveness is test precision and repeatability. At very low level of wear, the test is very sensitive to vibration and alignment. Statistically significant differences among lubricants or additives can only be obtained if the mechanical parameters, material homogeneity, and contact conditions are tightly controlled. For a pin-on-disk wear tester, this requirement can be very challenging. While the disk is relatively easy to fabricate and polish, the pin must be machined individually and polished for each specimen. A small variation in roundness and roughness can change the contact area and the asperity tip contact pressure significantly. Alignment and vibration are also significant issues especially at the beginning of the wear test. The loci of the pin-on-disk in creating a wear track can vary until the wear track is established. If the contact condition is relatively severe, alignment and vibration effects play relatively minor roles and can be controlled. However, if the wear level is very low (wear coefficient in the range of 10^{-8} or less as is typical in a well-lubricated system), alignment and vibration issues can introduce test repeatability problems. For this reason, wear testers for evaluating lubricants or additives are predominantly FBWTs or a variation of the FBWT such as the ball on three flats design (Figure 4.3).

Another measurement obstacle encountered by both types of wear test apparatus is referred to as "run-in." At the beginning of a test under boundary lubrication conditions, the contact pressure can be very high, on the order of 1–3 GPa. At this pressure, rapid wear and a considerable amount of surface damage due to abrasion and plowing by loosened wear particles can change the contacting surface characteristics. This alteration of the contact surface is highly undesirable since wear is a cumulative quantity, i.e. it is a function of the history of the contacting surfaces. Once the surfaces are severely damaged, the contact stresses and their distributions are changed inside the contact. This change alters the fundamental mechanical actions between the contacting asperities, hence producing entirely different wear processes and/or wear rates. The production of wear particles also changes the contact stress distributions, introducing third-body abrasive actions. These effects fundamentally alter the measurement results. Therefore, control of run-in is critical in lubricant and additive evaluations. Run-in, however, is unavoidable. Consequently, special test procedures to eliminate or minimize the effects of run-in are essential in wear measurement and must be considered as an integrated part of the test methodology.

4.5 Measurement Procedures

Four test procedures have been developed to measure different aspects of the lubricating films for a given set of materials and lubricant chemistry. Three of the four procedures are sequential wear testing procedures consisting of a series of steps. The first step in each case is a "run-in" procedure that serves to condition the surface to a standardized initial state. The subsequent steps in the sequence provide measures of wear and friction.

The three sequential procedures are based on the understanding provided by the lubrication model presented in Figure 4.4. The lubricant undergoes oxidation reactions driven by frictional heating and the stresses produced at the tribocontact. The breaking and forming of chemical bonds produce both lower and higher molecular weight reaction species. These products may escape through evaporation, participate in subsequent surface polymerization reactions, or both react and escape. As the reactions proceed, the products and the substrates interact to form a surface film protecting the surface against the action of the tribocontact.

The FBWT (Figure 4.3) consists of three balls held stationary in a ball pot plus a fourth ball held in a rotating spindle. In the most common configuration, the balls are 1.27 cm (0.5 in.) in diameter. Loads are applied by way of the spinning ball which presses into the center of the triangular formation of the three stationary balls. The load may be selected in the range from 1 to 180 kg, while the rotation speed may be chosen from 60 to 3000 rpm. The temperature of the sample chamber can be controlled by means of a heater attached to the ball pot. A gas line can be used to control the composition of the atmosphere. With the balls in place, the ball pot has sufficient capacity for 10 mL of lubricant.

The primary measurement made with an FBWT is wear. The wear produced on the three stationary balls is measured under a calibrated optical microscope and reported as the wear

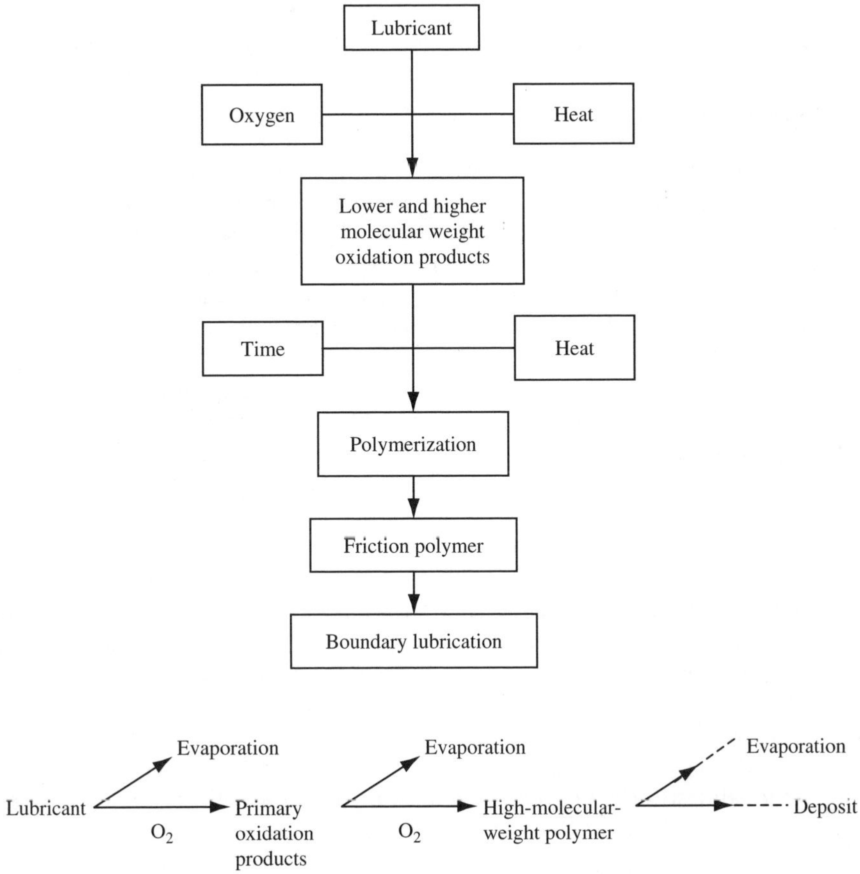

Figure 4.4 Lubricant degradation reaction pathway in a lubricated wear test

scar diameter or the calculated wear volume. To calculate a wear volume, it is usually assumed that the wear occurs only on the stationary balls. The missing material is assumed to come from spherical segments of the stationary balls that correspond to the net volume occupied by the rotating spherical ball that fits into the wear scar [8]. Experimental studies [9, 10] have shown that the measured wear volume and the calculated wear volume can differ greatly depending on the original location of the wear material. If the wear occurs predominantly on the rotating ball, rather than on the stationary balls, the actual wear volume can exceed the calculated wear volume by a factor as large as 10. Such a wear anomaly has been noted in several articles [11–13] and has been referred to as asymmetric wear. The anomaly is not unreasonable. It is clear that frictional heating can cause changes in hardness. It has also been shown that the temperature distribution found on the rotating ball can be different from the distribution on the stationary balls [14]. Both of these effects may lead to asymmetric wear. To avoid the confusion caused by the use of wear volume, the average wear scar diameter of the non-rotating balls is used in this chapter as the measure of wear. Further, wear for the "run-in" is reported as the wear scar minus the diameter of the Hertzian elastic contact. The subsequent incremental wear value is reported as the difference between the wear scar after the run-in step and the wear at the completion of the entire test.

4.5.1 Run-In Process

Run-in is a term that has been used to describe the early stages of operation of practical engineering systems such as automotive engines, gears, and bearings. During run-in, the system adjusts to reach a steady-state condition between contact pressure, surface roughness, interface layer, and the establishment of an effective lubricating film at the interface. These adjustments may include surface conformity, oxide film formation, material transfer, lubricant reaction product, martensitic phase transformation, and subsurface microstructure reorientation.

Run-in has a strong influence on conventional four-ball wear tests, which are commonly run at a high rate of change of wear scar during the early stage of the test. During this period, surface material is removed very rapidly through adhesion, abrasion, and a variety of other wear mechanisms. Because of the rapid wear process, the temperature of the wear contacts, the contact pressure, and the lubricant film thickness all vary significantly with time. These variations are usually reflected by variations in the frictional torque trace. For this reason, the FBWT (without a pre-wear-in step) often is not considered as an accurate friction-measuring device. As wear progresses to a nearly steady-state operation, the surface roughness of the tribocontacts becomes less variable, and the temperature and pressure of the contacts gradually approach their mean distribution values. The oil film thickness remains relatively constant throughout this process. After wear-in, then, the effective friction characteristic of the lubricant can be measured more accurately and consistently.

The improvement of friction determination when a run-in procedure is used is illustrated by Figure 4.5, which shows a comparison of friction traces for both conventional and run-in sequential four ball tests. The upper friction trace is for a conventional test run at 600 rpm

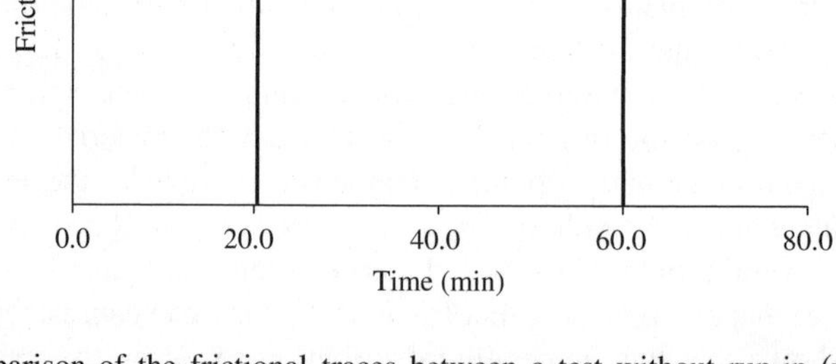

Figure 4.5 Comparison of the frictional traces between a test without run-in (upper curve) and a sequential test (lower curve) in which a run-in step precedes the friction test. μ is the coefficient of friction. Note especially the excessive noise in the measurement of the upper curve during the first 30 min and the continued variation out to 60 min. Contrast this result with the rapidly attained steady signal from the sequential test

and 40-kg load. Initially, the friction level is fairly high and erratic during the early high-wear stages of the test. As the test progresses, friction levels are reduced and seem to reach a steady-state value after approximately 60 min of test duration. The lower friction trace is from the second step of a sequential test. For this test, the surface was pre-conditioned by a run-in procedure using 10 mL of paraffin oil for 60 min at 40-kg load. The subsequent friction test used only 6 μL of lubricant to produce the lower trace. The friction level quickly obtained a steady-state value (in about 5 min) and was free of the erratic behavior seen in the upper friction trace.

To determine the consistency of the run-in process, experiments were conducted to study surface roughness characteristics. Profilometry was used to evaluate the surface texture of the worn halo on the rotating ball and the wear scars on the stationary balls. The profilometer trace was fed into a microprocessor and the resultant signal was displayed on a chart recorder providing surface roughness graphs as shown in Figures 4.6 and 4.7. The profilometer was used in a manner that eliminated the microscopic radius of curvature of the wear scar and which therefore displayed the surface of the halo and wear scar as a horizontal line as shown in the figures. The profilometer traces were taken perpendicular to the direction of sliding.

Figure 4.6 shows the roughness of the wear scars on the stationary lower balls for three run-in tests. The scale of the surface roughness is purposely magnified on the ordinate. While the tests were conducted independently, the traces exhibit similar surface roughness values.

Figure 4.7 shows traces of the wear scar surfaces of each of the upper rotating specimens in the three run-in tests. Again the central sections show a consistency in surface roughness. These data suggest that the run-in process produces a consistent wear scar surface that may be used as a well-defined baseline condition.

The consistency of the run-in process may also be assessed in terms of the quantitative repeatability of the wear scar produced by the run-in process. Experiments were conducted

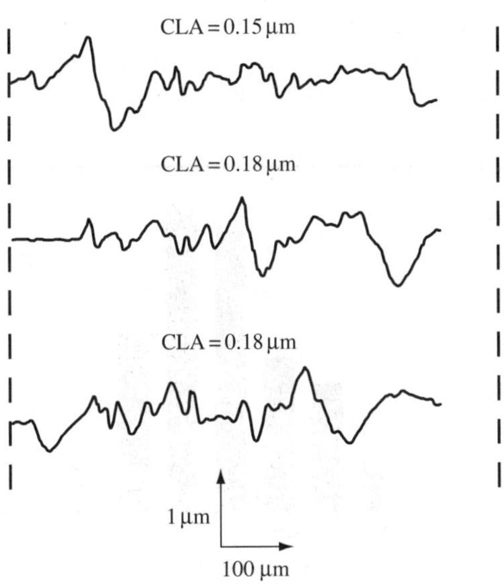

Figure 4.6 Consistency among tests of the surface roughness of the wear scar of a typical stationary steel ball in a four-ball wear test. CLA is the "center line average" measure of roughness

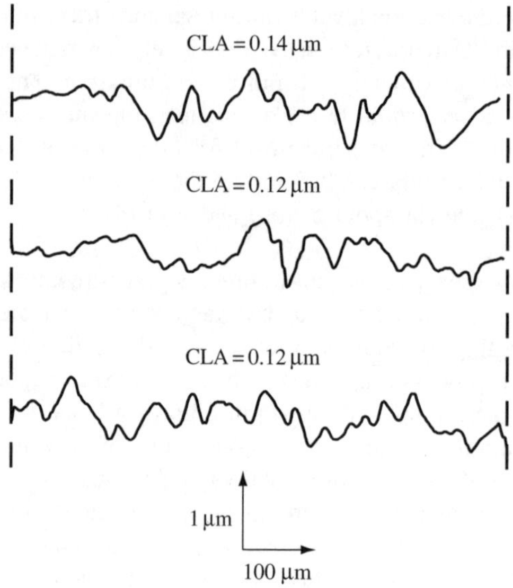

Figure 4.7 Consistency among tests of surface roughness of the wear track on the rotating ball in a four-ball wear test. CLA is the "center line average" measure of roughness

to determine the wear scar diameter of 52100 steel balls tested in the FBWT with 10 mL of purified paraffin oil at 600 rpm and 40-kg load. Surface roughness and friction traces were monitored to judge the validity of the tests. Over the course of a large number of experiments, only a small number of abnormalities were observed. Those few cases occurred with sufficiently large deviations from the norm that they could be considered as statistical outliers. When such cases were eliminated, the wear scar diameter was found to be in the range from 0.64 to 0.70 mm. Data collected from 45 repetitions of the run-in process produced the wear scar diameter frequency distribution shown in Figure 4.8. The distribution

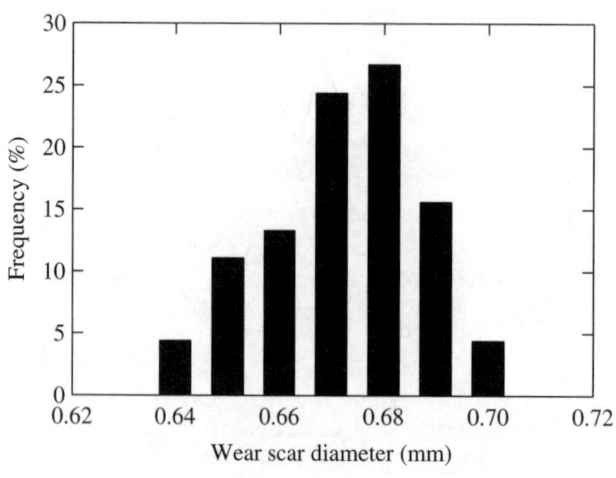

Figure 4.8 Frequency distribution of the wear scar diameter measurement for 45 replications of the run-in procedure of Figure 4.9 applied AISI 52100 steel balls

is somewhat asymmetric with a wider tail on the smaller diameter side. In general, under the conditions of this test, the run-in consistently produced a wear scar diameter of 0.67 ± 0.03 mm, yielding an estimated relative standard uncertainty of 5%.

4.5.2 General Performance Wear Test (GPT)

The functions of a lubricant include the reduction of friction and wear, the cooling of components, and the removal of wear debris. All of these factors can be involved in a wear test if a sufficient supply of lubricant is available and if the test is not so long in duration as to fully degrade the lubricant. Such conditions may be obtained in a general performance test (GPT) by exploiting the full capacity of the lubricant reservoir of the FBWT.

The ball pot of the FBWT is designed such that the tribocontact region can be flooded with 10 mL of lubricant. During operation, the rotating ball causes the lubricant to circulate within the pot. In the process, the lubricant removes any wear debris or soluble degradation products that occur in the contact region and cools the tribosurface. Further, if the duration of the test is not too extensive, the lubricant in the reservoir is comprised of largely unreacted oil. Thus, the lubricant interacting with the tribocontact may be continuously refreshed. The net result of such a test, therefore, would tend to probe the general performance characteristics of the test lubricant.

A GPT was developed in this manner for an FBWT using AISI 52100 steel balls. This test was designed to study the lubricating effectiveness of liquids on steel surfaces under boundary lubrication conditions in the presence of an air or oxygen environment.

The present GPT begins with a general cleaning of the test components followed by the run-in step as described in Figure 4.9 and Table 4.1. These cleaning steps are essential to achieve the precision described in this chapter. The FBWT is assembled with unused balls that are immersed in 10 mL of purified paraffin oil. A 40-kg load is applied by way of the ball pot while the rotating ball spins at 600 rpm. These conditions are maintained for 60 min. Upon completion of the run-in, the ball pot (with the three balls held in place) and the ball chuck (holding the fourth ball) are thoroughly cleaned with solvents. The wear scar diameters of the three balls in the ball pot are then measured. The average diameter should be in the range of 0.67 ± 0.03 mm. Following the run-in and pretest initialization measurements, 10 mL of the test lubricant is added to the ball pot, and the parts are reassembled. The wear test is then continued for 30 min at the desired test load. The friction level during this 30-min step and the increment in wear produced during this step form the primary measures of the performance of the test lubricant.

The application of the GPT requires an interpretive component to the procedure in addition to the mechanical step-by-step procedure outlined in Figure 4.10. In particular, a scale must be established showing a meaningful range of performance results. For the GPT, a baseline can be established by using unformulated lubricants, i.e. basestocks without additives. To calibrate the scale in a qualitative sense, a formulated lubricant considered good by industry standards, such as American Society of Testing Materials (ASTM) engine sequence tests, might be used. Table 4.1, for example, gives the results obtained for two base oils and two formulated lubricants. The base oils establish a reference or baseline wear change, δW, in the range of 0.07–0.09 mm. Hence, a formulated lubricant containing antiwear additives should be expected to have a value of δW that is smaller than about 0.08 mm. The reference lubricants used as calibrants indicate that a measure of good performance would be a value of δW in the range of 0–0.02 mm.

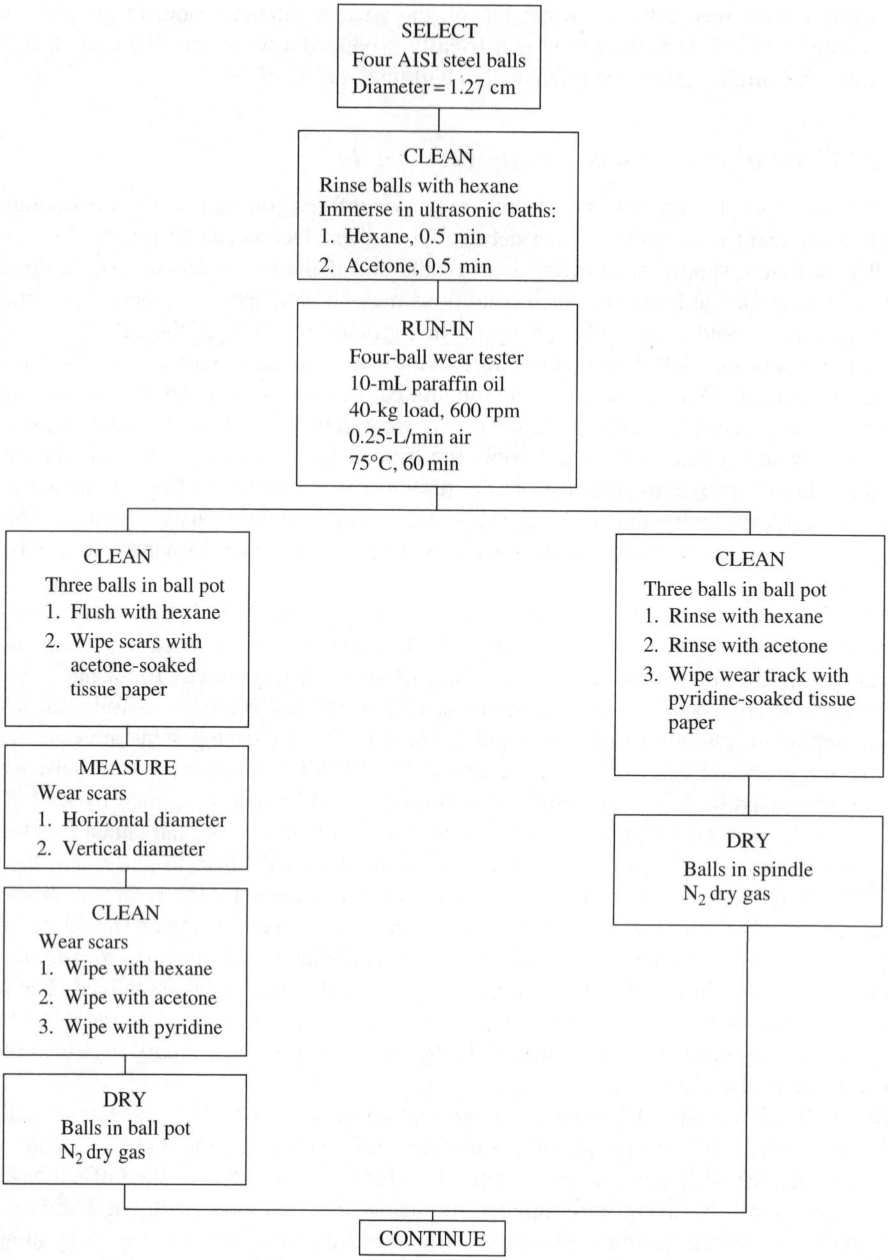

Figure 4.9 Pre-test specimen preparation procedure (run-in) for the general performance test and the enhanced oxidation test

Table 4.1 Friction and wear results obtained with the general performance wear test. In each case, the test was conducted at 60 rpm with a 40-kg load. Standard Uncertainties: wear scar change, 0.01 mm; coefficient of friction (COF), 0.005

Test fluid	Wear scar (mm)		Wear change (mm)	COF
	60 min	90 min		
150 neutral	0.68	0.77	0.09	0.105
600 neutral	0.68	0.75	0.07	0.085
Formulated automotive lube A	0.67	0.67	0.00	0.118
Formulated automotive lube B	0.65	0.67	0.02	0.113
Polyalphaolefin (8 cst)	0.70	0.76	0.06	0.084
Alkylated benzene	0.66	0.70	0.04	0.093
Polyolester	0.67	0.69	0.02	0.111
PAO + 1% ZDDP	0.68	0.68	0.00	0.094

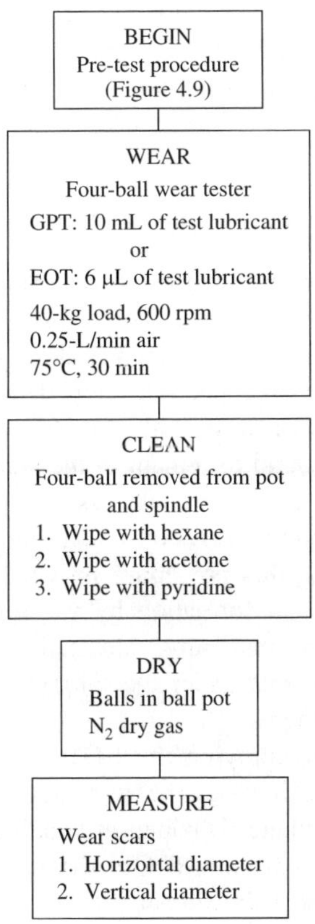

Figure 4.10 Wear test procedures for the general performance test and for the enhanced oxidation test

Such a scale may be used, for example, to study the lubricating effectiveness of various types of functional compounds that are found in hydrocarbon base oils. Analysis of the constituents of a base oil [15] indicates that saturate compounds comprise about 80% of the base fluid. Aromatic constituents account for about 19% of the oil, while polar compounds occur at approximately the 1% level. Table 4.1 includes data for three prototype compounds that correspond to the three major constituents in the base oil.

A polyalphaolefin (PAO) was selected to represent the saturate fraction; an alkylated benzene was chosen as an aromatic compound; and the polar fraction was represented by a polyolester. Each of the prototype fluids was tested by the present GPT. The results, given in Table 4.1, indicated that the polyolester provided significantly better wear protection than either the saturate or aromatic compounds. The alkylated benzene was substantially better than the conventional hydrocarbon base oil, while the PAO was found to be only somewhat better than the natural base oil as an antiwear agent.

These results are consistent with the practices observed in the petroleum industry. PAO, for example, is known to be a good base fluid from which a high-quality lubricant may be formed by use of appropriate additives. Table 4.1 concludes with an illustration of this effect. PAO with the additive zinc dialkyl dithiophosphate (ZDDP) at a 1% concentration produces excellent wear protection on the present GPT performance scale.

Variations on the GPT described in this work may be appropriate depending upon the goals of the performance testing. For example, increasing the second stage of the test from 30 min to a longer duration might allow a better resolution of the performance results among lubricants with very low wear. The essence of the test procedure, however, would remain unchanged.

4.5.3 Enhanced Oxidation Wear Test (EOT)

Under boundary lubrication conditions, lubricant molecules react with the surface to form a protective film at the interface. The formation of this film is, in part, influenced by the oxidation of the lubricant. A wear test specifically designed to study the combined wear and oxidation effects on lubricants has been developed. In this test, only 6 μL of the test lubricant is used. This 6 μL of lubricant is applied to the wear track that is produced during the run-in step. The lubricant is found to remain in the wear track region during the course of the subsequent wear experiment. As a result, essentially the entire lubricant sample on the tribosurface progresses through the series of oxidation/degradation reactions, from initial products to final deposits. Hence, this procedure provides an opportunity to examine the nature of the progression of deposit formation by varying the duration of the test or by selecting various speed and load combinations. Alternatively, the EOT may be used to study the relative effectiveness of lubricants under the same degree of oxidation or to study the consequences of different atmospheres.

Table 4.2 shows the results of applying the EOT to the same lubricants examined in Table 4.1 using the GPT. In each case, the test lubricant experienced 30 min of the wearing conditions. Under the enhanced oxidation conditions, the two base oils exhibited a substantially lower coefficient of friction (COF) than was found in the GPT. Indeed, the COF is sufficiently small that the tribocontact may have been converted from boundary lubricating conditions to elastohydrodynamic conditions. In other words, the 30-min tests converted the base oil into a deposit in such a way that the COF measured in the test was not that of 52100 steel lubricated by a base oil, but rather the COF characteristic of the deposited

Table 4.2 Friction and wear results using the enhanced oxidation wear test. In each case, the 6-μL test was conducted at 600 rpm with a 40-kg load. The 600-N fractions were obtained by chromatographic separation of the 600-N base oil [14]. Standard uncertainties: wear scar change, 0.01 mm; coefficient of friction (COF), 0.005

Test fluid	Wear scar (mm)		Wear change (mm)	COF
	60 min	90 min		
150 neutral	0.68	0.75	0.07	0.067
600 neutral	0.68	0.71	0.03	0.069
Formulated automotive lube A	0.66	0.66	0.00	0.102
Formulated automotive lube B	0.69	0.70	0.01	0.101
600-N saturated hydrocarbons	0.68	0.71	0.03	0.073
600-N aromatic hydrocarbons	0.67	0.73	0.06	0.085
600-N polars	0.70	0.74	0.04	0.107

material. In contrast, the two formulated lubricants produced comparable results for both GPT and EOT procedures. Both formulated lubricants contained antioxidant additives which inhibited the degradation of the lubricant and allowed the boundary lubricating condition to persist, as suggested by the retention of a COF of approximately 0.1.

A study of the saturate, aromatic, and polar fractions isolated from the 600-N base oil by high-performance liquid chromatography (HPLC) [15], also included in Table 4.2, shows that only the polar component retains a COF characteristic of boundary lubrication. The polar fraction also produces a relatively small wear change, as judged by the performance scale established by Table 4.1 for boundary lubrication conditions. Both the saturate and aromatic fractions show lower COF results, suggesting they contain effective friction modifiers or they form an effective friction-controlling layer in the process.

4.5.4 Boundary Film Persistence Test (BFPT)

An issue of considerable importance in boundary lubricated wear is the lifetime or 'tenacity' of the lubricating film. Additives in most formulated lubricants serve a variety of purposes, not all of which directly reduce wear. Antioxidant additives, for example, are intended to inhibit the degradation of the base oil rather than to enhance the formation of the lubricating film. Additives, however, do not perform their intended functions in isolation. Each may have an influence on the performance of the others, either by involvement with an interaction sequence or by affecting the availability of reactants to participate in a reaction sequence. Consequently, the formation of a lubricating film may be affected significantly by differing types and amounts of additives. The issue that arises from this concern is how well the film from a particular formulation protects the surface. How well does the film adhere to the surface and how durable is the film under the tribocontact conditions? These questions call for tests focused on the effectiveness of the film itself.

Theoretically, the boundary lubricating film strength is controlled by the adhesion between the film and the surface and the cohesive strength within the film. A recent study [16] has suggested that effective antiwear films are tenacious, consisting of a solid glassy component and a soft mobile high-molecular-weight "polymer." The solid glassy phase provides load support and the soft polymeric materials provide recirculation effecting self-repairing characteristics. Good effective films can survive contacts for a long time and therefore film lifetime is a critical parameter for designing good lubricants.

A test procedure for assessing the persistence of a boundary lubricating film is shown in Figure 4.11. The procedure consists of three major steps: a 30-min run-in with the test lubricant; a subsequent 30 min of steady-state wear with the test lubricant; and a third step of 30 min of wear using an additive-free white oil. The typical nature of the results from this test is illustrated in Figure 9.

The third step in this procedure has a twofold significance. First, it may provide a measure of the relative effectiveness of the particular boundary film formed by the test lubricant in the second 30-min test step. Second, the wear results may reveal insights into the manner in which selected lubricant additives function to protect the wearing surface.

In this procedure, the additive-free white oil (percolated in an activated alumina column just prior to test) is used as a baseline. Table 4.3 presents the data for a white oil along with data for three additive-containing oils. The results for the baseline white oil show relatively

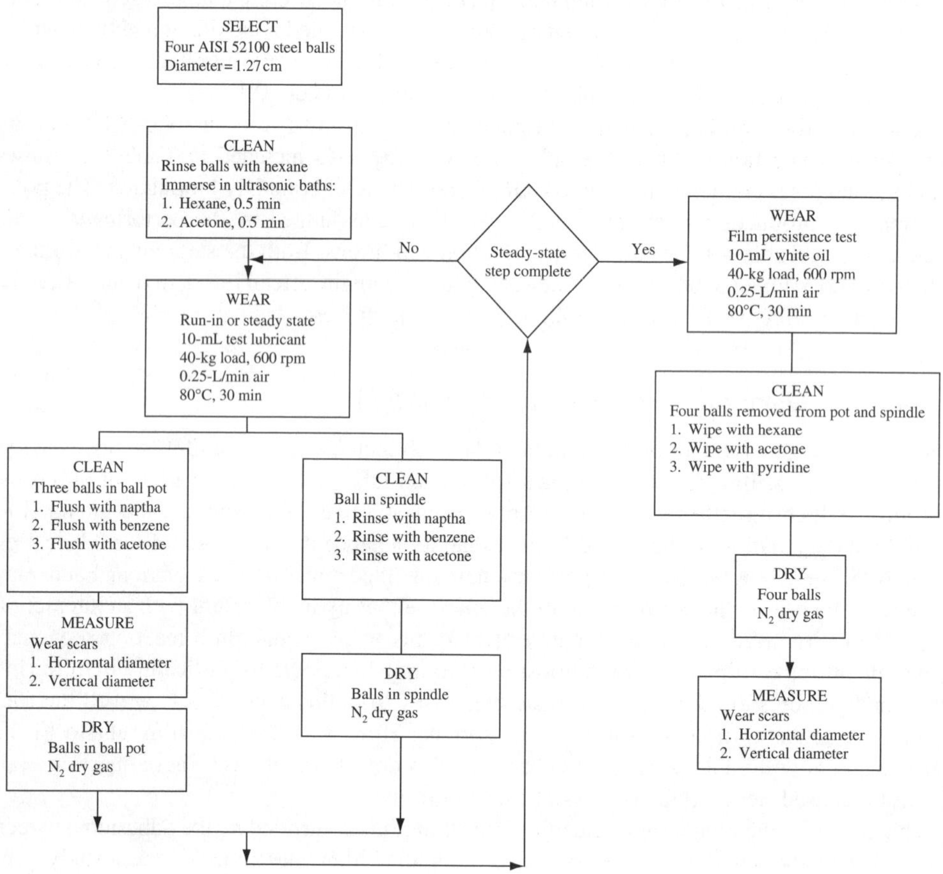

Figure 4.11 Wear test procedure for the boundary film persistence test

Table 4.3 Wear results obtained using the boundary film persistence test [6]. Each test was conducted at 600 rpm with a 40-kg load. Standard uncertainties: wear scar change, 0.02 mm

Test fluid	Wear scar (mm)			First 30-min run-in wear change (mm)[a]	Second 30-min steady-state wear change (mm)[b]	Final 30-min surface finish wear change (mm)[c]
	30 min	60 min	90 min			
White oil	0.52	0.63	0.73	0.22	0.11	0.10
Motor oil	0.36	0.38	0.44	0.06	0.02	0.06
EP industrial oil	0.56	0.71	0.80	0.26	0.15	0.09
Industrial oil	0.46	0.67	0.77	0.16	0.21	0.10

[a] Run-in wear equals 30-min wear scar minus Hertz elastic indentation.
[b] Steady-state wear equals 60-min wear minus 30-min wear.
[c] Surface finish wear equals 90-min wear minus 60-min wear.

large wear during the run-in stage followed by two stages of considerably less wear. It is important to note that the measures of wear in the second and third stages of the procedure are comparable values when the additive-free white oil is used. The wear results found for this white oil are typical of the results found for additive-free mineral oils and synthetic hydrocarbons in general.

In contrast, the presence of an antiwear additive changes the wear results significantly due to changes in the nature of the film. The formulated lubricant, a 10 W-30 multigrade motor oil, produces low wear in each stage of the test procedure. The first two stages of wear show that the additive package produces an antiwear effect compared to the white oil baseline. More importantly, the wear protection persists into the third stage of testing when the test lubricant is replaced with the additive-free white oil.

In contrast, boundary film persistence is not observed for either of the industrial oils tested. Both industrial oils show relatively high wear. The extreme pressure (EP) additive package may produce greater wear as a result of corrosion by the EP additives, while the other industrial oil has high wear as a result of the use of a rust inhibitor. The latter additive is a polar material that may dominate the surface chemistry and therefore influence the wear rate. In both cases, the wear results in the third stage of the test procedure are characteristic of the baseline white oil. Under the high-wear conditions of these test oils, no persistent boundary lubricating film is produced.

These test procedures, properly used, can help to identify some of the key properties of the lubricants in terms of wear. Conventional test procedures tend to confuse the issues and mix run-in behavior and steady-state behavior. In boundary lubricated wear, it is clear that we are measuring the formation tendency, effectiveness, and durability of the boundary lubricating films. Only proper test techniques can isolate the issues and provide useful answers.

4.5.5 Case Study with GPT and BFPT

Energy and environmental concerns have been a driving force for research on both improved lubricants and alternative engine designs. The US Department of Energy (DOE) has been a leading proponent of advances in these areas. With support from DOE, NIST has studied the feasibility of an integrated molecular-engineered lubricant that contains no metals, sulfur, and phosphorus elements. Such lubricants could provide superior performance in diesel engines while virtually eliminating particulate emissions. As a result, the need for particulate traps and other emission-control devices would be eliminated or greatly reduced, thereby allowing more fuel-efficient engine designs in anticipation of future emission targets.

A radical approach to lubricant design will be needed to achieve this goal. Current lubricants contain a large percentage of inhibitors with many functional groups. To eliminate those additives, a base oil molecule would need to be functionalized to provide the equivalent capability while avoiding many of the complex interactions resulting from the current lubricant formulations. Unfortunately, most synthetic base oils are not amenable to controlled functionalization due to the large number of isomers and the wide molecular weight distribution.

In response to this situation, research was focused on multiple alkylated cyclopentanes (MACs) that have molecular structures that are thermally stable and have high purity, low volatility, and good solubility. Most importantly, this family of molecular structures can easily be functionalized to incorporate various functional groups. The functionalized molecules exhibit a high degree of antioxidancy, thermal stability, and a mild antiwear function.

As part of this work, NIST examined a series of boron-containing chemical compounds. Studies based on both the GPT and the BFPT were conducted to provide comparisons among the new formulations and current antiwear additives. In the GPT study, Figure 4.12, base fluids exhibit wear curves parallel to the Hertzian contact diameter line over a significant range of load. Effective antiwear additives are designed to reduce wear and, therefore, are expected to produce wear scar diameters that approach the Hertzian curve. The degree to which an additive achieves results closer to the Hertzian contact diameter line provides a measure of antiwear effectiveness.

Results using 1% ZDDP in the base fluid, Figure 4.13, demonstrated the high degree of effectiveness for this well-established compound. The results obtained using 1% of a boron-containing additive (labeled AD6608) indicated that this functionalized MAC additive was almost as effective as ZDDP in this particular study. However, in the BFPT, Figure 4.14, the boundary film produced by AD6608 was clearly not as tenacious as the film produced by ZDDP. The combination of the results from the GPT and the BFPT established at once both

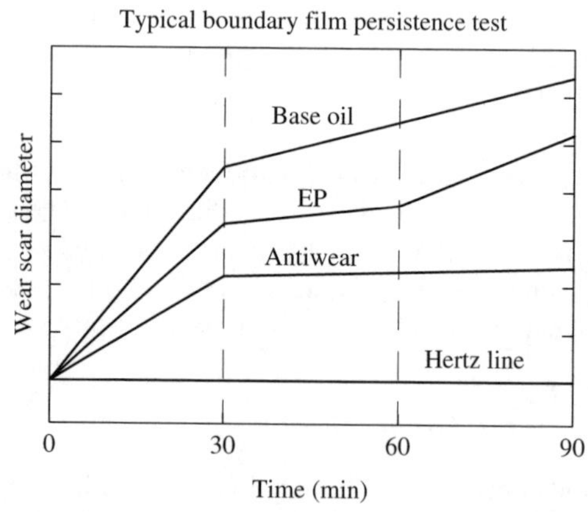

Figure 4.12 Typical wear behavior in a four-ball wear tester showing run-in during the first 30 min, steady-state wear during the second 30 min, and the film-only wear during the third 30 min. This sequence forms the basis of the boundary film persistence test

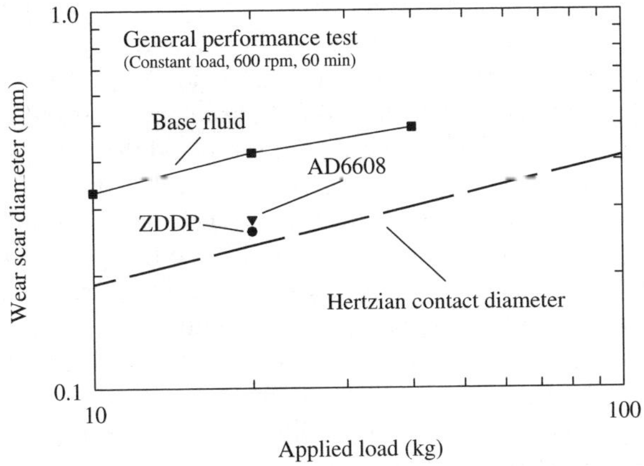

Figure 4.13 The general performance test reveals the overall relative effectiveness of lubricants with different additives. Result closer to the Hertzian contact diameter curve are considered better

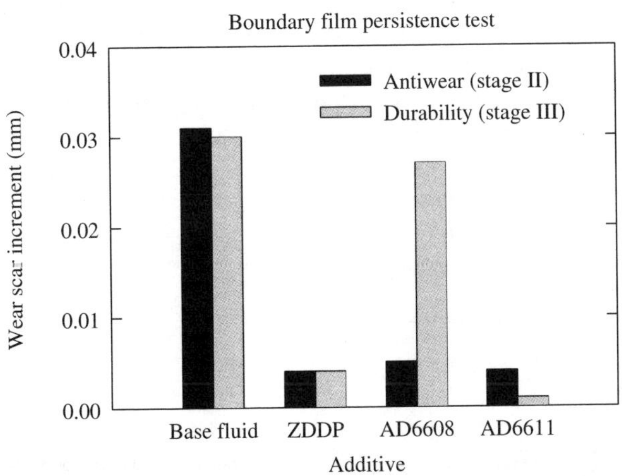

Figure 4.14 The boundary film persistence assesses the relative effectiveness of additives to form a durable lubricating film. Smaller wear scar increment indicates a more durable film

the potential for successful development of the new class of antiwear additive and the need for additional advances. Further work on another borate ester, labeled AD6611, showed that, indeed, the durability of the boundary film could be improved beyond the test performance level of the ZDDP additive.

4.5.6 Boundary Film Failure Test (BFFT)

An additional test has been developed to address the question of how well a lubricating film adheres to the surface. If the lubricating film fails, then scuffing and material wear can follow immediately. The material and material-pair factors that can be expected to affect the onset of film failure include the tensile strength of the film, the viscoelastic properties of the film, the adhesive strength of the bond between the film and the substrate, the surface roughness, and the dynamic COF at the tribocontact. Additionally, there are environmental factors, including the temperature and pressure at the contact and the chemical composition of the ambient atmosphere, that affect film failure. The temperature and pressure effects in a wear test may largely be controlled by the sliding speed and the applied load. Early studies either focused expressly on the temperature effect [17–20] or interpreted the film failure results as a consequence of the breaking of the physical or chemical bonds between the lubricant molecules or atoms and the substrate [21–25]. To focus on the mechanical aspects of film failure, recent work [26] has been directed toward the measurement of the maximum shear stress sustained by a lubricating film prior to failure.

The test apparatus for this work consists of a spherical ball that is fixed in position and an inclined plane that slides at slow speed under the ball. The condition of slow speed avoids the generation of high contact temperature, while the use of an inclined plane provides a contact pressure (and the resulting surface shear stress) that increases with the distance slid during the test. Quartz force transducers allow the vertical and horizontal contact forces to be monitored continuously. The apparatus allows an applied normal load from 1 to 800 N with an uncertainty of 0.01 N, while the data sampling rate can be selected from less than 1 Hz to more than 5000 Hz. To control the speed and position of the moving plane, *x*-, *y*-, and *z*-translational stages are used with uncertainties of 1 μm/s for linear speed and 1 μm for linear position. The inclination angle of the plane can be adjusted from −2.5° to 2.5° with a tilting table. A high-speed digital camera allowing up to 500 pictures per second is used to observe the contact interface *in situ* during sliding.

An immediate consequence of film failure is a loss of lubrication at the tribocontact and a corresponding rapid increase in the COF, as illustrated in Figure 4.15. This effect can be used as the signature characteristic indicating the onset of film failure.

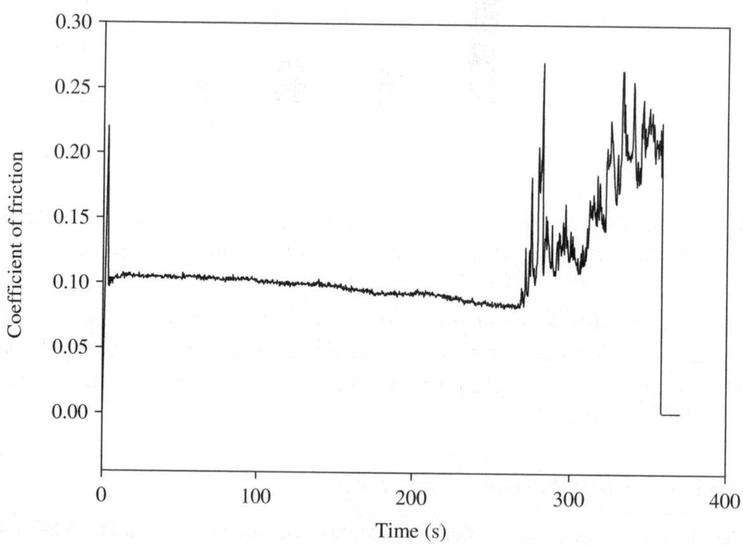

Figure 4.15 An illustration of the rapid increase in the coefficient of friction at film failure in the boundary film failure test

In the initial studies with this apparatus, the lubricant film was established on the plane specimen by a dipping procedure. After polishing the plane to remove any residual surface oxides, the surface was cleaned chemically and ultrasonically and then dipped into the test lubricant which had been diluted using hexane. The thickness of the resulting film was measured using laser light elipsometry. Each test was conducted with a clean ball (specified by the manufacturer as 52100 steel, 3.175 mm (1/8 in.) in diameter with Rockwell hardness Rc 62 and surface roughness Ra 12 nm). Each ball specimen was cleaned ultrasonically (2 min in hexane, 2 min in acetone, and 2 min in a detergent solution), rinsed in deionized water 10 times, and then dried with nitrogen gas before testing. The test ball was held firmly in position and without rotation using a screw-tightened holder, while the inclined plane specimen was driven by the high-precision stage to move against the stationary ball a distance of 5 mm at a speed of 0.2 mm/s. The angle of inclination of the plane was adjusted between 0.10° and 1.00° to produce a better resolution of the location of the onset of film failure. For this work, the data sampling speed was 20 Hz, and both vertical and horizontal forces were monitored. Two to five scratches were made on the same plane specimen at intervals of 2 mm, and the data from all of those scratches were averaged to obtain the mean result for the specimen. All of the tests were conducted in the laboratory atmosphere at 22°C and 50% relative humidity.

The maximum shear stress developed in the film prior to failure was used as the measure of the effectiveness of the lubricating film. The initial results for the film failure tests showed a reasonable correlation with the boundary persistence tests. Good antiwear additives, Figure 4.16, which previously showed good boundary film persistence, sustained the largest surface shear stresses. Quite reasonably, the results depended on the substrate, Figure 4.17,

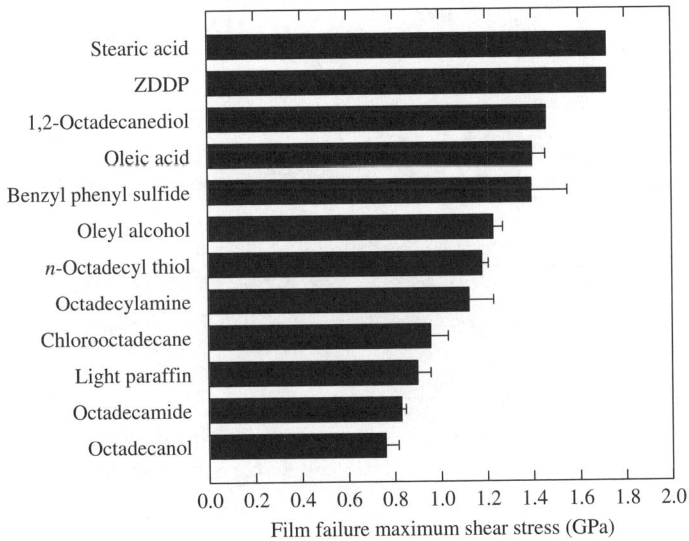

Figure 4.16 A comparison of results for the film failure maximum shear stress developed in the test lubricant films containing different additives. Ball: 52100 steel, Rc 62, Ra 12 nm. Plane: 52100 steel, Rc 62, Ra 85 nm. Film thickness = 2 nm. Sliding speed = 0.2 mm/s

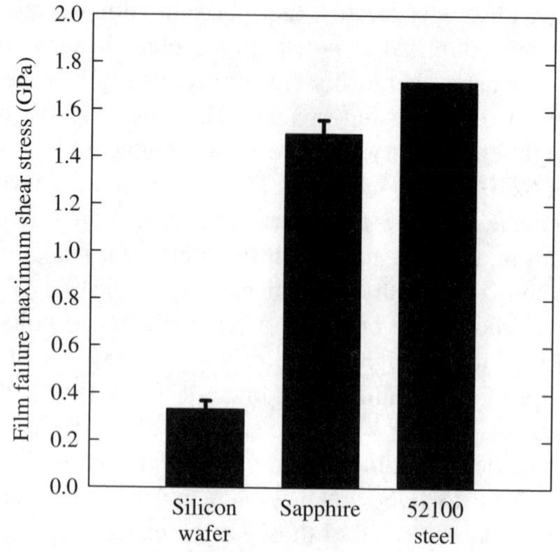

Figure 4.17 An illustration of the dependence of film failure on substrate. Ball: 52100 steel, Rc 62, Ra 12 nm. Planes: 52100 steel, sapphire and silicon wafer, Ra 2 nm. Film thickness = 8 nm. Sliding speed = 0.2 mm/s

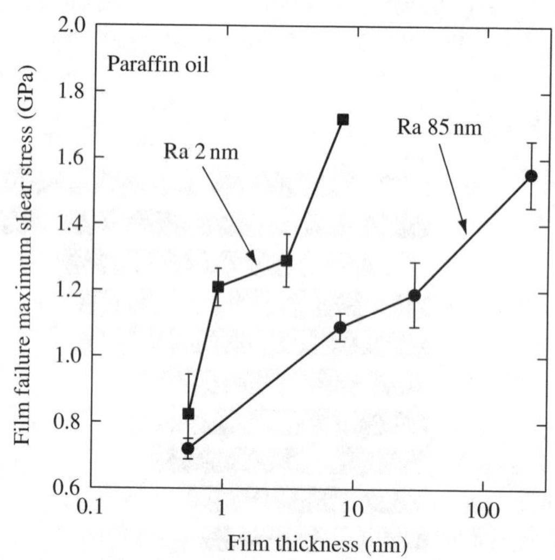

Figure 4.18 An illustration of the dependence of film failure on film thickness and surface roughness. Ball: 52100 steel, Rc 62, Ra 12 nm. Plane: 52100 steel, Rc 62. Sliding speed = 0.2 mm/s

as well as on the additive; thicker films, Figure 4.18, allowed larger shear stresses; and rougher surfaces, Figure 4.18, decreased the maximum sustained shear stress, presumably because of the more prominent intrinsic stress intensification that may occur at the tips of the surface asperities.

4.6 Wear Mechanisms Under Lubricated Conditions

The four wear test procedures (GPT, EOT, BFPT, and BFFT) provide substantial observations of the effectiveness of boundary lubricating films. Results from those tests lead to two significant questions. What is the nature of these boundary lubricating films and how do these films affect the boundary lubrication mechanisms?

A wear mechanism under boundary lubrication consists of two distinct parts: one part is mechanical and the other is chemical. The primarily mechanical wear mechanisms are plastic shear and abrasion [27]. Fatigue-induced third-body abrasion is also an important mechanism in a well-lubricated system under long cyclic stresses [28]. For most metals, wear particles are produced by the accumulation of plastic strain at and near the surface. Lubricating films effectively delay the process by redistributing the stresses over a larger area as well as by removing the strain as a result of film wear. The boundary lubricating film thickness required for a particular system depends on the relative surface roughness, mechanical properties of the surface materials, and the size of third-body particles that the film has to accommodate. For brittle solids and large particles (i.e. large grain size), thick films are needed. For materials with large elasticity and plasticity, thin films are sufficient. The wear prevention or wear rate, however, depends solely on the micromechanical properties of the film and its formation rate.

The chemical wear mechanism in boundary lubrication depends on the reactivity of the chemicals toward the surfaces. The ability of the molecule to react with the surface is also an indicator of a potential corrosion problem. In fact, one may describe the formation of an effective boundary lubricating film as a form of controlled corrosion. Because corrosion and organometallic chemistry are element specific, the composition of the surface is necessarily a very important factor in selecting the chemistry to protect that surface. For this reason, not all lubricant formulations work equally well on all surfaces.

In this context, the protective mechanisms of the boundary lubricating films need to be discussed. There are several mechanisms by which boundary lubricating films function: sacrificial layer, low-shear interlayer, friction-modifying layer, shear-resistant layer, and load-bearing glasses. The sacrificial layer is based on the fact that the reaction product layer is weakly bound and easily removed, thereby providing a low-shear interfacial layer against the rubbing. So, instead of the surface being worn away, it is the film layer that is removed. For such films to be effective in protecting the surface, the rate of film formation must be higher than the rate of film removal.

Another important consideration with respect to the effectiveness of a sacrificial layer is the shear resistance at the asperity level. It has long been known that large molecular weight entangled polymeric chains give superior cohesive strength and, hence, shear resistance. In a boundary lubricated contact, this effect is provided by the organometallic compounds. Oil-soluble metal-containing compounds were first identified as being generated in lubricants under oxidizing conditions [29]. These compounds were later identified to be high-molecular-weight organometallic compounds using gel permeation chromatography (GPC) coupled with atomic absorption spectroscopy (AAS) [30].

Figure 4.19 shows GPC–AAS analysis of results from an EOT. A super-refined mineral oil was used in an FBWT in the EOT configuration using a microsample of only 6 μL of lubricant. This small amount of lubricant allowed the reaction sequence and the resulting reaction products to be concentrated for ease of analysis. The net reaction product on the worn surface was extracted by a solvent (tetrahydrofuran). The solution was then injected

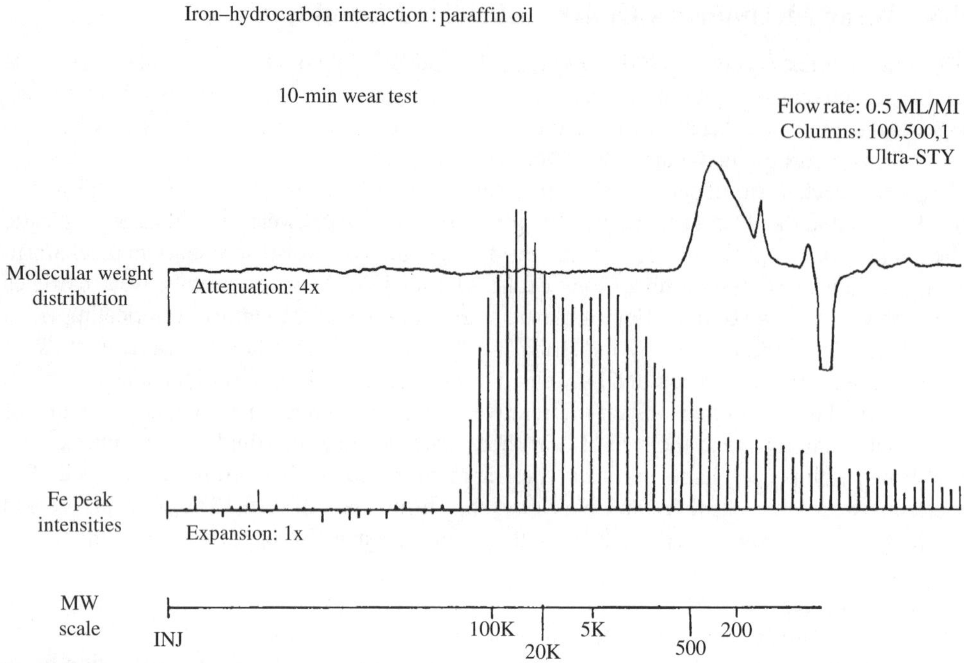

Figure 4.19 Post-test organometallic compound detection in a lubricant film after an enhanced oxidation test in a four-ball wear tester

into a GPC column for molecular size separation. The effluent flowed through two detectors, measuring refractive index and ultraviolet intensity. After the detectors, the effluent stream was collected in an autosampler vial to determine the metal content in the effluent by using AAS analysis. A broad spectrum of organo-iron compounds of various molecular weights were found, having molecular weights up to about 100,000. Molecular weight higher than 100,000 was not detected, suggesting that the solubility limit had been reached for this type of compound in the base oil. Optical pictures revealed that the boundary lubricating film was fully formed after only 1 min of wearing contact.

These organometallic compounds were also found on actual cam and tappet parts used in an ASTM engine dynamometer test, the sequence III oxidation wear test. Cam and tappet parts were taken from ASTM test stand calibration runs and analyzed for surface reaction products. Similar patterns, Figures 4.20 and 4.21, were observed. The discovery that these organometallic compounds were present in an engine component demonstrated that the same types of chemical processes occur in both full-scale engines and the microsample wear tests. It could be concluded, therefore, that these processes made fundamental contributions to the mechanisms of lubrication.

REO 76, Figure 4.20, is a low-wear reference oil, and the level of organometallic compound found is low. REO 77, Figure 4.21, is a high-wear reference oil, and the level of organometallic compound formation is very high. This observation suggested an optimum reactivity concept, illustrated in Figure 4.22. At very low reactivity, there would be

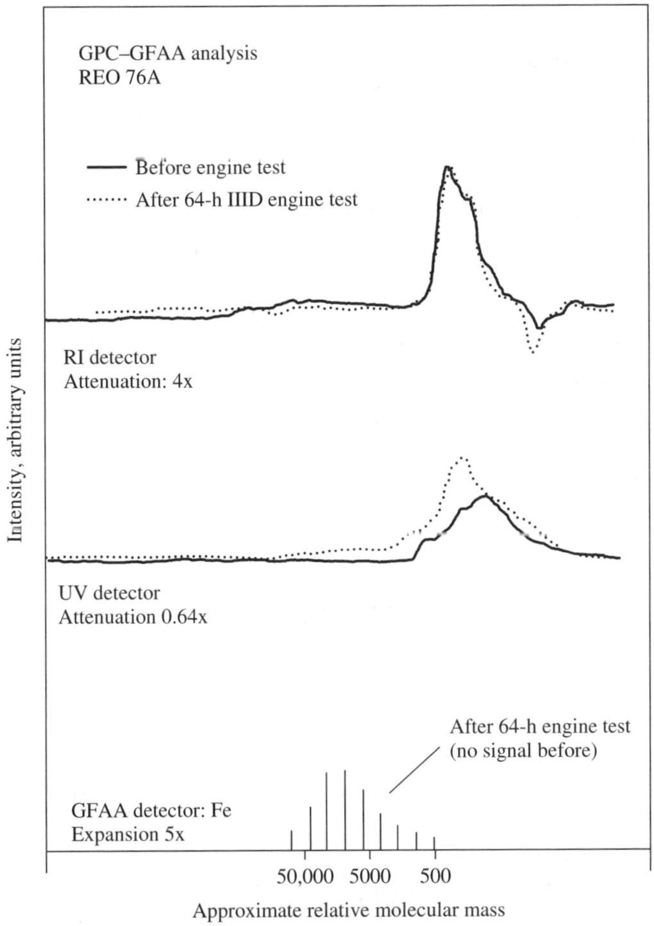

Figure 4.20 Organometallic compounds found on cam lifters after an ASTM engine dynamometer sequence test using an ASTM engine monitoring reference low-wear lubricant (GPC: gel permeation chromatography; GFAA: graphite furnace atomic absorption)

insufficient film formation to resist wear. At very high reactivity, the high chemical reaction rate would produce corrosive wear. Between these extremes, there should be an optimum reactivity condition providing optimum film formation. Direct experimental evidence of this concept is difficult to obtain because once the chemistry is changed, the system is fundamentally altered and many other factors need to be considered, such as oxidation resistance, film formation, adhesive strength, and cohesive strength. Cross-comparisons among widely different chemical systems may not be valid. However, in a base oil constituent fractionation study, various polar species were isolated and separated by HPLC. Tests on these fractions in an FBWT [16] tend to support this concept.

Recent data suggest that not all films formed in a boundary contact are capable of providing antiwear protection [31]. Because of the high temperature and reactive conditions in the contact (nascent metal surfaces, flash temperatures, and catalytic actions from

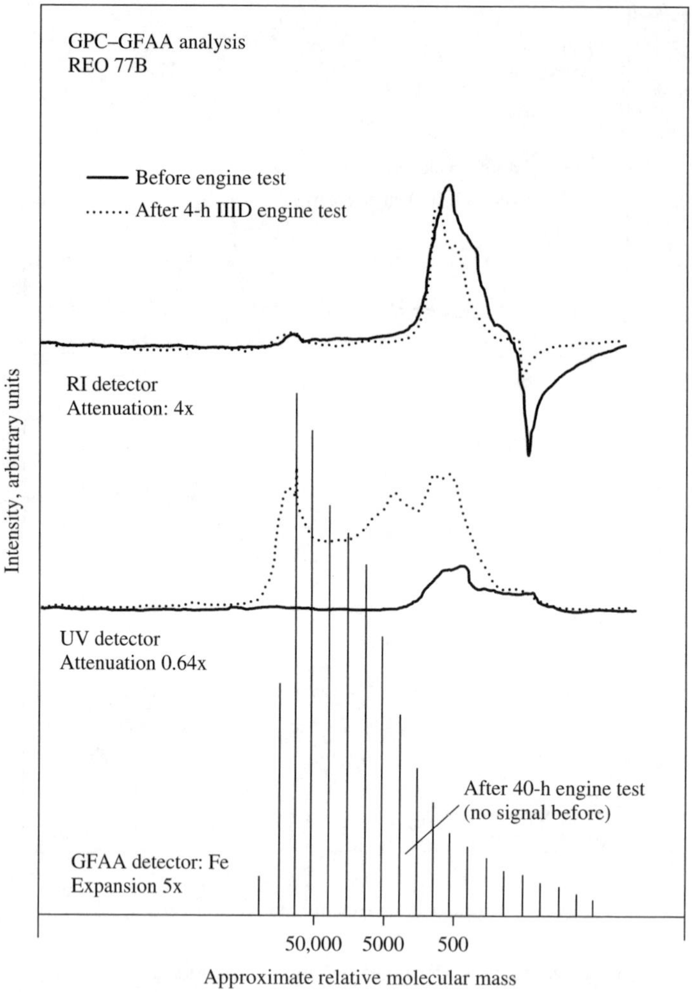

Figure 4.21 Organometallic compounds found on cam lifters undergone an ASTM engine dynamometer sequence test using an ASTM engine monitoring reference high-wear lubricant (GPC: gel permeation chromatography; GFAA: graphite furnace atomic absorption)

transitional metal ions and electrons), many chemical reactions may occur. In some cases, the reaction products can be degradation products that simply pass through the contact without attaching to the surfaces. This case is most likely when the surfaces are relatively chemically inert such as may occur with ceramics, diamond-like carbon, or other hard coatings. Consequently, depending on the nature of the solid surfaces, many different kinds of films can form. Some of the films may be protective (having suitable adhesive and cohesive strengths as a result of chemical bonds formed with the surfaces), some may be corrosive (having very high reactivity but weak adhesive strength), and some may be simply

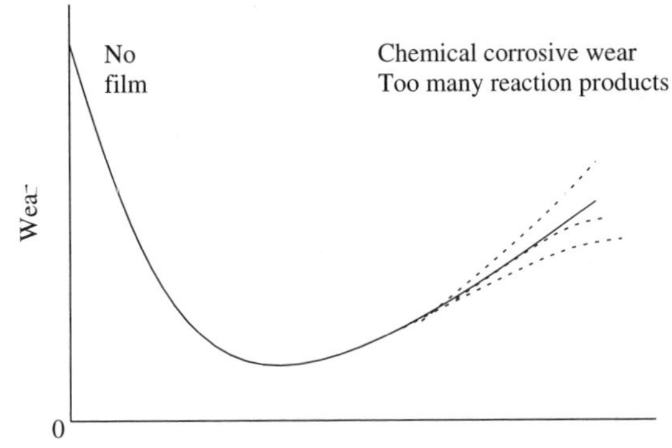

Figure 4.22 Optimum reactivity diagram for effective lubrication

degradation reaction products inside the wear scar (having very little influence on friction and wear).

4.7 Modeling of Lubricated Wear

Where do we stand in terms of having a predictive capability with respect to lubricated wear under boundary lubricated conditions? Given the materials pair, speed, load, surface roughness, lubricant type (including viscosity and additive chemistry), and duty cycles, can we predict length of service, amount of wear, time to scuffing, and seizure? In short, no; we currently do not have such models. At the same time, however, we can describe fairly well average film thickness, elastohydrodynamic support, and even some coarse estimate of wear.

4.7.1 Wear

Wear is a system function. Wear outcome depends on materials, surface roughness, lubricants, environment, operating conditions, temperatures, and other such factors. There are also many different wear mechanisms operating in different regimes delineated by combinations of load, speed, and environment. Each wear mechanism is governed by a set of principles, such as fracture mechanics, delamination, and/or plastic deformation. Therefore, to predict wear *a priori*, without experimental fitting constants, assumes *a priori* knowledge of the dominant wear mechanism for a given system. Understanding the dominant mechanism currently is the most significant barrier that tribologists need to overcome. For brittle materials, such as ceramics and plastics, significant advances have already been made, and predictive models capable of predicting the wear coefficient within one or two orders of magnitude have been developed on the basis of fracture mechanics [32].

4.7.2 Contact Area

In the boundary lubrication regime, one of the most critical initial system parameters is the real area of contact. This parameter controls the real load supported by the asperities and the subsequent stress/strain relationship beneath the contact because the fundamental wear process for most metals is controlled by the accumulation of strain. Greenwood and Williamson [33] proposed several models to describe the initial real area of contact which is usually a small fraction of the apparent area of contact between two engineering surfaces. One model, which assumes a distribution of peak heights, can be written as

$$A = \pi\eta\beta\sigma\ F_1(h)$$

where π is the transcendental number 3.141592654..., η number of asperities, β asperity radius, and σ the standard deviation of the peak height distributions, and

$$F_m(h) = \int_h^{\infty} (s-h)^m \phi^*(s)\,\mathrm{d}s$$

where $\phi^*(s)$ is a probability density.

However, once sliding takes place, the real area of contact changes with wear. Under a steady-state mild wear condition, the real contact area can be quite high if the two surfaces conform to one another [34].

4.7.3 Rheology

The next step is to model the rheology of the lubricant in the contact. Under boundary lubrication conditions, traditionally, bulk fluid viscosity does not play a role because the asperities are bearing the load. However, recent studies of fluid molecules under confined space exhibit radically different mechanical and flow properties as compared with those in the bulk [35–38]. The viscosity near the wall (within nanometers) is much higher than that of the bulk. This phenomenon has long been known to be important in high-pressure studies [39]. The clear implication is that viscosity of a fluid trapped among the asperities in a

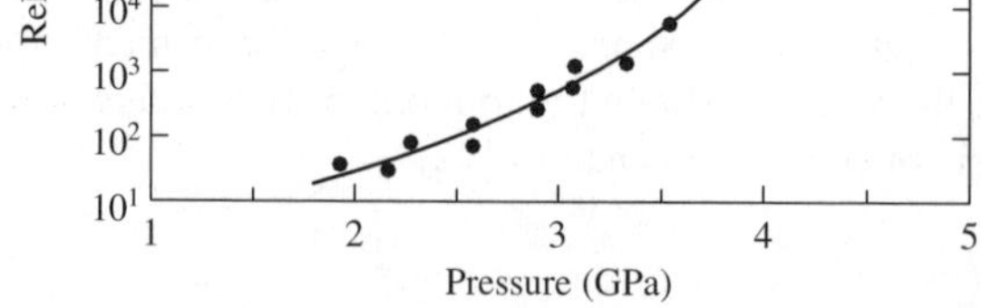

Figure 4.23 Typical pressure dependence of the viscosity of a liquid. Note the rapid, nearly critical point, increase in viscosity as the pressure-induced glass transition is approached

tribocontact may be higher than in the bulk liquid at ambient pressure; hence, there may be a significant elastohydrodynamic lift at the local site. For most liquids [40], a sufficient increase in pressure will cause a liquid to transform into a glassy state. The initial increase in the viscosity, Figure 4.23, is approximately exponential, but as the glass transition is approached, the rate of increase accelerates. While the glassy state represents a nonequilibrium condition, the transition is sufficiently rapid that a well-defined glass transition pressure, p_g, can be determined, and the increase of the viscosity with pressure can be related to p_g,

$$\eta/\eta(0) = (1 - p/p_g)^{-\upsilon}$$

where $\nu > 0$ is the exponent characterizing the nearly critical point behavior at the glass transition. For lubricants, the increase of viscosity with pressure is often offset by the exponential decrease in viscosity due to an increase of the temperature in the tribocontact.

4.7.4 Film Thickness

Once the rheology is defined, then fluid film thickness can be estimated by the equation given by Dowson and Higginson [41]:

$$h = 1.63\left[\frac{\eta_0^{0.7}\sigma_0^{0.54}V^{0.7}R^{0.43}}{L^{0.13}E'^{0.03}}\right]$$

where h is film thickness, η_0 the viscosity, σ_0 the pressure viscosity coefficient, V the velocity, R the effective radius of contacting bodies, L the load, and E the elastic modulus.

4.7.5 Contact Stress

There are several computer programs available to calculate the contact stresses, fluid flow, temperatures, and elastohydrodynamic lifts [42], but to link these calculations to wear prediction is a big step. The most advanced application, to date, appears to be that of Bell and coworkers [43, 44] who used component cam-follower wear data obtained from bench tests to correlate with oil film thickness and to calculate the oil film thickness using a finite element program. The wear contours and approximate amount of wear of the cam-follower contact in an engine were successfully simulated. Chemical film formation as a function of additive concentrations and temperatures was not taken into account but was simulated by the bench tests.

4.7.6 Flash Temperatures

Chemical reactions are key to successful boundary lubrication. So, prediction of the asperity temperature in a concentrated contact is very important. Since the chemical films are produced by asperity flash temperatures, if one is able to predict the asperity temperature in the contact, one can calculate the chemical reaction rate for the generation of such films. This would, in effect, predict lubrication effectiveness.

Naidu *et al.* [45] proposed a chemical kinetic rate model based on the consumption of lubricant under thin film oxidation conditions similar to those encountered in a sliding contact.

The model describes the primary oxidation step as well as the subsequent condensation polymerization step, which results in lubricant viscosity increase and insoluble sludge formation. These reactions can be described as follows:

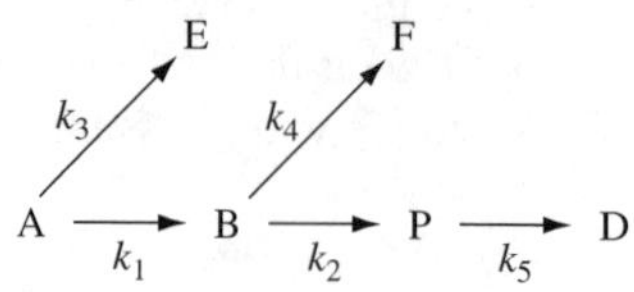

where A is original oil, E is evaporated original oil, B is low-molecular-weight oxidation products, F is evaporated low-molecular-weight products, P is high-molecular-weight liquid polymerization products, D is insoluble deposits, and k the reaction rate constants.

Using the results from a microsample four-ball wear test, Hsu *et al.* [46] were able to calculate the flash temperatures necessary to produce the observed amount of reaction products, 375 °C. They were also able to calculate the lubricant life at this temperature as a function of lubricant volume. The calculated lifetime value agreed with the experimental value reasonably well.

4.8 Summary

Boundary lubrication is the result of a sequence of chemical reactions involving the lubricant, the substrate, and the ambient atmosphere. The boundary film produced by this sequence is distinct from both the initial lubricant and the substrate. A set of four wear test procedures has been suggested as a means of investigating and probing different aspects of the reaction sequence and the resulting film characteristics. The set of tests consists of a general performance wear test, a procedure to enhance and emphasize oxidation effects, a test of the persistence or tenacity of the boundary lubricating film, and a test of the failure or rupture strength of the film.

Use of these test procedures together as a set represents a powerful approach to understanding the basic mechanisms of boundary lubricated wear. In the GPT, a collection of reference oils can be used to establish a scale of performance. Such a scale may be used to assess oils and to provide an approximate quantitative ranking among a collection of oils. One can effectively argue that most wear events under effective lubrication conditions are occurring at the asperity level. Conversely, lubrication means protecting the asperities.

When asperities are touching and sliding over each other, both mechanical events and chemical events occur. Material wear, however, is dominated by the mechanical events in the form of contact pressures, stresses, and strain accumulation rate. If the load is so high that fracture of the asperity is inevitable, lubrication has little impact.

For lower loads, the protective mechanisms are load-bearing interface; easily sheared layer; interfacial shear within a film; and adhesion barrier. Under high pressure, the lubricant or the chemical film behaves like a solid between the contact, enlarging the real area of contact and, hence, reducing the contact pressure. Redistributing the stresses, in turn, reduces the strain build-up rate for metals or reduces the stress intensity for brittle solids. Since wear of brittle solids depends primarily on stress intensity, lowering the stress reduces wear. If the film is easily sheared and has sufficient thickness, then the magnitude of the tensile stress imposed by the asperity will be significantly reduced. If the film is thick enough and can allow shear inside the film, such as solid lubricating films with weak planar attractive forces, the relative motion can be accommodated by interfacial shear within the film. If the molecules are large, have strong bonding to the surface to resist shear, and cover the asperity surface reasonably well, the presence of such films will prevent nascent surface contacts from forming cold welding or adhesion. Under current boundary lubrication technology, many of these protective mechanisms are being used to protect the surfaces.

Acknowledgments

Most of the data used in developing and refining the three sequential wear test procedures described in this chapter were generated over a period of years in our laboratory and involved a large number of individuals in various research projects. We especially acknowledge the contributions of D. Deckman, J. Yelletts, P. Pei, C. Ku, and K. Jewett.

This material has been based upon an article that first appeared in the *Journal of Engineering Tribology* – Proceedings Part J, 2002, Vol. 216, No. J6, ISSN 1350–6501, published by Professional Engineering Publishing. Permission is granted by the Institution of Mechanical Engineers.

References

1. Beerbower, A., 'Boundary Lubrication: A Scientific and Technical Applications Forecast,' final report, Army Office of Research, Contract Number DAHC-19-69-C-0033, 1972.
2. Hsu, S.M. and Klaus, E.E., 'Estimation of Molecular Junction Temperatures on Four-Ball Contacts by Chemical Reaction Rate Studies', *ASLE Transactions*, **21**(3), 1978, 201.
3. Hsu, S.M. and Klaus, E.E., 'Some Chemical Effects on Boundary Lubrication – Part 1: Base Oil–Metal Interaction', *ASLE Transactions*, **22**, 1979, 135.
4. Gates, R.S., Jewett, K.L. and Hsu, S.M., 'A Study on the Nature of Boundary Lubricating Film: Analytical Method Development', *Tribology Transactions*, **32**(4), 1989, 423.
5. Gates, R.S. and Hsu, S.M., 'Development of a Four-Ball Wear Test Procedure to Evaluate Automotive Lubricating Oils', *Lubrication Engineering*, **39**(9), 1983, 561.
6. Gates, R.S. and Hsu, S.M., 'Development of an Oxidation-Wear-Coupled Test for the Evaluation of Lubricants', *Lubrication Engineering*, **40**(1), 1984, 27.
7. Klaus, E.E. and Perez, J.M., 'Comparative Evaluation of Several Hydraulic Fluids in Operational Equipment, a Full Scale Pump Test and a Four-Ball Wear Tester', SAE, Special Publication 558, 1983, p. 25.
8. Feng, I.M., 'A New Approach in Interpreting the Four-Ball Wear Results', *Wear*, **5**, 1962, 275.
9. Bieber, N.E., 'Phosphate Esters and Their Function in Boundary Lubrication', Ph.D. thesis, Penn State University, 1965.
10. Bose, A.C., Klaus, E.E. and Tweksbury, E.J., 'Evaluation of Wear Products Produced by Some Chemical Reactions in Boundary Lubrication', *ASLE Transactions*, **19**(4), 1976, 287.
11. Willermet, P.A., Mahoney, L.R. and Kandah, S.K., 'Lubricant Degradation and Wear IV. The Effect of Oxidation on the Wear Behavior of Pentaerythrityl Tetraheptanoate', *ASLE Transactions*, **24**(4), 1981, 441.
12. Willermet, P.A. and Kandah, S.K., 'Wear Asymmetry – A Comparison of Wear Volumes of Rotating and Stationary Ball in the Four-Ball Wear Machine', *ASLE Transactions*, **26**(2), 1983, 173.
13. Willermet, P.A., Kandah, S.K., Diegl, W.O. and Chase, R.E., 'The Influence of Molecular Oxygen on Wear Protection by Surface Active Compounds', *ASLE Transactions*, **26**(4), 1983, 521.
14. Munro, R.G., 'Temperature Considerations in the Study of Surfaces Using a Four-Ball Wear Apparatus', *Journal of Applied Physics*, **57**(11), 1985, 4950.
15. Pei, P. and Hsu, S.M., 'Preparative Liquid Chromatographic Method for the Characterization of Minor Constituents of Lubricating Base Oils', *Journal of Liquid Chromatography*, **15**, 1986, 3311.
16. Hsu, S.M. and Gates, R.S., 'Boundary Lubricating Films: Formation and Lubrication Mechanism', *Tribology International*, **38**, 2005, 305–312.

17. Blok, H., 'Surface Temperatures Under Extreme Pressure Conditions', Congres Mondial du Petrole, Paris, 1937, pp. 471–486.
18. Enthoven, J.C., Cann, P.M. and Spikes, H.A., 'Temperature and Scuffing', *Tribology Transaction, STLE*, **36**(2), 1993, 258–266.
19. Horng, J.H., Lin, J.F. and Li, K.Y., 'Scuffing as Evaluated from the Viewpoint of Surface Roughness and Friction Energy', *Journal of Tribology, ASME*, **118**, 1996, 669–675.
20. Hsu, S.M., Shen, M.C., Klaus, E.E., Cheng, H.S. and Lacey, P.I., 'Mechano-Chemical Model: Reaction Temperatures in a Concentrated Contact', *Wear*, **175**, 1994, 209–218.
21. Spikes, H.A. and Cameron, A., 'A Comparison of Adsorption and Boundary Lubricant Failure', Proceedings of the Royal Society of London A, **336**, 1974, 407–419.
22. Camera, A., '*Basic Lubrication Theory*', Ellis Horwood Limited, London, UK, 1981.
23. Hsu, S.M. and Klaus, E.E., 'Some Chemical Effects in Boundary Lubrication Part I: Base Oil–Metal Interaction', *ASME Transactions*, **22**(2), 1979, 135.
24. Klaus, E.E., Nagarajan, R., Duda, J.L. and Shah, K.M., 'The Adsorption of Tribochemical Reaction Products at Solid Surfaces', Friction and Wear – 50 Years on, *Proceedings of the Institution of Mechanical Engineers, International Conference on Tribology*, 1987, 370.
25. Hsu, S.M. and Zhang, X.H., 'Lubrication: Traditional to Nano-Lubricating Films', in *Micro/Nanotribology and its Applications* (ed B. Bhushan), Kluwer Academic Publishers, Netherlands, 1977, pp. 399–414.
26. Wang, L.Y., Yin, Z.F., Zhang, J., Chen, C-I. and Hsu, S.M., 'Strength Measurement of Thin Lubricating Films' *Wear*, **237**, 2000, 55–162.
27. Ying, T.N. and Hsu, S.M., 'Effect of Friction on Subsurface Strain Distribution of Steel', *Tribology Transactions*, **40**(3), 1997, 420–435.
28. Ying, T.N., 'Wear Mechanisms for Ductile and Brittle Materials in a Micro-Contact', Ph.D. thesis, Engineering Materials, University of Maryland, College Park, MD, 1994.
29. Klaus, E.E. and Tewksbury, E.J., 'Microcorrosion Studies with Functional Fluids', *Lubrication Engineering*, **29**, 1973, 205.
30. Gates, R.S., Jewett, K.L. and Hsu, S.M., 'A Study on the Nature of Boundary Lubricating Film: Analytical Method Development', *Tribology Transactions*, **32**(4), 1989, 423–460.
31. Hsu, S.M., 'Boundary Lubrication of Advanced Materials', *MRS Bulletin*, **16**(10), 1991, 54.
32. Shen, M.C. and Hsu, S.M., 'A Modeling Wear of Ceramics Based on Wear Maps', in *The Advancing Frontier of Engineering Tribology* (ed Q. Wang, J. Netzel and F. Sadeghi), STLE, 1999, pp. 140–149.
33. Greenwood, J.A. and Williamson, J.B.P., 'Contact of Nominally Flat Rough Surfaces', *Proceedings of the Royal Society, London*, **A295**, 1966, 300.
34. Wang, F.X., Lacey, P.I., Gates, R.S. and Hsu, S.M., 'A Study of the Relative Surface Conformity Between Two Surfaces in Sliding Contact', *ASME Journal of Tribology*, **113**, 1991, 755.
35. Coy, R.C., 'Practical Applications of Lubrication Models in Engines', *Tribology International*, **31**(10), 1998, 563.
36. Granick, S., 'Motions and Relaxations of Confined Liquids', *Science*, **253**(5062), 1991, 1374.
37. Granick, S., 'Molecular Tribology', MRS Bulletin, **16**(10), 1991, 33.
38. Granick, S., 'Soft Matter in a Tight Spot', *Physics Today*, **52**(7), 1999, 26.
39. Munro, R.G., Piermarini, G.J. and Block, S., 'Wall Effects in a Diamond-Anvil Pressure-Cell Falling-Sphere Viscometer', *Journal of Applied Physics*, **50**, 1979, 3180.
40. Munro, R.G., Piermarini, G.J. and Block, S., 'Viscosities and Glass Transitions in Liquids at High Pressures', *Review of Physical Chemistry of Japan*, **50**, 1980, 79–96.
41. Dowson, D. and Higginson, G.R.A., 'Numerical Solution to the Elastohydrodynamic Problem', *Journal of Mechanical Engineering Science*, **1**(1), 1959, 6.
42. Lee, S.C. and Cheng, H.S., 'On the Relation of Load to Average Gap in the Contact Between Surfaces With Longitudinal Roughness', *Tribology Transactions*, **35**(3), 1992, 523.
43. Bell, J.C. and Colgan, T., 'Pivoted-Follower Valve Train Wear – Criteria and Modeling', *Lubrication Engineering*, **47**(2), 1991, 114.
44. Bell, J.C. and Willemse, P.J. 'Mid-Life Scuffing Failure in Automotive Cam-Follower Contacts', *Proceedings of the Institution of Mechanical Engineers, Part J, Journal of Engineering Tribology*, **212**(J4), 1998, 259.
45. Naidu, S.K., Klaus, E.E. and Duda, J.L., 'Evaluation of Liquid Phase Oxidation Products for Ester and Mineral Oil Lubricants', *Industrial and Engineering Chemistry Product Research and Development*, **23**, 1984, 613.
46. Hsu, S.M., Klaus, E.E. and Cheng, H.S., 'A Mechano-Chemical Descriptive Model for Wear Under Mixed Lubrication Conditions', *Wear*, **128**(3), 1988, 307.

5

Wear and Chemistry of Lubricants

A. Neville and A. Morina

5.1 Encountering Wear in Tribological Contacts

Wear occurs when moving surfaces under load, or surface asperities, come into contact. Plastic deformation results in a change in surface topography and/or removal of material and such 'wear' processes have severe consequences for component performance in a range of industrial machines. The phenomenon of wear has been widely studied over the years and much attention has focussed towards defining wear mechanisms, determining wear rates, modelling/predicting wear and developing measurement methods for wear. Avoiding wear, by applying surface coatings, modifying the surface topography, altering the surface structure or incorporating species into the surface, has led to the creation of a whole new field of surface engineering.

Wear occurs in dry as well as lubricated contacts. Here, the focus will solely be on lubricated wear. To understand wear, and in particular where wear may pose a problem in tribological contacts, it is important to understand the lubrication regime under which the tribocouple operates. In the Stribeck diagram [1], lubrication regimes are defined by the dimensionless bearing number $\eta\omega/p$ where η is the dynamic viscosity of the lubricant, ω the rotational speed and p the specific load. Three regimes are clearly defined in Figure 5.1(a). In Figure 5.1(b), a modified version of the Stribeck diagram is presented [2] where the lambda ratio (λ) replaces the $\eta\omega/p$ group. λ is defined as $\lambda = h_c \sigma_{rms}^{-1}$ where $\sigma_{rms} = \sqrt{(R_{q1}^2 + R_{q2}^2)}$, R_{q1} and R_{q2} are the rms roughness values of surfaces 1 and 2 in contact and h_c is the film thickness. The importance of using λ in the modified Stribeck diagram lies in the fact that the lubrication regime is inextricably linked to the nature of the surface topography. Figure 5.1(b) also describes where some components of an internal combustion would lie in terms of their lubrication regime. The components of Internal Combustion (IC) engines will be the primary focus of this chapter although reference will be made to additives and wear in other contexts.

Figure 5.1(a) and (b) defines the lubrication regimes in terms of their friction coefficient, and wear is not directly integrated into the Stribeck diagram. However, as it will become

Wear – Materials, Mechanisms and Practice Edited by G. Stachowiak

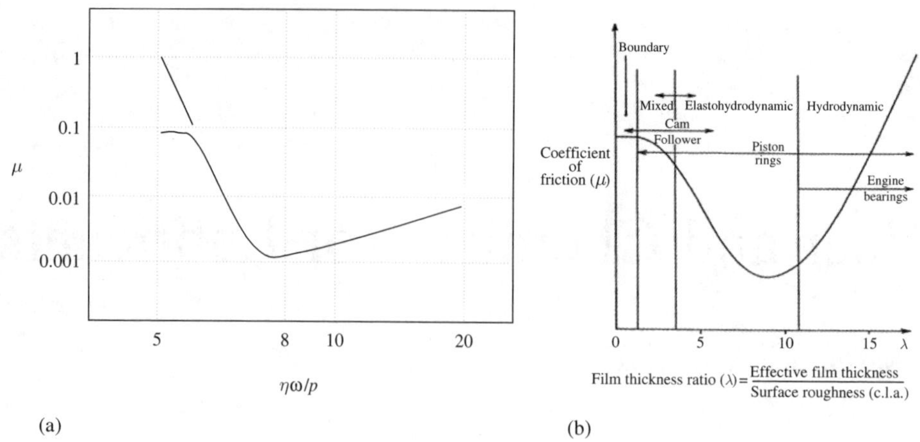

Figure 5.1 (a) Stribeck diagram and (b) modified Stribeck diagram [2] incorporating lubrication regimes for some IC engine components

clear later, the two important regimes from the point of view of durability of the contact (i.e. wear) are the boundary and mixed regimes where metal-to-metal contact (by asperity or surface contact) will occur and lubricants will be required to prevent excessive damage. There is no simple universal link between friction and wear and, as such, the Stribeck diagram does not give any appreciation of the level of wear in a system.

In recognition of the importance of the boundary and mixed lubrication regimes in engine operation, the newly developed Sequence VI test has been developed, with an increased operation cycle proportion being in these regimes [3]. It is appreciated that engine and lubricant performance will often be pushed to the limits of their capabilities where boundary lubrication conditions exist. In cases where full film lubrication is difficult to maintain, the physical properties of the lubricant such as viscosity are not as important as the chemical properties of the lubricant. The chemical additives added to the lubricant form a protective thin film between the rubbing surfaces which is intended to alleviate wear and control friction. The mechanisms by which this is done are described fully later in this chapter.

Wear processes can be divided into categories depending on the nature of material displacement or removal. The principal categories depend very much on the text being consulted as nomenclature can vary. In Reference [4] they are classified as adhesive wear, abrasive wear, contact fatigue and corrosive wear. Scuffing is described as the 'catastrophic wear occurring as a consequence of gradual starvation of the lubricating film under hard loading conditions' [5]. Initiation of scuffing can be by any of the wear processes mentioned above but is normally associated with adhesive wear.

Wear rates are often quoted in terms of a wear coefficient or a dimensionless wear coefficient as derived from the Archard treatment of wear processes [4]. The wear rate (ϖ) is calculated as shown in equation (1):

$$\varpi = \frac{KW}{H} \tag{1}$$

where K is the wear coefficient (dimensionless), W the load on the contact and H the hardness of the uppermost layer in the contact. Difficulties in precisely defining H have meant that often the dimensional wear coefficient (k) is used where $\varpi = kW$. Defining a wear coefficient is obviously useful to determine the extent of material deformation. The Archard treatment of wear has been shown to be valid for cases where mechanical influences are dominant but, as will be shown later, the influence of lubricant chemistry is not considered and often this can be important.

Although when engines are running, wear is inevitably occurring, in automotive engines the design, material selection and lubricant package have been finely tuned to ensure that wear is kept to a very low level. Wear rates of the order of 10^{-18} m^3/Nm are experienced in the cam and followers in normal running engines [6]. With current changes in legislation driving changes in lubricant formulation and the changes in engine materials for reducing fuel consumption, engine wear has to be optimized in every case. Wear in the piston ring/cylinder bore system is normally as a result of three main mechanisms: corrosion, adhesion and abrasion [7]. Corrosion occurs when the engine runs either very hot or very cold due to increasing deposition of acidic materials on the cylinder wall [8]. Abrasion occurs when hard particles (such as soot) are entrained in the contact and adhesion occurs when asperities are in contact due to breakdown of the lubricant film. Much has been discussed about the role of soot in wear and recent studies have concluded that soot size is primarily dictated by the quality of combustion and not by the type of oil as previously thought. Soot undoubtedly enhances the potential for wear and increased soot size leads to higher potential as it bridges the minimum film thickness [9]. Consequences of increased engine wear are increased oil consumption, blow by and eventually power loss [10].

Even with traditional engine materials and lubricants, in heavy-duty diesel engines (e.g marine) there is a much higher instance of wear than in the passenger car and light vehicle sector. On such large-scale components, the fine tolerances required on cylinders, which are generally adhered to in the automotive sector, cannot be met [11]. Wear in the piston ring/cylinder assembly near the top-dead-centre is often a limiting factor in the lifetime of two-stroke marine diesel engines.

5.2 Lubricant Formulations – Drivers for Change

It is perhaps surprising that a link can be made between claims from ecologists [12] that 1,000,000 species could be extinct in the next decade and wear. However, passenger and commercial light vehicles have been estimated [13] to account for 20% of the total CO_2 delivered into the atmosphere from hydrocarbon sources in the US. When the population of vehicles in the western world is considered, it is perhaps easier to see how a realistic link between wear and ecology can be made. According to recent UK government statistics [14], in 2000 in the UK alone there were 29 million motor vehicles registered for use on the public highway. International numbers are even more impressive, with the total number of road vehicles in service in Europe, Japan and the USA in 1999 being of the order of 530 million [14]. Bearing in mind that this total excludes all of Asia except Japan, former Soviet block countries, Australasia, Africa and South America, the staggering scale of the use of the reciprocating internal combustion engine becomes clear.

Reduction of CO_2 emissions is achieved by increasing fuel economy and, as summarized by Ukuno and Bessho [15], there are several means of achieving ‘good gas mileage’ (Figure 5.2).

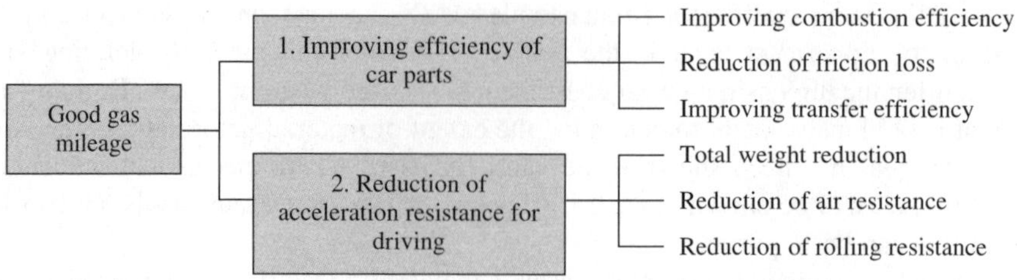

Figure 5.2 Techniques for improving fuel economy [15]

Advances are being made in both areas 1 and 2 as defined in Figure 5.2. With respect to point 2, weight reduction and reduced aerodynamic resistance [16] have been instrumental in achieving the 30% reduction in fuel consumption of the new Volkswagen Lupo 3L TDI. As material development continues at an alarming rate and ever-new alloys [17], composites and nano-engineered structures are developed and embraced by the automotive sector [18], it is expected that the trend will continue. Material selection and use in engine components brings its own challenges in relation to wear and selection of optimum lubricants as will be discussed later.

Of relevance to this chapter is the potential for fuel economy improvements in the moving engine parts as defined in point 1 which are important when wear and the chemistry of lubricants are concerned.

Fuel economy, important as it is for controlling harmful emissions, is also driven by the need to reduce worldwide energy consumption and reliance on fossil fuels. Jeremy Rifkin [19], in his recent book, presented the position on the world's energy supply and use. With the peak world oil production estimated to occur between 2010 and 2020, renewable sources of energy will have to succeed in substituting increasing proportions of the energy demand. It was estimated in 1999 that around 90% of the world's energy is supplied by fossil fuels [20].

Environmental concerns are just one factor influencing the development of lubricants. In a recent article by Korcek [21], the technology drivers in engine oil development were summarized, as shown in Figure 5.3. These can be broadly classified into drivers from customers, environmental protection and resource utilization.

The consequences of these pressures on lubricant formulators have meant a radical change has been required to occur (and is still occurring) in the allowable oil additive constituents. Table 5.1 shows the current European and US emission regulations for gasoline and diesel engines. Achieving these emission targets and the fuel economy targets imposed by CAFE[1] standards requires a radical change in lubricant formulation. Some of the major changes in engine oils, their consequences and also how future challenges are posed resulting from these are shown in Table 5.2.

The challenge in terms of P reduction is perhaps the most apparent and is certainly the most important when engine durability, and in particular wear, is considered. The acceptable level of P has decreased from 0.12 wt.% in 1993 for ILSAC[2] GF-1 oils to 0.1 wt.% in 1996 and 2001 (GF-2 and -3) and will be reduced again to 0.08 wt.% when GF-4 is introduced,

[1] Corporate average fuel economy.
[2] International Lubricant Standardization and Approval Committee.

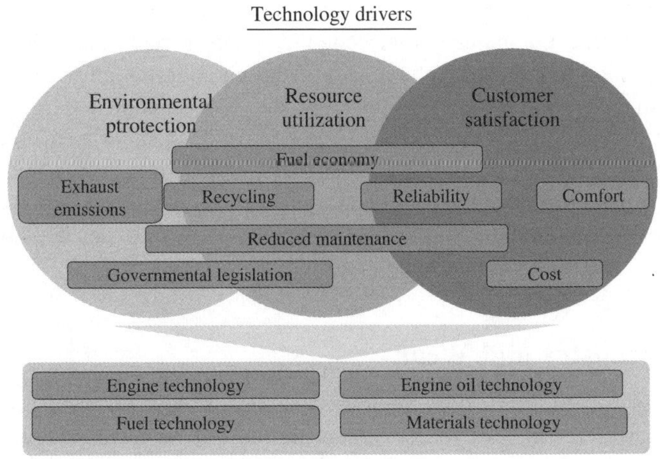

Figure 5.3 Original equipment manufacturer (OEM) drivers [21]

Table 5.1 Current and future emission regulations for Europe (a) and US (b) [22]

Gasoline engines				Diesel engines (HDD)				
g/km	CO	HC	NO_x	g/kW-h	CO	HC	NO_x	Particulates
(a) European emission regulation								
2000	2.3	0.2	0.15	2000	2.1	0.66	5.0	0.1
2005	1.0	0.1	0.08	2005	1.5	0.46	3.5	0.02
2008	?	?	?	2008	1.5	0.46	2.0	0.02
(b) US emission regulation								
Tier 0 (1987)	3.4	0.25	1.0	2000	20.8	1.7	5.4	0.13
Tier 1 – 1994–1997	3.4	0.41	0.4	2005	20.8	0.7	3.4	0.13
Tier 2 – 2004/2009	3.4	0.07[a]	0.07	2008	?	0.2	0.3	0.01

[a] NMOG value and not THC.

Table 5.2 Summary of the major changes in lubricant formulations, the major consequences and challenges

Lubricant change	Effects	Future challenge
Lower viscosity	– Improved fuel economy	– Improved antiwear technology
	– Risk of increased wear	– Durable surface development
Lower P	– Improved catalyst performance – Reduced emissions – Higher wear	– Novel zero P additives and low to zero S technology
Lower S	– Reduced EP performance – Reduced SO_x	– Alternative additives to deliver effective wear protection

which is estimated to be in 2005. Finding new lubricant additives, which can deliver adequate wear protection, whilst maintaining fuel economy poses a real challenge to formulators. As discussed later, lubricant manufacturers are focussing attention on replacement additives.

Environmental concerns are also influencing the use of additives in lubricants other than those for internal combustion engines. The control of wear in metal forming, cutting, wire drawing and other manufacturing processes requires additives to work under extreme conditions of temperature and pressure. Traditionally, this has been achieved through the use of Cl-, S- and P-containing additives. However, their link to ozone depletion [23] is also forcing new lubricants to be developed.

5.3 Tribochemistry and Wear

In boundary and mixed lubrication regimes, the occurrence of metal-to-metal contact, primarily at asperities, will lead to component failure due to local seizure. Avoidance of excessive wear and/or seizure in boundary lubricated contacts is achieved through formation of a protective film, derived from the complex additive packages which have been developed over the last 60–70 years.

One of the key aspects in boundary lubrication is that the performance of the lubricant (and more specifically the additives) is critically dependent on the environment within which it is used. This is not a simple case of assessing the bulk oil temperature and pressure – a much more complex situation arises where the 'environment' is defined as the local environment at the asperity contact when the asperities are in contact and then, as importantly, out of contact.

The word *tribochemistry* has been adopted to encompass aspects of chemical reactions and physical changes at the surface that can only occur when both tribological and chemical processes act together. The interaction of the tribological process and the chemical process when surfaces are in contact in lubricated conditions is vital since the combination of the two processes results in highly reactive surfaces being formed and subsequent formation of surface films [24]. As an example, the activation energy for formation of iron oxide through a thermochemical mechanism is 54 kJ/mol compared with a much lower value of 0.7 kJ/mol for the tribological or tribochemical pathway [25]. Kajdas [24] described the tribochemical reaction processes occurring at boundary lubricated surfaces in a schematic representation presented in Figure 5.4.

Because of the complexity and the transient nature of the conditions in the boundary lubricated contact, it has been difficult for the complete chemical and physical pathway for the reactions involved in film formation to be elucidated. There is uncertainty as to the temperature–time profile for individual asperities and the nature (temperature increment, rheology and chemical changes) of the local lubricant environment and this makes direct assessment of the layer formation difficult. Great strides have been made towards a full understanding of the action of lube additives at tribosurfaces with traditional lubricant additives and surfaces through extrapolation from post-surface analysis of reaction products and introduction of new methods for thin film analysis [26]. There are still uncertainties remaining about the exact nature of the tribofilm when it is in the contact. Recent emphasis has been on demonstrating the 'smart' properties of the film – having the ability to react to the severity of the conditions and adapt its form accordingly [27]. Also, new lubricants and surfaces are being introduced, as will be discussed later, and so there are challenges to be addressed in getting to the level of understanding we have with current lubrication

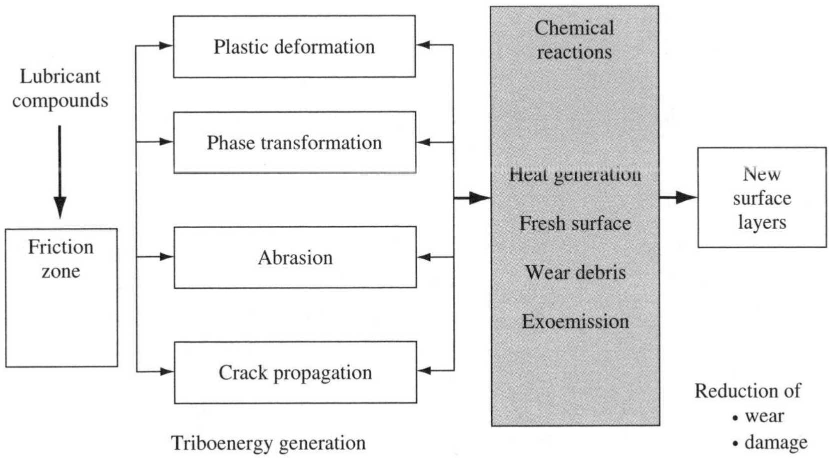

Figure 5.4 Major physical and chemical events in the boundary lubrication contact [24]

technologies for new additive and material systems. Progress will have to be made in some of the more complex areas of tribology, which involve probing the physical and chemical nature of the film when it is in the contact. This remains one of the principal outstanding challenges in this area.

5.4 Antiwear Additive Technologies

To deliver their antiwear functionality, antiwear additives, through tribochemical reactions, facilitate the formation of a very thin film (now commonly referred to as the tribofilm) on the surface of the asperities. In terms of assessing the role of wear processes on the stability and integrity of the film the following must be understood:

- the nature of the film (chemical, physical and mechanical properties);
- kinetics of formation; and
- removal and replenishment rate.

Because these are dependent on conditions, additive packages are normally only effective in certain operational windows [28].

5.4.1 Antiwear Technologies

The most common antiwear additives used in practice are organochlorine, organosulphur, organophosphorus [tricresyl phosphate (TCP) and dibutyl phosphite (DBP)] and organometallic (ZDDP, MoDTP, MoDTC) and organic borate compounds [28, 29]. In some instances antiwear additives are mentioned in the same context as extreme pressure (EP) additives since their function is primarily the same: to reduce metal-to-metal contact. The distinguishing feature is that antiwear functionality is normally associated with the mixed lubrication regime where some separating lubricant film still exists but the film is intermittently penetrated and asperity contact occurs. This is in contrast to EP conditions where collapse of the oil film leads to a high degree of metal contact and severe wear/high friction is encountered. In this chapter, the two will generally be treated separately, but in some instances a clear separation of the boundary and mixed regimes does not exist and so AW/EP (anti-wear/extreme pressure) functionality co-exist.

5.4.2 ZDDP – Antiwear Mechanism

In the category of organometallic additives, perhaps the most commonly used is zinc dialkyl dithiophosphate (ZDDP or ZDTP), whose structure is shown in Figure 5.5. Initially developed in the 1940s [30] to provide oxidation stability to engine oils, it was found to be effective in reducing wear and has become one of the most commonly used additives.

As mentioned previously, probing the chemical structure in the tribocontact is for now not possible using existing surface analysis techniques although in-contact (or 'buried' interface) spectroscopy is receiving attention and it is expected that progress will be made in this area soon [31]. The rapid progress made in the last two decades in the understanding of the mechanism of ZDDP action has been mainly due to the timely arrival of advanced surface analysis techniques. These include, as examples, ToF-SIMS, XPS, Auger analysis, P-NMR, H-NMR, XANES and AFM. Combining the information from all of these techniques has meant that the nature of the species of the tribofilm, formed from the decomposition of ZDDP and interaction with the surface, is accessible. There has been much debate relating to the nature of the film, once it is removed from the triboenvironment, and prepared for analysis in what are often UHV conditions. The effect of solvent washing was assessed in a recent article by Bec *et al.* [32], but the question regarding how realistic the surface is compared to when it is in motion in the contact still remains.

Two aspects of the ZDDP tribofilm are generally accepted, these being that the film has a complex structure and that it has a layered structure. In simplified terms, it is often regarded to comprise mixed sulphides and oxides near to the substrate surface and these are covered by short- and long-chain polyphosphate.

The generation of tribofilms from ZDDP is accepted to be as a result of its decomposition and, as such, the role of the decomposition products becomes important if a full understanding of the nature of the film is to be achieved. Although the species formed from tribochemical and thermal decomposition of ZDDP are known, the mechanisms by which they are obtained are still not fully understood. Proposed mechanisms have included *thermal decomposition, hydrolytic mechanisms* and *oxidative mechanisms*.

The important by-products in early research were postulated to be complex thiophosphates, phosphides and sulphides by consideration of the ZDDP molecule and possible reaction pathways as ZDDP composes. Later work, from the mid-1970s onwards, looking at tribochemical and thermally derived films, has advanced the understanding. The mechanisms proposed by different researchers are presented in the following paragraphs.

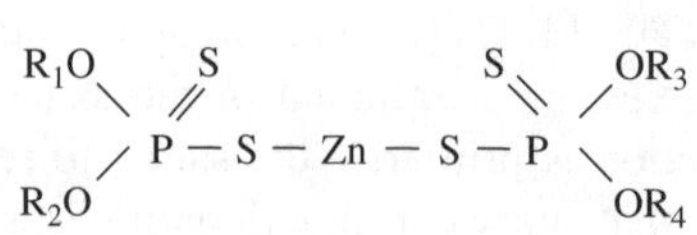

Figure 5.5 Molecular structure of ZDDP

Jones and Coy [33] heated samples of ZDDP-containing lubricant to induce thermal degradation and identified the decomposition products by 31 P NMR and 1H NMR. By correlating the structures of the products found (oil-insoluble solid, trialkyltetraphiophosphate, dialkyl sulphide, alkyl mercaptan and dialkyl disulphide) with the structure of the starting material, the mechanism of thermal decomposition of ZDDP was proposed. According to this mechanism, the decomposition is initiated by a general migration of the alkyl group from oxygen atoms to sulphur atoms. This process then is followed by the olefin elimination, leading to the formation of phosphorus acids. Nucleophilic substitution of one phosphorus species by another leads to formation of P-O-P-type structures and of zinc mercaptide $[Zn(SR)_2]$ as reaction intermediate. Reaction of mercaptide with an alkylating agent generates dialkyl sulphide, while reaction of mercaptide with dithiophosphate species leads to formation of trithiophosphates and eventually tetrathiophosphates. The oil-insoluble deposit is a mixture of zinc thiophosphates and zinc pyro- and polypyro-thiophosphates.

Spedding and Watkins [34] and Watkins [35] in their study of thermal decomposition followed by the role of surface reactions, proposed a hydrolytic mechanism of ZDDP decomposition in oil solution, forming zinc polyphosphate and a mixture of alkyl sulphides. They came to this conclusion by studying the effect of water removal from the reaction on the assumption that if water removal stopped the reaction, then water must play an integral part. This approach gave almost complete suppression of the decomposition, confirming that ZDDP breakdown is water-catalysed. In terms of surface reaction related to tribofilms [35], the antiwear effects of the tribofilm compositions formed from ZDDP breakdown were studied. Using Electron Spectroscopy Chemical Analysest (ESCA) to analyse cam and tappet components, they found that the tribofilm comprises zinc and phosphorus as phosphate and sulphur as iron sulphide. It is suggested that zinc phosphate is physically adsorbed on the surface oxide layer. The antiwear mechanism of zinc phosphate was postulated to be through its low melting point, glassy characteristics [36]. The fluid glass effectively lubricated the surface. Iron sulphide is formed via the oxide layer and elemental sulphur generated from alkyl sulphides and Fe_2O_3.

Willermet *et al.* [37] have proposed that under moderate wear conditions, the formation of an antiwear film from ZDDP proceeds by a thermo-oxidative mechanism. It was suggested that the antiwear and antioxidant chemistries of ZDDP are linked. The following steps are suggested for ZDDP tribofilm formation:

- Adsorption of ZDDP on the metal surface and formation of a ZDDP-rich near-surface region.
- Reaction of ZDDP with the metal surface to form phosphate/phosphothionic moieties chemically bonded to the metal.
- Formation of phosphate film precursors from the antioxidant reactions of ZDDP.
- Condensation of phosphates/phosphothionates and their esters, the growing phosphate chains being terminated by reaction with zinc-containing compounds and, in the presence of overbased detergents, by reaction with the metal overbase.

Under the 'moderate' wear conditions defined by Willermet *et al.* [37] no presence of sulphide is accounted for which is in line with the general understanding of their formation, linked to EP or severe wear conditions where the formation of debris plays a key role.

With the assumption that friction is necessary to produce the speciation of phosphorus and sulphur, Martin [38] proposed a mechanism for the film formation with metal dithiophosphates in which both chemical and mechanical aspects are linked together. It is based on a three-step mechanism:

Step 1: A thermo-oxidative reaction which leads to formation of phosphate precursor films. The result is mainly the formation of a long-chain zinc poly-thiophosphate film on the flat surface (a polymer-like material deposit) and the corresponding liberation of organic sulphur species in the lubricant phase. This reaction can be summarized as follows:

$$\text{ZnDTP(chemisorbed)} + \text{O}_2\text{(or ROOH)} \xrightarrow{100-200^\circ\text{C}} \text{Zn[PO}_x\text{S}_{3-x}]_n + \text{sulphur species (in the lubricant phase)}$$

Step 2: The *in situ* tribochemical reaction leading to P and S speciations. The zinc thiophosphate polymer reacts with crystalline iron oxides, which is generated from the surface during an induction period where abrasive wear is dominating, and a mixed iron and zinc phosphate glass matrix on one side and metal sulphide-embedded crystallites on the other side are formed. This reaction can be summarized as follows:

$$\text{Zn[PO}_x\text{S}_{3-x}]_n + \text{iron oxides } (\text{Fe}_2\text{O}_3, \text{Fe}_3\text{O}_4) \xrightarrow{\text{pressure, shear, contact temperature}} [\text{ZnO}, \text{P}_2\text{O}_5, \text{FeO/Fe}_2\text{O}_3] + \text{MeS}_x$$

Step 3: A self-organization system in which there is a selective transfer phenomena between the two friction surfaces which leads to friction reduction.

A schematic representation of the film structure as suggested by Martin [39] is shown in Figure 5.6.

Figure 5.6 is widely accepted as a true representation of the ZDDP tribofilm structure. It can be altered by incorporation of other additives (explained later) such as detergents and friction modifiers. The thickness of such tribofilms is still under debate and much of the uncertainty derives from not knowing how to define the limits of the tribofilm. Often the film on tribological

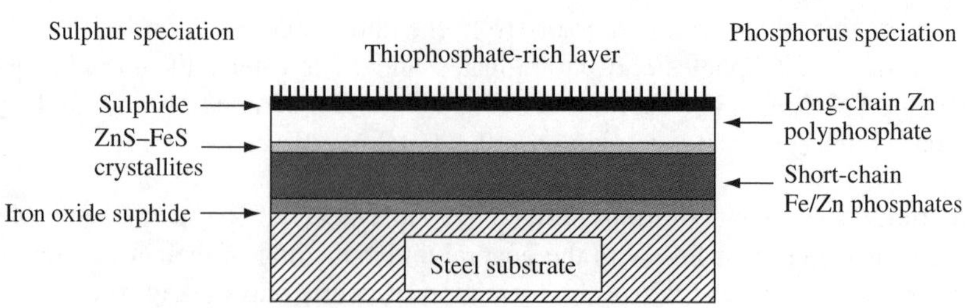

Figure 5.6 ZDDP film structure proposed by Martin [39]

surfaces can have an extended layer of C-rich material which has a structure not comparable with the bulk oil and is therefore often regarded as an integral part. How much of this layer is removed by solvent rinsing was addressed by Bec *et al.* [32] and the outer hydrocarbon-rich layer is removed, the extent depending on the solvent type and rinsing time. However, it is generally accepted that the film is in the order of several tens of nanometres thick [40].

The reaction of zinc thiophosphate with iron oxides is explained using hard and soft acids and bases (HSAB) principle as described by Pearson [41]. According to this principle, which is based on the role of chemical hardness in predicting the chemical reactions, hard acids react with hard bases while soft acids react with soft bases. So the reactions between phosphates and oxides stem from the fact that Fe^{3+} is a harder Lewis acid than Zn^{2+} and that the cation exchange is energetically favourable from the point of view of this principle. This reaction is shown below:

$$\begin{array}{ccc} 5Zn(PO_3)_2 + Fe_2O_3 & \longrightarrow & Fe_2Zn_3P_{10}O_{31} + 2ZnO \\ (ZnO, P_2O_5) & & (3ZnO, 5P_2O_5, Fe_2O_3) \end{array}$$

This process is called a 'digestion' of abrasive wear particles of oxides by the phosphates and is linked with the antiwear effectiveness of ZDDP. According to Martin [39], there are three ways by which ZDDP acts as an antiwear additive:

- By formation of long-chain zinc polyphosphate glasses on the surface. These glasses are known to have a relatively low transition melting point, so it is suggested that the rheological properties of the molten glass ensure some sort of elastohydrodynamic lubrication.
- By 'digesting' the iron oxide particles, known to be very hard and which can cause abrasive wear.
- In the cases where the phosphate film is disrupted the sulphur species react with the nascent surface forming a passivating film. The sulphides formed will prevent adhesion and the oxidation of nascent surfaces in very severe conditions.

According to the above, nascent iron and iron oxide, generated by wear processes occurring during sliding, are an integral part of the mechanism proposed for ZDDP tribofilm formation. In this mechanism, wear and film formation are inextricably linked. This issue assumes extreme importance when boundary lubrication of materials which are free from iron is considered. The question then arises as to how effective ZDDP will be in forming an effective tribofilm. This is dealt with later in the chapter. In addition, observations of films resembling tribofilms, formed under purely thermal conditions, have led to debate about how important the rubbing process and the debris it generates are. This debate continues.

Fuller *et al.* [42] combined a study of thermal decomposition with a study of tribofilm formation. The decomposition products were identified as polyphosphates with small amounts of sulphides. The proportion of long- and short-chain polyphosphate changed as the temperature changed. In comparing thermal films with tribofilms it was concluded that the thermal films formed at lower temperature (125°C compared with 200°C) were more akin to the structure of tribofilms. In later work, the presence of iron phosphate in the ZDDP antiwear films was not acknowledged. It was proposed that short-chain polyphosphates are formed by the hydrolysis of polyphosphates. A linkage isomer of ZDDP (LI-ZDDP) was proposed as an important precursor for film formation after analysis of thermal films and oil-insoluble ZDDP decomposition products. The modified mechanism of film formation from ZDDP is suggested and can be explained with the reactions below:

$$Zn[(RO)_2P_2]_2\,(\text{solution}) \Rightarrow Zn[(RO)_2PS_2]_2\,(\text{ZDDP adsorbed})$$

$$Zn[(RO)_2PS_2]_2\,(\text{solution}) \Rightarrow Zn[O_2P(SR)_2]_2\,(\text{LI-ZDDP in solution})$$

$$Zn[O_2P(SR)_2]_2\,(\text{solution}) \Rightarrow Zn[O_2P(SR)_2]_2\,(\text{LI-ZDDP adsorbed})$$

$$Zn(RO)_4P_2S_4 + O_2(\text{or ROOH}) \Rightarrow Zn(PO_3)_2\,(\text{polyphosphate}) + \text{sulphur species}$$

$$7Zn(PO_3)_2 + 6H_2O \Rightarrow Zn_7(P_5O_{16})_2 + 4H_3PO_4$$

$$2Zn(PO_3)_2 + 3H_2O \Rightarrow Zn_2P_2O_7 + 2H_3PO_4\ \text{short-chain polyphosphates}$$

An important point relating to the thickness of the ZDDP tribofilm was raised by Fuller *et al.* [42] and it was concluded that increased thickness is not necessarily beneficial in reducing wear. ZDDP tribofilms are primarily established to reduce wear and are effective in this respect under a wide range of operating conditions. However, the effect of film formation on friction has been shown to be complex. The literature shows that in oils containing ZDDP as the sole additive friction can increase [43–47], decrease [48, 49] or be unaffected [50]. From a fuel economy perspective, any adverse effects of antiwear formulations on friction must be managed. Hence, the ZDDP film optimized structure has to be achieved to gain reduction in wear whilst maintaining friction at an acceptable level.

In the work by Bell *et al.* [51], the antiwear films formed from ZDDP were analysed under both cryogenic and solvent-washed conditions using surface techniques such as SEM/EPMA, TOF-SIMS and for quantitative analysis X-ray photoelectron spectroscopy (XPS). Besides phosphates in polymeric and glassy states, evidence of a significant concentration of organic material in the furthest layers of the film was found. This layer is believed to be removed after solvent washing of the samples. A layered film structure model is proposed and is shown in Figure 5.7.

Although differences are apparent in the structure shown in Figure 5.6, there are generic similarities and several similar representations have been presented in mono- and multiple-additive systems.

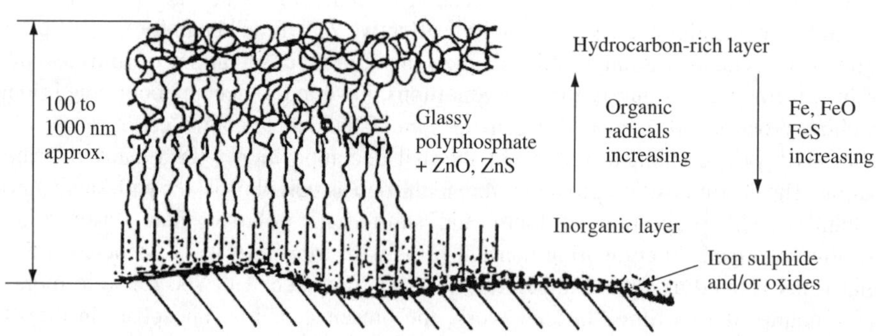

Figure 5.7 ZDDP film structure proposed by Bell *et al.* [51]

Much debate continues about the nature of the tribofilm in the contact and how the *idealized* layered structures, as shown schematically in several papers, actually relate to the load-bearing, antiwear films formed on the asperities. Undoubtedly, an important aspect with respect to the film stability and also the extent of wear associated with the tribological contact is the rate of film formation, removal and replenishment. Lin and So [52] recently summarized the important chemical and physical changes at the surface which cause changes in the surface film and these were presented in a simplified rule that governs antiwear performance.

In their analysis the growth rate of the effective film thickness, h_{effect}, was given by equation (2)

$$h_{\text{effect}} = h_{\text{phy}} + h_{\text{chem}} + h_{\text{react}} \tag{2}$$

and the ZDDP films exist if

$$h_{\text{effect}} \geq h_{\text{scrape}} \tag{3}$$

where h_{scrape} is the rate at which the film is scraped off (worn off), h_{phy} is the physisorbed film growth rate, h_{chem} is the growth rate of the chemisorbed film and h_{react} is the growth rate of the chemically reacting film.

As well as being chemically analysed, the mechanical properties of ZDDP films have also been investigated [32, 53]. With the introduction of the atomic force microscopy (AFM) in the late 1980s, tribologists have capitalized on the facility as a means of probing the near-surface features of tribofilms. One of the earliest studies on real tribofilms formed from additive-containing oils was performed by Pidduck and Smith [53] where they used the force–distance curve analysis to semi-quantitatively determine the hardness of the tribofilm. They showed clearly that the outer layer was viscous and the sub-layer was much harder than the outer layer. The structure and rheological properties of the antiwear films formed from ZDDP were also studied by Bec *et al.* [32]. It was shown in their work also that the film has varying structure and properties with depth. It was found that initial hardness increased with penetration depth which suggests that the ZDDP tribofilm accommodates the pressure applied by the indenter by increasing the hardness – thus exhibiting a degree of 'smartness'.

5.4.3 Interaction of ZDDP with Other Additives

Modern engine oils can contain up to 15 different additives each with their own specific function. In these multi-component systems there is potential for interactions between the different additive species. These interactions can be synergistic or antagonistic and so the formulator's challenge is to optimize the blend to address the performance and durability criteria. Table 5.3 shows the range of additives likely to be present in a commercial engine oil [54]. As discussed previously, ZDDP is one of the most common antiwear additives and other common species include PIBS (polyisobutylene succinimide) as a dispersant, MoDTC (molybdenum dithiocarbamate) as a friction modifier and Ca/Mg sulphonate/phenates as detergents.

In any review of the effect of lubricant chemistry on wear, it is therefore important to acknowledge the potential detrimental or positive effects on wear of other additives, especially when combined with ZDDP as the principal source of antiwear protection.

Table 5.3 Additives and typical concentrations in commercial engine oils [54]

Function	Component	Concentration (mass %)
Base oil (mineral and/or synthetic)		75–95
Control friction and wear	Viscosity index improver	0–6
	Antiwear additive	0.5–2
	Friction modifier	0–2
	Corrosion inhibitor	0–1
Reduces contamination and maintains cleanliness	Antioxidant	0–1
	Dispersant	1–10
	Detergent	2–8
Maintains fluid properties	Pour point depressant	0–0.5
	Anti-foam additive	0.001

So how can additives such as dispersants, detergents and friction modifiers potentially affect the wear performance in boundary lubricated contacts? There are three primary ways in which the antiwear additive performance can be disrupted/affected in the presence of additional species. These are

1. chemical reaction between the additives (e.g complexation [55, 56]) in the bulk oil phase;
2. by affecting the film formation by the antiwear additive [55, 57, 58];
3. by forming a film from the absorption and/or decomposition products derived from their own chemistry [59].

In the following paragraphs, the specific interactions between antiwear and other additives are reviewed, focussing on how wear performance is affected.

Yin *et al.* [60], in their study of the interactions between two detergents and a dispersant, concluded that calcium sulphonate interacts with the adsorbed ZDDP only at high concentration of the detergent (= 2 wt.%). The detergent species were found to be enriched at the outer part of the film which has also been reported in other work [59]. They found that calcium phenate detergents affect the film formation even at low concentrations – in the resulting film the chain length of the polyphosphate is reduced. In this study, no direct measurement of these changes and wear performance was made, but similar effects reported by Wan *et al.* [61] on the interaction of overbased metallic detergents with ZDDP resulted in the deterioration of the antiwear performance of the ZDDP.

Competition between ZDDP and metallic detergents for surface sites, thereby reducing the effective ZDDP surface concentration, has been postulated as one reason for the reduction of ZDDP antiwear properties in the presence of metallic detergents. In Wan *et al.*'s work [61], it was found that Ca^{2+} ions replace Zn^{2+} ions in the polyphosphate structure of tribofilms, forming short-chain polyphosphates.

This confirmed the previous work of Willermet *et al.* [62] which suggested that Ca partially replaced Zn as the cation in the amorphous phosphate film which leads, to the loss of the higher molecular weight phosphates in favour of ortho- and pyro-phosphates. It was found that the Ca cation can replace up to 50% of the Zn cation in the phosphate film. The phosphate chain length was found to be greater for ZDDP solutions than for the systems containing overbased detergents.

A clear interaction of ZDDP with Ca from detergents was shown in the work by Morina *et al.* [63]. When ZDDP was present in additive packages containing ZDDP, Ca was found as Ca phosphate, while in the additive package without ZDDP, Ca was found as $CaCO_3$. This affects the tribological properties – wear and friction are enhanced by the abrasive nature of the carbonate detergent species. These interactions were shown to be related to the lubrication conditions.

To further complicate additive systems, it has been reported that detergent–ZDDP interactions can display synergistic effects [57], and then on addition of a third additive (in this case a dispersant), the synergism is lost and the antiwear performance can also be lost. For dispersants added to ZDDP-containing oils a confusing picture arises. In some reports, the adsorption of ZDDP has been enhanced at the surface, leading to a reduction in wear [64], whereas in other studies [65] the formation of a chelating compound in ZDDP, in the presence of dispersant, decreases the adsorption. This points to the fact that tribofilms formed in the presence of ZDDP are themselves complex entities and addition of often surface-active additives can substantially complicate the system. There is no universal rule to predict additive–additive interactions and so much is dependent on the conditions and the nature of the additives.

In terms of fuel economy friction reduction is paramount and in terms of engine durability wear is of key importance and, as a result, the interactions between antiwear-type additives and friction modifiers (primarily MoDTC) have received considerable attention.

MoDTC, like ZDDP, delivers its performance through degradation compounds. The key to successful reduction in friction is the formation of MoS_2 (molybdenum disulphide) which has a hexagonal crystal structure in which each atom of molybdenum is surrounded at equal distances by six atoms of sulphur placed at the corners of a triangular right prism of 0.317-nm height [66]. The bonds between Mo and S are strong covalent and between the atoms of S are weak van der Waals. Hence, sliding along these planes occurs with little energy input and hence low friction. Figure 5.8 shows a schematic representation of the friction trace when MoDTC is used as the friction modifier. In most circumstances the reduction in friction is not immediate and there is an activation time (t_{act}) required to functionalize the additive. In recent work, it has been demonstrated that the chemical composition, and primarily the

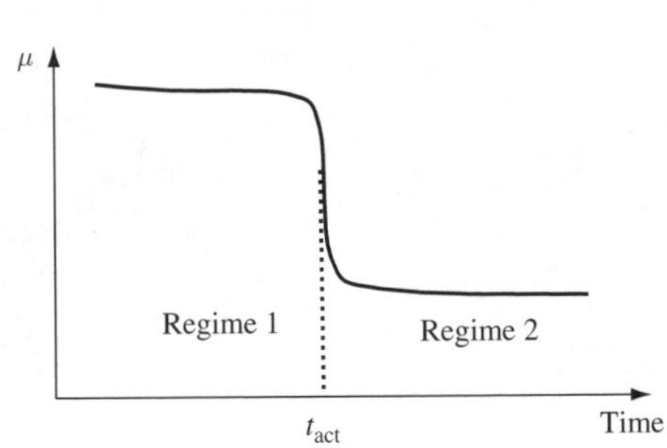

Figure 5.8 Schematic representation of the transient in friction observed when MoDTC is 'activated' at the tribosurface

MoS_2 content, of the tribo-layer evolves in going from regime 1 to regime 2. A significant enhancement in MoS_2 content is recorded at point 2. In some texts, it is written that MoDTC is not able to form MoS_2-rich friction layers on its own [66], but there is evidence in the literature to suggest that MoDTC can in fact work without the presence of ZDDP [67] and also friction reduction with molybdenum dialkyldithiophosphates (MoDTP) is achieved without the presence of ZDDP [68, 69]. However, as reported by Muraki and Wada [70], the formation of MoS_2 on an existing ZDDP-derived tribofilm is catalysed.

The surface topography as determined by AFM has been shown in recent work [71] to be contrasting when ZDDP alone and ZDDP/MoDTC are contained in the lubricants. In Figure 5.9, the different surface characteristics are presented. The MoS_2 significantly smooths the surface and the underlying phosphate-rich layer (offering wear protection) is covered. The layer produced by the friction modifier has been shown not to cause any significant disruption to the antiwear performance. As such, it can be concluded that the effect on the glassy antiwear phosphate layer is minimal.

It was found that in the presence of MoDTC/ZDDP mixture, a two-step reaction occurs: first, a reaction between phosphate and iron oxide, and second, a reaction between the nascent iron surface and a sulphide species. These reactions are found to be well explained by the HSAB principle [72]. The surface analysis made using transmission electron microscopy (TEM) and XPS showed two glassy phases in the tribofilm: mixed zinc/molybdenum phosphate glass and carbon-rich amorphous phase with MoS_2 sheets embedded in the phosphate glass [73]. Iwasaki [74] estimated that up to 40% of the S in MoS_2 can come from the ZDTP thus proving their intimate link. The MoDTC/ZDDP combination was found to be excellent in reducing both friction and wear; wear is reduced because of the elimination of MoO_3 and possible iron oxides when they react with zinc polyphosphate which at the same time preserves pure MoS_2 from oxidation [72]. Kasrai *et al.* [75] used XANES and XPS to obtain more information about the chemical compositions of the tribofilms formed when MoDTC and ZDDP/MoDTC systems were used. No effect of MoDTC on antiwear performance of ZDDP was seen.

MoDTP have been recognized as multifunctional additives, reducing both friction and wear [76]. Surface analysis using XPS showed that MoDTC formed a surface film composed of MoS_2, while MoDTP formed a surface film composed mainly of MoS_2 and $FePO_4$ at 120°C temperature and MoS_2, MoO_3 and $FePO_4$ at 200°C [68]. Unnikrishnan *et al.* [67] concluded that MoDTCs and MoDTP give the same performance in terms of friction reduction but in

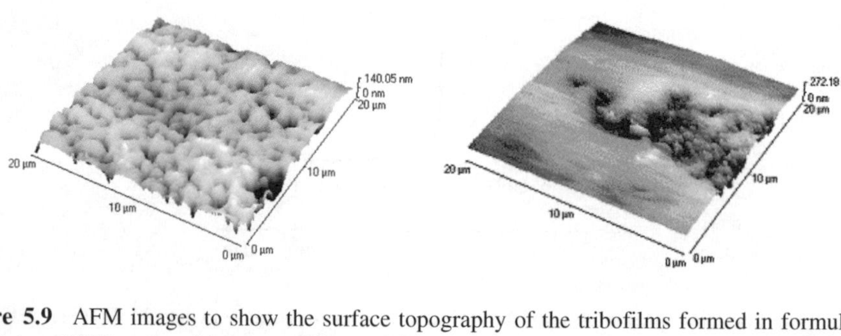

Figure 5.9 AFM images to show the surface topography of the tribofilms formed in formulated oil containing ZDDP and no friction modifier and (b) ZDDP with friction modifier (MoDTC) [71]

terms of wear reduction MoDTP was found to possess better wear prevention characteristics. In parallel with ZDDP, the formation of a glassy phosphate glass as a protective layer over the asperities is the principal mechanism. However, the AW properties of the MoDTC can be improved by combination with ZDDP which is not the same with MoDTP/ZDDP systems.

Interactions between additives in lubricants for control of wear in manufacturing processes are important. Control of wear in deep drawing processes has also been provided by ZDDP and incorporation of solid lubricants (such as CeF_3 [77]) has been shown to reduce wear. Such synergies are therefore not restricted to engine oils.

5.4.4 New Antiwear Additive Technologies

The presence of P in ZDDP is known to contaminate the emission systems in vehicles and hence have an impact on the environment. The demand for reduction of P content in lubricants has made the need for new additives ever-pressing. Phosphorous is not the only species in additives regarded as being detrimental to the environment and equal pressure for a reduction in sulphur in lube oils, parallel with fuels, exists. It has been shown that almost 100% of S in ZDTP is converted to SO_x, which poisons the NO_x trap [78]. The use of ZDDP is not going to stop overnight and in fact there is a large remaining market for ZDDP in industrial lubricants, not legislated for to the same extent as in the automotive sector, especially in countries like China and India [79]. However, ZDDP will continue to be phased out in new vehicles and hence the need for a replacement strategy is urgent. To obtain the effectiveness of ZDDP but with no or significantly less impact on the environment presents a real challenge. The main consideration is replacing the antiwear functionality of ZDDP [78]. Although it also offers resistance to oxidation, there are a number of other species which can offer antioxidancy. These include ashless antioxidants which also have the added benefit that they reduce ash loadings in diesel particulate filters (DPFs). Also, group II basestocks offer enhanced resistance to oxidation in their own right.

Where additive technology can go in the next decade is a major topic for debate as lubricant manufacturers strive to make step-changes in their technology. Developments are likely in two main areas:

1. *Soluble additive technology* which, in a manner similar to ZDDP, forms wear-resistant layers on asperities. The film formation occurs mainly through additive decomposition.
2. *Additives of dispersed nanoparticles* which function by presenting themselves in the asperity regions and forming lubricant layers and some load-bearing capacity.

Organic boron compounds are possible candidates to replace ZDDP. Boron-containing additives are not new and patents were filed as early as the 1960s [80]. Formation of a borate glass, in a manner similar to phosphate glass (in the presence of ZDDP or other P-containing additives), is often regarded as the key mechanism by which they provide wear protection.

Boron-containing additives of different types have been synthesized and tested as potential replacements for the P-containing ZDDP. In general, the boron-containing compounds are borates of mixed alcohols, amides, hydroxyl esters, ethoxylated amides and their mixtures [81]. Limited success has been reported but there is clearly a scope for continued development. Stanulov *et al.* [82] investigated the potential synergies between boric additives and traditional ZDTP additives using the environmental reduction/replacement of ZDTP as the driver for the study. Whilst it was found that there was a narrow region of tribological conditions in which the boric and P-containing additives acted synergistically, it demonstrated that there is an economic and ecological benefit in partial replacement. Other works have investigated full B-containing additives, fully replacing the P-additive system. Whilst these have shown some success in laboratory-based studies, there is a long way to go to reach the widespread use in real engines. One further concern with some of the B-containing species is that they contain relatively high amounts of S and, as such, do not solve the entire environmental problem. Also, it is not clear from the laboratory-based studies reported to date what the mechanisms of action of the B-type additives are. Boric acid, as a surface treatment (through surface films on boron carbide) and as a solid lubricant, has recently been heralded [83] as an exciting low-friction surface and industries were recently criticized in the editorial comments in *Lubrication Engineering* (issue January 2003) for not realizing its potential since the earlier works by Erdemir [84, 85].

A drawback of these additives is noted to be the ability to hydrolyse in the presence of water and by that to form hard granules of oxides or hydroxides which contribute to noise, damage to bearings and deposits near seals [81]. Junbin [86] and Zheng *et al.* [87] improved the hydrolytic stability of novel borate esters by the formation of a stable six-member ring of B and N coordination in a borate ester molecule. Surface analysis showed mainly adsorbed borate esters, BO and boron nitride on the surface. Under boundary lubrication, calcium borate overbased salicylate micelles were shown to form a thin amorphous film of a borate glass material at the microscopic level that acts as antiwear layer similar to P in ZDDP tribofilms [88]. Although this additive gave lower friction than ZDDP, it was not as effective in wear reduction as ZDDP [89]. The antiwear properties of this borate additive were improved when used together with ZDDP, giving the same performance on wear reduction as ZDDP alone [89]. Surface analytical techniques showed the tribofilm to be calcium and zinc borophosphate glass. Addition of calcium borate to ZDDP/MoDTC reduced friction even further and had no effect on antiwear action of ZDDP [90].

Nano-dispersed particles in lubricants offer potential to solve some of the problems associated with 'soluble' additives. A range of nanometre-sized particles have been suggested and tested, including titanium borate [91] and cobalt hydroxide [92], and have shown some promise for reduction in friction and wear. Such lubricant additives can be used where solubility problems (as are often associated with Mo-containing additives) and also more specifically where hydrolytic stability for B-containing additives are an issue.

5.5 Extreme Pressure Additives

Pressures are also being exerted on industry to radically change the chemistry of additives used in metal working, machining, wire-drawing and other manufacturing processes where EP functionality is needed. Currently, the most common EP additives contain chlorine, sulphur or phosphorus [23]. Organochlorine additives are known to give tribofilms of $FeCl_2$ and $FeCl_3$ and organosulphur $FeS/FeSO_4/Fe_2O_3$ composite films [93] through a complex sequence of thermal and physical changes at the metal/additive interface. It is acknowledged by Gao *et al.* [23] that because the nature of the action of the additive is very much surface dependent, the replacement of lubricant additives may well need to consider the type of surface being lubricated.

5.6 Lubricating Non-Fe Materials

Advanced material technology is being embraced in the transport sector; material and surface treatments are being developed at a rapid rate to achieve weight savings in vehicles, improve durability, improve structural integrity, etc. In Table 5.4 some of the different solid materials used in engine components are presented. The list is not exhaustive but is meant to illustrate the range of materials used. There has been a general move away from fully ferrous systems which generally comprise bearing steels and cast iron, used extensively in tribological applications. This move brings with it some associated challenges for tribologists – that being how to effectively lubricate non-Fe-based surfaces. In this chapter, the mechanisms of action of lubricant additives have been discussed. In antiwear technology, the important role of Fe_2O_3 for the development of the wear film has been discussed in several works. In addition, cationic exchange of Fe and Ca/Zn species in antiwear and detergent systems has been demonstrated thus emphasizing the role of chemical and physical interactions between the surface and the lubricant additives. It is not at all reasonable to assume that when lubricating non-Fe-based surfaces, the traditional technologies that are known to be effective will have the same functionality. Indeed, several recent workers in tribology have recently turned their attention to optimizing lubricant additive and surface compatibility for reduced friction and wear [94].

The antiwear ability of metal surfaces can be strengthened by different treatments such as carburizing, nitriding, carbonitriding, sulphonitriding, sulphirizing, pre-oxidation, boriding, phosphating, and electrolytic and electroless plating [93]. In addition, the application of different surface coatings has become widely used in lubricated contacts and includes materials such as thermally sprayed molybdenum, ceramic Cr_2O_3 plasma coatings, cermet coatings applied by HVOF, etc. [95]. Recent interest in more exotic coatings such as diamond-like carbon (DLC) [96–99] and modified versions of DLC incorporating elements such as Ti has made the whole area of surface engineering for lubricated contacts reach a new era. Although some aspects of lubrication of new surfaces, whether as solid materials or as surface coatings, have been addressed, it is fair to say that little is known about surface/lubricant interactions in new, non-ferrous and perhaps exotic materials. With the increased use of

Table 5.4 Materials used for engine components

Component	Materials
Valves	• γ-Titanium aluminide (with nitriding for wear protection) • Al–SiC MMCs • Ti alloys (especially in F1)
Pistons	• C–C composites (eliminates need for piston rings) • Mg alloys (1921 Indianapolis 500) • PSZ, Al2O3 (crown)
Engine block	• Mg (through USAMP – Magnesium Powertrain Cast Components Project) • Al – Lupo

non-ferrous materials for improved durability, it is vital that lubricant/surface compatibility is achieved.

Taking as an example Al–Si alloys, there has been increased use of these alloys in internal combustion engine components such as pistons, due to their excellent resistance to corrosion, good thermal conductivity, low cost and ease of fabrication by casting. However, these materials are susceptible to high wear and friction under loaded, sliding conditions. The tribological behaviour of Al–Si alloys under dry sliding has been studied extensively but there is little published on the lubrication of Al–Si alloys and in particular surface/lubricant additive interactions. Preliminary research on Al–Si alloys against steel [100–103] has shown that commercially available engine oils are not as effective at lubricating this system as they are in steel against steel contacts. In boundary lubrication of aluminium, the chemical interaction, especially the chemical reaction of the lubricant with the rubbing surface, is very important. Results by Wan and coworkers [101, 102] and Hah *et al.* [104] suggest that some chemical reaction takes place on the surface and therefore some chemical wear or corrosion takes place resulting in larger wear volumes. These tests indicate that conventional P- and S-based antiwear additives are not effective in preventing seizure of the aluminium.

Whilst some surface coatings are essentially inert when used outside tribological applications, it is now recognized that the chemical and mechanical interactions involved when lubricated surfaces come into contact can throw up some surprises in terms of the chemical reactivity of surfaces. On DLC tribocouples, there have been conflicting reports on the level of triboactivity in lubricated contacts. Podgornik *et al.* [97] reported a 25% reduction in friction and wear in conventional EP/AW additive systems where DLC was intimately involved in the formation of the tribofilm. In contrast, when the lubricant was fully formulated no such benefit was recorded. These examples, of which there are plenty emerging, confirm two things; firstly, that there are complex surface/additive reactions involved in tribochemical film formation on non-Fe surfaces, and secondly, that a new approach and lateral thinking are needed to provide integrated lubricated systems where materials and lubricant additives are going to deviate from standard traditional materials.

References

1. Bhushan, B. and Gupta, B.K., '*Handbook of Tribology. Materials, Coatings and Surface Treatments*', Krieger Publishing Company, Florida, 1997.
2. Taylor, C.M., 'Automobile Engine Tribology – Design Considerations for Efficiency and Durability', *Wear*, **221**(1), 1998, 1–8.
3. Korcek, S., Jensen, R.K., Johnson, M.D. and Sorab, J., 'Fuel Efficient Oils, Additive Interactions, Boundary Friction, and Wear', in *Lubrication at the Frontier* (ed D. Dowson), 1999, pp. 13–24.
4. Williams, J.A., '*Engineering Tribology*', Oxford Science Publications, Oxford University Press, New York, 1994.
5. Jensen, M.F., Bottiger, J., Reitz, H.H. and Benzon, M.E., 'Simulation of Wear Characteristics of Engine Cylinders', *Wear*, **253**, 2002, 1044–1056.
6. Bell, J.C., 'Gasoline Engine Valve Train Design Evolution and the Antiwear Requirements of Motor Oils', *Proceedings of the Institution of Mechanical Engineers Part J*, **212**, 1998, 243–257.
7. Becker, E.P. and Ludema, K.C., 'A Qualitative Empirical Model of Cylinder Bore Wear', *Wear*, **225–229**, 1999 387–402.
8. Lyyrnena, J., Jokiniemia, J., Kauppinena, E.I. and Silvonen, A., 'Corrosion Studies with a New Laboratory-Scale System Simulating Large-Scale Diesel Engines Operating with Residual Fuels, Part II. Particle and deposit characteristics', *Fuel Processing Technology* 86, 2004, 329–352.
9. http://www.fpc1.com/tests/ftc/ftp1/tb108-97.htm

10. Tung, S.C. and Huang, Y., 'Modeling of Abrasive Wear in a Piston Ring and Engine Cylinder Bore System', Proceedings of the 2003 ASME/STLE Joint International Tribology Conference, Ponte Vedra Beach, Florida, USA, October 26–29, 2003.
11. Jensen, M.F., Bottinger, J., Reitz, H.H. and Benzon, M.E., 'Simulation of Wear Characteristics of Engine Cylinders', *Wear*, **253**, 2002, 1044–1056.
12. The Independent, 'Revealed: how global warming will cause extinction of a million species', January 8, 2004.
13. http://www.ems.org/energy_policy/cate.html
14. DTLR (Department of Transport, Local Government and the Regions), Transport Statistics Great Britain 2001, The Stationery Office, London, 2001.
15. Ukuno, K. and Bessho, T., 'Need for Environmentally Friendly Surface Modification Technology in the Japanese Automotive Industry', in *Surface Modification Technologies XIV* (ed T.S. Sudershan and M. Jeandin), ASM International, 2001, pp. 135–140.
16. Curro, S., 'Economical and Earth-friendly', The Lupo 3L TDI, AL Alluminio E Leghe, April 200, pp. 119–121.
17. Powell, R., 'The USAMP Magnesium Powertrain Cast Components Project', *JOM*, February, 2002, 49–51.
18. Presting, H. and Konig, U., 'Future Nanotechnology Developments for Automotive Applications', *Materials Science and Engineering*, **C 23**, 2003, 737–741.
19. Rifkin, J., 'The Hydrogen Economy: The Creation of the Worldwide Energy Web and Redistribution of Power on Earth; J.P. Tarcher/Putman, New York, 2002.
20. Isenberg, G., 'Assessment of Automotive Fuels', *Journal of Power Sources*, **84**, 1999, 214–217.
21. Korcek, S., 'Future Challenges in Engine Oil Development – OEM Views', Ford Motor Company, Presentation at OMV, Schwechat, Austria, September 6, 2001.
22. Bovington, C.H. and Castle, R., 'Lubricant Chemistry Including the Impact of Legislation', Proceedings of the 28th Leeds-Lyon Symposium on Tribology, Vienna, 2001.
23. Gao, F., Kotvis, P.V. and Tysoe, W.T., 'The Surface and Tribological Chemistry of Chlorine- and Sulphur-Containing Additives', *Tribology International*, **37**, 2004, 87–92.
24. Kajdas, C., 'Tribochemistry', Proceedings of the World Tribology Congress, Vienna, 2000.
25. Heinicke, G., 'Tribochemistry', Akademie Verlag, Berlin 1984.
26. Spikes, H., 'Advances in the Study of Thin Film Lubricant Films', Invited paper at the World Tribology Congress, London, 1997, New Directions in Tribology, MEP Ltd, London, September 1997, pp. 353–369.
27. Williams, J.A., 'The Behaviour of Sliding Contacts Between Non-Conformal Rough Surfaces Protected by "Smart" Films', *Tribology Letters*, **17**(4), 2004, 765–778.
28. Xue, Q. and Liu, W., 'Tribochemistry and the Development of AW and EP Oil Additives – A Review', *Lubrication Science*, **7**(1), 1994, 81–92.
29. Yao, J. and Xu, Z., 'A Proposed Mechanism for Boundary Layer Formation from Organic Borates and cd(II) Compound', *Tribology Letters*, **3**, 1997, 277–281.
30. Grossiord, C., Martin, J.M., Mogne, Th.Le and Palermo, Th., 'In-situ MoS_2 Formation and Selective Transfer from MoDTP Films', *Surface and Coatings Technology*, **108–109**, 1998, 352–359.
31. Williams, C.T. and Beattie, D.A., 'Probing Buried Interfaces with Non-Linear Optical Spectroscopy', *Surface Science*, **500**, 2002, 545–576.
32. Bec, S., Tonck, A., Georges, J.M., Coy, R.C., Bell, J.C. and Roper, G.W., 'Relationship Between Mechanical Properties and Structures of Zinc Dithiophosphate Anti-Wear Films', *Proceedings of the Royal Society of London A*, **455**, 1999, 4181–4203.
33. Jones, R.B. and Coy, R.C., 'The Chemistry of the Thermal Degradation of Zinc Dialkyldithiophosphate Additives', *ASLE Transactions*, **24**(1), 91–97.
34. Spedding, H. and Watkins, R.C., 'The Antiwear Mechanism of ZDDP's Part I', *Tribology International*, **15**, 1982, 9–12.
35. Watkins, R.C., 'The Antiwear Mechanism of ZDDP's Part II', *Tribology International*, **15**, 1982, 13–15.
36. Ray, N.H., Laycock, J.N.C. and Robinson, W.D., Oxide Glasses of Very Low Softening Point', *Glass Technology*, **14**(2), 1973, 55–59.
37. Willermet, P.A., Dailey, D.P., Carter III, R.O., Schmitz, P.J., Zhu, W., 'Mechanism of Formation of Antiwear Films from Zinc Dialkyldithiophosphates', *Tribology International*, **28**(3), 1995, 177–187.
38. Martin, J.M., 'Lubricant Additives and the Chemistry of Rubbing Surfaces: Metal Dithiophosphates Triboreaction Films Revisited', *Japanese Journal of Tribology*, **42**, 1997, 9.
39. Martin, J.M., 'Antiwear Mechanisms of Zinc Dithiophosphate: A Chemical Hardness Approach', *Tribology Letters*, **6**, 1999, 1–8.

40. Martin, J.M., Grossiord, C., Le Mogne, T., Bec, S. and Tonck, A., 'The Two Layer Structure of Zndtp Tribofilms Part 1: AES, XPS and XANES Analyses', *Tribology International*, **34**, 2001, 523–530.
41. Pearson, R.G., in *Chemical Hardness: Applications from Molecules to Solids*, Wiley, VCH, Weinhein, Germany, New York, 1997, 1–26.
42. Fuller, M.L.S., Kasrai, M., Bancroft, G.M., Fyfe, K. and Tan, K.H., 'Solution Decomposition If Zinc Dialkyldithiophosphate and Its Effect on Anti-Wear and Thermal Film Formation Studied by X-ray Absorption Spectroscopy', *Tribology International*, **31**(10), 1998, 627–644.
43. Holinski, R., 'The Influence of Boundary Layers on Friction', *Wear*, **56**, 1987, 147–154.
44. Kennedy, S. and Moore, L.D., 'Additive Effects on Lubricant Fuel Economy', SAE Paper 872121, 1987.
45. Kubo, K., Kibukawa, M. and Shimakawa, Y., 'Effect on Friction of Lubricants Containing Zinc Dithiophosphate and Organo-Molybdenum Compound', Combustion Engines – Reduction of Friction and Wear, Institute of Mechanical Engineers Conference Publication, 1985, 121–131.
46. Tripaldi, G., Vettor, H. and Spikes, H.A., 'Friction Behaviour of ZDDP Films in the Mixed, Boundary/EHD Regime', SAE Paper 962036, 1996.
47. Taylor, L., Dratva, A. and Spikes, H.A., 'Friction and Wear Behaviour of Zinc Dialkyldithiophosphate Additive', *Tribology Transactions*, **43**, 2000, 469–479.
48. Cann, P., Spikes, H.A. and Cameron, A., 'Thick Film Formation by Zinc Dialkyldithiophosphates', *ASLE Transactions*, **26**(1), 1996, 48–52.
49. Cann, P. and Cameron, A., 'Studies of Thick Boundary Lubrication-Influence of ZDDP and Oxidized Hexadecane', *Tribology International*, **17**(4), 1984, 205–208.
50. Sheasby, S., Caughlin, T.A., Blahey, A. and Laycock, K.F., 'A Reciprocating Wear Test for Evaluating Boundary Lubrication', *Tribology International*, **23**(5), 1990, 301–307.
51. Bell, J.C., Delargy, K.M. and Seeney, A.M., 'The Removal of Substrate Material Through Thick Zinc Dithiophosphate Anti-Wear Films', Proceedings of Leeds–Lyon Symposium, *Tribology Series*, **21**, *Wear Particles from the Cradle to the Grave* (eds D. Dowson *et al.*), Elsevier Amsterdam, 1992, 387–396.
52. Lin, Y.C. and So, H., 'Limitations on Use of ZDDP as an Anti-Wear Additive in Boundary Lubrication', *Tribology International*, **37**, 2004, 25–33.
53. Pidduck, A.J. and Smith, G.C., 'Scanning Probe Microscopy of Automotive Anti-Wear Films', *Wear*, **212**, 1997, 254–264.
54. Smith, G.C., 'Surface Analytical Science and Automotive Lubrication', *Journal of Physics D: Applied Physics*, **33**, 2000, R187–R197.
55. Inoue, K. and Watanabe, H., 'Interactions of Engine Oil Additives', *ASLE Transactions*, **26**, 1983, 189–199.
56. Kapsa, Ph., Martin, J.M., Blanc, C. and Georges, J.M., 'Antiwear Mechanism of ZDDP in the Presence of Calcium Sulphonate Detergent', *Transactions ASME, Journal of Tribology*, **103**, 1981, 486–496.
57. Rounds, F., 'Changes in Friction and Wear Performance Caused by Interactions Among Lubricant Additives', in Proceedings of the 5th International. Colloquium On Additives for Lubricants and Operational Fluids, Germany 1986, pp. 4.8-1–4.8-21.
58. Silver, H.B., 'The Interaction Between Corrosion Inhibitors and Load Carrying Additives in Mineral Oil', *Tribology International*, **11**, 1978, 185–188.
59. Reyes, M.E. and Neville, A., 'The Effect of Anti-Wear Additives, Detergents and Friction Modifiers in Boundary Lubrication of Traditional Fe-Base Materials', Tribological Research and Design for Engineering Systems, Proceedings of the 29th Leeds–Lyon Symposium on Tribology, Leeds, 2002, Elsevier, Amsterdam, 2003.
60. Yin, Zh., Kasrai, M., Bancroft, G.M., Fyfe, K., Colaianni, M.L. and Tan, K.H., 'Application of Soft X-ray Absorption Spectroscopy in Chemical Characterization of Antiwear Films Generated by ZDDP Part II: The Effect of Detergents and Dispersants', *Wear*, **202**, 1997, 192–201.
61. Wan, Y., Suominen Fuller, M.L., Kasrai, M., Bancroft, G.M., Fyfe, K., Torkelson, J.R., Hu, Y.F. and Tan, K.H., 'Effects of Detergent on the Chemistry of Tribofilms from ZDDP: Studied by X-ray Absorption Spectroscopy and XPS', in *Boundary and Mixed Lubrication: Science and Application* (eds D. Dowson *et al.*), *Tribology Series*, **40**, 2002, 155–166.
62. Willermet, P.A., Dailey, D.P., Carter III, R.O., Schmitz, P.J., Zhu, W., Bell, J.C. and Park, D., 'The Composition of Lubricant-Derived Surface Layers Formed in a Lubricated Cam/Tappet Contact II. Effects of Adding Overbased Detergent and Dispersant to a Simple ZDTP Solution', *Tribology International*, **28**(3), 1995, 163–175.

63. Morina, A., Green, J.H., Neville, A. and Priest, M., 'Surface and Tribological Characteristics of Tribofilms Formed in the Boundary Lubrication Regime with Application to Internal Combustion Engines', *Tribology Letters*, **15**(4), 2003, 443–452.
64. Forbes, E.S., Groszek, A.J. and Neustadter, E.L., 'Adsorption Studies of Lubricating Additives', *Journal of Colloid and Interface Science*, **33**, 1970, 629.
65. Harrison, P.G., Brown, P. and McManus, J., 'P-31 NMR Study of the Interaction of a Commercial Succinimide-Type Lubricating Oil Dispersant with Zinc(II) bis (o,o′-di-iso-butyldithiophosphate)', *Wear*, **156**, 1992, 345–349.
66. Bovington, C.H., 'Friction, Wear and the Role of Additives in Their Control', in *Chemistry and Technology of Lubricants* (eds Mortier R.H. and Orszulik, S.T.), Blackie Academic and Professional, London, 1997.
67. Unnikrishnan, R., Jain, M.C., Harinarayan, A.K. and Mehta, A.K., 'Additive–Additive Interaction: An XPS Study to the Effect of ZDDP on the AW/EP Characteristics of Molybdenum Based Additives', *Wear*, **9022**, 2001, 1–10.
68. Yamamoto, Y. and Gondo, S., 'Friction and Wear Characteristics of Molybdenum Dithiocarbamate and Molybdenum Dithiophosphate', *Tribology Transactions*, **32**(2), 1989, 251–257.
69. Lin, Y., 'Antifriction and Antiwear Characteristics of Molybdenum Dithiophosphate in Engine Oils', *Journal of the Society of Tribologists and Lubrication Engineers*, October, 1995, 855–860.
70. Muraki, M. and Wada, H., 'Influence of the Alkyl Group of Zinc Dialkyldithiophosphate on the Frictional Characteristics of Molybdenum Dialkyldithiocarbamate Under Sliding Conditions', *Tribology International*, **35**, 2002, 857–863.
71. Green, J.H., Morina, A., Priest, M. and Neville, A., 'Evolution of Tribofilms Under Lubrication Conditions Experienced in Engine Valve Trains', *Transit Progression in Tribology*, accepted for publication in Proceedings of the 30th Leeds–Lyon Symposium, Lyon, September 2003, 2004, 97–108.
72. Martin, J.M., Grossiord, C., Le Mogne, Th. and Igarashi, J., 'Transfer Films and Friction Under Boundary Lubrication', *Wear*, **245**, 2000, 107–115.
73. Martin, J.M., Grossiord, C., Varlot, K., Vacher, B. and Igarashi, J., 'Synergistic Effects in Binary Systems of Lubricant Additives: A Chemical Hardness Approach', *Tribology Letters*, **8**, 2000, 193–201.
74. Iwasaki, H., 'TOF-SIMS Analysis of MoS_2 Formed on the Rubbing Surface for MoDTC and ZnDTP Synthesized by Using ^{34}S Isotope', in Proceedings of Japanese Society of Tribologists Tribology Conference, Takamatsu, 1999, 359–360.
75. Kasrai, M., Cutler, J.M., Gore, K., Canning, G., Bancroft, G.M. and Tan, K., 'The Chemistry of Antiwear Films Generated by the Combination of ZDDP and MoDTC Examined by X-ray Absorption Spectroscopy', *Tribology Transactions*, **41**, 1998, 69–77.
76. Lansdown, A.R., 'Molybdenum Disulphide Lubrication' (ed D. Dowson), *Tribology Series*, **35**, 1999.
77. Lian, Y., Xue, Q., Zhang, X. and Wang, H., 'The Mechanism of Synergism Between ZDDP and CeF_3 Additives', *Lubrication Science*, **7**(3), 1995, 261–272.
78. Peckham, J., 'Lubes Sulfur, ZDTP Second only to ULSD Impact on NOx Traps', Diesel Fuel News, July 7, 2003. http://www.findarticles.com/cf_dls/m0CYH/12_7/105617532/p2/
79. Lloyd, B., 'Watchful Eye on ZDDP', *Lubes and Greases Magazine*, August, 2003.
80. Samuel, M.D., 'Lubricating Oils Containing Borate Compound', US Patent 2975135, 1961.
81. Khorramain, B.A., Iyer, G.R., Kodali, S., Natarajan P. and Tupil, R., 'Review of Antiwear Additive for Crankcase Oils', *Wear*, **169**, 1993, 87–95.
82. Stanulov, K.G., Harhara, H.N. and Cholakov, G.S., 'An Opportunity for Partial Replacement of Phosphates and Dithiophosphates in EP Packages with Boron-Containing Additives', *Tribology International*, **31**(5), 1998, 257–263.
83. Dvorak, S.D., Wahl, K.J. and Singer, I.L., 'Friction Behaviour of Boric Acid and Annealed Boron Carbide Coatings Studied by In-situ Raman Tribometry', *Tribology Transactions*, **45**, 2002, 354–362.
84. Erdemir, A., Bindal, C., Zuiker, C. and Savrun, E., 'Tribology of Naturally Occurring Boric Acid Films on Boron Carbide', *Surface Coatings and Technology*, **86–87**, 1996, 507–510.
85. Erdemir, A., Eryilmaz, O.L. and Fenske, G.R., 'Self-Replenishing Solid Lubricant Films on Bulk Borides and Their Coatings', *Surface Engineering*, **15**, 1999, 291–295.
86. Junbin, Y., 'Antiwear Function and Mechanism of Borate Containing Nitrogen', *Tribology International*, **30**(6), 1997, 387–389.
87. Zheng, Zh., Shen, G., Wan, Y., Cao, L., Xu, X., Xue, Q. and Sun, T., 'Synthesis, Hydrolytic Stability and Tribological Properties of Novel Borate Esters Containing Nitrogen as Lubricant Additives', *Wear*, **222**, 1998, 135–144.

88. Normand, V., Martin, J.M., Ponsonnet, L. and Inoue, K., 'Micellar Calcium Borate as an Antiwear Additive', *Tribology Letters*, **5**, 1998, 235–242.
89. Varlot, K., Martin, J.M., Grossiord, C., Vargiolu, R., Vacher, B. and Inoue, K., 'A Dual-Analysis Approach in Tribochemistry: Application to ZDDP/Calcium Borate Additive Interactions', *Tribology Letters*, **6**, 1999, 181–189.
90. Varlot, K., Kasrai, M., Bancroft, G.M., Yamaguchi, E.S., Ryason, P.R. and Igarashi, J., 'X-ray Absorption Study of Antiwear Films Generated from ZDDP and Borate Micelles', *Wear*, **249**, 2001, 1029–1035.
91. Hu, Z.S. and Dong, J.X., 'Study on Anti-Wear and Reducing Friction Additive of Nanometer Titanium Borate', *Wear*, **216**, 1998, 87–91.
92. Chen, G.X., Hu, Z.S., Dong, J.X., Wang, L.G., Peng, Y., He, T. and Lai, R., 'Study on Antiwear and Reducing Friction Additive of Nanometer Cobalt Hydroxide', *Lubrication Engineering*, **54**(4), April, 2001, 36–39.
93. Wei, D., 'Future Directions of Fundamental Research in Additive Tribochemistry', *Lubrication Science*, **7**(3), 1995, 211–232.
94. Voong, M., Neville, A. and Castle, R., 'The Compatibility of Crankcase Lubricant–Material Combinations in Internal Combustion Engines', *Tribology Letters*, **15**(4), 2003, 431–441.
95. Erdemir, A., Erck, R.A., Fenske, G.R. and Hong, H., 'Solid/Liquid Lubrication of Ceramics at Elevated Temperatures', *Wear*, **203–204**, 1997, 588–595.
96. Gahlin, R., Larsson, M. and Hedenqvist, P., 'ME-C:H Coatings in Motor Vehicles', *Wear*, **249**, 2001, 302–309.
97. Podgornik, B., Jacobson, S. and Hogmark, S., 'Influence of EP and AW Additives on the Tribological Behaviour of Hard Low Friction Coatings', *Surface and Coatings Technology*, **165**, 2003, 168–175.
98. Podgornik, B., Jacobson, S. and Hogmark, S., 'DLC Coating of Boundary Lubricated Components – Advantages of Coating One of the Contact Surfaces Rather than Both or None', *Tribology International*, **36**, 2003, 843–849.
99. Ahn, H-S., Chizhik, S.A., Dubravin, A.M., Kazachenko, V.P. and Popov, V.V., 'Application of Phase Contrast Imaging Atomic Force Microscopy to Tribofilms on DLC Coatings', *Wear*, **249**, 2001, 617–625.
100. Wan, Y., Cao, L. and Xue, Q., 'Friction and Wear Characteristics of ZDDP in the Sliding of Steel Against Aluminium Alloy', *Tribology International*, **30**(10), 1998, 767–772.
101. Wan, Y., Xue, Q. and Cao, L., 'Boundary Lubrication of Aluminium Alloy with Cl-Containing Anti-Wear and Extreme Pressure Additives', *Wear*, **208**, 1997, 57–60.
102. Wan, Y. and Xue, Q., 'Effect of Antiwear and Extreme Pressure Additives on the Wear of Aluminium Alloy in Lubricated Aluminium-on-Steel contact', *Tribology International*, **28**(8), 1995.
103. Hu, L., Chen, J., Liu, W., Xue, Q. and Kajdas, C., 'Investigation of Tribochemical Behaviour of Al–Si Alloy Against Itself Lubricated by Amines', *Wear*, **243**, 2000, 60–67.
104. Hah, S.R., Fischer, T.E., Gruffel, P. and Carry, C., 'Effect of Grain Boundary Dopants and Mean Grain Size on Tribomechanical Behaviour of Highly Pure α-Alumina in the Mild Wear Regime', *Wear*, **181–183**, 1995, 165–177.

6

Surface Chemistry in Tribology

A.J. Gellman and N.D. Spencer

Abstract

Surface chemistry is key to the understanding of tribological phenomena in the absence of a thick lubricant film. Progress in the development of surface-analytical techniques has opened a new window onto tribochemical phenomena, which holds the promise of a better understanding of many critically important tribological processes. In this review, we survey the areas in which surface chemistry has played an important role in enhancing our tribological understanding. These include boundary lubrication, surface-additive interactions, the anomalous tribological behavior of quasicrystals, and the lubrication of hard disks.

6.1 Introduction

In general, the twentieth century marked the first major incursion of chemists into the world of tribology. Driven partly by the need for improved lubricants for new machines developed in the first few decades of the century, and partly by new analytical capabilities, lubricant chemists and tribochemists were to become full partners in the science of tribology. In this review, we cover a number of topics in surface chemistry in tribology, which serve both to illustrate some of the successes in tribochemical research over the past decades and to show that much work remains to be done before true, first-principle, tribochemistry-based lubricant and tribopair design can begin.

6.2 Boundary Lubrication and Oiliness Additives

6.2.1 Introduction

In an ideal world, lubricated sliding surfaces are separated by a stable fluid film of lubricant, which effectively lowers friction and prevents wear, holding the surfaces apart by virtue of its

Wear – Materials, Mechanisms and Practice Edited by G. Stachowiak

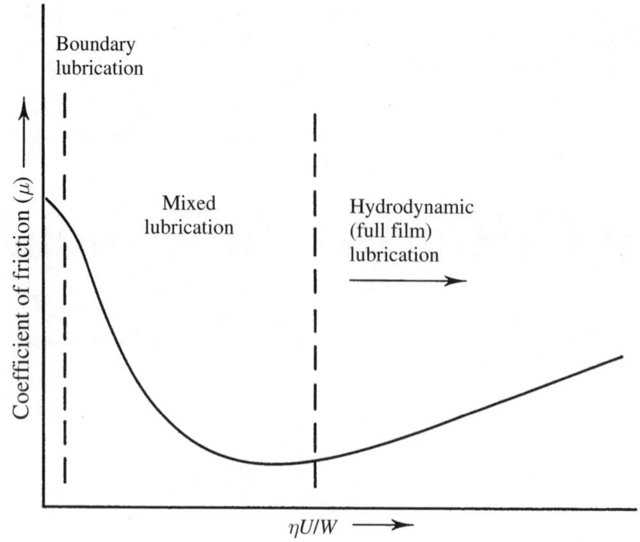

Figure 6.1 Stribeck curve: the variation in frictional drag (expressed as the coefficient of friction, μ) with the quantity $\eta U/W$ for a lubricated sliding bearing. After Hutchings [1]

fluid-mechanical properties. The friction coefficient in such a hydrodynamically lubricated bearing is proportional to $\eta U/W$, where U is the relative sliding speed of the surfaces, W the normal load supported by the bearing and η the Newtonian viscosity [1]. At low sliding speeds (e.g. upon startup) and/or higher forces, however, the friction coefficient varies in a more complex way (Figure 6.1), and at the lowest speeds and friction forces can be two orders of magnitude higher than under hydrodynamic lubrication conditions. This is the regime of "boundary lubrication," where fluid films have given way to thin layers that serve as the active lubricants, and the realm of pure engineering (hydrodynamic lubrication) has given way to the complex world of surface chemistry under confinement. Boundary lubrication is clearly of vital importance, since it is the mechanism in operation under the most extreme of tribological conditions. Understanding boundary lubrication, however, has been a challenge to tribochemists over the last century, with light appearing at the end of the tunnel only quite recently.

6.2.2 *Monolayers, Multilayers and Soaps*

It was noticed in the 1920s that certain lubricants possessed a quality designated "oiliness," which, independent of their viscosity, led to better lubrication at low sliding speeds [2–4]. Generally, these oily lubricants tended to be those of natural origin, such as castor oil, containing long-chain oxygenated organic compounds (esters and acids), which have surfactant properties, and are not usually present in their mineral oil counterparts. It was shown that these oxygenated molecules could, by themselves, reduce friction considerably between sliding surfaces, their effectiveness depending on their molecular weight [5] (Figure 6.2). Hardy developed a monolayer theory of boundary lubrication (Figure 6.3),

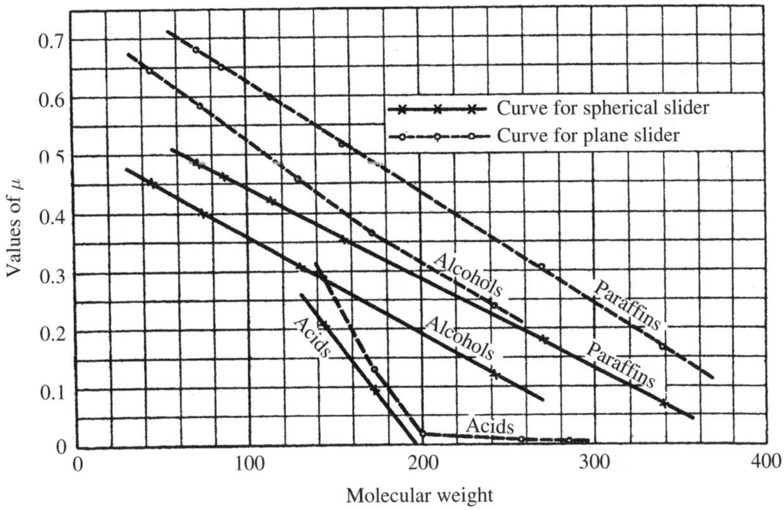

Figure 6.2 Dependence of friction coefficient on molecular weight for various homologous series (reprinted with permission from the Royal Society of London) [5]

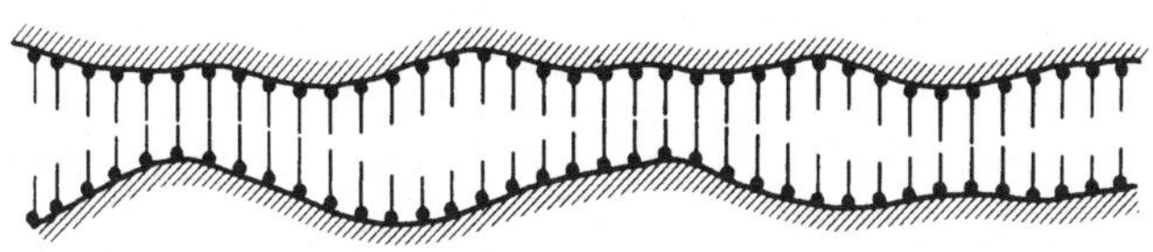

Figure 6.3 Hardy boundary model, adapted from Reference (reprinted with kind permission from Hodder Education) [1]

whereby he postulated that, under boundary conditions, the sliding surfaces were held apart by adsorbed, oriented monolayers of polar molecules, which form a plane of low shear strength, thus lowering friction and affording protection of the surfaces.

Stanton [3] and Deeley [4], at around the same time, advanced an alternative hypothesis, however, involving a thick film consisting, essentially, of an organo-iron compound as the agent of surface oiliness. The monolayer versus multilayer controversy was to continue for many decades, a cogent set of arguments against the monolayer theory being presented by Allen and Drauglis [6].

Gregory was to lend support to the monolayer theory by investigating the friction between clean cadmium surfaces lubricated by pure paraffin oil [7]. In the absence of additives, he measured μ (static) to be 0.6, which was reduced by an order of magnitude in the presence of 1% dodecanoic (lauric) acid. In the presence of only 0.01% dodecanoic acid, μ was still only 0.1, while a 0.001% acid concentration showed a μ value (0.45) close to that of the pure paraffin oil, dropping, however, over a period of 12 h to a value of 0.26, suggesting the slow adsorption of a sub-monolayer quantity of acid onto the sliding surfaces. By using a 0.1% dodecanoic acid concentration and slowly extending the lubricated area, Gregory could calculate the thickness of lubricating layer that apparently yielded significantly reduced friction. This was calculated at somewhere between one and two molecular layers.

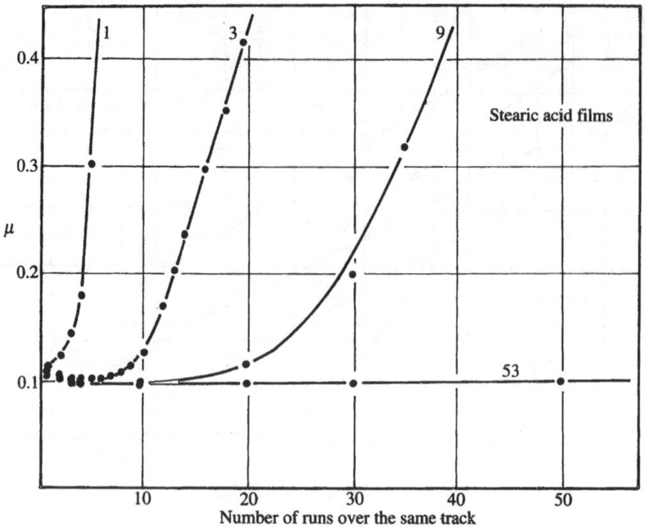

Figure 6.4 The friction of octadecanoic (stearic) acid films deposited on a stainless steel surface (adapted with permission from Oxford University Press) [7, 8]

Bowden and Tabor took a tremendous step forward in the scientific treatment of boundary lubrication with the publication of their book *Friction and Lubrication of Solids* in 1950 [7]. They showed that while a monolayer of octadecanoic (stearic) acid deposited on stainless steel by means of the Langmuir–Blodgett (LB) technique could indeed produce a low μ of 0.1 against a stainless steel slider, this value increased to 0.3 after five passes (Figure 6.4). The friction-reduction effect of three LB-deposited octadecanoic acid monolayers was more durable, with the $\mu = 0.1$ value persisting for some half-dozen passes. Nine monolayers performed even better, with $\mu = 0.3$ only being reached after 35 passes, and 53 monolayers provided an apparently constant value of $\mu = 0.1$. Thus, it was shown that while a single monolayer of a polar molecule can reduce friction, it is far from robust, and requires continuous replenishment. This is the situation, of course, in an oil containing the molecule as an additive at a reasonable concentration. Bowden and Tabor also found that molecular structure played a role in boundary lubrication ability, with nonanoic acid (C9) being the shortest acid that could show any boundary lubrication ability. Looking at this in terms of a modern interpretation of the Hardy model, this would correspond to the minimum chain length that, by virtue of inter-chain van der Waals forces, could maintain itself in an ordered, tails-up arrangement. Similar arguments can be used to explain why branched molecules are less effective boundary lubricants than their straight-chain counterparts: the side chains simply impair close packing of the molecules [9, 10].

The substrate dependence of lubrication by oil additives can also provide useful insights. In Table 6.1, the effect on μ of 1% dodecanoic acid in paraffin oil used with nickel, chromium, platinum, and silver surfaces is shown to be slight. On copper, cadmium, zinc, and magnesium surfaces, however, friction reduction by an order of magnitude was observed. This study suggests that reaction of the additive with the substrate (presumably to form a metal "soap") has occurred for the latter group. Temperature dependence is also revealing (Figure 6.5):

Table 6.1 Coefficients of sliding friction at room temperature [7, 8]

Surfaces	Lubricant: paraffin oil	Paraffin oil + 1% dodecanoic acid
Nickel	0.3	0.28
Chromium	0.3	0.3
Platinum	0.28	0.25
Silver	0.8	0.7
Copper	0.3	0.08
Cadmium	0.45	0.05
Zinc	0.2	0.04
Magnesium	0.5	0.08

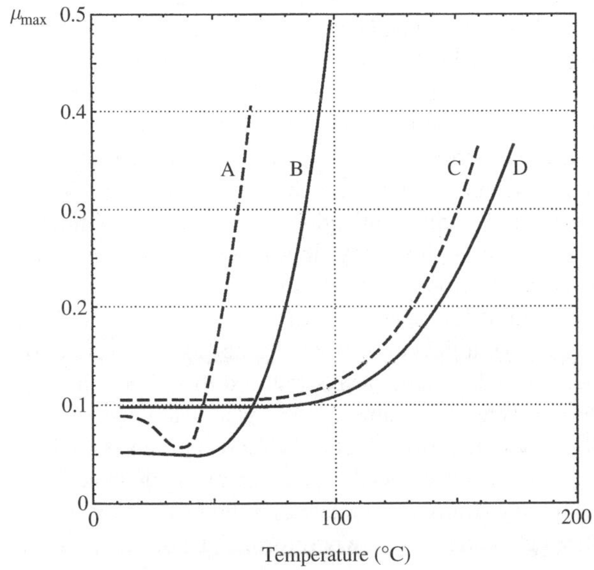

Figure 6.5 Effect of temperature on friction of platinum and copper surfaces, lubricated by docosane and fatty acids. Curve A: solid docosane (m.p. = 44°C) on Pt; B: solid octadecanoic acid (m.p. = 69°C) on Pt; C: solid copper dodecanoate (softening point = 110°C) on Pt; D: 1% dodecanoic acid in paraffin oil on Cu (adapted with permission from Oxford University Press) [7, 10]

Many solid lubricants, such as lead or solid alkanes, will effectively lubricate surfaces upto their melting points, after which they can no longer hold the sliding surfaces apart. While docosane (C22, m.p. = 44°C) and octadecanoic acid (m.p. = 69°C) effectively lubricate platinum surfaces up to their melting points, dodecanoic acid lubricates copper surfaces up to 90°C, which corresponds to the softening point of copper dodecanoate. Moreover, dodecanoic acid on copper shows lubrication behavior similar to copper dodecanoate on platinum! This might suggest that copper dodecanoate is involved in both cases. It is also interesting to note that sodium octadecanoate (m.p. = 260°C) lubricates steel surfaces up to 280°C.

Clearly, the lubrication ability of octadecanoic and dodecanoic acids on reactive metals involves more than a simple adsorbed molecular film, although it is also clear from results described earlier that a simple molecular layer can serve as the basis for boundary lubrication. More light has been shed on this apparently confusing issue by a recent set of experiments. Fischer *et al.* [11] used X-ray absorption spectroscopy to show that octadecanoic acid adsorbs on copper in a mixture of bidentate and monodentate states, with the chains oriented close to the surface normal. At this stage, the spectra appear very similar to those of an octadecanoic acid standard. Tribological treatment (rubbing in air against a clean Cu surface in a reciprocating motion for 10 passes with a normal force of 1 N) induces a conversion of some of the monodentate states into the bidentate form, and an enhanced orientation toward the surface normal, while showing no loss of the organic layer. X-ray absorption measurement of the tribologically treated surface reveals a spectrum that is very similar to that of copper octadecanoate. Thus it seems as if the octadecanoic acid monolayer is converted almost instantaneously into the corresponding octadecanoate upon rubbing, although the chain orientation is still very much as Hardy envisaged.

Possibly the results that provide the most insight into the monolayer/multilayer controversy have been generated by Ratoi *et al.* [12], who used the method of ultrathin film interferometry [13] to monitor the lubricant film thickness during rolling contact of bearing steel on glass. When the lubricant used is purified hexadecane, the logarithm of the film thickness varies linearly as the logarithm of the rolling speed (Figure 6.6(a)), as predicted from elastohydrodynamic lubrication theory, down to a thickness of around 1 nm (interestingly, due to elastic conformality, this holds despite the 12-nm roughness of the contacts). In the presence of carboxylic acids, the situation is more complex, and a deviation from linearity is observed at low speeds. In the case of dry 0.1 wt.% octadecanoic acid in hexadecane (Figure 6.6(b)), a deviation from the hexadecane results of around 2 nm was observed at the lowest speeds, although this reduced to < 1 nm upon stopping the relative motion of the surfaces. When the same system was examined in the presence of water, however, the results were quite different (Figure 6.6(c)): a much thicker film was formed at low speeds. The film thickness was speed dependent, reaching about 8–12 nm thickness at its maximum, but disappearing almost completely as rolling speed was increased. Upon reducing the rolling speed once again, the film slowly reformed. Upon stopping the machine, a film of some 2–4 nm remained in the contact region. Similar wet and dry behavior was observed for eicosanoic (C20, arachidic) acid. Interestingly, when stainless steel was used in the contact instead of bearing steel, no film was found to form upon testing with wet octadecanoic acid in hexadecane, suggesting that the above observations are due to a corrosive reaction of the additive with the steel substrate.

The above observations can be explained in terms of the water-induced oxidation of iron to Fe(II)/Fe(III), which then goes on to form the carboxylate thick films. Similar behavior has been observed previously with copper carboxylates in mineral oil on a steel surface [14], where a 40-nm boundary film forms slowly at low speeds, collapsing at high speeds. In general, redox reactions play an important role in the formation and removal of these thicker films [15], with the films first forming in the Fe(II) state, which is insoluble in oil, and slowly oxidizing to the oil-soluble Fe(III) state. Thus, as a system moves from boundary to elastohydrodynamic conditions, the boundary film is no longer being formed and dissolves away. Extending this approach, by judicious addition of various metal carboxylates, Ratoi *et al.* [15] found that carboxylates of metals below iron in the electrochemical series

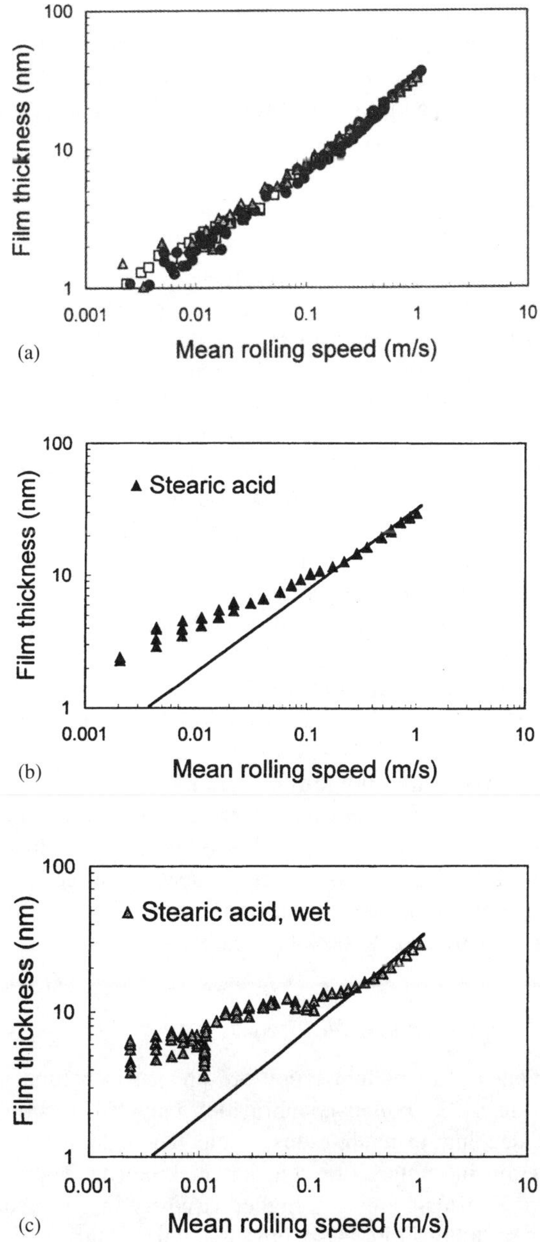

Figure 6.6 Elastohydrodynamic film formation measured by ultrathin film interferometry: (a) purified hexadecane; (b) dry 0.1% octadecanoic (stearic) acid in hexadecane (the solid line is pure hexadecane); (c) wet 0.1% octadecanoic acid in hexadecane (reprinted with kind permission from Elsevier) [12]

(e.g. copper) reacted to form iron carboxylate boundary films, while carboxylates of metals above iron (e.g. zinc, magnesium) formed no boundary films at all. In summary, ultrathin film interferometry has shown that *either monolayers or multilayers can be formed from carboxylic acid additives,* depending on the prevailing tribological conditions. The relative lubricating efficiency of the two possibilities is currently under investigation.

6.2.3 Viscous Near-Surface Layers

In the previous section, we have seen that multilayers of carboxylates can form in additive-modified lubricants in the presence of moisture. These lubricate the rolling contact by means of their locally high viscosity. However, even in the absence of additives, there have been suggestions or reports over the years of a locally high viscosity at the surface, starting with Deeley's suggestion in the 1920s of a "friction surface, which is a compound of oil and metal" [3, 4]. Çavdar and Ludema [16–18] studied the lubrication of steel by additive-free mineral oils by means of *in situ* ellipsometry and found that after some 20 min, a 5–10 nm thick layer of organo-iron compounds was formed on an oxide/carbide underlayer. Hsu [19] has actually analyzed oil samples that have been used to lubricate sliding contacts. Using gel-permeation chromatography, he was able to show that organo-iron compounds with molecular weights up to 100,000 were present in the case of a highly refined mineral oil lubricant. Hsu hypothesized that metal-catalyzed oil oxidation leads to a polymerization mechanism.

Results from the surface force apparatus (SFA) suggest that liquids in the immediate vicinity of surfaces are strongly influenced by that surface, with a phase behavior that is distinct from that of the bulk liquid. Israelachvili *et al.* [20] and Granick [21] have shown that the effective viscosities of thin liquid films can be several orders of magnitude higher than bulk values within one or two molecular layers of the surface, and, very recently, SFA refractive-index studies by Heuberger *et al.* [22] have indicated fluctuations between phases resembling solids, liquids, and gases when cyclohexane is confined between mica surfaces at room temperature. Although interesting, the significance of these SFA observations under real tribological conditions remains to be established.

6.2.4 Boundary Lubrication in Natural Joints

Although many different modes of lubrication are present in natural joints, such as those in the hip and in the knee [23], boundary lubrication plays an important role after periods of inactivity, such as sleeping, or reading this article. The natural lubricant, synovial fluid, is similar to blood serum in composition, but has a somewhat higher level of hyaluronic acid, which plays a role in maintaining a higher viscosity in the medium – useful in the hydrodynamic modes of lubrication. Also present in the fluid are many different kinds of proteins, lipids, and inorganic salts. While there is some discussion in the literature as to whether lipids play an important tribological role, it is now commonly accepted that glycoproteins are important boundary lubricants.

Lubrication in the natural joint is highly relevant for understanding the progress of certain joint disorders, such as arthritis. The sliding surfaces consist of cartilage, which contains glycoproteins and polysaccharides, such as chondroitin sulfate. In artificial joints, however, which are being increasingly implanted, as a consequence of the aging population and of improved technology, sliding surfaces generally consist of Co–Cr alloys, ceramics, or ultrahigh-molecular-weight polyethylene (UHMWPE), all of which have certain advantages and disadvantages. Wear of UHMWPE has been an important issue in recent years, since PE particles can activate the body's immune system, leading to a response that ultimately causes loosening of the implant in the bone, requiring a further operation.

Widmer *et al.* [24] have shown that the surface hydrophilicity of the sliding surfaces in a hip implant can have a major effect on both protein adsorption and frictional behavior. Using oxygen plasma treatment to render a polyethylene surface more hydrophilic, they were able to show a reduction of dynamic friction by around 50% in protein-containing solutions. In salt solutions (Ringer's), this effect did not occur. It was also observed that the plasma treatment led to an increase in the amount of protein that adsorbed on the polyethylene surface, lending support to the hypothesis that proteins denature and spread out on hydrophobic surfaces, and are less able to function efficiently as boundary lubricants. On hydrophilic surfaces, they maintain a more compact shape, can therefore pack better onto the surface, and function better as boundary lubricants.

6.2.5 Summary

In a sense, modern analytical methods have not so much led to the confirmation of one particular proposed mechanism for boundary lubrication, but rather have shown that several different mechanisms, from Hardy's monolayers, through the formation of Deeley's organo-iron compounds and other thick reaction layers, to near-surface layers of pure lubricant with enhanced viscosity, are all possible under different tribological situations. The relative importance of each mechanism under different sets of real conditions is not yet completely clear, but is currently under investigation by a number of research groups.

6.3 Zinc Dialkyldithiophosphate

6.3.1 Background

There are few lubricant additives that have received as much attention in the literature as zinc dialkyldithiophosphate ("ZnDTP," "ZDTP," "ZDDP," or "ZDP"). Originally added to lubricating oil as an antioxidant [25], it was rapidly discovered that it also functioned as a highly effective antiwear and extreme-pressure additive, and is an essential ingredient in the vast majority of current lubricant formulations. An important role of ZnDTP in such lubricants is to protect surfaces under conditions where elastohydrodynamic lubrication breaks down, such as in counterformal contacts present in valve-train systems between cams and followers [26]. In order to function effectively, films of this kind must possess a shear strength that is sufficiently low to ensure that the shear plane resides within the protective layer itself, while being high enough to maintain layer integrity [27]. Clearly, the rate of formation of the film must also keep pace with any film wear that occurs, while not being so great that it unnecessarily corrodes the substrate. Furthermore, the films should inhibit adhesion between the contacting surfaces. All in all, this is a tough set of requirements for an additive, but has been met by ZnDTP with great success.

One of the reasons for the effectiveness of ZnDTP is its ability to function in different ways, depending on the nature and severity of the tribological conditions. It is in this sense a

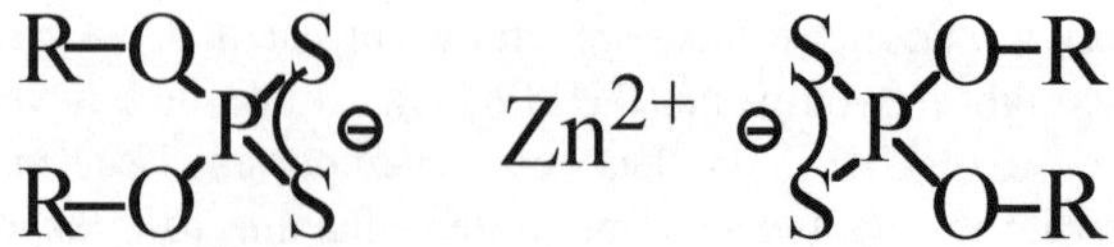

Figure 6.7 Schematic of ZnDTP. R can be a primary or secondary alkyl group, or aryl functionality

"smart" material [27]. Thus, the many apparently conflicting results obtained from studies of this system have often been illustrating different facets of the rich thermal and tribochemical behavior of this molecule, or rather family of molecules, since many variations on the basic chemistry (Figure 6.7), as well as their combinations, are commercially employed.

In addition to academic curiosity, there are good practical reasons for trying to understand the mechanism by which ZnDTP functions. Among these is the undesirable propensity for engines that are exposed to these additives to produce zinc- and phosphorus-containing emissions, which are potentially deleterious to both downstream catalyst performance and the environment. Clearly, a better understanding of the ZnDTP surface-reaction mechanisms should assist in the development of more benign alternatives.

6.3.2 Analytical Approaches

One particularly daunting aspect of the generally daunting ZnDTP literature is that virtually every weapon in the surface-analysis arsenal has been brought to bear upon the problem, including Auger electron spectroscopy (AES) [28, 29], X-ray photoelectron spectroscopy (XPS) [30–35], near-edge X-ray absorption fine structure (NEXAFS, or XANES) [30, 36–43], Fourier transform infrared (FTIR) spectroscopy [33, 44–46], time-of-flight secondary-ion mass spectroscopy (ToF-SIMS) [35, 47], scanning force microscopy (SFM) [48, 49], the SFA [27, 50], transmission electron microscopy (TEM) [43], scanning electron microscopy (SEM) [51], extended X-ray absorption fine structure (EXAFS) [40], and profilometry [51], as well as *in situ* tribometry methods such as the direct-observation wear machine (DOWM) [52] and ultrathin film interferometry [53].

6.3.3 Summary of Film-Formation Mechanism

By the end of the twentieth century, a general picture of the mechanism for alkyl-ZnDTP molecules with steel surfaces had emerged. The molecule appears to interact weakly with the steel surface (possibly through the sulfur atoms [54], according to quantum-chemical calculations) at room temperature [50], starting to catalytically decompose around 50°C in the presence of the iron oxide substrate [55]. Upon reaching around 100°C, either through simple thermal treatment or by tribo-induced flash heating, the ZnDTP starts to undergo thermo-oxidative decomposition, due to the presence of either oxygen or peroxide radicals in the oil [56]. There is evidence from XANES studies [39] as well as from ^{31}P-NMR [57, 58] that the molecule first undergoes a rearrangement to form an O/S exchange isomer of ZnDTP in solution:

$$\underset{\text{ZnDTP}}{Zn[(RO)_2PS_2]_2} \rightarrow \underset{\text{O/S exchange isomer}}{Zn[O_2P(SR)_2]_2}$$

This species, following reaction with unchanged ZnDTP, goes on to form a long-chain glassy zinc polyphosphate and polythiophosphate coating on the surface, as inferred from optical microscopy [51], XANES [37], and TEM of wear debris [56]. Under mild conditions, simple phosphates rather than polyphosphates can be formed [33–35]. An organosulfur species is released into the lubricant, which, under extreme conditions, reacts with nascent iron surfaces to form FeS, which has been detected within the polyphosphate matrix in TEM studies (Figure 6.8) [56] and immediately next to the steel substrate using AES [29]. Partially reacted, semipolymerized alkylphosphates seem to form a viscous layer above the harder polyphosphates, as determined by ToF-SIMS [47] and SFA [27] studies.

Many of the reactions involved in dialkyldithiophosphate decomposition can be rationalized using Pearson's theory of hard acids and bases [56, 59], where the $P = S$ group, for example, is a typical soft base (low nuclear charge, high polarizability), which is more likely to react with a soft acid (e.g. tetravalent carbon), as in the O/S exchange isomer formation reaction described above, while a hard base (high nuclear charge, low polarizability, e.g. $P = O$) is likely to react preferentially with hard acids (e.g. tetravalent P, or H^+), leading to the reaction [58]

$$=P=O + =P-SR \rightarrow =P^+-O-P=SR^-$$

Similarly, looking at the iron oxide "digestion" process, since Fe^{3+} is a harder acid than Zn^{2+}, ferric oxide will react with (hard) zinc phosphates to produce zinc iron phosphate and zinc oxide. Zinc oxide may go on to react with organosulfur compounds to produce zinc sulfide, which has the advantage of being mechanically softer than its oxide counterpart.

6.3.4 *Studies of Film Structure, Composition, and Thickness*

Many studies have focused on the thickness of the films produced by ZnDTP. Gunsel *et al.* [53] used ultrathin film interferometry to monitor ZnDTP film formation at various

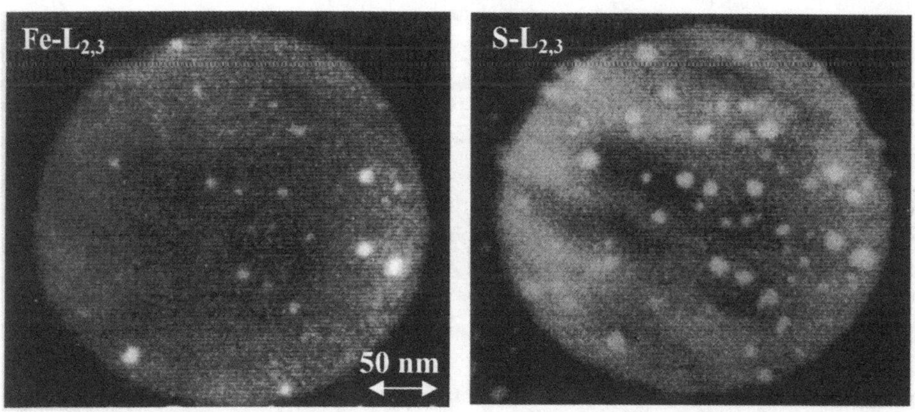

Figure 6.8 Energy-filtering TEM image of iron and sulfur in a wear particle originating from severe plane-on-plane wear tests in the presence of ZnDTP. The images show the existence of nm-scale precipitates of iron sulfide in the phosphate-based matrix (reprinted with kind permission from Springer Science and Business Media) [56]

temperatures. A solid film appeared to be forming, under rolling conditions, once a threshold temperature of 130°C was reached. At 200°C, the film grew, over a period of an hour, to a saturation thickness of just over 40 nm. Using a similar method, which enabled reaction films to be observed outside the contact in a mixed rolling/sliding regime, Taylor *et al.* [60] were able to show that the films could be as thick as 200 nm, but depended strongly on the slide–roll ratio. XANES studies [42] have shown that the films are of the order of 40–100 nm thickness, depending on the concentration of ZnDTP in solution.

XANES has proved to be a particularly useful tool for examining the growth of polyphosphate chains on the steel surface in the presence of ZnDTP. Using specially synthesized zinc polyphosphate chains of various lengths, Kasrai *et al.* [30] were able to show that the fine structure in the phosphorus L-edge in the XANES spectrum is a sensitive indicator of polyphosphate chain length (Figure 6.9), while the height of the corresponding K-edge can serve as a reliable indicator of polyphosphate film thickness [42].

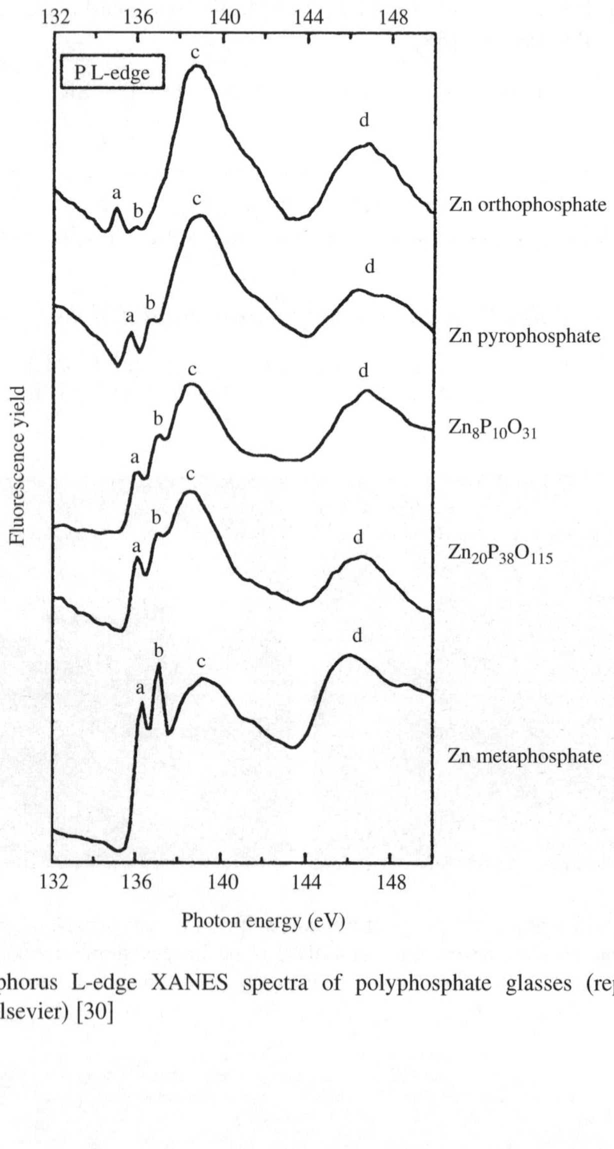

Figure 6.9 Phosphorus L-edge XANES spectra of polyphosphate glasses (reprinted with kind permission from Elsevier) [30]

XANES spectra can be monitored by either total electron yield (TEY) or fluorescence yield (FY). These two approaches are complementary, the former having a surface sensitivity of some 5 nm, while the latter probes 50 nm (at the P L-edge, and 1μm for the K-edge) into the sample. By using both methods on the same sample, an indication of the relative composition of the outer and inner regions of the film may be obtained. Bancroft *et al.* [38] were thus able to monitor the fate of tribo- and thermally generated films. Immediately after formation, the outer surface (from TEY) consisted of long-chain polyphosphates, while the bulk (from FY) shows only short-chain phosphates. Upon rubbing the films in pure base oil (i.e. without the possibility of film replenishment), they appear to remain on the surface for many hours, although the XANES fine structure shows that the chain length decreases markedly with time. It is thought that the reactions are of the general form:

$$5Zn(PO_3)_2 + Fe_2O_3 \rightarrow Fe_2Zn_3P_{10}O_{31} + 2ZnO$$

or

$$Zn(PO_3)_2 + 2FeO \rightarrow Fe_2Zn(PO_4)_2$$

Long-chain phosphate   Short-chain phosphate

In other words, the film is intergrown with the oxide on the steel, which presumably also enhances mechanical stability [38]. The mechanism also shows that the glass can function as a digestion agent for iron oxides, which are abrasive and therefore enhance wear. The tribo-generated films appear to last for days when being rubbed in base oil, and are still some 30 nm thick after 6 h. The thermal films appear to be somewhat less robust.

All of the studies just described measure average properties over a macroscopic area. However, there is considerable evidence that the antiwear films produced are highly uneven [60]. Sheasby and Nisenholz Rafael [51] detected the existence of 20-μm-diameter "pads" of ZnDTP-derived antiwear film by means of SEM. Graham *et al.* [49] went on to extend this work, using SFM to examine antiwear films derived from both alkyl and aryl ZnDTPs. The alkyl additives indeed yielded 10–20-μm-diameter pads with flattened tops, elongated in the sliding direction and surrounded by lower, smaller pads. They also applied the interfacial force microscope (IFM) to examine the mechanical properties of these pads. At the center of the large pads, indentation modulus values as high as 209 GPa were measured (compared to 220 GPa for the substrate 52100 steel!), as well as a very elastic response to indentation. The aryl-ZnDTP-derived films, which display greatly inferior wear resistance, did not appear to contain large pads; neither did they display such high indentation moduli. Using spectromicroscopy (spatially resolved XANES) on similarly produced films, Canning *et al.* [41] showed that the large pads produced from the alkyl ZnDTP consisted of long-chain polyphosphates, while the smaller pads, and all features in the aryl-derived films, contained short-chain phosphates only.

Eglin and co-workers have studied the effect of load on the tribochemical fate of ZnDTP by a combinatorial approach [34, 35]. By creating a "library" of different tribological conditions in a single pin-on-disk experiment [61] and subsequently analyzing the surface by means of XPS, they were able to correlate trends in friction coefficient and chemistry as load was increased. Increasing load up to 5 N at room temperature correlated with a decreasing friction coefficient, an increasing phosphate signal (Figure 6.10), and a decreasing P:S ratio, corresponding to sulfur depletion in the film.

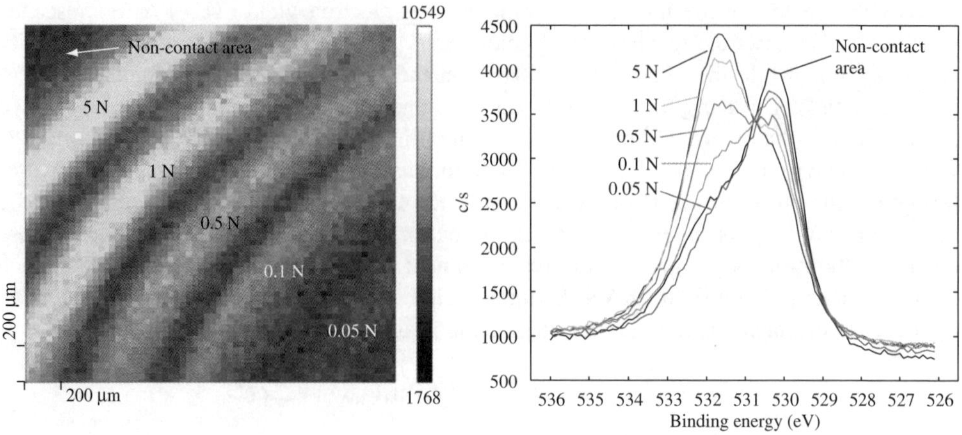

Figure 6.10 Left-hand side: O(1s) XPS chemical-state map of five concentric contact areas created with different loads in a pin-on-disk experiment (52100 steel, di-isopropyl ZnDTP (1 wt.%) in decane). The map shows the correlation of the tracks with phosphate-type spectra (extracted from a contact area), and thus the distribution of phosphate-type species. Right-hand side: O(1s) spectra extracted from the chemical-state map. The spectra show an increasing peak at 531.7 eV with increasing load. This peak can be assigned to oxygen bound in a phosphate structure (reprinted with kind permission from Elsevier) [35]

In many of the analytical approaches used to investigate this system, it has been necessary to rinse the sample in organic solvent prior to analysis. Depending on the aspect of the films under investigation, some workers have found that solvent washing does not affect the outcome of the analysis [43], while others have noticed a significant effect [27, 47]. Bec *et al.* [27] used a unique version of the SFA to investigate fully formed films both with and without solvent washing. They found that in the absence of solvent washing, a viscous layer, several hundred nanometers thick and displaying a hardness in the MPa range, was present at the outer surface. This presumably corresponds to a partially reacted ZnDTP layer, consisting of alkylphosphates [47], which serves both as a reserve for polyphosphate formation and as a low shear-strength layer. Solvent washing apparently removed the alkylphosphate layer, leaving the polyphosphate layer, with hardness in the GPa range, largely intact. This layer was patchy, as found by many others, with features some 10 μm across. It seems unlikely that there is a clear interface between the alkylphosphate and polyphosphate layers, but rather a gradual transition. An interesting observation by Bec *et al.* [27] was the similarity in the hardness value of the polyphosphate layer to the mean pressure applied during the previous tribological testing. This suggests an accommodation on the part of the polyphosphate film to the contact pressure to which it is exposed. The overall structure of the produced films is summarized in Figure 6.11.

As can be seen from the figure, the formation, structure, and function of ZnDTP-derived antiwear films are highly complex. Thanks to the developments in surface-analytical techniques over the last 30 years, and their judicious combination, a fairly comprehensive picture has emerged. The challenge for the additive chemist is now the application of this knowledge in the design of improved antiwear additives with greater environmental compatibility.

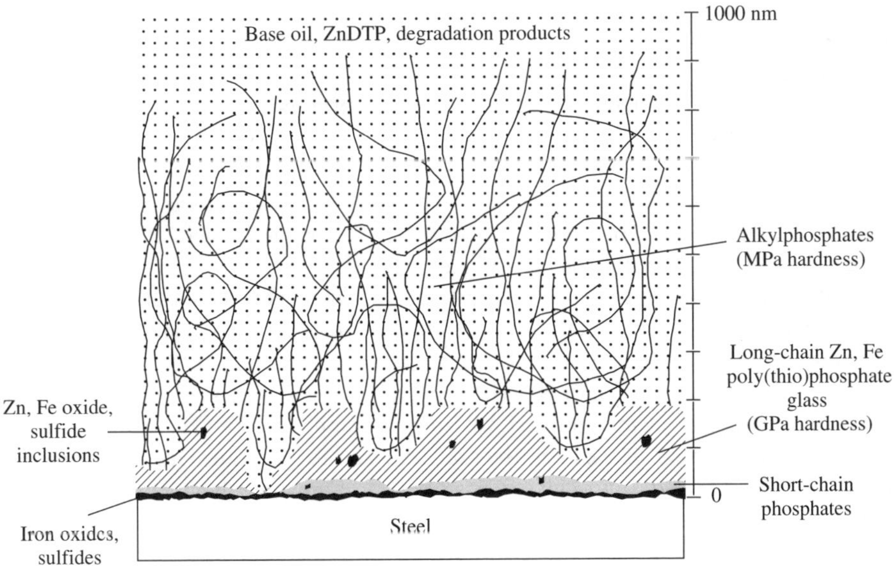

Figure 6.11 Schematic of ZnDTP antiwear film structure

6.4 Hard Disk Lubrication

The majority of information stored in computers is stored in hard disk drives. At the heart of these hard disk drives is the head–disk interface (HDI). The disk, of course, is the medium on which the data are written and from which they are read. The recording head is the device that does the reading and the writing. Although they come in several shapes and sizes, the most commonly shipped drives currently have disks that are 65 mm in diameter and are fabricated from glass or an AlMg alloy. The size of the disk has continuously decreased over the years and as the technology is developing extremely rapidly it will, no doubt, continue to do so. The data are written into a thin (~200 Å) layer of magnetic material that is sputtered onto the surface of the disk. As the disk spins at speeds on the order of 10,000 rpm, the read-write head flies over its surface on an air-bearing at a height of ~100 Å. With such incredibly fine tolerances, periodic contacts between the head and the disk surface are unavoidable and the need to lubricate the disk surface in order to protect it from damage creates one of the most interesting problems in modern tribology [62–64].

To get some idea of the magnitude of the problem associated with protecting the HDI from damage, consider a simple scaling of the device up to more familiar macroscopic dimensions. The read-write head is ~1 mm in length. If this is scaled linearly ($\times 10^4$) to the size of a small airplane, 10 m in length, then this airplane is flying at ~400,000 kph at an altitude of ~0.1 mm! Furthermore, the problems associated with operating these devices at such fine tolerances will become more severe in the future. There is a constant market drive to increase the density of data stored on magnetic media. Products currently being shipped have areal densities of about 40 Gbit/in.^{2}. This number has been increasing by 60% per year for the past decade and will continue to do so for the foreseeable future. The highest demonstrated recording densities are 130 Gbit/in.^{2} and current targets are set at 1 Tbit/in.^{2}

Achieving higher recording densities requires smaller bits and this in turn requires the head to fly even closer to the disk surface in order to be able to sense the magnetic field from magnetic domains that are so small [65, 66].

Needless to say, the protection of magnetic data poses one of the most high-tech and most demanding problems in modern tribology. The surface of the disk is protected from the read-write head by a thin (30–50 Å) film of sputtered carbon, which, in turn, is coated with an even thinner film (5–20 Å) of lubricant. These serve the dual purpose of providing both mechanical protection and corrosion protection for the magnetic media. The composition and structure of the HDI is illustrated in Figure 6.12.

The sputtered carbon films that are used to protect magnetic media are usually of two types: amorphous hydrogenated carbon (a-CH_x) and amorphous nitrogenated carbon (a-CN_x). In addition, there are, of course, hybrid materials: a-CH_xN_y. The carbon is sputtered from targets in the presence of H_2 and/or N_2. These overcoats are quite complex materials and the details of their structure at the atomic level are not completely known. Although they are sputtered onto the surface in vacuum, they are exposed to air immediately afterwards and thus their surfaces are partially oxidized. Many analysis methods have been used to probe the composition and structure of these surfaces. XPS has been used extensively to determine O and N concentrations at the surface and Rutherford back-scattering is used to measure the H content. Raman spectroscopy has been used quite widely to characterize the bulk of the carbon films. Analysis of the "G" and "D" peaks can be used to estimate the fractions of sp^2- and sp^3-bonded carbon in the films. Finally, electron spin resonance has been used as a means of determining the concentrations of dangling bonds. These all reveal a quite complex material containing carbon atoms with a mixture of hybridizations and various types of partial oxidation and nitrogenation [67–71].

The surface of the amorphous carbon film on the hard disk is coated with an ultrathin film (<20 Å) of lubricant. There are enormous constraints on the types of materials that might be considered for use as lubricants in such an application. The films of <20 Å thickness must survive for periods of up to five years in environments that reach operating temperatures

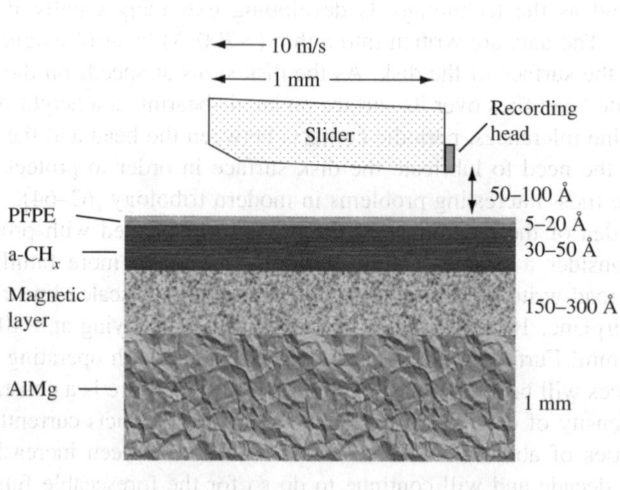

Figure 6.12 Illustration of the head–disk interface in a hard disk drive

of ~80°C and cover a wide range of humidity. This requires that the lubricant be thermally and chemically stable, and that it have a very low rate of evaporation or, in other words, a very low vapor pressure. The most commonly used lubricants belong to the Fomblin® family of fluids, which are perfluoropolyalkyl ethers (PFPEs). They have the general structure

$$R - CF_2O - (CF_3CF_2O)_n - (CF_2O)_m - CF_2 - R$$

These are high-molecular-weight polymers (1000–8000 amu) that are liquid and meet all the criteria listed above. There are a number of different end groups that can be used, including R = F, CH_2OH, and $CH_2OCH_2CH(OH)CH_2OH$, although by far the most commonly used has been Fomblin® Zdol, which is terminated by $-CH_2OH$ groups. The PFPEs are long polymeric molecules that have a diameter of roughly 7 Å. Thus, the lubricant films are only two to three monomolecular layers in thickness. After the disk surfaces have been sputter coated with magnetic media and carbon overcoats, they are removed from vacuum and the lubricant is applied by dip-coating the disks into a solution of the lubricant in a fluorinated solvent [62, 63, 69].

At production scale simply determining the lubricant thickness and uniformity on the surface of the disk is a challenging problem for quality control. This is done most commonly by using ellipsometry measurements or FTIR absorption in a reflection mode to examine the absorption intensity from the CF_2 stretch modes. Calibration of the coverage has been done using XPS [72–74]. The fact that the lubricant films on the carbon surface are roughly two monolayers in thickness means that their characteristics are truly determined by the surface chemistry of the adsorbed molecules and probably have little to do with the bulk properties of the fluid. Surface science is clearly needed in order to understand the tribological properties of these films – issues such as the nature of the bonding of the lubricant to the surface are important. Once coated on the surface, the lubricant appears to evolve into two forms: a "bonded" fraction and a "non-bonded" fraction. The evolution of the coverages of the two components and the evaporation of the lubricant with time have been measured using FTIR [74]. The nature of the "bonded" species is not known, although it is clear that it is associated with the end groups of the molecule, since the CF_3-terminated Fomblin® Z does not form a bonded fraction while the hydroxyl-terminated Fomblin® Zdol does. The non-bonded component is thought to be the fraction of the lubricant that is mobile on the surface and can diffuse into regions of the surface at which the lubricant has been displaced by head–disk contact. The mobility of the films has been studied in some detail using scanning microellipsometry. These studies have revealed that the hydroxyl-terminated Fomblin® family can form layered structures on the disk surface, while the CF_3-terminated species do not [75]. Finally, the nature of the chemical interactions of Fomblin® lubricants with the a-CH_x surface has been studied using model compounds and thermal desorption methods [67, 76]. These have shown that the ether linkages of the Fomblins® interact with the surface of the carbon film through a dative electron-donation mechanism, while the hydroxyl end groups are hydrogen-bonded to the surfaces.

The fact that the lubricant films used for protection of magnetic data storage media are so thin mandates the need for a deep understanding of their surface chemistry. Surface-analysis methods have already played an important role in this endeavor and will continue to be necessary as the evolution of data storage technology places ever greater constraints and demands on the tribological performance of these systems.

6.5 Vapor-Phase Lubrication

Devices and applications requiring lubrication at extreme temperatures present tribological problems that cannot be solved using traditional, fluid-based lubricants. Although some of the PFPEs used for hard disk lubrication are, in fact, stable fluids at temperature as high as $\sim 400\,^{\circ}C$, even this is insufficient for some applications. High-efficiency gas turbine engines may require lubrication at temperature in excess of $600\,^{\circ}C$ and one can find applications needing lubrication at even higher extremes. One of the obvious solutions is the use of solid films, such as graphite or MoS_2. However, under conditions of high or even moderate wear, these films must be periodically replaced. What is needed is a lubrication scheme based on thin solid films that can be replenished continuously and *in situ*.

One of the methods proposed and successfully tested for use in high-temperature gas turbine engines is known as vapor-phase lubrication (VPL) [77–79]. In this scheme, the lubricant is vaporized and added to a hot gas stream flowing through the engine. On contact with the hot surfaces of the engine components, it reacts to deposit a thin solid film, which lubricates and protects the engine surfaces from wear. This solid film is continuously worn away during engine operation, but is continuously replenished from the vapor phase. The compounds tested most commonly as vapor-phase lubricants have been arylphosphates, such as tricresylphosphate (now referred to as TCP) shown below, and some alkylphosphates [80–82].

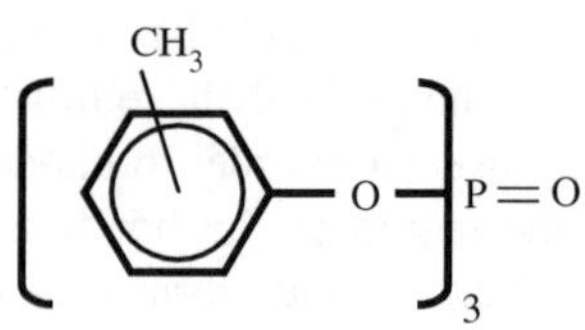

The mechanism by which these vapor-phase lubricants react on the surface to form lubricating films is undoubtedly quite complex and is certainly not well understood. It does, however, pose an interesting problem for surface science investigations.

VPL films have been examined by SEM and those that are effective as solid lubricating films have structures consisting of micrometer-sized nodules [83–85]. These are observed to form on some metals such as steels and Cu but not on ceramics such as SiC or some Ni-based alloys. The films that showed poor lubricating characteristics had flaky structures that did not appear to adhere well to the substrate.

The thin lubricating films deposited by VPL using TCP are thought to consist of polyphosphate glasses containing significant amounts of graphitic carbon [86, 87]. A schematic representation of this thin film is illustrated in Figure 6.13. A number of surface-analytical measurements have been made in order to gain insight into the composition and structure of these films. Analysis of the film composition using XPS revealed the presence of phosphorus, carbon, and oxygen. The use of FTIR and Raman spectroscopy showed that the structure of these films is quite complex and includes bonding consistent with the presence of a polyphosphate glass containing small graphitic particles. In addition, there is some evidence of the presence of P–O–C bonds. It is thought that the polyphosphate serves as a binder for the graphitic carbon [86, 88]. Depth profiling using Auger spectroscopy

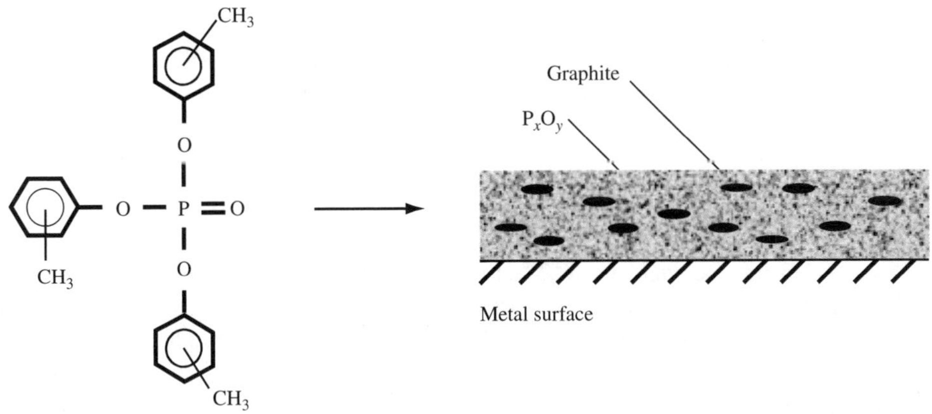

Figure 6.13 A schematic representation of the reaction of a vapor-phase lubricant (TCP) to form a thin film of polyphosphate glass with embedded graphite

to determine composition has shown that there is a significant amount of Fe present in the films [85]. In fact, it is thought that the ability of Fe to diffuse through and to the surface of the film is a key to the success of the use of TCP as a vapor-phase lubricant with TCP.

The formation of lubricating films by vapor deposition must be the result of a complex, multi-step reaction beginning with TCP and leading to the formation of both the polyphosphate film and the embedded graphite. Besides what is known about the composition of the films themselves, there is evidence that the *aryl*phosphates serve as better vapor-phase lubricants than the *alkyl*phosphates [81, 82, 89]. The root cause of this difference must lie in differences in the surface chemistry of these two classes of compounds. Arylphosphates $[(RO)_3P = O]$, such as TCP, are complex molecules and one can imagine a number of initial reaction steps or mechanisms which might initiate their decomposition to form lubricating films of polyphosphate glasses. The initial reaction steps that appear to be most likely include the cleavage of P–O bonds to form adsorbed aryloxy groups [RO–] or the cleavage of C–O bonds to form adsorbed aryl groups [R–]. There have been several studies that have tried to shed light on this issue by using the model compound trimethylphosphite, $[(CH_3O)_3P]$, the simplest organophosphorus compound containing these basic structural linkages [90–92]. These studies used temperature-programmed desorption (TPD) and XPS on the Cu(111), Ni(111), and Fe(111) surfaces and all suggest that it is the P-O bonds that are cleaved in the case of trimethylphosphite to produce adsorbed methoxy groups $[CH_3O–]$. Studies on Fe foils using TCP itself are a little bit harder to interpret but suggest that it is the C–O bonds that dissociate to produce adsorbed toluyl groups $[CH_3(C_4H_4)-]$ [93, 94]. These then decompose to deposit carbon onto the surface.

The nature of the ligand on the phosphates used for VPL can have a significant effect on their performance [81, 82, 89]. This can be understood in terms of the chemistry of the fragments that are left on the surface by scission of the P–O or C–O bonds in the phosphate. Alkyl ligands leave either alkyl or alkoxy groups on the surface, as illustrated below with cyclohexanoxy and cyclohexyl groups.

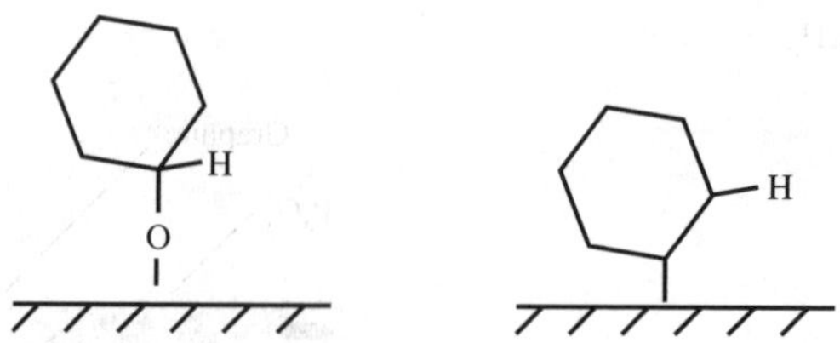

These species are characterized by β-CH bonds, which tend to dissociate readily on most metal surfaces, releasing the adsorbed groups as either aldehydes or ketones in the case of alkoxy groups, or as olefins in the case of alkyl groups. These products of β-hydride elimination can desorb from the surface quite readily and thus remove much of the surface carbon. Aryloxy and alkyloxy ligands have very different surface chemistry. The arlyoxy ligands dissociate from the phosphates by cleavage of either the P–O or the C–O bonds to produce either aryloxy or aryl groups on the surface. Since they have no β-CH bonds, they cannot react by β-hydride elimination to generate products that are easily desorbed

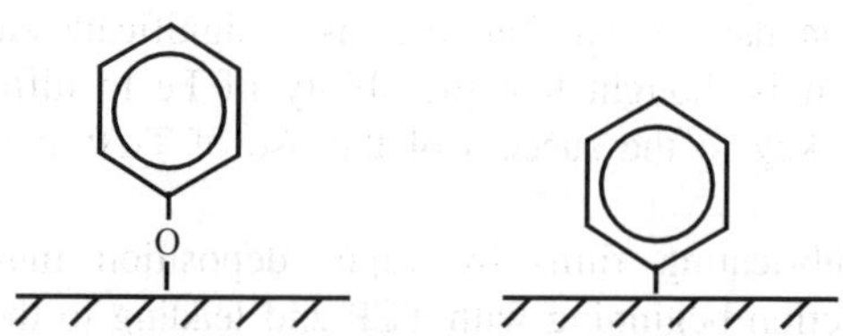

into the gas phase. Instead, they react by complete decomposition to deposit large amounts of carbon onto the surface in the form of graphite. This chemistry has been observed and studied using TPD to detect desorbing reaction products produced by scission of the β-CH bonds in the alkyl and alkyloxy groups, and AES to detect the carbon left on the surfaces due to decomposition of the aryl and aryloxy groups [92, 94, 95]. The fact that the aryloxy ligands decompose completely on the metal surfaces to deposit graphite suggests that this is an important component of the surface chemistry of VPL and that the efficient deposition of graphite onto the surface or into the lubricating films is critical to the performance of the film as a solid lubricant.

Although ceramics are ideally suited to use in many high-temperature applications, their use in engines and as machine components is hampered by the fact that they are brittle and it is very difficult to find lubricants compatible with ceramic materials. Typical vapor-phase lubricants such as TCP do not react on the surfaces of ceramics such as SiC or Si_3N_4 and thus cannot be used as vapor-phase lubricants. Some means is needed to activate ceramic surface for TCP decomposition. One approach that has been demonstrated quite recently and that has some promise is activation of the surface by exposure to high vapor pressure Fe-containing compounds such as $Fe(CO)_5$. At high temperatures, $Fe(CO)_5$ will decompose to deposit a thin Fe film on the surface. Subsequent or concurrent exposure of the Fe-modified surface to TCP has shown that it will decompose on the modified surface to deposit a film containing both phosphorus and carbon [96].

6.6 Tribology of Quasicrystals

Quasicrystals constitute a class of alloys that have the unusual property of being ordered in the sense that their structures are deterministic but do not have periodicity. One of the most interesting consequences of this is that they have bulk symmetry elements such as five- and tenfold rotation axes, which cannot be found in periodic lattices. They were first discovered, roughly 20 years ago, by Shechtman *et al.* [97]. Since then, many ternary and higher-order alloys have been found which have stable quasicrystalline structures.

The extraordinary structural properties of quasicrystals have motivated numerous measurements of material properties that might be directly influenced by their quasicrystallinity. One such property that might lead to important commercial and technological applications of these materials is their tribological behavior. Interestingly, there have been a number of reports of apparently low friction measured on the quasicrystal surfaces [98–104]. From a tribological science perspective, however, the interesting question is whether such behavior is a direct consequence of quasicrystallinity.

There are several possible origins of the low frictional properties of quasicrystals. One, of course, is that because of their inherent lack of periodicity, quasicrystalline surfaces can never come into commensurate contact with one another or with any periodic surface. While the connection between commensurability and friction is far from being clearly resolved at the experimental level, there are several theoretical articles that predict such a connection [105, 106]. Another obvious potential source of low friction during most friction measurements is the possible presence of thin adsorbed films. If the surfaces of quasicrystals are coated with thin films of contaminants that effectively serve as lubricants, the hardness of the quasicrystals results in low contact area and thus low friction. Hardness may be a direct result of the quasicrystalline structure and in that sense serves as the link between friction and quasicrystallinity [98]. A similar idea is that thin oxide films on air-exposed quasicrystal surfaces serve as lubricants because they are only weakly adherent to the quasicrystal substrate and delaminate under shear [107]. Current work in the field of quasicrystal tribology is aimed at unraveling the competing theories and finding links between experimental tribological results and quasicrystallinity.

Making the connection between the macroscopic material properties of quasicrystals and their atomic-level structure is a complicated problem. As always, there is a need for good experimental measurements of friction under well-defined conditions. This is always an issue in the discussion of tribological phenomena, since they are inherently surface-related properties and are extremely sensitive to the presence of surface contamination [107, 108]. The second hurdle to any experimental test of the connection between quasicrystallinity and material properties is that it is not possible to experimentally vary the relevant parameters in a truly independent manner. Many of the properties of such materials are not truly independent and thus cannot be varied independently. Changes in one parameter cause changes in others, to which they are physically coupled. Finally, in such problems one is faced with the fact that it is often impossible to change the relevant parameters in a continuous fashion. For example, although the composition of an alloy may be varied continuously, its structure is dictated by a phase diagram that does not allow continuous variation. As a result of these numerous problems, it is difficult to find reliable measurements that allow one to make unambiguous statements about the influence of quasicrystallinity on macroscopic material properties.

There have been a few studies of the frictional properties of quasicrystalline surfaces that have been performed under ultrahigh vacuum conditions, which allow the controlled preparation and analysis of quasicrystal surfaces [109, 110]. These have made use of the

$Al_{70}Pd_{21}Mn_9$ quasicrystal and $Al_{48}Pd_{42}Mn_{10}$, an approximant of the quasicrystal. The latter has a composition that is similar to that of the quasicrystal, but has a CsCl structure that is periodic. The friction measurements were performed in a ultrahigh vacuum (UHV) surface-analysis chamber that allowed both controlled surface preparation and surface analysis before and after friction measurements. Figure 6.14 shows a low-energy electron diffraction pattern of the $Al_{70}Pd_{21}Mn_9$ quasicrystal taken from a fivefold symmetric plane. This surface was sputtered and annealed to generate a truly clean surface for friction measurements.

Prior to cleaning in vacuum, the surfaces of the $Al_{70}Pd_{21}Mn_9$ quasicrystals were analyzed using AES and shown to be contaminated by a layer of carbon, sulfur, chlorine, and oxygen. The friction coefficient between these two contaminated surfaces was $\mu_s = 0.11 \pm 0.02$. After cleaning in vacuum, the friction coefficient rose to $\mu_s = 0.60 \pm 0.08$. Clearly, the friction of these surfaces is heavily influenced by the presence of air-borne contaminants, and without making measurements in vacuum it is not possible to probe the frictional properties intrinsic to the quasicrystal itself.

The use of UHV methods allows the preparation of truly clean surfaces for friction measurements. It also allows the controlled preparation of adsorbed layers on such surfaces.

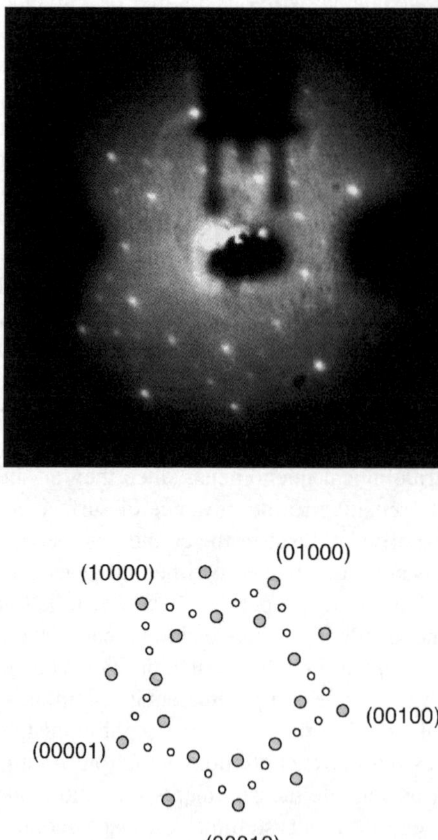

Figure 6.14 A low-energy electron diffraction pattern of the truly clean $Al_{70}Pd_{21}Mn_9$ quasicrystal on its fivefold symmetric surface. Two of these truly clean surfaces were used for subsequent friction measurements in vacuum (reprinted with kind permission from Elsevier) [109]

The $Al_{70}Pd_{21}Mn_9$ quasicrystal surface was modified by controlled exposures to O_2 and to H_2O in order to grow thin oxide films. AES shows that exposure to O_2 results in the formation of a thin film of aluminum oxide that passivates the surface against further oxidation. This is similar to the oxidation of the surface of aluminum itself [109, 110]. These films have an influence on the frictional properties of the surface. This is revealed in Figure 6.15, which shows the friction coefficient measured as a function of the degree of oxidation by both O_2 and H_2O. It is quite apparent that the presence of oxide films reduces the friction between the quasicrystal surfaces.

In order to test the importance of the bulk crystal structure and quasicrystallinity on the friction of these alloys, similar measurements have been taken with the $Al_{48}Pd_{42}Mn_{10}$ approximant phase [110, 111]. As with the quasicrystal phase, the presence of airborne contaminants on the surface of the approximant phase has a significant effect on friction. The interesting thing is that once these contaminants have been removed from the surface and the surface is clean, as determined by AES, the friction is roughly twice that of the quasicrystal. Furthermore, Figure 6.15 reveals that the friction remains higher than that of the quasicrystal at all levels of oxidation.

The results of the comparison between the frictional properties of the quasicrystals and their approximants clearly show that there are differences between them. Unfortunately, this is still insufficient to prove that the quasicrystalline structure is the determining factor. Nonetheless this is a very interesting result and one that could not have been obtained without the combined use of surface-analytical approaches and tribological measurements.

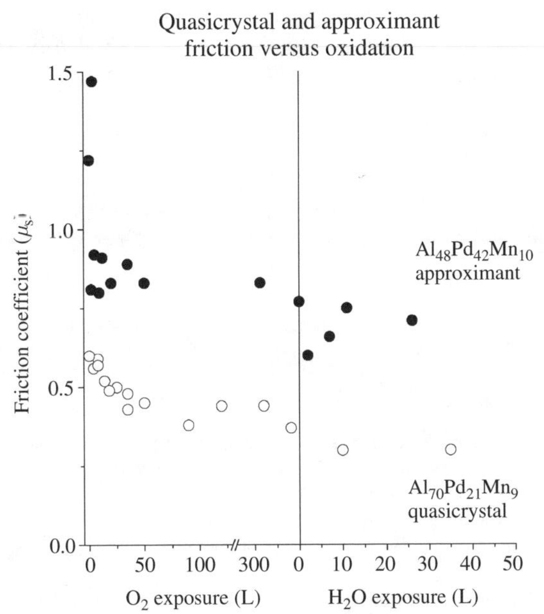

Figure 6.15 Friction coefficient between two Al_{70} Pd_{21} Mn_9 quasicrystal surfaces and two Al_{48} Pd_{42} Mn_{10} approximant surfaces as a function of surface exposure to O_2 and H_2O. The friction clearly drops as a function of the level of oxidation. It is also apparent that the friction between the approximant surfaces is roughly twice that of the quasicrystals at all levels of oxidation

6.7 Conclusions

In pursuit of our original objective – a survey of surface chemistry in tribology – we have illustrated a number of tribological problems and technologies in which surface chemistry is a critical component. As technologies push the need for tribological knowledge to new frontiers and demand solutions to tribology problems in new environments, such as the human body or the vacuum of space, a greater and deeper understanding of the influences of surface chemistry on tribology is needed. As our examples illustrate, the types of chemistry that are implicated in tribological phenomena are broad and varied. It is probably fair to say that at this time there is no single theory or concept that broadly describes the role of molecules in reducing friction at interfaces. Much of the understanding that we have gained over the past century has been specific to the problems at hand. In fact, what is needed to successfully approach the study of tribological surface chemistry and to solve real tribological problems is a broad understanding of organic chemistry, solid-state chemistry, materials chemistry, and, of course, the mechanical aspects of tribology and contact mechanisms.

Broadly speaking, however, one can make a few general statements about the role of surface science as an approach to solving tribological problems. There is no doubt that the tools developed by physicists and chemists in the field of surface science have had a major impact on our understanding of tribology. Electron microscopy, Auger spectroscopy, and the many other tools of the trade have provided tribologists with a molecular or atomic scale understanding of the world of the solid–solid interface. However, in addition, the field of surface science has provided the tools and the means for preparing highly controlled and highly characterized interfaces between solids. These are clearly needed in order to separate the many variables that can influence friction and the role of lubricants at interfaces. While the complexities of these problems are undisputed, inspiration, creativity, and hard work, both in the laboratory and by theoreticians, have pushed forward our understanding of lubrication and will continue to do so for the foreseeable future.

Acknowledgments

This material has been based upon an article that first appeared in the *Journal of Engineering Tribology* – Proceedings Part J, 2002, Vol. 216, No. J6, ISSN 1350–6501, published by Professional Engineering Publishing. Permission is granted by the Institution of Mechanical Engineers.

References

1. Hutchings, I.M., '*Tribology—Friction and Wear of Engineering Materials*', Arnold, London, 1992.
2. Spikes, H.A., 'Boundary Lubrication and Boundary Films', in *Thin Films in Tribology* (eds D. Dowson *et al.*), Elsevier Science Publishers B.V., Amsterdam, 1993, pp. 331–346.
3. Stanton, T.E., '*Friction*', Longmans, Green & Co., London, 1923.
4. Deeley, R.M., 'Oiliness and Lubrication. Discussion on Lubrication', *Proceedings of the Physical Society*, **32**, 1919, 1s–11s.
5. Hardy, W.B. and Bircumshaw, I., 'Boundary Lubrication—Plane Surfaces and the Limitations of Amontons' Law', *Proceedings of the Royal Society of London, A*, **108**, 1925, 1–27.
6. Allen, C.M. and Drauglis, E., 'Boundary Layer Lubrication: Monolayer or Multilayer', *Wear*, **14**, 1969, 363–384.
7. Bowden, F.P. and Tabor, D., '*Friction and Lubrication of Solids*', Oxford University Press, Oxford, UK, 1950.
8. Williams, J.A., '*Engineering Tribology*', Oxford University Press, Oxford, UK, 1994.

9. Studt, P., 'The Influence of the Structure of Isometric Octadecanols on Their Adsorption from Solution on Iron and Their Lubricating Properties', *Wear*, **70**, 1981, 329–334.
10. Stachowiak, G.W. and Batchelor, A.W., '*Engineering Tribology*', Elsevier, Amsterdam, 1993.
11. Fischer, D.A., Hu, Z.S. and Hsu, S.M., 'Molecular Orientation and Bonding of Monolayer Stearic Acid on a Copper Surface Prepared in Air', *Tribology Letters*, **3**, 1997, 41–46.
12. Ratoi, M., Anghel, V., Bovington, C.H. and Spikes, H.A., 'Mechanisms of Oiliness Additives', *Tribology International*, **33**, 2000, 241–247.
13. Spikes, H.A., 'Direct Observation of Boundary Layers', *Langmuir*, **12**, 1996, 4567–4573.
14. Anghel, V., Bovington, C.H. and Spikes, H.A., Thick Film Formation by Friction Modifier Additives', *Lubrication Science*, **11**, 1999, 313–335.
15. Ratoi, M., Bovington, C.H. and Spikes, H.A., 'Mechanism of Metal Carboxylate Friction Modifier Additive Behavior', International Tribology Conference, Nagasaki, 2000.
16. Çavdar, B. and Ludema, K.C., 'Dynamics of Dual Film Formation in Boundary Lubrication of Steels. Part I Functional Nature and Mechanical Properties', *Wear*, **148**, 1991, 305–327.
17. Çavdar, B. and Ludema, K.C., 'Dynamics of Dual Film Formation in Boundary Lubrication of Steels, Part 2: Chemical Analyses', *Wear*, **148**, 1991, 329–346.
18. Çavdar, B. and Ludema, K.C., 'Dynamics of Dual Film Formation in Boundary Lubrication of Steels, Part 3: Real Time Monitoring with Ellipsometry', *Wear*, **148**, 1991, 347–361.
19. Hsu, S.M., 'Boundary Lubrication: Current Understanding', *Tribology Letters*, **3**, 1997, 1–11.
20. Israelachvili, J.N., McGuiggan, P.M. and Homola, A.M., 'Dynamic Properties of Molecularly Thin Liquid Films', *Science* (Washington, D.C., 1883), **240**, 1988, 189–191.
21. Granick, S., 'Motions and Relaxations of Confined Liquids', *Science* (Washington, D.C., 1883), **253**, 1991, 1374–1379.
22. Heuberger, M., Zäch, M. and Spencer, N.D., 'When is a Fluid Not a Fluid?', *Science* (Washington, D.C., 1883), **292**, 2001, 905–908.
23. Persson, B.N.J., '*Sliding Friction, Physical Principles and Applications*', Springer-Verlag, Berlin Heidelberg, 1998.
24. Widmer, M.R., Heuberger, M. and Spencer, N.D., 'Influence of Polymer Surface Chemistry on Frictional Properties Under Protein-Lubrication Conditions', *Tribology Letters*, **10**, 2001, 111–116.
25. Habeeb, J.J. and Stover, W.H., 'The Role of Hydroperoxides in Engine Oil and the Effect of Zinc Dialkylthiophosphates', *ASLE Transactions*, **30**, 1987, 419–426.
26. Roper, G.W. and Bell, J.C., '*Review and Evaluation of Lubricated Wear in Simulated Valve Train Contact Conditions SAE 952473. Recent Snapshots and Insights into Lubricant Tribology SP-1116*', Society of Automotive Engineers, Warrendale, PA, 1995, pp. 67–83.
27. Bec, S., Tonck, A., Georges, J.M., Coy, R.C., Bell, J.C. and Roper, G.W., 'Relationship Between Mechanical Properties and Structures of Zinc Dithiophosphate Anti-Wear Films', *Proceedings of the Royal Society of London, A*, **455**, 1999, 4181–4203.
28. Debies, T.P. and Johnson, W.G., 'Surface Chemistry of Some Antiwear Additives as Determined by Electron Spectroscopy', *ASLE Transactions*, **23**, 1980, 289–297.
29. Jahanmir, S., 'Wear Reduction and Surface Layer Formation by a ZDDP Additive', *Journal of Tribology*, **109**, 1987, 577–586.
30. Kasrai, M., Fuller, M., Scaini, M., Yin, Z., Brunner, R.W., Bancroft, G.M., Fleet, M.E., Fyfe, K. and Tan, K.H., 'Study of Tribochemical Film Formation Using X-ray Absorption and Photoelectron Spectroscopies', in Proceedings of the 21st Leeds/Lyon Symposium, Lubricants and Lubrication (eds D. Dowson *et al.*), Elsevier Science B.V., Amsterdam, 1995, pp. 659–669.
31. Rhodes, K.L. and Stair, P.C., 'The Effect of ZDP's on the Surface Chemistry of Hydrocarbon Films on Oxidized Iron Substrates', *Tribology Transactions*, **36**, 1993, 27–34.
32. Eglin, M., Rossi, A., Piras, F.M. and Spencer, N.D., 'Zinc Diisopropyl Dithiophosphate by XPS', *Surface. Science Spectra*, **8**, 2001, 97–104.
33. Piras, F.M., Rossi, A. and Spencer, N.D., 'Combined In Situ (ATR FT-IR) and Ex Situ (XPS) Study of the ZnDTP–Iron Surface Interaction', *Tribology Letters*, **15**, 2003, 181–191.
34. Eglin, M., Rossi, A. and Spencer, N.D., 'X-ray Photoelectron Spectroscopy Analysis of Tribostressed Samples in the Presence of ZnDTP: A Combinatorial Approach', *Tribolog Letters*, **15**, 2003, 199–209.
35. Rossi, A., Eglin, M., Piras, F.M., Matsumoto, K. and Spencer, N.D., 'Surface Analytical Studies of Surface-Additive Interactions, by Means of In Situ and Combinatorial Approaches', *Wear*, **256**, 2004, 578–584.

36. Yin, Z., Kasrai, M., Bancroft, G.M., Laycock, K.F. and Tan, K.H., 'Chemical Characterization of Antiwear Films Generated on Steel by Zinc Dialkyl Dithiophosphate Using X-ray Adsorption Spectroscopy', *Tribology International*, **26**, 1993, 383–388.
37. Yin, Z., Kasrai, M., Fuller, M., Bancroft, G.M., Fyfe, K., Yamaguchi, E.S., Ryason, P.R., Willermet, P.A. and Tan, K.H., Application of Soft X-ray Absorption Spectroscopy in Chemical Characterization of Antiwear Films Generated by ZDDP: Part I—The Effect of Physical Parameters', *Wear*, **202**, 1997, 172–191.
38. Bancroft, G.M., Kasrai, M., Fuller, M. and Yin, Z., 'Mechanisms of Tribological Film Formation: Stability of Tribo-Thermally-Generated ZDDP Films', *Tribology Letters*, **3**, 1997, 47–51.
39. Suominen Fuller, M.L., Kasrai, M., Bancroft, G.M., Fyfe, K. and Tan, K.H., Solution Decomposition of Zinc Dialkyl Dithiophosphate and Its Effect on Antiwear and Thermal Film Formation Studied by X-ray Absorption Spectroscopy', *Tribology International*, **31**, 1998, 627–664.
40. Ferrari, E.S., Roberts, K.J., Sansone, M. and Adams, D., 'A Multi-Edge X-ray Absorption Spectroscopy Study of the Reactivity of Zinc Di-alkyl-di-thiophosphates Anti-Wear Additives 2. In Situ Studies of Steel/Oil Interfaces', *Wear*, **236**, 1999, 259–275.
41. Canning, G.W., Suominen Fuller, M.L., Bancroft, G.M., Kasrai, M., Cutler, J.N., De Stasio, G. and Gilbert, B., 'Spectromicroscopy of Tribological Films from Engine Oil Additives. Part I. Films from ZDDP's', *Tribology Letters*, **6**, 1999, 159–169.
42. Suominen Fuller, M.L., Rodriguez Fernandez, L., Massoumi, G.R., Lennard, W.N., Kasrai, M. and Bancroft, G.M., 'The Use of X-ray Absorption Spectroscopy for Monitoring the Thickness of Antiwear Films from ZDDP', *Tribology Letters*, **8**, 2000, 187–192.
43. Varlot, K., Kasrai, M., Martin, J.M., Vacher, B., Bancroft, G.M., Yamaguchi, E.S. and Ryason, P.R., 'Antiwear Film Formation of Neutral and Basic ZDDP: Influence of the Reaction Temperature and of the Concentration', *Tribology Letters*, **8**, 2000, 9–16.
44. Willermet, P.A., Carter, R.O. and Boulos, E.N., 'Lubricant-Derived Tribochemical Films – An Infra-Red Spectroscopy Study', *Tribology International*, **25**, 1992, 371–380.
45. Piras, F.M., Rossi, A. and Spencer, N.D., 'In Situ Attenuated Total Reflection (ATR) Spectroscopic Analysis of Tribological Phenomena', in Proceedings of the 28th Leeds/Lyon Symposium, Boundary and Mixed Lubrication (eds D. Dowson *et al.*), Elsevier Science B.V., 2002, pp. 371–380.
46. Piras, F.M., Rossi, A. and Spencer, N.D., 'Growth of Tribological Films: In Situ Characterization Based on Attenuated Total Refection Infrared Spectroscopy', *Langmuir*, **18**, 2002, 6606–6613.
47. Bell, J.C., Delargy, K.M. and Seeney, A.M., 'The Removal of Substrate Material Through Thick Zinc Dithiophosphate Anti-Wear Films', in Proceedings of the 18th Leeds/Lyon Symposium, Wear Particles: From the Cradle to the Grave (eds D. Dowson *et al.*), Elsevier Science Publishers B.V., 1992, pp. 387–396 .
48. Pidduck, A.J. and Smith, G.C., Scanning Probe Microscopy of Automotive Anti-Wear Films', *Wear*, **212**, 1997, 254–264.
49. Graham, J.F., McCague, C. and Norton, P.R., 'Topography and Nanomechanical Properties of Tribochemical Films Derived from Zinc Dialkyl and Diaryl Dithiophosphates', *Tribology Letters*, **6**, 1999, 149–157.
50. Georges, J.M., Tonck, A., Poletti, S., Yamaguchi, E.S. and Ryason, P.R., 'Film Thickness and Mechanical Properties of Adsorbed Neutral and Basic Zinc Diisobutyl Dithiophosphates', *Tribology Transactions*, **41**, 1998, 543–553.
51. Sheasby, J.S. and Nisenholz Rafael, Z., 'Antiwear Characteristics of a Commercial Secondary ZDDP Additive', *Tribology Transactions*, **36**, 1993, 399–404.
52. Sheasby, J.S. and Caughlin, T.A., 'The Direct Observation of the Anti-Wear Action of ZDDP', in Proceedings of the 21st Leeds/Lyon Symposium, Lubricants and Lubrication (eds D. Dowson *et al.*), Elsevier Science B.V., 1995, pp. 399–408.
53. Gunsel, S., Spikes, H.A. and Aderin, M., 'In-Situ Measurement of ZDDP Films in Concentrated Contacts', *Tribology Transactions*, **36**, 1993, 276–282.
54. Armstrong, D.R., Ferrari, E.S., Roberts, K.J. and Adams, D., 'An Examination of the Reactivity of Zinc Di-alkyl-di-thiophosphate in Relation to Its Use as an Anti-Wear and Anti-Corrosion Additive in Lubricating Oils', *Wear*, **217**, 1998, 276–287.
55. Bovington, C.H. and Dacre, B., 'The Adsorption and Reaction of Decomposition Products of Zinc Di-isopropyldithiophosphate on Steel', *ASLE Transactions*, **27**, 1984, 252–258.
56. Martin, J.M., 'Antiwear Mechanisms of Zinc Dithiophosphate: A Chemical Hardness Approach', *Tribology Letters*, **6**, 1999, 1–8.
57. Coy, R.C. and Jones, R.B., 'The Thermal Degradation and EP Performance of Zinc Dialkyldithiophosphate Additives in White Oil', *ASLE Transactions*, **24**, 1980, 77–90.

58. Jones, R.B. and Coy, R.C., 'The Chemistry and Thermal Degradation of Zinc Dialkyldithiophosphate Additives', *ASLE Transactions*, **24**, 1981, 91–97.
59. Ho, T.-L., 'The Hard Soft Acids Bases (HSAB) Principle and Organic Chemistry', *Chemical Reviews*, **75**, 1975, 1.
60. Taylor, L., Dratva, A. and Spikes, H.A., 'Friction and Wear Behavior of Zinc Dialkyldithiophosphate Additive', *Tribology Transactions*, **43**, 2000, 469–479.
61. Eglin, M., Rossi, A. and Spencer, N.D., 'A Combinatorial Approach to Elucidating Tribochemical Mechanisms', *Tribology Letters*, **15**, 2003, 193–198.
62. Gellman, A.J., 'Lubricants and Overcoats for Magnetic Storage Media', *Current Opinion in Colloid and Interface Science*, **3**, 1998, 368–372.
63. Mate, C.M., 'Picking the Best Lubricant for Contact Recording', *Data Storage*, 1997, 45–48.
64. Mate, C.M. and Homola, A.M. (eds), '*Micro/Nanotribology and Its Applications, Molecular Tribology of Disk Drives*', Kluwer Academic Publishers, Netherlands, 1997, pp. 647–661.
65. Kryder, H.M., 'Ultrahigh-Density Recording Technologies', *Materials Research Society Bulletin*, **21**, 1996, 17–22.
66. Thompson, D.A., 'The Role of Perpendicular Recording in the Future of Hard Disk Storage', *Journal of Magnetics Society of Japan*, **21**, 1997, 9–15.
67. Shukla, N. and Gellman, A.J., 'Interaction of Alcohols with a-CHx Films', *Journal of Vacuum Science and Technology*, **18**, 2000, 2319–2326.
68. Kobayashi, A., Yoshitomi, D., Yoshihara, O., Imayoshi, T., Kinbara, A., Fumoto, T. and Ueno, M., Carbon Films for the Protection of Magnetic Recording Media', *Surface and Coating Technology*, **72**, 1995, 152–156.
69. Cho, N.-H., Krishnan, K.M., Veirs, D.K., Rubin, M.D., Hopper, C.B. and Bhushan, B., Chemical Structure and Physical Properties of Diamond-Like Amorphous Carbon Films Prepared by Magnetron sputtering', *Journal of Material Science*, **5**, 1995, 2543–2554.
70. Tsai, H. and Bogy, D., 'Critical Review: Characterization of Diamond-Like Carbon Films and Their Application as Overcoats on Thin-Film Media for Magnetic Recording', *Journal of Vacuum Science and Technology, A*, **5**, 1987, 3287–3312.
71. Yanagisawa, M., 'Adsorption of Perfluoropolyethers on Carbon Surfaces', *Tribology and Mechanics of Magnetic Storage System*, **9**, 1994, 36.
72. Toney, M.F., Mate, C.M. and Pocker, D.J., 'Calibrating ESCA and Ellipsometry Measurements of Perfluoropolyether Lubricant Thickness', *IEEE Transactions on Magnetics*, **34**, 1998, 1774.
73. Gao, C. and Dai, P., 'Molecular Orientation of PFPE Lubricant Films and Its Quantification', *IEEE Transactions on Magnetics*, **33**, 1997, 3118–3120.
74. Tyndall, G.W. and Waltman, R.J., 'Thermodynamics of Confined Perfluoropolyether Films on Amorphous Carbon Surfaces Determined from the Time-Dependent Evaporation Kinetics', *The Journal of Physical Chemistry B*, **104**, 2000, 7085–7095.
75. Ma, X., Gui, J., Smoliar, L., Grannen, K., Marchon, B., Bauer, C.L. and Jhon, M.S., 'Complex Terraced Spreading of Perfluoropolyalkylether Films on Carbon Surfaces', *Physical Review E*, **59**, 1999, 722–726.
76. Cornaglia, L. and Gellman, A.J., 'Fluoroether Bonding to Carbon Overcoats', *Journal of Vacuum Science and Technology A*, **15**, 1997, 2755–2765.
77. Van Treuren, K.W., Barlow, D.N., Heiser, W.H., Wagner, M.J. and Forster, N.H., 'Investigation of Vapor-Phase Lubrication in a Gas Turbine Engine', *Journal of Engineering for Gas Turbines and Power*, **120**, 1998, 257.
78. Rao, A.M.N., 'Vapor-Phase Lubrication: Application-Oriented Development', *Lubrication Engineering*, **52**, 1996, 856–862.
79. Placek, D.G. and Freiheit, T., 'Progress in Vapor Phase Lubrication Technology', *Journal of Engineering for Gas Turbines and Power*, **115**, 1993, 700–705.
80. Forster, N.H. and Trivedi, H.K., 'Rolling Contact Testing of Vapor Phase Lubricants – Part II: System Performance Evaluation', *Tribology Transactions*, **40**, 1997, 493–499.
81. Forster, N.H. and Trivedi, H.K., 'Rolling Contact Testing of Vapor Phase Lubricants – Part I: Material Evaluation', *Tribology Transactions*, **40**, 1997, 421–428.
82. Groeneweg, M., Hakim, N., Barber, G.C. and Klaus, E., 'Vapor Delivered Lubrication of Diesel Engines – Cylinder Kit Rig Simulation', *Lubrication Engineering*, **47**, 1991, 1035–1039.
83. Makki, J. and Graham, E., 'Formation of Solid Films from the Vapor Phase on High Temperature Surfaces', *Lubrication Engineering*, March, 1990, 199–206.
84. Makki, J.F. and Graham, E.E., 'Vapor Phase Deposition on High Temperature Surfaces', *Tribology Transactions*, **33**, 1990, 595–603.
85. Hanyaloglu, B.F., Graham, E.E., Oreskovic, T. and Hajj, C.G., 'Vapor Phase Lubrication of High Temperature Alloys', *Lubrication Engineering*, **51**, 1994, 503–508.

86. Forster, N.H., 'Rolling Contact Testing of Vapor Phase Lubricants – Part III: Surface Analysis', *Tribology Transactions*, **42**, 1999, 1–9.
87. Hanyaloglu, B. and Graham, E.E., 'Vapor Phase Lubrication of Ceramics', *Lubrication Engineering*, **50**, 1994, 814.
88. Morales, W., Hanyaloglu, B.F. and Graham, E.E., Infrared Analysis of Vapor Phase Deposited Tricresyl Phosphate (TCP), NASA, TM106423, 1994.
89. Forster, N.H., personal communication, 2000.
90. Holbert, A.W., Batteas, J.D., Wong-Foy, A., Rufael, T.S. and Friend, C.M., 'Passivation of Fe(110) via Phosphorus Deposition: The Reactions of Trimethylphosphite', *Surface Science Reports*, **401**, 1998, L437.
91. Ren, D., Zhou, G. and Gellman, A.J., 'The Decomposition Mechanism of Trimethylphosphite on Ni(111)', *Surface Science Reports*, **475**, 2001, 61–72.
92. Ren, D. and Gellman, A.J., 'The Carbon Deposition Mechanism in Vapor Phase Lubrication', *Tribology Transactions*, **43**, 2000, 480–488.
93. Wheeler, D.R. and Fuat, O.D., 'The Adsorption and Thermal-Decomposition of Tricresylphosphate (TCP) on Iron and Gold', *Application of Surface Science*, **18**, 1984, 106.
94. Sung, D. and Gellman, A.J., 'The Surface Chemistry of Alkyl- and Aryl-Phosphate Vapor Phase Lubricants on Fe Foil', *Tribology International*, **35**(9), 2002, 579–590.
95. Ren, D. and Gellman, A.J., Reaction Mechanisms in Organophosphate Vapor Phase Lubrication of Metal Surfaces', *Tribology International*, **34**, 2001, 353–365.
96. Ren, D., Sung, D. and Gellman, A.J., 'Activation of the SiC Surface for Vapor Phase Lubrication by Fe chemical vapor deposition from $Fe(CO)_5$', *Tribology Letters*, **10**, 2001, 179–185.
97. Shechtman, D., Blech, I., Gratias, D. and Cahn, J.W., Metallic Phase with Long-Range Orientational Order and No Translational Symmetry', *Physical Review Letters*, **53**, 1984, 1951.
98. Kang, S.S., Dubois, J.M. and von Stebut, J., 'Tribological Properties of Quasi-Crystalline Coatings', *Journal of Materials Research*, **8**, 1993, 2471–2481.
99. Dubois, J.M., Kang, S.S. and Perrot, A., 'Towards Application of Quasi-Crystals', *Materials Science and Engineering*, **A179/A180**, 1994, 122–126.
100. Wittmann, R., Urban, K., Schandl, M. and Hornbogen, E., 'Mechanical Properties of Single Quasi-Crystalline AlCuCoSi', *Journal of Materials Research*, **6**, 1991, 1165–1168.
101. Dubois, J.M., Kang, S.S. and von Stebut, J., 'Quasi-Crystalline Low-Friction Coatings', *Journal of Material Science Letters*, **10**, 1991, 537–541.
102. von Stebut, J., Soro, J.M., Plaindoux, P. and Dubois, J.M. (eds), 'New Horizons in Quasicrystals Research and Applications', *Friction Behavior of Pure Quasicrystalline Materials*, World Scientific Publishing, Singapore, 1997, pp. 248–255.
103. Gavatz, M., Rouxel, D., Pigeat, P., Weber, B. and Dubois, J.M., Surface Oxidation of the $Al_{62}Cu_{25.5}Fe_{12.5}$ Icosahedral Quasicrystal', *Philosophical Magazine A*, **80**, 2000, 2083.
104. Pinhero, P.J., Anderegg, J.W., Sordelet, D.J., Besser, M.F. and Thiel, P.A., Surface Oxidation of Al–Cu–Fe Alloys: A Comparison of Quasicrystalline and Crystalline Phases', *Philosophical Magazine B*, **79**, 1999, 91–110.
105. He, G., Muser, M.H. and Robbins, M.O., 'Adsorbed Layers and the Origin of Static Friction', *Science*, **284**, 1999, 1650.
106. Sokoloff, J.B., 'Theory of Energy Dissipation in Sliding Crystal Surfaces', *Physical Review B*, **42**, 1990, 760–765.
107. Gellman, A.J., 'Lubrication by Molecular Monolayers at Ni–Ni Interfaces', *Journal of Vacuum Science Technology A*, **10**, 1992, 180.
108. McFadden, C.F. and Gellman, A.J., 'Ultra-High Vacuum Boundary Lubrication of the Cu–Cu Interface by 2,2,2-Trifluoroethanol', *Langmuir*, **11**, 1995, 273.
109. Ko, J.S., Gellman, A.J., Jenks, C., Lograsso, T. and Thiel, P.A., 'Friction Between Single Grain $Al_{70}Pd_{21}Mn_9$ Quasicrystal Surfaces', *Surface Science Reports*, **423**, 1999, 243–255.
110. Mancinelli, C., Ko, J.S., Jenks, C.J., Thiel, P.A., Ross, A.R., Lograsso, T.A. and Gellman, A.J., 'Comparative Study of the Tribological and Oxidative Properties of AlPdMn Quasicrystals and Their Cubic Approximants in Quasicrystals – Preparation, Properties and Applications' (eds E. Belin-Ferré, P.A. Thiel, A-P. Tsai, K. Urban), *MRS Proceedings*, **643**, 2001, K8.2.1–K8.2.10.
111. Mancinelli, C., Jenks, C.J., Thiel P.A. and Gellman A.J., 'Tribological Properties of a B2-Type AlPdMn Quasicrystal Approximant', *Journal of Materials Research*, **18**(6), 2003, 1447–1456.

7

Tribology of Engineered Surfaces

K. Holmberg and A. Matthews

Abstract

Wear and friction behaviour in moving contacts depend on the properties of the surfaces of the two interacting solids and the material, usually a lubricant, between them. The properties of the surfaces can be modified and changed by surface treatment techniques and by applying one or several layers of different thickness on one or both of the surfaces. In this chapter, we discuss the possibilities that the rapidly developing deposition techniques, such as physical and chemical vapour deposition, offer to change the tribological properties by applying very thin hard or soft surface coatings. The wear and friction mechanisms are described on the macro-, micro- and nanoscale and their effect is discussed in sliding, abrasive, impact, fatigue, fretting, chemical dissolution as well as in lubricated conditions. Thin hard coatings such as TiN, TiAlN and Al_2O_3 are excellent for wear reduction in severe rubbing conditions, while both the very hard diamond-like carbon and the lamellar MoS_2-based coatings can produce conditions of ultra-low friction coefficients down to 0.001 and below. Multilayers, nanocomposites and functionally graded, hybrid and otherwise structured surfaces have been tailored to optimize tribological performance in special situations such as cutting and forming tools, fuel pumps, gears and rolling bearings, and some of these applications are described.

7.1 Introduction

The introduction of a fluid between the surfaces is the traditional and most common way to change the tribological behaviour in a contact with two surfaces moving relative to each other. This approach has shortcomings in many applications due to aspects such as lubricant supply, ageing, degradation, contamination, maintenance and environment. An increasingly popular alternative which does not suffer from these limitations is to apply a thin surface

Wear – Materials, Mechanisms and Practice Edited by G. Stachowiak

layer or coating on one or both of the surfaces. This approach is promoted by the new coating techniques that have been developed over the last few decades. In particular, plasma-assisted physical vapour deposition (PVD) and chemical vapour deposition (CVD) offer considerable possibilities to tailor the thin surface layers with regard to their composition and structure as well as their mechanical and chemical properties.

In PVD, the samples to be coated are placed in a vacuum chamber at a pressure typically ranging from 0.3 to 1.3 Pa. First, the surfaces are cleaned by argon bombardment in a glow discharge plasma. Then, the deposition takes place by atomization or vaporization of the coating material from a solid source. The ionized species are attracted to the negatively biased sample surface. The deposition temperature can be from room temperature up to over 500°C.

In CVD, precursor gases are utilized containing the elements to be deposited and these are introduced into a reaction chamber to condense onto a substrate at a very high temperature, typically at about 800–1500°C and at a pressure ranging from atmospheric down to about 1 Pa. There are various means of assisting the process, such as through the use of laser or electron beams, or by ion bombardment of the growing films. Today, there is a large variety and many combinations of the PVD and CVD processes. They are described in more detail by Holmberg and Matthews (1994) and Rickerby and Matthews (1991).

The first major commercial application of tribological thin surface coatings deposited by plasma-assisted PVD (PAPVD) was the use of titanium nitride and titanium carbide coatings on cutting tools. Other early applications were the use of lead coatings for dry lubrication of rolling bearings in near-vacuum conditions in space, aluminium coatings for corrosion protection of aircraft fasteners and gold coatings for electrical contacts. Following the early rapid adoption of advanced coatings on cutting tools, there has been a gradual but accelerating interest in the use of vapour deposition technologies in other application sectors, such as automotive and aerospace, consumer products, medical devices and in other branches of manufacturing industry, such as food processing (Leyland and Matthews, 1994; Cselle and Barimani, 1995; Zabinski and Prasad, 1996; Enomoto and Yamamoto, 1998; Theunissen, 1998; Vercammen *et al.*, 1999; Lampe *et al.*, 2003; Merlo, 2003).

One factor that has constrained the wider adoption of coatings has been their relatively high cost; another is concern over repeatability of properties such as thickness, composition, hardness and adhesion to the substrate. Such reliability issues are now being solved through improvements in process monitoring and control. Developments in continuous and semi-continuous high-throughput coating equipment have reduced the coating cost per component and opened up high-volume markets such as automotive engine and lighting components and consumer products such as cellular phones and shaving razor blades. The automotive parts include components used in fuel injection systems, air conditioning units, engines and even transmission systems in high-performance vehicles. Thus, the nature of the tribological contacts to be protected by the coatings is becoming ever more wide-ranging and demanding. Hence, there is a need for a full and systematic understanding of both the mechanisms occurring and the response behaviour of coatings to these challenging applications.

In this chapter, our aim was to present an overview of the present understanding of the tribological mechanisms occurring on engineered surfaces and to elucidate how advanced coatings and treatment can provide protection. This text complements our previous reviews of the subject, Holmberg and Matthews (1994, 2001) and Holmberg *et al.* (1998).

7.2 Definition of an Engineered Surface

An engineered surface can be defined as a surface of a solid that has been designed and technologically modified or coated with new materials in such a way that the resulting surface has a performance that neither of the involved materials can provide alone. We will concentrate on thin tribological coatings and define them as those that are sufficiently thin that the substrate material plays a role in determining the friction and wear performance. These coatings are typically in the thickness range of 0.01–10 μm. Thus, we exclude coatings that are so thick that there is little or no substrate influence on the tribological behaviour – the coating in effect acts as a bulk material. We also exclude through-thickness treatment methods where the properties of the entire solid are changed. Both thick coatings and bulk treatment methods have been described elsewhere (Cartier, 2003).

7.3 Tribomechanisms of Coated Surfaces

7.3.1 Scales of Tribology

The tribological process in a contact in which two surfaces are in relative motion is very complex, since it involves simultaneously friction, wear and deformation mechanisms at different scale levels and of different types. In the literature, the tribological phenomena have been described with a characteristic set of parameters and characteristic scientific approaches typically only for one scale level, which often may be remarkably different from the next or the previous scale level. An attempt to illustrate the tribological scale levels from the very basic friction and wear phenomena of atomic dimensions to even global and universal dimensions has been presented by Holmberg (2001), as shown in Figure 7.1.

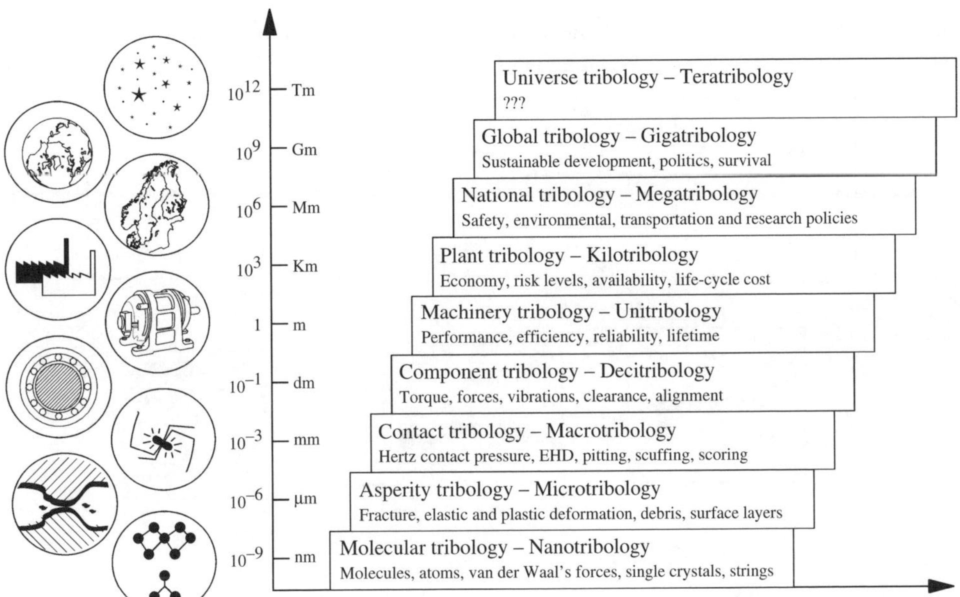

Figure 7.1 Scales of tribology from nanotribology to teratribology

Nanotribology or molecular tribology includes investigations that concentrate on phenomena related to the interaction between molecules and atoms, such as the effects of van der Waals forces and single crystal structures of materials.

Microtribology or asperity tribology was introduced by Bowden and Tabor (1950) with their studies of friction, wear and adhesion that take place at the peaks of the surface topography. Phenomena such as fracture, elastic and plastic deformation, debris formation, surface layer formation and topography effects are of central importance.

Macrotribology or contact tribology was in focus in the research at the beginning of the twentieth century. This work was related to contacts between gears, bearing elements and rollers; and phenomena like Hertzian contact pressure, elastohydrodynamic lubrication, and wear mechanisms clearly observable to the naked eye (scuffing, scoring and pitting) are of interest.

Component tribology or decitribology is related to defining and measuring typical parameters originating from the interaction of components and related to their performance such as torque, forces, vibrations, clearance and alignment.

The tribology scale presentation is of course a broad one where the higher levels are reliability engineering oriented and the very highest levels are of a more speculative or philosophical nature. However, when trying to understand the tribological mechanisms and interactions in engineered surfaces in a holistic way it is very useful to separately analyse the tribological changes on three different scales, the macro-, micro- and nanoscale, and to study separately the mechanical and chemical changes as well as material transfer taking place in the contact, as shown in Figure 7.2.

7.3.2 *Macromechanical Friction and Wear*

The macromechanical tribological mechanisms describe the friction and wear phenomena by considering the stress distribution and deformations in the whole contact, the total elastic

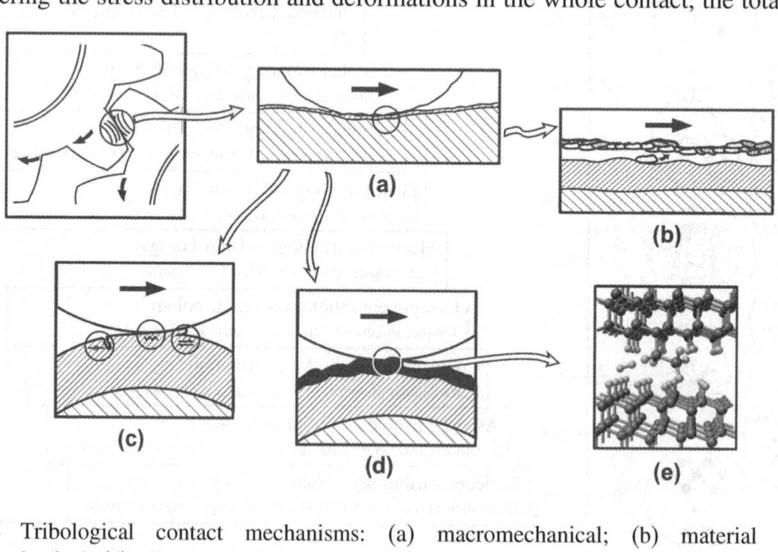

Figure 7.2 Tribological contact mechanisms: (a) macromechanical; (b) material transfer; (c) micromechanical; (d) tribochemical and (e) nanophysical (Part (e) from Krim, 1996 and Harrison *et al.*, 1998)

and plastic deformations they result in and the total wear particle formation process and its dynamics. In contacts with one or two coated surfaces, four main parameters that control the tribological process have been defined (Holmberg and Matthews, 1994). They are the coating and bulk deformability (hardness and elasticity), coating thickness, surface roughness and debris and tribolayers in the contact.

The influence on the macromechanical friction mechanisms of coating and substrate hardness, coating thickness, surface roughness and debris present in the contact is illustrated in Figure 7.3 and the corresponding wear mechanisms in Figure 7.4. It is important to notice that the three geometrically related parameters, the film thickness, the surface roughness and the wear debris, all typically appear in the same dimension range from 0.01 to 10 μm. This means that the interrelation between these dimensions for each real case can be considered to have a dominating effect on both friction and wear on macroscale. The mechanisms involved are very different depending on whether the coating and the substrate is soft or hard (Holmberg and Matthews, 1994; Donnet, 1995; Ramalingam and Zheng, 1995). We will discuss both these cases separately in the case when a sphere is sliding over a coated surface.

In this chapter, we use the words *soft* and *hard* for coatings in a general sense, meaning their deformability and not only their hardness. The loading conditions, both in vertical and transverse direction, will influence the stresses and strains in a surface and its friction and wear behaviour. Hardness is one important parameter but not the only one. The influence of elasticity in combination with hardness gives a more comprehensive and accurate

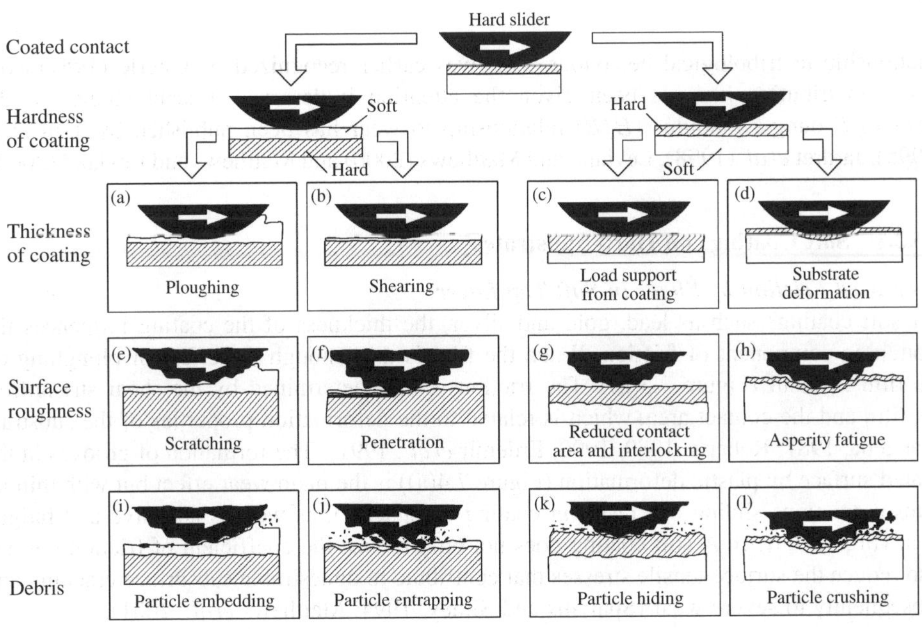

Figure 7.3 Macromechanical contact conditions for different mechanisms that influence friction when a hard spherical slider moves on a coated flat surface

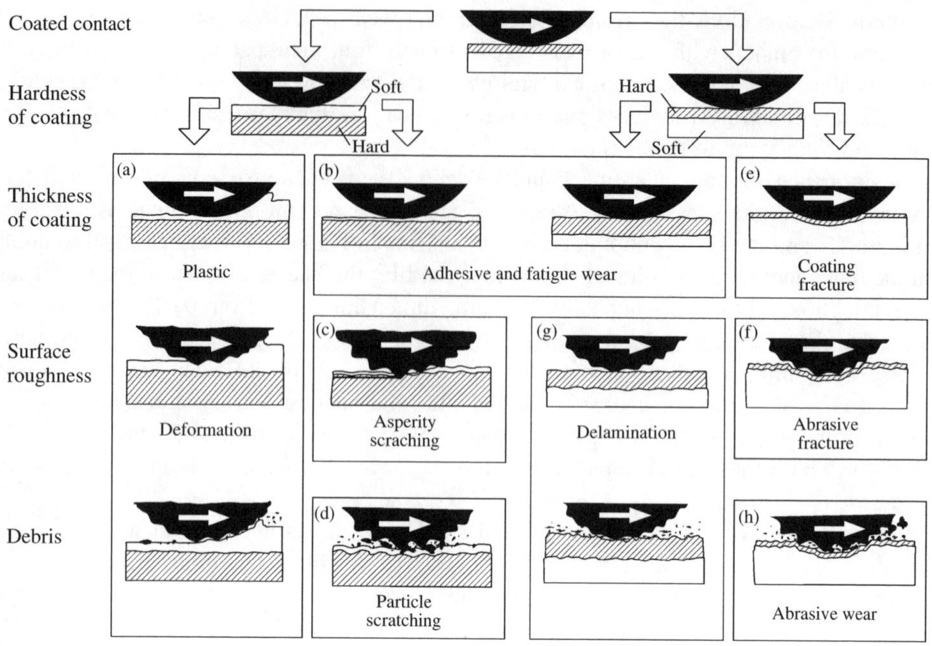

Figure 7.4 Macromechanical contact conditions for different mechanisms that influence wear when a hard spherical slider moves on a coated flat surface

relationship to tribological behaviour. This was earlier recognized by Oberle (1951) even if his contribution has not been given the attention it deserves. Recent studies of the Hardness/Young's modulus (H/E) relationship to wear has been published by Tsui *et al.* (1995), Jardret *et al.* (1998), Leyland and Matthews (2000) and Matthews and Leyland (2002).

7.3.2.1 Soft Coatings on Hard Substrates

Decreased Friction by Shear in Soft Top Layer

For soft coatings such as lead, gold and silver, the thickness of the coating influences the ploughing component of friction. When the film is thin enough the effect of ploughing on the film is small (Figure 7.3(b)). The friction is thus determined by the shear strength of the film and the contact area, which is related to the deformation properties of the substrate (Tangena, 1987; Roberts, 1989, 1990; Erdemir *et al.*, 1991). The formation of grooves in the coated surface by plastic deformation (Figure 7.4(a)) is the main wear effect but with thinner films continuous sliding can result in coating compaction, as well as adhesive and fatigue wear (Figure 7.4(b)). A soft coating does not only reduce the coefficient of friction but can also reduce the surface tensile stresses that contribute to undesirable subsurface cracking and subsequently to severe wear (Spalvins and Sliney, 1994; Matthews *et al.*, 2001).

Ploughing Friction

The friction usually increases with coating thickness for soft coatings due to plastic or elastic deformation of the film and due to the increased contact area at the interface between the sliding counterface and the coating where the shear takes place (Figure 7.3(a)). This increase in friction has been experimentally demonstrated for lead films (Tsuya and Takagi, 1964; Sherbiney and Halling, 1977), for gold films (Takagi and Liu, 1967), for silver films (El-Sherbiny and Salem, 1986; Yang *et al.*, 1999) and for MoS_2 films (Aubert *et al.*, 1990; Wahl *et al.*, 1999). For very thick soft coatings, the mechanism of ploughing will be very similar to that of soft bulk materials scratched by a hard indenter. These tribological mechanisms have been described by several authors, e.g. Hokkirigawa and Kato (1988), who in addition to the ploughing mechanism, also identify a wedge forming and a cutting mechanism depending on the degree of penetration and the shear strength of the contact interface.

Influence of Surface Roughness

The substrate surface roughness has an almost negligible influence on friction if the roughness is considerably smaller than the thickness of the soft coating and the coating is stiff enough to carry the load, as shown in Figure 7.3(e). However, when the roughness of the slider is higher than the coating thickness, coating penetration will take place (Figure 7.3(f)) and the friction is considerably increased due to scratching of the substrate material. This has been described both experimentally and theoretically by Sherbiney and Halling (1977) and El-Sherbiny and Salem (1984). A model for predicting the wear rate when hard asperities penetrate the soft coating and produce grooves in the substrate (Figure 7.4(c)) was developed by El-Sherbiny and Salem (1986).

Particle Embedding

In the sliding situation shown in Figure 7.3(i) hard particles are present in the contact, the particles having a diameter considerably smaller than the thickness of the soft coating on a hard substrate. The particles are pressed into the soft coating and embedded into it without any further contact with the slider as long as the soft coating remains thicker than the particle diameter. In this case, the particles have no great effect on friction, which is mainly controlled by the ploughing mechanisms described earlier. The slider will produce a main groove by ploughing in the soft coating and the surface asperities or trapped debris may cause microploughing and microgrooves within the main groove (Hwang *et al.*, 1999). El-Sherbiny and Salem (1979) showed with a theoretical model that the wear appears to be dependent on the surface texture of the system rather than on material properties during the initial wear when hard conical asperities are ploughing a soft surface coating.

Particle Entrapping

For thin surface coatings where the dimensions of the particles are of the same magnitude or bigger than the coating thickness and the surface roughness, as shown in Figure 7.3(j), their influence on friction can be considerable. If the particles are harder than the coating but softer than the substrate, then they are easily caught by the roughness of the counterface or partly sink into it and scratch grooves in the soft coating, as in the case of asperity penetration. The friction will increase because of particles ploughing the coating (Hwang *et al.*, 1999). The influence of ploughing wear particles on friction in sliding wear contacts has been shown to be considerable (Kim and Suh, 1991; Komvopoulos, 1991a,b). An increase in friction may follow if the slider and the substrate are of equal hardness and the loose particles in the contact have a higher hardness (Suh, 1986). Then the particles may partly sink through the coating into the substrate and also into the counterface and by a kind of anchoring mechanism resist motion. The wear rate depends on the particle size (Sin *et al.*, 1979).

If the particles in the contact are soft, then their tribological effect is quite different. Soft particles with low shear strength trapped in the contact can carry part of the load and inhibit direct substrate-to-counter-surface contact, thus reducing both wear and friction (Grill, 1997; Voevodin *et al.*, 1999b). A similar effect has been observed by Yamada *et al.* (1990) who found that polymer particles can act very effectively to reduce wear and friction. In contacts with MoS_2 coatings or when Mo and S are present, MoO_2 wear debris are produced that have a very low shear stress and thus can act as solid lubricants reducing the friction and wear (Donnet, 1998; Singer *et al.*, 2003). The tribological behaviour of molybdenum disulphide coatings is treated in more detail in Section 7.3.7.2.

7.3.2.2 Hard Coatings on Softer Substrates

The use of hard thin coatings on softer substrate materials is today very popular in many tribological applications. The hard coating can provide good wear protection and with a suitable choice of material and surface layer design the friction can be very low as well.

Hard Coating Can Reduce Wear

With a very thin hard layer on top of a softer substrate (Figure 7.4(e)) it may be that neither the coating nor the substrate is able to support the load. However, the function of the coating is to separate the substrate from the counter-surface and to prevent ploughing by increasing the hardness of the top layer of the surface. The latter effect has been considered to be very important by Shepard and Suh (1982), Suh (1986) and Bull and Rickerby (1990). The increase in wear resistance with increased coating hardness has recently been reported by Kodali *et al.* (1997) and Wiklund (1999). Increased substrate hardness results in a decreased contact area, where the shear takes place, and decreased friction, which is indicated in the results from Ronkainen *et al.* (1998b, 1999). Decreased surface roughness normally results in lower friction (Ahmed *et al.*, 2003; Svahn *et al.*, 2003).

Hard Coatings Without Microfilms Can Have High Friction

The prevention of ploughing reduces both friction and wear; but the higher shear strength introduced at the contact interface by the hard coating can, on the other hand, have the effect of increasing friction in sliding if no microfilms are formed (Figure 7.3(c)). That is why very high coefficients of friction often occur in sliding contacts with hard coatings (Holmberg and Matthews, 1994; Voevodin *et al.*, 1995b; Kullmer *et al.*, 2003; Tricoteaux *et al.*, 2003; Tuszynski *et al.*, 2003). The increase in friction caused by increased shear strength generally seems to be more dominant than the reduction in friction due to decreased ploughing. Here the recently developed diamond and diamond-like carbon (DLC) coatings are an exception. The friction may then be extremely low because of the low shear conditions produced at the surface top layer (Erdemir, 2002).

Thick Hard Films are Better Able to Carry the Load

Increased loads can also be resisted with thicker hard coatings because of their load-carrying capacity which reduces deflection as shown in Figure 7.3(c) (Rabinowicz, 1967; Roth *et al.*, 1987). A thick hard coating will have a modifying effect on the size and the shape of the stress zone beneath the coating, as has been shown for a hardness indenter by Burnett and Rickerby (1987a).

Deflection Influence on Stresses and Cracks
For very soft substrates, the indentation deformation will be considerable and this will thus add a ploughing or hysteresis effect on friction (Figure 7.3(d)). The deflection increases the stresses in the coating and also at the interface between the coating and the substrate, possibly resulting in fracture or fatigue cracks that may harm the coating or the substrate material. With a soft substrate, cracks may occur in the coating both within the contact area and outside at the substrate material pile-up area (Burnett and Rickerby, 1987b; Page and Knight, 1989; Guu *et al.*, 1996a,b; Lin *et al.*, 1996a,b; Bull, 1997; Yuan and Hayashi, 1999). The harder the substrate is, the higher the loading which the coating can resist without failure by fracture (Hedenqvist *et al.*, 1990).

Particle Hiding
The introduction of small particles into the sliding contact of a hard and rough surface, as shown in Figure 7.3(k), does not necessarily make the tribological contact more severe. The particles can be hidden in the valleys formed by the asperities while the sliding takes place at the asperity tops. Thus, the particles will have no great effect on either friction or wear. It is important to notice that reduced surface roughness of the surfaces may increase both friction and wear if the particles cannot hide in the valleys and instead interact with the surfaces by scratching and interlocking.

Particle Crushing
When particles, large in relation to the surface roughnesses, are introduced between two hard surfaces the result can be particle crushing, scratching or rolling as shown in Figure 7.3(l). If the particles have lower hardness than the surfaces, then they will be crushed and destroyed under the load of the contact, with smaller debris and some increase in friction as a result. If the particles have a higher hardness than the surfaces, they will be caught by the roughness of the surfaces resulting in ploughing and scratching. The scratching particles carry part of the load which results in concentrated pressure peaks on both surfaces as they try to penetrate them. The high-pressure peaks may well be the origin of crack nucleation in the coating (De Wit *et al.*, 1998; Tricoteaux *et al.*, 2003).

The presence of hard particles between hard surfaces may in some cases even reduce the coefficient of friction. If the particles are fairly round in shape, hard enough to carry the load, and at least one of the surfaces is smooth, the particles may act as rollers and reduce the friction (Blomberg, 1993; Fu *et al.*, 1998). The debris in the contact may also change in its properties during the sliding action due to mechanical hardening or chemical reaction (De Wit *et al.*, 1998; Huq and Celis, 1999).

The sliding process will have an influence on the material properties of the counterface surface material. Because of work hardening, phase transitions or third-body formation, the microhardness of a steel wear surface can be about three times larger than the initial bulk hardness. In sliding abrasive contacts with uncoated steel surfaces most of the wear particles are often smaller than 1 μm in size (Kato, 1992).

7.3.3 Micromechanical Mechanisms

The origin of the friction and wear phenomena that we observe on the macro-level is found in the mechanisms that take place at the micro-level. The integration of all the micromechanical mechanisms results in the macromechanical mechanisms discussed above.

The micromechanical tribological mechanisms consider the stress and strain at an asperity level, the crack generation and propagation, material liberation and single-particle formation. In typical engineering contacts, these phenomena are at the size level of about 1 μm or less.

Shear and fracture are two basic mechanisms for the first nucleation of a crack and for its propagation until it results in material liberation and formation of a wear scar and a wear particle (Figure 7.2(c)). These mechanisms have been discussed by, for example, Argon (1980), Suh (1986), Kato *et al.* (1991) and Gerde and Marder (2001), but still today there is only a poor understanding of these quite fundamental phenomena. Another approach is to study the tribological micromechanical mechanisms by using the velocity accommodation concept developed by Berthier *et al.* (1989) and Singer *et al.* (2003). This has been extended to an energy accommodation approach to the micromechanical tribological mechanisms by Holmberg and Matthews (2001). A third and very promising approach is to model the stresses and strains in a contact and to simulate the different parameter interactions in the contact. This is described in more detail in the following text.

7.3.4 Modelling Stresses and Strains in a Coated Microcontact

The basic micromechanical mechanisms can be studied by using the geometry of a small spherical tip sliding on a flat surface. When loaded a stress field is distributed in and around the contact. The material will react to the stresses either elastically, plastically or by fracture. The elastic and plastic responses have been modelled (Djabella and Arnell, 1992, 1993; Mao *et al.*, 1995; Zheng and Ramalingam, 1995) in coated surfaces but the fracture behaviour has been more difficult to approach by modelling. In the literature, the fracture behaviour has been described mainly by crack pattern descriptions and analysis (Buckley, 1981; Je *et al.*, 1986; Hedenqvist *et al.*, 1990; Hills *et al.*, 1990; Bull, 1997). Advanced two-dimensional fracture calculation analyses have been published by Beuth (1992), Diao *et al.* (1994), Oliveira and Bower (1996), Nastasi *et al.* (1999), Malzbender and de With (1999, 2000a,b,c), Bouzakis *et al.* (2003) and Ye and Komvopoulos (2003a,b).

A comprehensive three-dimensional finite element method model for the stress and strain distribution in the contact of a rigid sphere sliding on a thin hard coating on a softer substrate has recently been published by Holmberg *et al.* (2003). The model is elasto-plastic and takes into account strain hardening effects and the fracture behaviour. The fracture toughness of the coating can be calculated on the basis of this model.

The first simulations were carried out with the scratch test geometry, that is a 200-μm-radius diamond tip sliding with increasing load from 5 to 50 N on a 2-μm-thick TiN coating on a steel substrate. The contact situation and the stress distribution results are shown in Figure 7.5.

The model illustrates in an elegant way the very complicated stress fields and material behaviour, elastic, plastic and fracture, in the coating/substrate system. This is thus a tool with great potential for surface design and parameter optimization. The simulations show the compressional stresses below the tip, the tensile stresses behind the tip originating from the friction by shear and ploughing, and the compressional pile-up stresses growing in front of the sliding tip. The stress field is at an early stage star shaped with four arms and changes to form a stress concentration at the border of the scratch groove in the tail part which is the location for the first crack to appear in empirical comparison. The load-carrying capacity of the coating/substrate system can be observed by considerable material deflection. The fracture toughness of the TiN coating was calculated to be $K_c = 7$ MPa m$^{0.5}$ (Holmberg *et al.*, 2003).

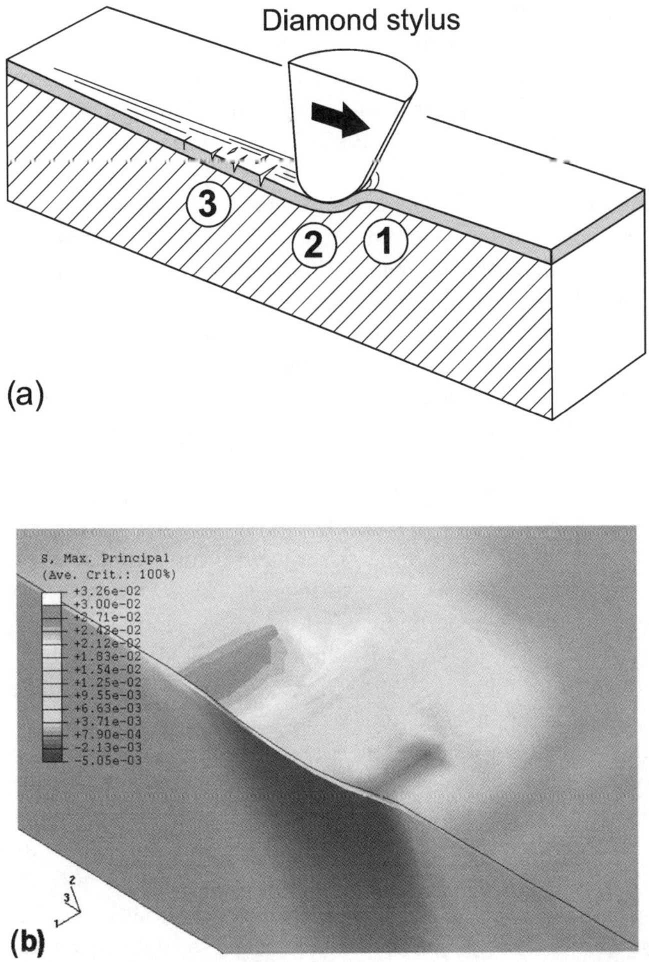

Figure 7.5 (a) The contact geometry and (b) the stress distribution when a diamond tip is sliding over a 2-μm TiN coating on steel at a load of 15 N after a sliding distance of 2.3 mm. The sliding tip is invisible in the figure

7.3.5 Tribochemical Mechanisms

The chemical reactions taking place at the surfaces during sliding contact, and also during the periods between repeated contacts, change the composition of the outermost surface layer and its mechanical properties. This has a considerable influence on both friction and wear because they are determined to a great extent by the properties of the surface, where phenomena such as shear, cracking and asperity ploughing take place (Gee and Jennett, 1995). The chemical reactions on the surfaces are strongly influenced by the high local pressures and the flash temperatures, which can be over 1000°C, at spots where asperities collide.

Low coefficients of friction, down to 0.1, have been reported for a hard titanium nitride coating sliding against itself (Mäkelä and Valli, 1985) and even lower values, down to about 0.01 but more typically 0.05, have been measured for DLC coatings sliding against different counterface materials (Donnet, 1995; Holmberg *et al.*, 1994; Erdemir *et al.*, 2000; Voevodin and Zabinski, 2000; Erdemir and Donnet, 2001; Teer, 2001; Andersson *et al.*, 2003b; Bandorf *et al.*, 2003; Singer *et al.*, 2003) and diamond coatings sliding against diamond and ceramics (Hayward *et al.*, 1992; Gardos, 1994; Habig, 1995; Erdemir *et al.*, 1996b). This can be explained by chemically or physically formed low shear microfilms on the hard coating or perhaps only on the asperity tips of the coating.

Thus, if we consider such a contact on a microscale, there is effectively a soft, easy-to-shear coating on a hard substrate (Figure 7.3(b)), although now the coating (e.g. TiN or diamond) plays the role of the hard substrate and the soft microfilm formed plays the role of a coating. It is obviously advantageous if the substrate under the hard coating is as hard as possible, to avoid fracture of the brittle coating by deformation, to improve the load support and to decrease the real area of contact. The very low coefficients of friction of polished diamond and diamond-like coatings are further explained by the extreme smoothness of the surface, excluding effects such as interlocking and asperity ploughing as well as of the hard coating reducing the ploughing component of friction (Erdemir, 2002).

In a tribological contact, the sliding process brings energy and often high local temperatures into the contacts and at the same time the wear process results in exposure of pure uncontaminated material surfaces to the environment. This situation is very favourable for chemical reactions to take place on the newly formed or deformed surfaces. The chemical reactions are dependent on what kind of additional material is brought to the contact in gaseous or liquid form (Ruff *et al.*, 1995; Singer *et al.*, 1996b; Ronkainen *et al.*, 1998a; Andersson *et al.*, 2003a,b; Eglin *et al.*, 2003).

In environments containing oxygen, such as air, a thin (about 1–10 nm thick) oxide layer is formed very quickly on most metal surfaces. Some oxide layers, like copper oxide, are sheared more easily than a metal, while others, such as aluminium oxide, form a very hard layer. Erdemir *et al.* (1998) have studied the shear properties and lubricity of a number of oxides for high-temperature tribosystems. They conclude that a complex system, including the kinetics of oxidation and cation diffusion rates, heats of formation, electrostatic electronegativity, surface energy and other fundamental crystallochemical properties, may influence the adhesion and shear rheology of an oxide film forming on a sliding surface. Mechanistically, the crystallochemical properties relate strongly to the melting point of an oxide and its viscosity, the lowering of the surface energy and melting point of an oxide when a second oxide is present in the system, and the solubility, chemical interactions and compound-forming tendencies between two or more oxides. Tribologically, these phenomena can play a significant role in shear rheology, adhesive interactions, and hence the frictional properties in a sliding contact.

There are a number of reports where different aspects of oxide layer formation on tribological properties have been studied. In contacts with alumina (Al_2O_3) surfaces, Gee and Jennett (1995) found that the tribologically formed hydroxide films are much softer than the remaining alumina and that the films are liable to fracture under concentrated loading conditions. The hydroxide films were composed of small particles, about 20–50 nm in size. Erdemir *et al.* (1995, 1997) investigated the formation and self-lubricating mechanism of thin boron oxide (B_2O_3) and boric acid (H_2BO_3) films on the surfaces of boron oxide substrates. They developed a short-duration annealing procedure resulting in the formation of a glass-like boron oxide layer for which they measured very low coefficients of friction down to 0.05

when sliding against a steel ball. The tendency for formation of an oxide layer on the surface of different (Ti, Al, Zr, Si)N coatings has been reported by Rebouta *et al.* (1995).

Experiments with steel sliding against a chromium nitride (CrN) coating resulted in the formation of a chromium oxide (Cr_2O_3) surface film with very good wear resistance (Lin *et al.*, 1996b). Increases in the applied load or the sliding velocity helped to form a thicker Cr_2O_3 film, thus reducing the wear. However, the growth of a thick TiO_2 surface film in a TiN-coated contact under the same conditions did not help to reduce the wear loss.

7.3.6 Nanoscale Mechanisms

Emerging technologies such as atomic force microscopy and other surface analysis methods (Israelachvili and Tabor, 1972; Bhushan, 1999) have opened the possibility to study friction and wear phenomena on a molecular scale and to measure frictional forces between contacting molecules at the nano-Newton level. Increased computational power has made it possible to study friction and associated phenomena by molecular dynamic simulations of sliding surfaces and to investigate the atomic scale contact mechanisms, as shown in Figure 7.2(e) (Landman *et al.*, 1992; Zhang and Tanaka, 1997; Gao *et al.*, 2003).

Only some aspects of these complex nanophysical phenomena have so far been investigated such as the friction that arises from slippage between solid to solid interfaces (Thompson and Robbins, 1989) and between closely packed films in sliding contact (McClelland and Glosli, 1992). The atomic scale mechanisms of friction when two hydrogen-terminated diamond surfaces are in sliding contact have been studied and the dependence of the coefficient of friction on load, crystallographic sliding direction, roughness and methane molecules as third bodies in the contact has been investigated (Harrison *et al.*, 1992, 1993, 1995, 1998; Perry and Harrison, 1996). Work on molecular scale viscoelastic effects and viscous flow has been reported (Wahl and Unertl, 1998; Zhang and Tanaka, 1998).

Diamond and DLC surfaces have been of major interest in nanoscale studies because of their structure, inertness and low friction and wear properties. Gao *et al.* (2003) showed that the three-dimensional structure of an amorphous DLC film is paramount in determining its mechanical properties. Particular orientations of sp^2 ring-like structures create films with both a high sp^2 content and large elastic constants. Films with graphite-like top layers parallel to the substrate have lower elastic constants than films with large amounts of sp^3-hybridized carbon. Several of the most recent studies of DLC coatings indicate that the general explanation for their extremely low coefficient of friction, down to $\mu = 0.001$ and below (Erdemir *et al.*, 2000; Erdemir, 2002), which has been measured is related to three effects. The surfaces are extremely smooth so asperity interlocking effects are eliminated, the surfaces are hard so ploughing effects are eliminated and the dangling bonds of the carbon structure are attached to hydrogen atoms, creating inert hydrophobic surfaces that exhibit almost no frictional resistance (Erdemir, 2002; Harrison *et al.*, 1995). Actually, no internal material shear is taking place at all, the hydrogen planes just 'fly' over each other.

The improved understanding of the origin of friction at the atomic scale and even why friction exists has resulted in an examination of the relationship between the commonly used laws of friction at a macroscale and the molecular frictional behaviour on a nanoscale. There have been suggestions that friction arises from atomic lattice vibrations occurring when atoms close to one surface are set into motion by the sliding action of atoms in the opposing surface (Celis, 1987). Thus, some of the mechanical energy needed to slide one surface over the other would be converted to sound energy, which is then eventually transformed into heat (Krim, 1996, 2002a,b; Robbins and Krim, 1998).

In an interesting investigation, Tervo (1998) found that friction correlates to some extent with the Rayleigh surface waves. This would suggest that friction arises mainly from the elastic interactions, i.e. lattice vibrations, on the surfaces due to sliding motion. The velocity of a Rayleigh surface wave is dependent on Poisson's ratio, shear modulus and density. Tervo found that nitrogen alloying of stainless steel slightly increases the friction.

In a molecular level study of Amonton's law, Berman *et al.* (1998) found that the projected area is not necessarily proportional to the load and that the shear strength is not constant. Despite this, Amonton's laws are obeyed and the friction force is still proportional to load on a macro-level. They offer a physical model, based on intermolecular forces and thermodynamic considerations, which explains why the friction force is proportional to the applied load, and why the case of adhering surfaces – where the friction force is found to be proportional to the molecular contact area – is quite different from that of non-adhering surfaces.

Today, we are only at the very beginning of the understanding of the nanomechanical tribological contact effects that explain the origin of friction and wear and there is no doubt that in the near future many new theories and explanations for the origin of tribological phenomena will become available. The scaling up of the nanomechanical explanations of contact mechanisms to practically useful conclusions on a macroscale is a most challenging and complex task and will take many years. Already there are practical applications on a nanoscale where the increasing knowledge of tribological nanomechanisms can be used. This has resulted in the development of micro-electro mechanical systems (MEMS) such as motors, transducers, gears and bearings of sizes in the micrometre range (Bhushan, 1998). For these extremely small components, silicon has been used in the early applications for production reasons but studies have shown that tribological improvements can be achieved by using polycrystalline diamond or MoS_2 thin coatings or hydrogenated DLC coatings (Donnet *et al.*, 1995; Beerschwinger *et al.*, 1995; Gardos, 1996).

7.3.7 Debris Generation and Transfer Layers

7.3.7.1 Crack Generation and Debris Formation

The process of wear and friction results in both geometrical and structural changes in the surfaces of the contacting bodies. These changes will influence future contact conditions and friction and wear generated in the same contact. The changes range from nanoscale to macroscale. At the nanoscale outer surface molecules are released and they react chemically with adjacent molecules, at the microscale cracks are initiated and debris released and at the macroscale wear products are agglomerated and surface layers formed and deformed.

The initiation of cracks at the surface or in the material is the starting point of a process that may result in material detachment, debris generation and the formation of transfer layers. In most wear situations, in addition to hardness, the ability to elastically recover from deformation and the fracture toughness of the material are very important parameters (Zum Gahr, 1998; Wiklund, 1999; Holmberg, 2000). The modern fracture mechanics approach to the initiation and generation of cracks is presented by Anderson (1995). For the case of a loaded very thin coated surface, a model for crack initiation or generation has been presented by Holmberg *et al.* (2003) and is described in Section 7.3.4. A review of the basic mechanisms related to crack initiation and propagation is presented by Holmberg and Matthews (2001).

Debris that have been generated by the wear process or loose particles originated from the surrounding environment may be present in a tribological contact. It has been observed that the coefficient of friction rises significantly once wear particles are formed at the sliding surface and particles present in the contact affect the instantaneous coefficient of friction and the wear (Wang and Kato, 2003a,b). The coefficient of friction can be altered by removing wear particles or by inserting particles in the interface as shown by Hwang *et al.* (1999). They also found that the particle size influences friction but not so much the number of particles present in the contact. In experiments with sliding surfaces of materials with different hardnesses (Pb, Zn, Al, Cu, Ni, Ti and AISI 1045 steel), they found a clear difference between soft and hard surfaces. Soft and ductile surfaces produced larger wear particles with a stronger tendency to agglomerate, while hard surfaces produced smaller wear particles with weaker agglomeration tendency.

For coated surfaces, the influence of debris in the contact on friction and wear is in some conditions considerable, depending on the particle size and shape, coating thickness and surface roughness relationship, and the particle, coating and substrate hardness relationship (Figures 7.3 and 7.4).

7.3.7.2 Transfer Layers

Particles that have been liberated from a surface by wear may still have an influence on the future friction and wear behaviour of the contact. In sliding contacts with materials of different hardness (lead, zinc, aluminium, copper, nickel, titanium and steel), Hwang *et al.* (1999) observed different wear particle agglomeration behaviour depending on hardness and sliding direction. Smaller wear particles had a stronger tendency to join together and form larger ones, and particles of soft and ductile metals had a stronger tendency to agglomerate than those of hard and brittle metals. There was a difference in agglomeration behaviour depending on whether the sliding was unidirectional or reciprocating, and lower friction and wear were measured in the cases of reciprocating sliding. Particle agglomeration was not just limited to one location but occurred simultaneously over a distributed area and the observed particle or flake sizes were in the range of 100–600 μm. The mechanism of agglomeration is complicated and probably due to adhesion and/or mechanical interlocking. The agglomerated particles can act as larger particles in the contact, attach to either of the surfaces or be rejected from the contact (Yuan and Hayashi, 1999).

On the basis of observations from sliding contacts with MoS_2 and TiN coatings, Singer (1998) presents a three-stage nanoscale model for the formation of reaction layers and for the build-up of transfer layers. First a thin layer is removed from the coating and transferred to the counterface. Meanwhile, both the surface and transfer layers can react with possible surrounding gases, forming new compounds. The first films to be transferred to the countersurface may be very thin, only of molecular dimensions. As the transfer film thickens, it is extruded from the contact area and may break up to form new wear debris. This process is then repeated and a layer-by-layer build-up takes place. It has been observed for MoS_2 coatings that particle detachment appears to be preceded by deformation, reorientation and sometimes crack propagation within the first 20–50 nm of the coating.

It is important to note that even though the transfer films originate from one of the sliding surfaces they do not always have the same phase or composition as the parent material. Examples of this are given in the excellent transfer layer study by Singer *et al.* (2003).

Amorphous Pb–Mo–S transforms to crystalline MoS_2 layers and Al_2O_3 (sapphire) sliding against TiN and TiC has been observed to result in the expected TiO_2 (rutile) phase as steel counterface produced ternary phases. There is very little knowledge on how the transfer films attach to the surfaces but both mechanical trapping and chemical bonding have been suggested. Similarly, the reason why some surfaces form reservoirs or platforms of wear debris while others eject wear debris is not understood.

Direct evidence of third body control of friction and wear have been shown by Singer *et al.* (2003) for three kinds of coated surfaces; amorphous Pb–Mo–S that formed a MoS_2 transfer film, DLC that formed a graphite-like carbon film and annealed boron carbide forming a mix of H_3BO_3 and carbon or carbon alone. They showed that for a MoS_2 film the shear took place as interfacial sliding and not, as earlier suggested, as intergranular shear of the coating material or its debris. A direct correlation between the coefficient of friction and the area in the contact covered by transfer film in a sapphire sliding against a boron carbide coating has been shown by Dvorak *et al.* (2002). This correlation can be expressed in terms of the rule of mixtures.

The material transfer mechanism (Figure 7.2(b)) is well known for polymers, e.g. polytetrafluoroethylene (PTFE), sliding on steel. Surface material from the polymer will wear off and attach by adhesion to the steel counterface to form an extensive PTFE film. This means that after some time of sliding, the tribological pair is actually PTFE sliding against a thin PTFE film on steel, which has very low friction. Similar mechanisms have been observed for sputtered, less than 1 μm thick PTFE coatings on steel substrates, resulting in film-like PTFE wear debris transfer to the steel counterface, and for sputtered polyimide (PI)-coated steel, resulting in fine flake-like wear debris transfer to form a polymer layer on the steel counterface (Yamada *et al.*, 1990).

A typical building-up process of a transferred surface layer in contacts with steel and hard coatings such as TiN, CrN and (TiAl)N has been described by Huang *et al.* (1994). As a result of the ploughing action of hard coating asperities, slider material fragments were first removed and then adhered to some preferential sites on the sliding track of the coating. The preferential sites were the highest asperities that made earliest contact with the counter-surface. Repeated sliding resulted in accumulation of fragments, which then united and formed discontinuous layers on the coating surface. After some sliding, the highest asperities were covered with transferred material, which was deformed and flattened under continuous sliding and a transferred layer was built up. Similar processes of transfer layer build-up have been observed and reported for different ceramic and steel contacts, e.g. by Andersson and Holmberg (1994).

Depending on the contact condition, the formation of transfer films in the sliding contact of a titanium nitride coating against a steel slider may be very complex. Sue and Troue (1990) have described the process of minute wear fragments adhering to both surfaces, plastic deformation and strain hardening of the transferred layers and patches, cracking and oxidation of the layers, removal of the layers and patches by fracture and finally the formation of very thin films, possibly oxides, on both surfaces. In pin-on-disc experiments with a steel ball sliding on TiN coatings, Wilson and Alpas (1998) observed a load effect. At low loads of about 20 N, transfer and build-up of oxidized counterface material on the coating surface and minimal damage to the coating took place, and at higher loads of up to 100 N increased debris transfer, polishing and brittle spallation occurred, but at the higher loads plastic deformation and ductile ploughing or smearing of the TiN coating prevailed.

Badisch *et al.* (2003) have found that for plasma-assisted CVD TiN films, the presence of above about 3.2% chlorine in the coating can modify considerably the wear mechanism and the transfer layer formation, resulting in significant reductions in the friction coefficient in dry sliding against ball-bearing steel in the normal laboratory atmosphere, though this benefit was not observed in dry nitrogen conditions.

The material transfer layers generated in sliding coated contacts are generally a few micrometre in thickness but may vary in the range of 0.01–50 μm (Ko *et al.*, 1992; Blomberg, 1993; Scharf and Singer, 2003a,b,c).

7.4 Contact Types

This chapter has so far mainly considered the basic tribological contact, one surface sliding over another, and the tribological characteristics related thereto. However, there are also other surface contacting conditions where some of the contact parameters otherwise considered as not important may have a completely dominating influence on friction and wear. These are *abrasion*, where a hard sharp counterface ploughs in the surface; *impact*, where the load is momentary and mainly perpendicular to the surface; *fatigue*, where a low load is continuously repeated; *fretting*, where the movement is reciprocating with a small amplitude; *chemical dissolution*, where e.g. high temperature changes material behaviour at the surface; and *lubricated sliding*, when an external fluid is introduced into the contact as shown in Figure 7.6 (Franklin, 1993; Matthews *et al.*, 1998). The dominating contact features, the required surface properties and some ideas about suitable coating choices for those contact types are listed in Table 7.1.

7.4.1 *Sliding*

Sliding is the most common tribological contact condition. The tribological mechanisms involved are fairly well known and are largely described, with different coating examples, in the earlier text. In case of conformal surfaces, e.g. flat sliding on flat such as in many seal solutions, the contact is characterized by low pressure, but high contact temperature is generated and removal of wear particles is not so easy and may cause high wear and friction. The aim of the coating is then to minimize wear and create a long lifetime and robust contact

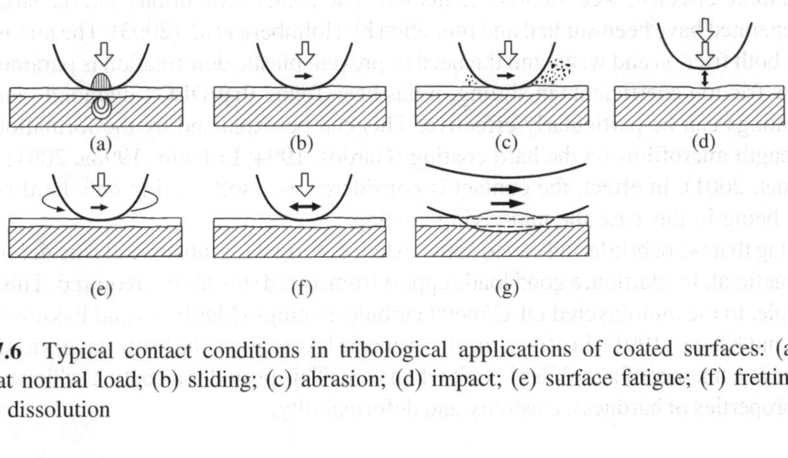

Figure 7.6 Typical contact conditions in tribological applications of coated surfaces: (a) contact stresses at normal load; (b) sliding; (c) abrasion; (d) impact; (e) surface fatigue; (f) fretting and (g) chemical dissolution

Table 7.1 Guideline for coating selection

Contact type	Dominating contact conditions	Required surface properties	Recommended coatings
Sliding	Low friction sliding Mild wear	Low shear strength at surface top layer Good load support	DLC, MoS_2, diamond
Abrasion	(a) Third-particle indentations (b) Two-body ploughing	(a) Good-microtoughness and load support (b) High hardness to resist plastic deformation	TiN, TiAlN, TiC, Al_2O_3, CrN
Impact	Concentrated impacting stress waves Abrasive wear	Good macrotoughness Good elasticity	Multilayer
Fatigue	Continuous large stress waves	Good macrotoughness Good load support	TiN, DLC Multilayer
Fretting	High-frequency large stress waves Wear debris in contact continuously	Good elasticity Low shear strength surface layer Not producing hard wear debris	MoS_2, Cu–Ni–In Multilayer
Chemical dissolution	High temperature	Non-soluble Thermally conductive	TiN, TiAlN, TiC, WC, CrAlN, DLC, diamond
Lubricated sliding	Coating giving load support for lubricant film and acting as emergency layer	Interaction with lubricant additives Texturing to support lubricant prevalence	DLC, TiN, TiC, CrN

situation. This is often related to the build-up of suitable transfer layers on the surface that allow the shear from sliding to take place with insignificant wear (Section 7.3.7.2).

In counter formal contacts, such as shown in Figure 7.6(b), the contact conditions are characterized by high contact pressure, continuous temperature generation only on the smaller body and more effective wear debris evacuation. The contact conditions and the stresses and strains generated have been studied and modelled by Holmberg *et al.* (2003). The aim is usually to reduce both friction and wear, and the need to prevent plastic deformation is paramount.

For low friction coefficients in sliding, it has been found that DLC-, diamond- and MoS_2-based coatings can be particularly effective. This can be explained by the formation of low shear strength microfilms on the hard coating (Gardos, 1994; Erdemir, 1996a, 2001; Erdemir and Donnet, 2001). In effect, the contact is considered as a soft coating on a hard substrate, the latter being in this case the underlying coating.

A coating that is not brittle and can to some extent deform with substrate elastic deformations can be beneficial. In addition, a good load support from a hard substrate is required. This applies, for example, to the multilayered DLC/metal carbide coatings (Matthews and Eskildsen, 1994; Dimigen and Klages, 1991). In effect, the more elastic layers allow the brittle layers to slide over each other in a manner of a multileaf book when bent. This provides a coating which combines optimal properties of hardness, elasticity and deformability.

7.4.2 *Abrasion*

Contact conditions involving abrasion have long been recognized as being some of the most significant in terms of the amount of material loss (Figure 7.6(c)). The theoretical understanding of abrasion, and related phenomena, such as erosion, has advanced considerably (Hutchings, 1992; Adachi and Hutchings, 2003; Gee *et al.*, 2003). In abrasion, a hard counterbody moves against and partly ploughs through a softer material and thus the hardness of the softer material is of paramount importance. The aim of the surface material choice is to inhibit large plastic deformation, such as grooves, on the surface. Good results for this have been achieved by applying thick hard coatings (Cartier, 2003).

The performance of advanced thin coatings, such as PVD TiN, has been less remarkable under abrasive conditions than under sliding conditions. This is partly due to the thickness limitations of hard PVD and CVD coatings, their lack of toughness and the need for effective load support, especially under three-body abrasion with sharp and hard abrasive particles. The load support requirement, to prevent adverse macromechanical mechanisms, can be fulfilled by utilizing interlayers such as electroless nickel (Matthews and Leyland, 1995).

It has been found that by multilayering Ti and TiN films it is possible to produce a composite coating which is both hard and tough, thus controlling micromechanical mechanisms and performing well in erosive and abrasive conditions (Leyland and Matthews, 1994). A coating with similar abrasion resistance can be formed by depositing stainless steel under nitrogen plasma conditions and then post-coat plasma nitriding the coating (Raehle, 1988; Matthews *et al.*, 1995; Matthews, 1997).

7.4.3 *Impact*

In impact contact situations (Figure 7.6(d)), a coated surface must possess a high toughness in order to absorb the impacts that result in considerable instantaneous and repeated stress fields. The coating must also be sufficiently elastic to be able to accommodate any substrate deformation that may occur under impact. Recent work has shown that multilayered coatings that alternate hard inelastic and soft elastic layers can provide excellent performance in impact conditions (Voevodin *et al.*, 1995a; Iwai *et al.*, 2001), i.e. they prevent the occurrence of micromechanical mechanisms such as microscratches and crack growth. Erosion wear is a combination of impact and abrasive wear.

7.4.4 *Surface Fatigue*

In a surface fatigue contact the surface is repeatedly loaded, as shown in Figure 7.6(e). This results in deterioration of the surface strength and crack initiation and propagation at or below the surface. The aim of a surface coating is mainly to distribute and decrease the surface stresses and to inhibit crack propagation. This condition occurs, for example, in ball and roller bearings and in gears (Ding *et al.*, 1996; Murakami *et al.*, 1997).

The fatigue life of a thin coating may be considerably longer than that of a thick coating for different reasons. Under similar deformation conditions, the thicker coating will experience higher bend stress levels. Since the coating typically has columnar growth morphologies, any crack normal to the surface will be large in a thick coating, and may exceed the critical crack length, whereas in a thin coating this may not be the case. It has been shown experimentally that in rolling contact fatigue tests hard TiN coatings with a film thickness well below 1 μm have up to two orders of magnitude longer lifetime than 2–3 μm thick similar coatings (Erdemir and Hochman, 1988; Chang *et al.*, 1990, 1992; Erdemir, 1992).

On the other hand, Polonsky *et al.* (1997) claim that they have been able to demonstrate that the rolling contact fatigue life increase achieved with less than 1-μm TiN coatings was entirely due to polishing of the steel loading balls by the significantly harder TiN coating. They changed the test procedure and eliminated the polishing effect and observed a somewhat negative effect of the TiN coating on the rolling contact fatigue life. In their article, they present a theoretical model of near-surface rolling contact fatigue initiation in coated rolling contact elements. They conclude that a truly effective hard coating that can resist near-surface rolling contact fatigue must be relatively thick (>3 μm), adherent to the substrate and have a fine microstructure to resist cohesive fatigue failure in the coating.

The presence of interfaces interrupting the columnar structure of PVD TiN coatings causes the cracks propagating from the surface to form kinks, which can significantly slow down crack propagation. Polonsky and Keer (2002) show by simulation that high-endurance coatings that are required in rolling contact fatigue can be produced by a multilayer alternating 250-nm TiN layers with much thinner interlayer coatings of another material.

Some laboratory rolling contact fatigue tests have been carried out with DLC coatings and the results seem to indicate that also a hard DLC coating on steel can increase the fatigue life. Metal mixed Me-C:H coatings on rollers in roller bearings considerably increased the bearing life and significantly reduced the particles generated in the contact (Olsson *et al.*, 2000; Sjöström and Wicksröm, 2001). Rosado *et al.* (1997) found no difference between fatigue life for DLC-coated and uncoated M50 steel specimen in rolling contact fatigue tests while the coating significantly improved the fatigue life even at a temperature of 177°C.

7.4.5 Fretting

Fretting is a special case of fatigue wear at the surface, where the distance of reciprocating sliding is typically smaller than the contact length, often less than 1 mm (Figure 7.6(f)). In the contact, the reciprocating friction load produces surface stresses that can result in cracks and fretting fatigue. Fretting wear appears when the cracks at the surface result in wear particles. Then the released wear products stay for some period in the reciprocating contact and influence the contact conditions crucially, e.g. concentrating the surface load due to the released wear particles and increasing the surface stresses under them. On steel surfaces, the contact process wears off the oxide layers on the surface, which is exposed to chemical reactions. Often the temperature is simultaneously increased, which speeds up chemical reactions (Waterhouse, 1981; Vingsbo and Söderberg, 1988; Vincent *et al.*, 1992; Fouvry *et al.*, 1996, 2003; Fouvry and Kapsa, 2001).

The aim of coatings in fretting contacts can be to increase surface elasticity and/or decrease the friction and thus reduce surface stresses, to increase the surface toughness and reduce the crack initiation and propagation or to increase the surface inertness and thus reduce the unfavourable chemical reactions. It is possible in some cases to build up favourable transfer of reaction layers on the surfaces that decrease the shear and friction, as discussed in Section 7.3.7.2.

Fretting wear is often most efficiently reduced by effective macromechanical design measures, e.g. to control the displacements and stresses induced in the contact. However, it is also possible to modify the frictional conditions to reduce friction, e.g. by means of a solid lubricant coating such as MoS_2 applied by PVD, or a soft electroplated metal. Thermomechanical treatments can also be beneficial, especially if compressive residual stresses are induced. Copper-based thick coatings can be beneficial in fretting wear contacts (Fridrici *et al.*, 2003).

7.4.6 Chemical Dissolution

A typical chemical dissolution dominated tribochemical application is workpiece material moving over a cutting tool, as shown in Figure 7.6(g). The aim of a coating on the cutting tool surface is to resist the high-temperature conditions, to reduce friction and thus heat generation, to inhibit dissolution of the surface material and to reduce scratches produced by hard elements in the workpiece material (Holmberg and Matthews, 1994). To be successful, the coating material should exhibit the following characteristics: low adhesion to the workpiece material but high adhesion to the tool material, good abrasive resistance, high chemical and thermal stability and high toughness. Thin hard coatings of nitrides and carbides match these requirements well and have been used successfully.

In metal-cutting applications, the resistance of TiN films to oxidation has been improved by employing TiAlN composites (Ronkainen *et al.*,1991, 1992; Bouzakis *et al.*, 2003) and by advanced multiphase ceramics incorporating yttrium, which helps to form a stable oxide on the surface (Luo *et al.*, 2001). Coatings in cutting applications are discussed in Section 7.6.

However, other contacts where chemical dissolution and reaction are involved are those subject to high-temperature diffusion and oxidation conditions. The early performance of TiN films was not regarded as outstanding in any of these regards, as e.g. pin-hole defects made such coatings vulnerable to aqueous corrosion. This situation was considerably improved by utilizing interlayer films such as electroless nickel (Leyland *et al.*, 1993; Bin-Sudin *et al.*, 1996). Another promising coating in this regard is a carbon-doped tungsten film which is both dense and hard (Rebholz *et al.*, 1998a).

7.4.7 Lubricated

Most of the current research on surface coatings deals with their performance in dry conditions, some even in vacuum or in an inert gas. However, there are many applications, e.g. related to engines, transmissions, seals, etc., where the coating is expected to perform in a liquid, often in oil. Therefore, in general, the oil lubricant carries the load and separates the surfaces from each other and the shear takes place in the lubricant. The role of the coating is then to act as a safety element that comes into action when the lubricant film fails for one reason or another. Thus, the coated surface is expected to have good tribological performance in boundary lubricated conditions where surfaces are partly in direct contact with each other and surrounded by oil.

In boundary lubrication, a reaction film is formed on top of the surface and the shear takes place in this film. The film is typically formed by chemical reaction between oil additive elements such as Zn, S, Cl or P and the steel surface and is enhanced by high temperature (Holmberg and Matthews, 1994). For a coating to be effective in such conditions, it is necessary that either it has the ability to react chemically with the additives and form a strong low shear reaction film, or it has the ability to itself produce a strong and low shear surface layer that protects the contact from destruction.

Ronkainen *et al.* (1998c) showed how a reaction layer is built up on a steel ball sliding against a steel surface in lubrication oil with additives. Very similar tribological performance was achieved with the ball sliding in dry conditions, against a DLC coating and then a transfer layer was formed on the steel ball. When the steel ball was sliding against the DLC layer in oil-lubricated conditions, no visible transfer layer was formed but the tribological performance was in the same range as in the previous cases. This indicates that a DLC coating does not really improve the friction and wear in an oil-lubricated contact, but can act as a safety layer (Ronkainen, 2001).

There is a difference in the behaviour of DLC coatings in lubricated conditions depending on their structure. Ronkainen *et al.* (1998c) showed in a comparison of three kinds of DLC films that hydrogen free (a-C) and hydrogenated (a-C:H) carbon films had a low friction coefficient in dry sliding conditions (0.15–0.22), which was further decreased by 10–40% under boundary lubrication. The a-C:H(Ti) films exhibited good self-lubricated properties ($\mu = 0.10$) in dry sliding conditions and the a-C films had the lowest friction coefficient in water- (0.03) and oil-lubricated (0.08) conditions. The a-C films showed excellent wear resistance in dry, water- and oil-lubricated conditions. The performance of a-C:H films could be improved by titanium alloying. In dry sliding conditions, the tribolayer formation of the DLC films influences the friction and wear performance, but in oil-lubricated conditions, boundary layers were formed, which governed the tribological mechanisms in the contact.

The effect of humidity and water on DLC coatings was clearly shown by Andersson *et al.* (2003a,b). Water molecules destroyed the conditions for superlow lubricity of a highly hydrogenated carbon film. The friction coefficient of a hydrogen-free DLC (ta-C) film was halved from its vacuum level, $\mu = 0.6$, to a water pressure of 1 Pa (0.04% RH, relative humidity) and it decreased one order of magnitude at a water vapour of 100 Pa (4% RH).

Jacobson and Hogmark (2001) have studied DLC-coated lubricated components and conclude with the following recommendations. It is tribologically advantageous to apply DLC coating only on one of the surfaces because it improves sliding in. The substrate of the coated surface should be harder than the counter material. The substrate surface roughness should not exceed 0.1 μm. If both surfaces are coated, the recommendations are less straightforward.

There is another mechanism by which coatings can influence friction and wear in coated contacts. Ortmann *et al.* (2003) deposited structured creviced CrN coatings by PAPVD which produced micro-reservoirs on the surface that accumulated the lubricant in the top layer of the surface for critical lubrication and also acted as traps for wear debris. This improved the friction and wear performance in boundary lubrication conditions.

7.5 Advanced Coating Types

Whilst early thin tribological coatings were dominated by single-element metallic or ceramic compositions, in recent years there has been a significant move towards multiphase coatings having a range of compositions and structures (Hogmark *et al.*, 2000). In this section, we overview some of the significant developments, under the following sub-headings:

1. Hard binary compound coatings. These include the ceramic coatings already referred to, as well as new 'doped' metal and modified DLC coatings.
2. Multilayer coatings. These include so-called 'superlattice' and nanolayered coatings, as well as macrolayered structures which can arrest crack propagation.

3. Nanocomposite (nc) coatings. These offer significant opportunities to control and enhance mechanical properties and include metal/ceramic, ceramic/ceramic and ceramic/metal composites.
4. Hybrid and duplex coatings. These refer to combined coatings or treatments which can give enhanced properties to the composite substrate/coating system.

7.5.1 Hard Binary Compound Coatings

Early hard and thin tribological coatings were dominated by TiN. Now a number of coating developments are presently being pursued, which range from advanced multiphase systems to coatings based on single metal or binary compounds or, for example, chromium can be doped with nitrogen and tungsten with carbon – thereby influencing the relative hardness and elasticity. The performance has been compared to that of fully stoichiometric ceramic phases of these mixtures. This work has revealed some remarkable wear performance from coatings which might previously have been considered non-optimal in terms of hardness (Rebholz *et al.*, 1999a,b). For example, a coating with a relatively low hardness and low elastic modulus is nitrogen-doped chromium metal. In trials under impact and sliding contacts, and under three-body abrasion conditions, these coatings performed better than harder, ceramic-phase chromium nitride. This is illustrated for impact conditions in Figure 7.7(a) and (b) (Rebholz *et al.*, 1999a). It is the lack of cracking around the indentation, rather than the impact crater volume, which reveals the benefits of low nitrogen coatings in this test – in that they remain well-adhered and comparatively undamaged, despite the quite severe plastic deformation of the underlying substrate material.

Optimum Cr(N) coatings, in terms of wear resistance, indicate a hardness of about 15 GPa, and an elastic modulus of 240 GPa in a chromium metal, single-phased coating, with ~10–12 at.% N in supersaturated solid solution; i.e. a high H/E value of 0.06. Similarly, coatings comprised of tungsten doped with small amounts of carbon performed better than those with greater carbon contents in rubber wheel abrasion tests, as shown in Figure 7.8, even though their hardness was only about 20 GPa, compared to 40 GPa for the hardest coating produced

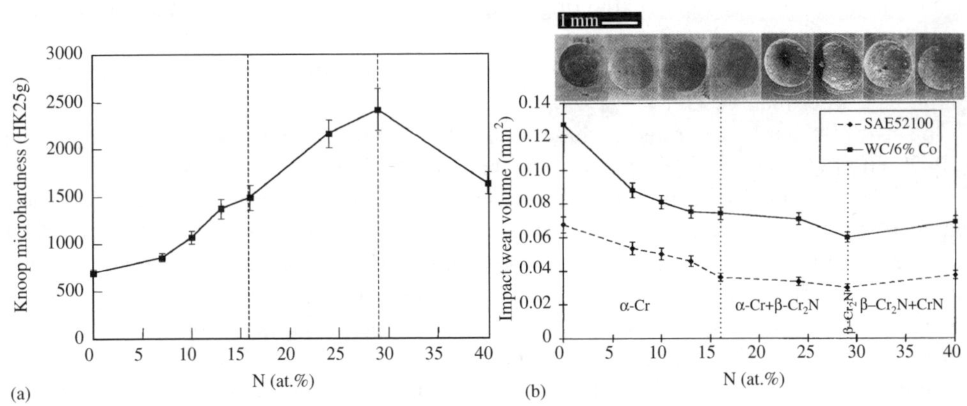

Figure 7.7 (a) Hardness levels and (b) impact wear volumes and the indentation crater appearance of chromium–nitrogen coatings

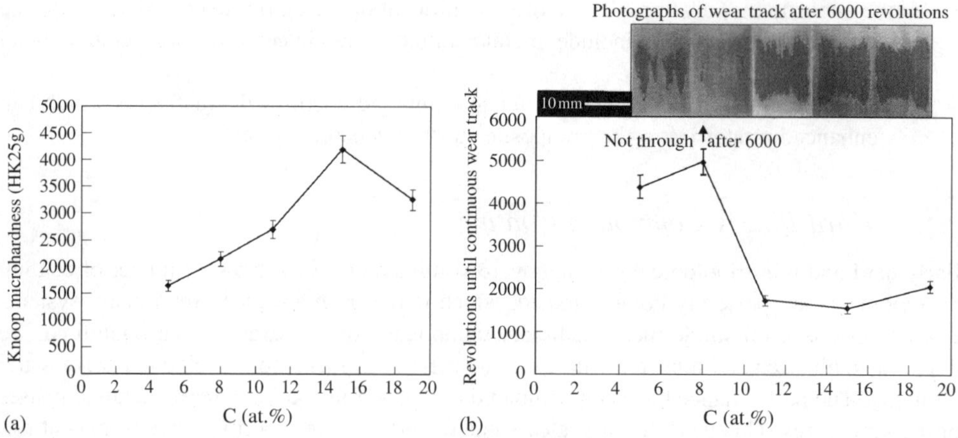

Figure 7.8 (a) Hardness levels and (b) abrasive wear wheel test results showing the wear scars, for tungsten–carbon coatings

(Rebholz *et al.*, 1998a). In this case, due to the rather high elastic modulus of tungsten, about 400 GPa, the H/E value is about 0.05. This perhaps explains why even better results can be obtained with tungsten-doped DLC films, i.e. utilizing the lower elastic modulus of the DLC material. However, in that case, it is important to ensure that the excess carbon is not present in graphitic form, since this creates a low elastic limit of strain, i.e. a low H/E (Neuville and Matthews, 1997).

7.5.2 Multilayer Coatings

Much of the scientific interest in PVD coating techniques has centred on attempts to produce 'ultra-hard' coatings, with hardnesses in excess of 70 GPa. In this regard, a distinction has to be made between coatings exhibiting 'intrinsic' hardness and those which derive their hardness from 'extrinsic' effects, such as their morphological structure and grain size. Veprek (1999) has used the generic term 'superhard', stating that this refers to materials with a hardness greater than 40 GPa.

Intrinsically hard materials include diamond (70–100 GPa), cubic boron nitride (50–70 GPa) and some other ternary compounds in the B–C–N system. Veprek states that an excellent example of a group of materials which derive extrinsic hardness from their microstructure is the heterostructures such as the superlattices. They have an artificial arrangement of epitaxial layers, with layer thicknesses in the nanometre range, and provide over a narrow range of layer thickness additional hardening effects that are not predicted by conventional metallurgical theory. Koechler (1970) reported this approach as a means of producing a strong solid. His work and that of Lehoczy (1978a,b) indicated that dislocations were unable to multiply or propagate from layer to layer in nanolayered heterostructures of binary metal systems (e.g. Al/Cu and Al/Ag).

Explanations for this were founded on strain coherency effects at layer interfaces, though other mechanisms have been presented. Barnett (1993) and co-workers have contributed significantly to the level of understanding of the behaviour of such superlattice coatings and other research groups, e.g. at Linköping University and the University of Illinois (Helmersson *et al.*, 1987), have been able to demonstrate the laboratory-based deposition of such coatings. While nanolayered, so-called superlattice coatings present great theoretical possibilities, the technicalities of coating real engineering components with well-controlled epitaxial nanolayers can present practical problems. However, there are considerable mechanical property benefits to be obtained from polycrystalline and even amorphous multilayer coatings as demonstrated, for example, by Springer and co-workers (Springer and Catlett, 1978; Springer and Harford, 1982).

One of the earliest exponents of the multilayering concept for tribological coatings was Holleck (1986). He first used multilayers of different ceramic materials, such as TiC and TiB_2, selected on the basis of their dominant bonding mechanism: metallic, covalent or ionic. Holleck demonstrated improvements in hardness, indentation toughness, adhesion and wear performance under optimized layer thicknesses (Holleck *et al.*, 1990; Holleck and Shier, 1995) and attributed these improvements, in part, to the crack deflection and stress relaxation mechanisms for the TiC/TiB_2 system. Several research groups (Dimigen and Klages, 1991; Chu *et al.*, 1992; Bewilogua and Dimigen, 1993; Donohue *et al.*, 1995, 1997; Sproul, 1996) have demonstrated that multilayer coatings in the nanometre and submicrometre layer-spacing range can potentially be extremely effective in mitigating wear in a range of metal-cutting and sliding-contact applications.

The authors and their collaborators have described how multilayer stacks with alternating high and low modulus layers can provide particular benefits, especially when substrate deformation occurs (Leyland and Matthews, 1994; Holmberg *et al.*, 1998; Matthews *et al.*, 1998, 2001). For thick, hard and brittle deposits, the bending stresses created in the coating can be high, whereas thin coatings bent to the same radius do not experience the same bending stress levels. By alternating high and low modulus, usually this means hard and soft materials, the hard layers can effectively slide over each other, and this prevents the build-up of high-bending stresses. The 'soft' layers act as shear zones to permit this sliding in a manner which is in some ways analogous to the leaves of a book if it is bent. This is illustrated in Figures 7.9 and 7.10. Coatings with such alternating properties can, in effect, combine high hardness with an ability to accommodate deformation, and have been shown to perform well in a range of wear conditions, including erosion and abrasion, for which Ti/TiN multilayers are very effective. Also, metal carbide/DLC multilayers are effective in bearing and gear applications, where substrate deformation has to be accommodated by the coating (Matthews and Eskildsen, 1994).

The bending model of hard/soft multilayer coatings does not rely on a discrete interface between the layers. Indeed, it could be argued that a gradual transition in composition and mechanical properties is to be preferred – as that is less likely to result in inter-layer separation or delamination of layer boundaries, which can occur for discrete layer systems under deformation conditions (Matthews, 1994). This is particularly true for systems using layers with very different chemical compositions. Such a system could be a MoS_2/metal or MoS_2/ceramic multilayer. In such systems, it has been desirable to intermix the metal or ceramic phase with the MoS_2 over the transition region.

The benefits in multilayering of relatively high modulus and relatively low modulus layers have been demonstrated in cyclic impact tests (Bantle and Matthews, 1995; Voevodin *et al.*, 1995, 1996). Such work has confirmed that a multilayer stack providing a transition in

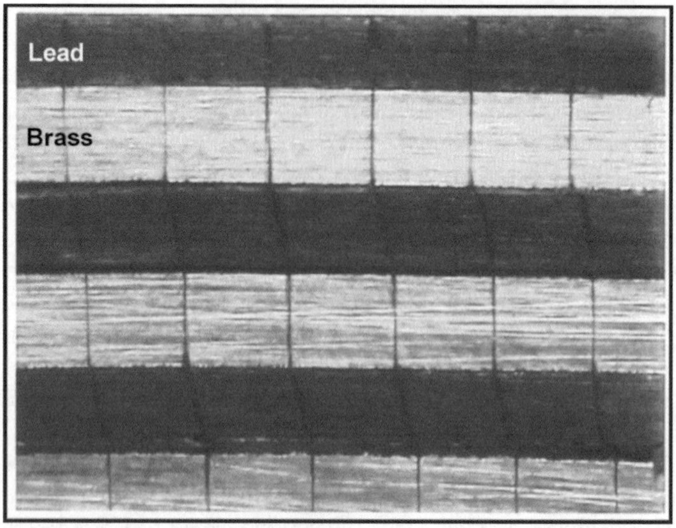

Figure 7.9 An illustration of how, in a multilayer hard–soft sandwich structure subject to bending, the shear occurs in the softer layers. The dark layer is lead and the light layer is brass

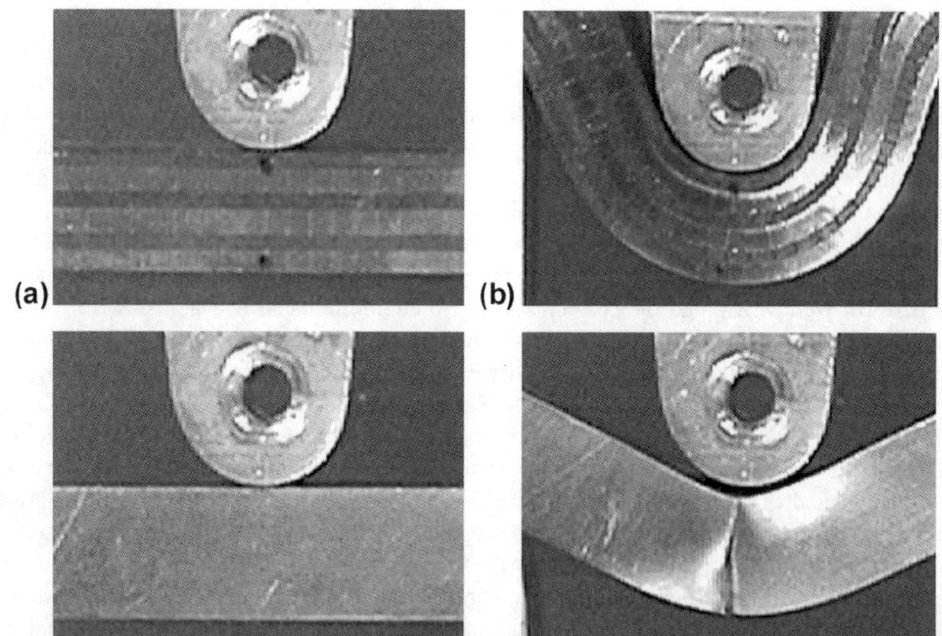

Figure 7.10 Illustrating how a multilayer structure shown in Figure 7.9 allows to bend without fracture ((a) and (b)), when a single hard brass layer ((c) and (d)) will crack

properties from the substrate to the surface can be particularly beneficial. In effect, this provides a functionally graded composition, as shown schematically in Figure 7.9.

7.5.3 Nanocomposite Coatings

Another class of materials which derives its hardness from 'extrinsic' influences is that which relies on grain-size effects. Veprek (1997, 1999), Musil *et al.* (1999b), Musil (2000), Mitterer *et al.* (1998), Gissler (1994) have been major advocates of this approach to the deposition of superhard coatings.

Musil (2000) has traced the developments in hard and superhard films, identifying the important steps which have occurred. His view is that superlattices represent a very important milestone, but the inherent problems of controlling layer thicknesses and avoiding interdiffusion of elements at high service temperatures can be avoided by using nanocomposite coatings. He discusses methods to control the size and orientation of grains, through ion bombardment and mixing processes, and describes the key differences between nanocrystalline alloy films and nanocomposite films based on nitrides of binary metal alloys. Examples of the latter are ZrCu–N (Musil *et al.*, 1999a), CrNi N (Musil and Regent, 1998) and TiNi–N (Misina *et al.*, 1998). In such films, one phase, usually metallic, segregates to the grain boundaries of the second, usually ceramic, phase and this limits grain growth, provided that the deposition conditions are appropriate. Hardnesses quoted are mostly 16–20 GPa, but some coatings achieve 50 GPa. Musil states that there are still questions concerning the thermal stability of such films.

The next group of nanocomposite films which Musil discusses are the superhard nanocomposites. He states that these can be produced not only with two hard phases, as proposed by Veprek *et al.* (1998), but also with one hard and one soft phase present such as for a nanocrystalline (nc)-ZrN/Cr ceramic/metal nanocomposite (Musil *et al.*, 1999a).

Musil thus cites two groups of superhard nanocomposite coatings: nc-MeN/nitride (e.g. TiN + Si_3N_4) and nc-MeN/metal (e.g. transition metal nitride + Cu, Ni, Y, Ag, Co, etc.). He claims that the hardness of both of these groups can be continuously varied from values of about 10 GPa up to 50–70 GPa. Indeed, following recent work by Veprek (1999), it seems that around 100 GPa can be obtained (e.g. nc-TiN/a-SiN_x) under certain conditions.

It is interesting that Musil recognizes that a low modulus is desirable in preventing plastic deformation. Using a derivation which has similar origins to the plasticity index, he cites the ratio H^3/E^2 as a parameter that controls the resistance of metals to plastic deformation. Musil provides data from the literature on the H^3/E^2 values for different coatings having a range of hardnesses. However, he does not elaborate on how the nanocomposite coating concept might be used to achieve the desirable goal of high H and low E.

Some of the earliest work on nanocomposite tribological coatings was based on the Ti–B–N and/or Ti–B–C ternary phase systems (Ronkainen *et al.*, 1991; Gissler, 1994; Mitterer *et al.*, 1999). These offer considerable potential to control the hardness and elastic properties, as demonstrated using both plasma-assisted electron beam PVD and sputter deposition (Rebholz *et al.*, 1999c,d,f). In particular, the controlled addition of aluminium to Ti–B–N coatings can give reduced E values, while maintaining a reasonable hardness. Such coatings exhibit excellent sliding wear properties, as illustrated in Figure 7.11 (Rebholz *et al.*, 1998b).

It is interesting to calculate the H/E values for the best of these coatings, i.e. approximately 30 GPa divided by 330 GPa, which gives an H/E value of about 0.09. This is better than

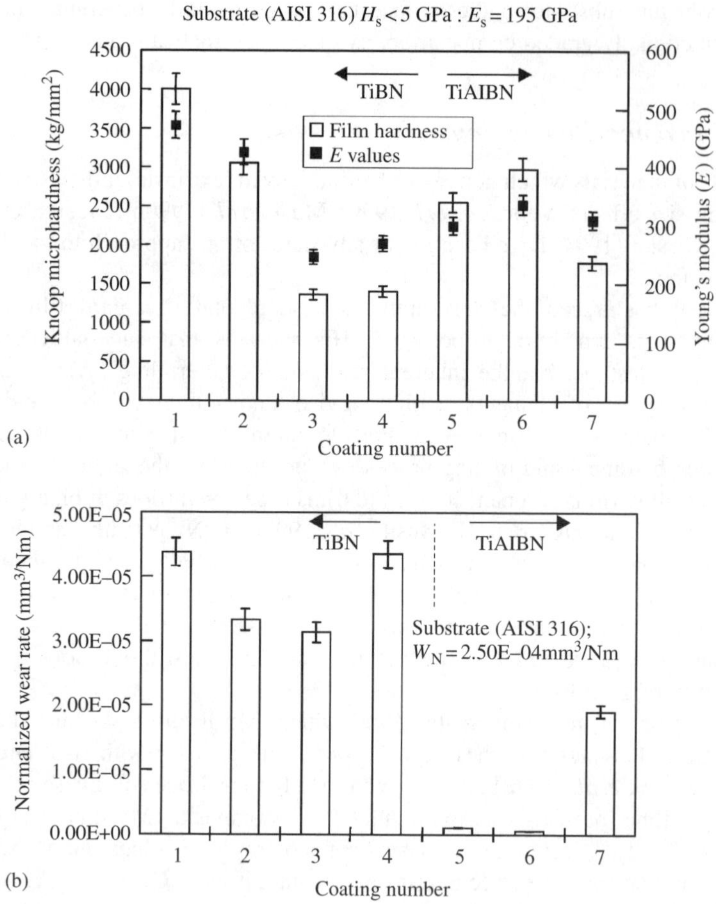

Figure 7.11 (a) Hardness and elastic modulus results and (b) pin-on-disc sliding wear rates (W_N) for TiBN and TiAlBN coatings produced by electron beam plasma-assisted PVD against cemented tungsten carbide

diamond, and earlier work, with controlled additions of titanium to adjust the metal content, has shown that the H/E value can be increased further, to 0.12 or higher (Rebholz *et al.*, 1999e). Moreover, considering the high temperature stability and hot hardness of Ti–Al–B–N, the H/E ranking is likely to be even more favourable for applications involving high-temperature environments, or situations in which heat is generated in the contact, such as metal cutting, especially in view of the enhanced oxidation resistance of this material.

The emphasis of this section so far has been on the wear properties of coatings, rather than their frictional behaviour. Usually low friction coatings do not exhibit low wear rates – since the achievement of low friction is often associated with a shear mechanism at the interface between the coating and the counterface surface. Such shearing may be accompanied by wear. However, the nanocomposite coating concept allows unique combinations of properties in a single coating.

This has been well employed by researchers at the Wright Patterson Air Force Laboratory, who have developed several tribological coatings for aerospace systems (Zabinski *et al.*, 1996; Voevodin *et al.*, 1999a). Applications for reaction wheels, gyroscopes, antenna drives, gears, pumps, etc. cover a broad range of contact stresses, from 10^7 to 10^{10} Pa, and sliding speeds from 0 to about 20 $m\,s^{-1}$. They can experience low-frequency launch vibrations and high peak loads (Voevodin *et al.*, 1999a). Furthermore, the operating environmental conditions can be extreme. Thus, the demands placed on surfaces are considerable – and this has led to the development of a number of innovative composite coatings, especially for situations where low friction is required in the absence of conventional lubricants.

Transition metal dichalcogenides such as MoS_2, WS_2 and $NbSe_2$ have been used for some time as solid lubricants in space applications (Voevodin *et al.*, 1999a). Typically, these are applied as powders mixed with various binders or deposited by spray or vacuum-based methods. However, they are easily abraded and prone to oxidation in air. Multilayering with metal or ceramic layers can reduce these deficiencies, and the production of nanocomposite mixed phases can provide similar, if not greater, benefits. Zabinski studied the co-deposition of a number of materials, Ni, Fe, Au and Sb_2O_3 with MoS_2, and evaluated their effect on film chemistry, crystallinity, microstructure and tribology (Zabinski *et al.*, 1995). The presence of dopants caused film densification and affected grain size, resulting in a reduction in the friction coefficient and wear rate. Presently this is an area of fairly intense research activity. This is due, in part, to the fact that undoped MoS_2 films are known to be susceptible to undesirable oxidation and water absorption effects which can adversely influence their in-service performance and shelf-life (Singer *et al.*, 1996a). Recently, further observations have been made on the improvements obtained by metal additions, following on from the work of Zabinski and others, and the fact that even in 1981 Stupp showed that the co-deposition of nickel and other metals improved the performance of MoS_2 coatings (Stupp, 1981).

The addition of ceramic phases is arguably of even greater significance, since this route can provide a means to create a very hard low friction composite with a relatively low elastic modulus, i.e. a high H/E. Goller reports a nano-indentation hardness measurement of 29 GPa for TiN, and 30 GPa for a TiN and 8 mol% MoS_2 coating, the latter giving low friction and low wear (Goller *et al.*, 1999). This unexpected result is probably related to the *in situ* formation of MoS_2 at the coating surface during sliding; i.e. the coating is 'adaptive' to the operating conditions. These coatings will probably be suitable for a wide range of tribological contact types – e.g. from dry sliding to metal cutting (Gilmore *et al.*, 1998).

The concept of 'adaptive' or 'chameleon' coatings is one which has been given considerable attention by Zabinski and his co-workers. They have perfected PbO+MoS_2 coatings which can provide low friction across a wide range of operating temperatures (Zabinski *et al.*, 1992; Zabinski and Donley, 1994; Voevodin and Zabinski, 2000). They also produced WC.DLC/WS_2 nanocomposites that demonstrate low friction and wear in tests performed in high vacuum, dry nitrogen and humid air (Voevodin *et al.*, 1999a). Such performance would not be achievable with one of the phases alone.

7.5.4 Hybrid and Duplex Coatings

A recurring theme in this section has been the need to develop coatings that can elastically deform under load and accommodate substrate surface deflections. A corollary to that is to enhance the substrate surface hardness and elastic modulus in order to reduce the deflections occurring; in other words to provide improved load support for the coating. A number of approaches to achieve this have been investigated. These are generally known as hybrid or duplex processes, since they normally involve a hybrid version of, for example, a vacuum-based plasma process and/or the duplex combination of two or more different, usually sequential, processes of which one is also usually plasma based (Bell, 1997).

In the hybrid category, one might include plasma thermochemical treatments such as plasma nitriding or carburizing coupled with plasma-assisted PVD coating (Leyland *et al.*, 1993; Matthews and Leyland, 1995), whereas a duplex technique might involve a coating such as electroless nickel, to act as a corrosion-resistant and load-supporting interlayer between the substrate and a subsequently deposited PVD coating (Leyland *et al.*, 1993; Matthews *et al.*, 1995; Matthews and Leyland, 1995; Bin-Sudin *et al.*, 1996). A recent novel innovation is to employ a technique known as plasma electrolysis, which can be used to produce an oxide or nitrobarburized layer, especially in the case of oxide films on light metal alloys of aluminium, titanium or magnesium (Yerokhin *et al.*, 1999, 2000; Nie *et al.*, 2001).

Our results show that this is particularly effective when used in combination with a low-friction top layer, such as MoS_2 or DLC. The latter material can be deposited, after a pre-deposition implantation stage, using another innovative plasma process known as plasma immersion ion implantation (PI^3). This method, which harnesses a combination of PI^3 and plasma-assisted CVD features, is becoming known as plasma-immersion ion-assisted deposition (PIAD), which adds to the already considerable armoury available to process developers to achieve precisely tailored surface properties (Nie *et al.*, 2000a,b; Matthews *et al.*, 2003).

7.6 Applications

The early uses for hard PVD and CVD ceramic coatings were dominated by metal-cutting and forming tool applications; there are several reasons for this. The first is that the conventional CVD deposition temperature (typically over 900°C) meant that it was best suited to the coating of WC-Co ‘cemented’ carbides which were widely used in the manufacture of turning tools.

When PVD ceramic coating processes emerged, with a typical deposition temperature of 450°C, they were seen as well suited for the coating of hardened high-speed steels. Such materials could provide the necessary load support for the hard coating and were suitable for immediate use on existing tools, such as drills, milling cutters and gear-cutting hobs. Such tools were well characterized in terms of their wear lives and cutting performance, and it was relatively easy to demonstrate the economic benefit of applying a coating.

Early trials with PVD titanium coatings on tools such as drills revealed significant scatter in the results; they also demonstrated that the tool geometries, which had been optimized for the uncoated condition, were not necessarily ideal for coated tools. Additionally, it was soon realized that the main benefit to be derived from hard ceramic coatings was not simply that they could make tooling last longer but that the coated tools could be operated at much higher feeds and speeds, and therefore the productivity obtained from the cutting machinery itself could be significantly increased. This recognized the fact that the main costs in many manufacturing operations are not those of tooling but of capital equipment and labour and that coatings allow these costs to be reduced per component produced.

In the UK, a new drill design was developed (designated ‘ADX’) which, when combined with a hard PVD TiN coating, allowed peripheral drill speeds to be doubled and feed rates to be increased (Dormer Ltd). Over the past 20 years or so, the use of PVD TiN coatings has become ubiquitous in the cutting tool sector. There is hardly a tool in current industrial use which does not have a PVD ceramic coating applied to it.

The adoption of PVD ceramic tribological coatings in engineering components – especially mass-marketed, high-volume parts, such as on automotive products – was, until recently, very limited. There are various reasons for this, not least the need for a hard underlying substrate material, which can provide the necessary load support. Such materials, if they are to withstand the PVD deposition temperature without softening, tend to be rather expensive, e.g. high-speed and hot-working tool steels. There has recently been extensive work to develop multilayered and nanocomposite coatings, which are more elastic and do not require the same load support attributes from the substrate. Similarly, there have been great strides in the development of techniques to harden the surface of lower-cost substrate materials, to make them suitable for PVD coating. However, these developments were being adopted in the market place at a rather slow pace. There are now an increasing number of impressive examples of the use of second- and third-generation PVD coatings in diverse tribological applications, some of which are mentioned below.

One such coating is a multilayer WC/C coating with excellent friction and wear properties ('Balinit®C' by Balzers; Grischke *et al.*, 2001). It is used, for example, on fuel injection pump components of gears. Another new coating from that company combines a TiAIN coating with a WC/C outer layer ('Balinit® Hardlube'). Significant improvements are claimed over earlier coatings, in drill tests. An alternative to this is to utilize the mixed-phase nanocomposite coatings described earlier.

There are today many carbon-based coatings on the market, fulfilling different requirements. One of the main advantages of carbon coatings is that they can be engineered to provide different properties (Neuville and Matthews, 1997; Schaefer *et al.*, 2000). For example, Schaefer and co-workers show that a steel ball valve can survive adverse environments when coated with a metal-containing DLC (Me-DLC), and forming and stamping dies can resist sticking in metal forming with that coating, whilst dies used for the production of plastic covered cables are best coated with silicon-, oxygen- and nitrogen-modified DLC. So, whilst metal cutting remains the main domain of crystalline diamond and some DLC films, new applications are continuously emerging for carbon-based films with different compositions and structures. For example, Schuelke (2000) has demonstrated the benefits of metal/DLC (Diamor™) films in both sliding and rolling contacts.

Carbon-based coatings are not the only ones that are leading the emergence of vapour deposition processes out of the cutting tool sector. Chromium nitride is also in increasing use, for example in vehicle applications. An article by Hawbaker and Taylor (2003) describes how PVD chromium nitride coatings have proved to be extremely effective in improving the abrasive wear resistance of Caterpillar track bushings. A track bushing is an integral part of the undercarriage of any track-type machine. Bushings, along with track pins and links, create the 'chain' that ties together the tracks. The application is very demanding and Hawbaker and Taylor describe how a force of well over 1 MN may be transferred through the track components on a fully loaded tractor. The application of a CrN coating has reduced the occurrence of a particular bush wear problem and thereby extended the productive life of the track bushing, and therefore the tracked vehicle. Clearly, this is a most demanding application, potentially involving several of the contact mechanisms described earlier, such as sliding wear and abrasion, and even tribo-corrosion.

Thin films are well known in the electronic and optical industrial sectors. Indeed, semiconductors consist entirely of a sequence of thin films. Similarly, many optical devices require multilayers of different phases and compositions. It is often forgotten that many of these functional layers must also possess excellent tribological properties if they are to meet the required operational performance. Even some coatings which at first sight appear to have no tribological function are actually tested for their scratch and sliding wear resistance before they can be adopted. Consider, for example, the coatings applied to mobile phones to enhance their electro-magnetic interference suppression. Although their main function is electro-magnetic, they are also scratch resistant. Many such coatings are extensively modified, not to meet their main functional needs, but to ensure that they survive the mechanical, tribological and environmental conditions under which they operate. Another good example is the scratch-resistant optical lens. Similarly, many decorative finishes, although selected and applied for their aesthetic appearance, usually colour, are in fact sold on an ability to function for the lifetime of the product. In this case, the tribological performance demands frequently exceed the functional demands for which they were originally applied. Examples in the latter category are gold-coloured ZrN and TiN films, which have an oxide outer layer deposited to enhance their resistance to exposure to cleaning chemicals and abrasion. Indeed, the testing regimes for such coatings are extremely rigorous (Van der Kolk *et al.*, 1998).

It can be seen from the above examples of the applications that PVD tribological coatings are starting to be more widely used outside their traditional domain of cutting and forming tools. With the further optimization of PVD coating mechanical properties, as described earlier, and the development of the techniques mentioned to improve the surface hardness and the load support characteristics of substrates, these coatings will find an ever wider range of applications, to which they will bring previously unachievable levels of durability, wear and corrosion resistance, and friction control. Importantly, these functionality improvements will be achieved with more predictable and statistically reliable overall lifetimes, and therefore with better quantified life-cycle costs. Such improvements in statistical lifetime analysis, including the influence of geometrical and surface coating effects, have been achieved in the tooling sector (Dowey and Matthews, 1998; Dowey *et al.*, 1999) and will be necessary for other application sectors if the goal of optimized tribological performance is to be achieved.

7.7 Conclusions

By considering tribological interactions at several levels of scale, and adopting a macromechanical approach to the categorization of surface interactions, it is possible to achieve a clearer insight into the friction and wear behaviours of tribological coatings. This helps in the identification of the requirements from coatings and the gaps in current knowledge – both in the modelling of behaviour and in the development of new coatings. The advanced plasma-based PVD and CVD processes facilitate the deposition of ceramic, binary alloy, multilayered, nanocomposite, hybrid, duplex and functionally graded surfaces which hold the potential to fulfil any desired operating requirement, and this will lead to the adoption of these surface engineering processes in an increasingly diverse range of demanding applications.

References

Adachi, K. and Hutchings, I., 'Wear-Mode Mapping for the Micro-Scale Abrasion Test', *Wear*, **255**, 2003, 23–29.

Ahmed, S., Bregliozzi, G. and Haefke, H., 'Microfrictional Properties of Diamond-Like Carbon Films Sliding Against Silicon, Sapphire and Steel', *Wear*, **254**, 2003, 1076–1083.

Anderson, T., '*Fracture Mechanics – Fundamentals and Applications*', 2nd edn, CRC Press, Boca Raton, Florida, 1995.

Andersson, P. and Holmberg, K., 'Limitations on the Use of Ceramics in Unlubricated Sliding Applications due to Transfer Layer Formation', *Wear*, **175**, 1994, 1–8.

Andersson, J., Erck, R. and Erdemir, A., 'Frictional Behaviour of Diamond-Like Carbon Films in Vacuum and Under Varying Water Vapour Pressure', *Surface and Coatings Technology*, **163–164**, 2003a, 535–540.

Andersson, J., Erck, R. and Erdemir, A., 'Friction of Diamond-Like Carbon Films in Different Atmospheres', *Wear*, **254**, 2003b, 1070–1075.

Argon, A.S., 'Mechanical Properties of Near-Surface Material in Friction and Wear', in *Fundamentals of Tribology* (eds N.P. Suh and N. Saka), MIT Press, London, 1980, pp. 103–125.

Aubert, A., Nabot, J.P., Ernoult, J. and Renaux, P., 'Preparation and Properties of MoS_2 Films Grown by d.c. Magnetron Sputtering', *Surface and Coatings Technology*, **41**, 1990, 127–134.

Badisch, E., Stoiber, M., Fontalvo, G.A. and Mitterer, C., 'Low Friction PACVD TiN Coatings: Influence of Cl Content and Testing Conditions on the Tribological Properties', *Surface and Coatings Technology*, **174–175**, 2003, 450–454.

Bandorf, R., Luthje, H., Wortmann, A., Steadtler, T. and Wittorf, R., 'Influence of Substrate Material and Topography on the Tribological Behaviour of Submicron Coatings', *Surface and Coatings Technology*, **174–175**, 2003, 461–464.

Bantle, R. and Matthews, A., 'Investigation into the Impact Wear Behaviour of Ceramic Coatings', *Surface and Coatings Technology*, **74/75**, 1995, 875–868.

Barnett, S.A., 'Deposition and Mechanical Properties of Superlattice Thin Films', in *Physics of Thin Films* (eds M. Francombe and J.A. Vossen), **17**, Academic Press, San Diego, USA, 1993.

Beerschwinger, U., Albrecht, T., Mathieson, D., Reuben, R.L., Yang, S.J. and Taghizadeh, M., 'Wear at Microscopic Scales and Light Loads for MEMS Applications', *Wear*, **181–183**, 1995, 426–435.

Bell, T., 'Realising the Potential of Duplex Surface Engineering', in *New Directions in Tribology* (ed I.M. Hutchings), Mechanical Engineering Publications, London, 1997, pp. 121–133.

Berman, A., Drummond, C. and Israelachvili, J., 'Amonton's Law at the Molecular Level', *Tribology Letters*, **4**, 1998, 95–101.

Berthier, Y., Godet, M. and Brendle, M., 'Velocity Accommodation in Friction', *Tribology Transactions*, **32**(4), 1989, 490–496.

Beuth, J., 'Cracking of Thin Bonded Films in Residual Tension', *International Journal of Solid Structures*, **29**(13), 1992, 1657–1675.

Bewilogua, K. and Dimigen, H., 'Preparation of W-C-H Coatings by Reactive Magnetron Sputtering', *Surface and Coatings Technology*, **61**, 1993, 144.

Bin-Sudin, M., Leyland, A., James, A.S., Matthews, A., Housden, J. and Garside, B., 'Substrate Surface Finish Effects in Duplex Coatings of PAPVD TiN and CrN with Electroless Nickel-Phosphorus Interlayers', *Surface and Coatings Technology*, **81**, 1996, 215–224.

Bhushan, B. (ed), 'Tribology Issues and Opportunities in MEMS', *Proceedings of NSF/AFOSR/ ASME Workshop on Tribology Issues and Opportunities*, MEMS, 9–11 November 1997, Columbus, USA, Kluwer Academic Publishers, London, 1998.

Bhushan, B., 'Nanoscale Tribophysics and Tribomechanics', *Wear*, **225–229**, 1999, 465–492.

Blomberg, A., 'Friction and Wear of Ceramics', Ph.D. thesis, Uppsala University, Material Science Division, Uppsala, Sweden, 1993, p. 47.

Bouzakis, K., Hadjiyiannis, S., Skordaris, G., Anasopoulos, J., Mirisidis, I., Michailidis, N., Efstatiou, K., Knotek, O., Erkens, G., Cremer, G., Rambadt, S. and Wirth, I., 'The Influence of Coating Thickness on Its Strength Properties and on the Milling Performance of PVD Coated Inserts', *Surface and Coatings Technology*, **174–175**, 2003, 393–401.

Bowden, F.P. and Tabor, D., '*Friction and Lubrication of Solids. Part I*', Oxford University Press, Oxford, 1950, p. 321.

Buckley, D., '*Surface Effects in Adhesion, Friction, Wear, and Lubrication*', Elsevier, Amsterdam, 1981, p. 631.

Bull, S., 'Failure Mode Maps in the Thin Film Scratch Adhesion Test', *Tribology International*, **30**(1), 1997, 491–498.

Bull, S.J. and Rickerby, D.S., 'Compositional, Microstructural and Morphological Effects on the Mechanical and Tribological Properties of Chromium Nitrogen Films', *Surface and Coatings Technology*, **43/44**, 1990, 732–744.

Burnett, P.J. and Rickerby, D.S., 'The Mechanical Properties of Wear-Resistant Coatings I: Modelling of Hardness Behaviour', *Thin Solid Films*, **148**, 1987a, 41–50.

Burnett, P.J. and Rickerby, D.S., 'The Relationship Between Hardness and Scratch Adhesion', *Thin Solid Films*, **154**, 1987b, 403–416.

Cartier, M. (ed), '*Handbook of Surface Treatments and Coatings*', Tribology in Practice Series, Professional Engineering Publishers, London, 2003, p. 412.

Celis, B., 'Theoretical Analysis of Dry Friction in Brittle and Ductile Materials', *Wear*, **116**, 1987, 287–298.

Chang, T.P., Cheng, H.S. and Sproul, W.D., 'The Influence of Coating Thickness on Lubricated Rolling Contact Fatigue Life', *Surface and Coatings Technology*, **43/44**, 1990, 699–708.

Chang, T.P., Graham, M.E., Sproul, W.D. and Cheng, H.S., 'Scuffing Behaviour of TiN-Coated Steel Rollers Under Rolling and Sliding Conditions', *Surface and Coatings Technology*, **54/55**, 1992, 490–495.

Chu, X., Wong, M.S., Sproul, W.D., Rohde, S.L. and Barnet. S.A., 'Deposition and Properties of Polycrystalline TiN/NbN Superlattice Coatings', *Journal of Vacuum Science and Technology*, **A10**(4), 1992, 1604.

Cselle, T. and Barimani, A., 'Today's Applications and Future Developments of Coatings for Drills and Rotating Cutting Tools', *Surface and Coatings Technology*, **76–77**, 1995, 712–718.

De Wit, E., Blanpain, B., Froyen, L. and Celis, J.-P., 'The Tribochemical Behaviour of TiN-Coatings During Fretting Wear', *Wear*, **217**, 1998, 215–224.

Diao, D., Kato, K. and Hayashi, K., 'The Maximum Tensile Stress on a Hard Coating Under Sliding Friction', *Tribology International*, **27**(4), 1994, 267–272.

Ding, Y., Jones, R. and Khunell, B., 'Elastic–Plastic Finite Element Analysis of Spall Formation in Gears', *Wear*, **197**, 1996, 197–205.

Djabella, H. and Arnell, R., 'Finite Element Analysis of the Contact Stresses in an Elastic Coating on an Elastic Substrate', *Thin Solid Films*, **213**, 1992, 205–219.

Djabella, H. and Arnell, R., 'Finite Element Analysis of Contact Stresses in Elastic Double-Layer Systems Under Normal Load', *Thin Solid Films*, **223**, 1993, 98–108.

Dimigen, H. and Klages, C.-P., 'Microstructure and Wear Behaviour of Metal-Containing Diamond-Like Coatings', *Surface and Coatings Technology*, **49**, 1991, 543.

Donnet, C., 'Tribology of Solid Lubricant Coatings', *Condensed Matter News*, **4**(6), 1995, 9–24.

Donnet, C., 'Problem Solving Methods in Tribology with Surface-Specific Techniques', Chapter 16, in *Handbook of Surface and Interface Analysis* (eds J.C. Riviera and S. Myhra), Marcel Decker, New York, 1998, pp. 697–745.

Donnet, C., Martin, J.-M., Le Mogne, T. and Belin, M., 'How to Reduce Friction in the Millirange by Solid Lubrication', *Proceedings of the International Tribology Conference*, Yokohama, 1995, p. 6.

Donohue, L.A., Cawley, J., Lewis, D.B., Brooks, J.S. and Munz, W.S., 'Investigation of Superlattice Coatings Deposited by a Combined Steered-Arc Evaporation and Unbalanced Magnetron Sputtering Technique', *Surface and Coatings Technology*, **76**, 1995, 149.

Donohue, L.A., Munz, W.D., Lewis, D.B., Cawley, J., Hurkmans, T., Trinh, T., Petrov, I. and Greene, J.E., 'Large-Scale Fabrication of Hard Superlattice Thin Films by Combined Steered-Arc Evaporation and Unbalanced Magnetron Sputtering', *Surface and Coatings Technology*, **93**, 1997, 69–87.

Dowey, S.J. and Matthews, A., 'Taguchi and TQM: Quality Issues for Surface Engineered Applications', *Surface and Coatings Technology*, **110**, 1998, 86–93.

Dowey, S.J., Zhang, J., Doyle, E.D. and Matthews A., 'Life Analysis of Coated Tools Using Statistical Methods', *Surface and Coatings Technology*, **116–119**, 1999, 54–661.

Dvorak, S., Wahl, K. and Singer, I., 'Friction Behaviour of Boric Acid and Annealed Boron Carbide Coatings Studied by In Situ Raman Tribometry', *Tribology Transactions*, **45**, 2002, 354–362.

Eglin, M., Rossi, A. and Spencer, N., 'A Combinatorial Approach to Elucidating Tribochemical Mechanisms', *Tribology Letters*, **15**(3), 2003, 193–198.
El-Sherbiny, M.G. and Salem, F.B., 'Initial Wear Rates of Soft Metallic Films', *Wear*, **54**, 1979, 391–400.
El-Sherbiny, M. and Salem, F., 'A Wear Equation for Solid Lubricant Films', *3rd International Conference on Solid Lubricants*, ASLE, Denver, 1984, 44–49.
El-Sherbiny, M. and Salem, F., 'Tribological Properties of PVD Silver Films', *ASLE Transactions*, **29**(2), 1986, 223–228.
Enomoto, Y. and Yamamoto, T., 'New Materials in Automotive Tribology', *Tribology Letters*, **5**, 1998, 13–24.
Erdemir, A., 'Rolling Contact Fatigue and Wear Resistance of Hard Coatings on Bearing-Steel Substrates', *Surface and Coatings Technology*, **54/55**, 1992, 482–489.
Erdemir, A., 'Solid Lubricants and Self-Lubricated Films', in *Modern Tribology Handbook* (ed B. Bhushan), Vol. 2/22, CRC Press, Boca Raton, USA, 2001, 787–825.
Erdemir, A., 'Friction and Wear of Diamond and Diamond-Like Carbon Films', *Proceedings of the Institution of Mechanical Engineers, Part J: Engineering Tribology*, **216**, 2002, 387–400.
Erdemir, A. and Donnet, C., 'Tribology of Diamond, Diamond-Like Carbon and Related Films', in *Modern Tribology Handbook* (ed B. Bhushan), Vol. 2/24, CRC Press, Boca Raton, USA, 2001, pp. 871–908.
Erdemir, A. and Hochman, R.F., 'Surface Metallurgical and Tribological Characteristics of TiN-Coated Bearing Steels', *Surface and Coatings Technology*, **36**, 1988, 755–763.
Erdemir, A., Bindal, C., Pagan, J. and Wilbur, P., 'Characterization of Transfer Layers on Steel Surfaces Sliding Against Diamond-Like Hydrocarbon Films in Dry Nitrogen', *Surface and Coatings Technology*, **76–77**, 1995, 559–563.
Erdemir, A., Bindal, C., Fenske, G., Zuiker, C. and Wilbur, P., 'Characterisation of Transfer Layers Forming on Surfaces Sliding Against Diamond-Like Carbon', *Surface and Coatings Technology*, **86/87**, 1996a, 692–697.
Erdemir, A., Bindal, C., Fenske, G.R., Zuiker, C., Csencsits, R., Krauss, A.R. and Gruen, D.M., 'Tribological Characterization of Smooth Diamond Films Grown in Ar-C_{60} and Ar-CH_4 Plasmas', *Diamond Films and Technology*, **6**(1), 1996b, 31–47.
Erdemir, A., Busch, D.E., Erck, R.A., Fenske, G.R. and Lee, R., 'Ion-Beam-Assisted Deposition of Silver Films on Zirconia Ceramics for Improved Tribological Behaviour', *Lubrication Engineering*, **47**(10), 1991, 863–872.
Erdemir, A., Eryilmaz, O. and Fenske, G., 'Synthesis of Diamond-Like Carbon Films with Superlow Friction and Wear Properties', 46th International Symposium of the American Vaccine Society, 25–29 October 1999, Seattle, USA, Manuscript revised March 2000 (ID#1625, Conference Paper #TF+VM-MoA3), 2000, p. 14.
Erdemir, A., Halter, M. and Fenske, G., 'Preparation of Ultralow-Friction Surface Films on Vanadium Diboride', *Wear*, **205**, 1997, 236–239.
Erdemir, A., Halter, M. and Fenske, G., 'New Oxide-Based Lubricants for Sliding Friction and Wear Applications at Extreme Temperatures', *World Ceramics Congress and Forum on New Materials – CIMTEC*, 14–19 June 1998, Florence, p. 16.
Fouvry, S. and Kapsa, P., 'An Energy Description of Hard Coating Wear Mechanisms', *Surface and Coatings Technology*, **138**, 2001, 141–148.
Fouvry, S., Kapsa, P. and Vincent, L., 'Quantification of Fretting Damage', *Wear*, **200**, 1996, 186–205.
Fouvry, S., Liskiewicz, T., Kapsa, P., Hannel, S. and Sauger, E., 'An Energy Description of Wear Mechanisms and Its Application to Oscillating Sliding Contacts', *Wear*, **255**, 2003, 287–298.
Franklin, S., Surface Modifications for the Control of Wear and Friction, Report CTR 545-93-0036/7.4.1993, CTF technology, Philips Centre for Manufacturing Technology, 1993.
Fridrici, V., Fouvry, S. and Kapsa, P., 'Fretting Wear Behaviour of a Cu–Ni–In Plasma Coating', *Surface and Coatings Technology*, **163–164**, 2003, 429–434.
Fu, Y., Batchelor, A.W. and Loh, N.M., 'Study on Fretting Wear Behaviour of Laser Treated Coatings by X-ray Imaging', *Wear*, **218**, 1998, 250–260.
Gao, G., Mikulski, P., Chateauneaf, G. and Harrison, J., 'The Effects of Film Structure and Surface Hydrogen on the Properties of Amorphous Carbon Film', *The Journal of Physical Chemistry, B*, **107**(40), 2003, 11082–11090.

Gardos, M.N., 'Tribology and Wear Behaviour of Diamond', in *Synthetic Diamond: Emerging CVD Science and Technology* (eds K.E. Spear and J.P. Dismukes), John Wiley & Sons, New York, 1994, pp. 419–502.

Gardos, M.N., 'Tribological Behaviour of Polycrystalline Diamond Films', NATO Advanced Research Workshop, *Protective Coatings and Thin Films*, 30 May–5 June 1996, Portimao, Algarve, Portugal, 1996.

Gee, M. and Jennett, N.M., 'High Resolution Characterization of Tribochemical Films on Alumina', *Wear*, **193**, 1995, 133–145.

Gee, M., Gant, A., Hutchings, I., Bethke, R., Schiffman, K., Van Acker, K., Poulat, S., Gachon, Y. and von Steubut, J., 'Progress Towards Standardisation of Ball Cratering', *Wear*, **255**, 2003, 1–13.

Gerde, E. and Marder, M., 'Friction and Fracture', *Nature*, **413**, 2001, 285–288.

Gilmore, R., Baker, M.A., Gibson, P.N., Gissler, W., Stoiber, M., Losbichler, P. and Mitterer, C., 'Low Friction TiN MoS_2 Coatings Produced by dc Magnetron Co-Deposition', *Surface and Coatings Technology*, **108/109**, 1998, 345–351.

Gissler, W., 'Structure and Properties of Ti–B–N Coatings', *Surface and Coatings Technology*, **68/69**, 1994, 556–563.

Goller, R., Torri, P., Baker, M.A., Gilmore, R. and Gissler, W., 'The Deposition of Low Friction TiN–MoS_2x Coatings by a Combined Arc Evaporation and Magnetron Sputtering Process', *Surface and Coatings Technology*, **120/121**, 1999, 453–457.

Grill, A., 'Tribology of Diamond-Like Carbon and Related Materials: An Updated Review', *Surface and Coatings Technologies*, **94–95**, 1997, 507–513.

Grischke, M., Herb, R., Massler, O., Karner, J. and Eberle, H., 'High Vacuum Based Deposition of Carbon-Based Films for Industrial Applications', in *Proceedings of the 44th Annual Technical Conference of the Society of Vacuum Coaters*, SVC, Albuquerque, Philadelphia, USA, 21–26 April 2001.

Guu, Y.Y., Lin, J.F. and Ai, C.F., 'The Tribological Characteristics of Titanium Nitride Coatings, Part I: Coating Thickness Effects', *Wear*, **194**, 1996a, 12–21.

Guu, Y.Y., Lin, J.F. and Ai, C.F., 'The Tribological Characteristics of Titanium Nitride Coatings, Part II: Comparisons of Two Deposition Processes', *Wear*, **194**, 1996b, 22–29.

Habig, K.-H., 'Fundamentals of Tribological Behaviour of Diamond, Diamond-Like Carbon and Cubic Boron Nitride Coatings', *Surface and Coatings Technology*, **76–77**, 1995, 540–547.

Harrison, J.A., Colton, R.J., White, C.T. and Brenner, D.W., 'Effect of Atomic Scale Surface Roughness on Friction: A Molecular Dynamics Study of Diamond Surfaces', *Wear*, **168**, 1993, 127–133.

Harrison, J., Stuart, S. and Perry, M., 'The Tribology of Hydrocarbon Surfaces Investigated Using Molecular Dynamics', in *Tribology Issues and Opportunities in MEMS* (ed B. Bhushan), Kluwer Academic Publishers, The Netherlands, 1998, pp. 285–299.

Harrison, J.A., White, C.T., Colton, R.J. and Brenner, D.W., 'Molecular-Dynamics Simulations of Atomic-Scale Friction of Diamond Surfaces', The American Physical Society, *Physical Review B*, **46**(15), 1992, 9700–9708.

Harrison, J., White, J., Colton, R. and Brenner, D., 'Investigation of the Atomic-Scale Friction and Energy Dissipation in Diamond Using Molecular Dynamics', *Thin Solid Films*, **260**, 1995, 205–211.

Hawbaker, R. and Taylor, S., 'PVD Wear Coatings: Anatomy of a Component Coating Application', in *Proceedings of the 46th Annual Technical Conference of the Society of Vacuum Coaters*, SVC, Albuquerque, San Francisco, USA, 3–8 May 2003.

Hayward, I.P., Singer, I.L. and Seizman, L.E., 'Effect of Roughness on the Friction of Diamond on CVD Diamond Coatings', *Wear*, **157**, 1992, 215–227.

Hedenqvist, P., Olsson, M., Jacobson, S. and Söderberg, S., 'Failure Mode Analysis of TiN-Coated High Speed Steel: In Situ Scratch Testing in the Scanning Electron Microscope', *Surface and Coatings Technology*, **41**, 1990, 31–49.

Helmersson, U., Todorova, S., Barnett, S.A. and Sundgren, J.-E., 'Growth of Single-Crystal TiN/VN Strained-Layer Superlattices with Extremely High Mechanical Hardness', *Journal of Applied Physics*, **62**(2), 1987, 481.

Hills, D.A., Nowell, D. and Sackfield, A., 'A Survey of Cracks in Layers Propelled by Contact Loading', in *Mechanics of Coatings* (eds D. Dowson *et al.*), Tribology Series 17, Elsevier, Amsterdam, 1990, pp. 203–208.

Hogmark, S., Jacobson, S. and Larsson, M., 'Design and Evaluation of Tribological Coatings', *Wear*, **246**, 2000, 20–33.

Hokkirigawa, K. and Kato, K., 'An Experimental and Theoretical Investigation of Ploughing, Cutting and Wedge Formation During Abrasive Wear', *Tribology International*, **21**(1), 1988, 51–58.

Holleck, H., 'Material Selection for Hard Coatings', *Journal of Vacuum Science and Technology*, **A4**, 1986, 2661–2669.

Holleck, H. and Shier, V., 'Multilayer PVD Coatings for Wear Protection', *Surface and Coatings Technology*, **76/77**, 1995, 328–336.

Holleck, H., Lahres, M. and Woll, P., 'Multilayer Coatings – Influence of Fabrication Parameters on Constitution and Properties', *Surface and Coatings Technology*, **41**, 1990, 179–190.

Holmberg, K., 'The Basic Material Parameters that Control Friction and Wear of Coated Surfaces Under Sliding', *Tribologia – Finnish Journal of Tribology*, **19**(3), 2000, 3–18.

Holmberg, K., 'Reliability Aspects of Tribology', *Tribology International*, **34**(12), 2001, 801–808.

Holmberg, K. and Matthews, A., '*Coatings Tribology – Properties, Techniques and Applications in Surface Engineering*', Elsevier Tribology Series, 28, Elsevier Science B.V., The Netherlands, 1994, p. 442.

Holmberg, K. and Matthews, A., 'Tribological Properties of Metallic and Ceramic Coatings', in *Modern Tribology Handbook* (ed B. Bhushan), Vol. 2/23, CRC Press, Boca Raton, USA, 2001, pp. 827–870.

Holmberg, K., Koskinen, J., Ronkainen, H., Vihersalo, J., Hirvonen, J.-P. and Likonen, J., 'Tribological Characteristics of Hydrogenated and Hydrogen-Free Diamond-Like Carbon Coatings', *Diamond Films and Technology*, **4**(2), 1994, 113–129.

Holmberg, K., Laukkanen, A., Ronkainen, H., Wallin, K. and Varjus, S., 'A Model for Stresses, Crack Generation and Fracture Toughness Calculation in Scratch TiN-Coated Steel Surfaces', *Wear*, **254**, 2003, 278–291.

Holmberg, K., Matthews, A. and Ronkainen, H., 'Coatings Tribology – Contact Mechanics and Surface Design', *Tribology International*, **31**, 1998, 1–3, 107–120.

Huang, Z.P., Sun, Y. and Bell, T., 'Friction Behaviour of TiN, CrN and (TiAl)N Coatings', *Wear*, **173**, 1994, 13–20.

Huq, M.Z. and Celis, J.-P., 'Fretting Wear of Multilayered (Ti,Al)N/TiN Coatings in Air of Different Relative Humidity', *Wear*, **225–229**, 1999, 53–64.

Hutchings, I., '*Tribology – Friction and Wear of Engineering Materials*', Arnold, London, 1992.

Hwang, D.H., Kim, D.E. and Lee, S.J., 'Influence of Wear Particle Interaction in the Sliding Interface on Friction of Metals', *Wear*, **225–229**, 1999, 427–439.

Israelachvili, J.N. and Tabor, D., 'The Measurement of van der Waals Dispersion Forces in the Range 1.5 to 130 nm', *Proceedings of the Royal Society*, **A331**, 1972, 19–38.

Iwai, Y., Honda, T., Yamada, H., Matsubara, T., Larsson, M. and Hogmark, S., 'Evaluation of Wear Resistance of Thin Hard Coatings by a New Solid Particle Impact Test', *Wear*, **251**, 2001, 861–867.

Jacobson, S. and Hogmark, S., 'On the Tribological Character of Boundary Lubricated DLC Coated Components', *2nd World Tribology Congress*, Vienna, Austria, 3–7 September 2001.

Jardet, V., Zahouani, H., Loubet, J. and Mathia, T., 'Understanding and Quantification of Elastic and Plastic Deformation During a Scratch Test', *Wear*, **218**, 1998, 8–14.

Je, J.H., Gyarmati, E. and Naoumidis, A., 'Scratch Adhesion Test of Reactively Sputtered TiN Coatings on a Soft Substrate', *Thin Solid Films*, **136**, 1986, 57–67.

Kato, K., 'Micro-Mechanics of Wear – Wear Modes', *Wear*, **153**, 1992, 277–295.

Kato, K., Diao, D. and Tsutsumi, M., 'The Wear Mechanism of Ceramic Coating Film in Repeated Sliding Friction', in *Wear of Materials*, ASME, 1991, pp. 243–248.

Kim, D.E. and Suh, N.P., 'On Microscopic Mechanisms of Friction and Wear', *Wear*, **149**, 1991, 199–208.

Ko, P.L., Robertson, M. and Magel, E.E., 'Wear Particle Formation in Lubricated Sliding Between a Hardened Sphere and a Flat Surface', in *Wear Particles: From the Cradle to the Grave* (eds D. Dowson *et al.*), Elsevier Tribology Series 21, Amsterdam, 1992, pp. 81–90.

Kodali, P., Walter, K.C. and Nastasi, M., 'Investigation of Mechanical and Tribological Properties of Amorphous Diamond-Like Carbon Coatings', *Tribology International*, **30**(8), 1997, 591–598.

Koechler, J.S., 'Attempt to Design a Strong Solid', *Physical Review*, **B2**, 1970, 547–551.

Komvopoulos, K., 'Sliding Friction Mechanisms of Boundary-Lubricated Layered Surfaces: Part I – Basic Mechanical Aspects and Experimental Results', *Tribology Transactions, Journal of STLE*, **34**(2), 1991a, 266–280.

Komvopoulos, K., 'Sliding Friction Mechanisms of Boundary-Lubricated Layered Surfaces: Part II – Theoretical Analysis', *Tribology Transactions, Journal of STLE*, **34**(2), 1991b, 281–291.

Krim, J., 'Friction at the Atomic Scale', *Scientific American*, October, 1996, 48–56.

Krim, J., 'Friction at Macroscopic and Microscopic Length Scales', *American Journal of Physics*, **70**(9), 2002a, 890–897.

Krim, J., 'Surface Science and the Atomistic-Scale Origins of Friction: What Once was Old is New Again', *Surface Science*, **500**, 2002b, 741–758.

Kullmer, R., Lugmair, C., Figueras, A., Bassas, J., Stoiber, M. and Mitterer, C., 'Microstructure, Mechanical and Tribological Properties of PACVD Ti(B,N) and TiB_2 Coatings', *Surface and Coatings Technology*, **174–175**, 2003, 1229–1233.

Lampe, Th., Eisenberg, S. and Rodriguez Cabeo, E., 'Plasma Surface Engineering in the Automotive Industry – Trends and Future Prospectives', *Surface and Coatings Technology*, **174–175**, 2003, 1–7.

Landman, U., Luedtke, W. and Ringer, E., 'Atomistic Mechanisms of Adhesive Contact Formation and Interfacial Processes', *Wear*, **153**, 1992, 3–30.

Lehoczy, S.L., 'Strength Enhancement in Thin-Layered Al–Cu Laminates', *Journal of Applied Physics*, **49**, 1978a, 5479.

Lehoczy, S.L., 'Retardation of Dislocation Generation and Motion in Thin-Layered Metal Laminates', *Physical Review Letters*, **41**, 1978b, 1814.

Leyland, A. and Matthews, A., 'Thick Ti/TiN Multi-Layered Coatings for Abrasive and Erosive Wear Resistance', *Surface and Coatings Technology*, **70**, 1994, 19–25.

Leyland, A. and Matthews, A., 'On the Significance of the H/E Ratio in Wear Control: A Nanocomposite Coating Approach to Optimized Tribological Behaviour', *Wear*, **246**, 2000, 1–11.

Leyland, A., Bin-Sudin, M., James, A.S., Kalantary, M.R., Wells, P.B. and Matthews, A., 'TiN and CrN PVD Coatings on Electroless Nickel Coated Steel Substrates', *Surface and Coatings Technology*, **60**, 1993, 474–479.

Lin, J.F., Liu, M.H. and Wu, J.D., 'Analysis of the Friction and Wear Mechanism of Structural Ceramic Coatings, Part 1: The Effect of Processing Conditions', *Wear*, **194**, 1996a, 1–11.

Lin, J.F., Liu, M.H. and Wu, J.D., 'Analysis of the Friction and Wear Mechanism of Structural Ceramic Coatings, Part 2: The Effect of Operating Conditions and Substrate Material', *Wear*, **198**, 1996b, 7–14.

Luo, Q., Leyens, C., Hovsepian, P., Lewis, D., Constable, C. and Munz, W., 'Oxidation Mechanism of PVD TiAlCrYN Coatings Observed by Analytical TEM', *Electron Microscopy and Analysis 2001: Institute of Physics Conference Series*, **168**, 2001, 369–372.

Mäkelä, U. and Valli, J., 'Tribological Properties of PVD TiN and TiC Coatings', *Finnish Journal of Tribology*, **4**(2), 1985, 74–85.

Malzbender, J. and de With, G., 'Sliding Indentation, Friction and Fracture of a Hybrid Coating on Glass', *Wear*, **236**, 1999, 355–359.

Malzbender, J. and de With, G., 'Cracking and Residual Stress in Hybrid Coatings on Float Glass', *Thin Solid Films*, **359**, 2000a, 210–214.

Malzbender, J. and de With, G., 'Friction Under Elastic Contacts', *Surface and Coatings Technology*, **124**, 2000b, 66–69.

Malzbender, J. and de With, G., 'Modeling of the Fracture of a Coating Under Sliding Indentation', *Wear*, **239**, 2000c, 21–26.

Mao, K., Sun, Y. and Bell, T., 'A Numerical Model for the Dry Sliding Contact of Layered Elastic Bodies with Rough Surfaces', *Tribology Transactions*, **39**(2), 1995, 416–424.

Matthews, A., Modulated Composition Composites Produced by Vapour Deposition', UK Patent, GB 2242442 B, 1994.

Matthews, A., 'Plasma Based Vacuum Deposition Processes to Enhance Wear and Corrosion Performance', in *Protective Coatings and Thin Films* (eds Y. Pauleau and P. Barna), Kluwer Academic Publishers, Dordecht, 1997.

Matthews, A. and Eskildsen, S.S., 'Engineering Applications for Diamond-Like Carbon', *Diamond and Related Materials*, **3**, 1994, 902–911.

Matthews, A. and Leyland, A., 'Hybrid Techniques in Surface Engineering', *Surface and Coatings Technology*, **71**, 1995, 88–92.
Matthews, A. and Leyland, A., 'Hard Tribological Coatings: Developments and Applications', in *Total Tribology: Towards an Integrated Approach* (eds I. Sherrington, W. Rowe and R. Wood), IMechE 'Tribology in Practice' Series, PEP Ltd, Bury St Edmunds, UK, 2002, pp. 39–63.
Matthews, A., Jones, R. and Dowey, S., 'Modelling the Deformation Behaviour of Multilayer Coatings', *Tribology Letters*, **11**(2), 2001, 103.
Matthews, A., Leyland, A., Dorn, B., Stevenson, P.R., Bin-Sudin, M., Rebholz, C., Voevodin, A.A. and Schneider, J.M., 'Plasma-Based Surface Engineering Processes for Wear and Corrosion Protection', *Journal of Vacuum Science and Technology*, **A13**(3), 1995, 1202–1207.
Matthews, A., Leyland, A., Holmberg, K. and Ronkainen, H., 'Design Aspects for Advanced Tribological Coatings', *Surface and Coatings Technology*, **100–101**, 1998, 1–6.
Matthews, A., Leyland, A. and Wilson, A., 'Plasma Immersion Ion Implantation as a Technique in Duplex and Hybrid Processing', *Vacuum*, **68**, 2003, 57–64.
McClelland, G.M. and Glosli, J.N., in *Proceedings on Fundamentals of Friction*, NATO ASI (eds I.L. Singer and H.M. Pollock), Kluwer, Dordrecht, 1992, pp. 405–426.
Merlo, A.M., 'The Contribution of Surface Engineering to the Product Performance in the Automotive Industry', *Surface and Coatings Technology*, **174–175**, 2003, 21–26.
Misina, M., Musil, J. and Kadlec, S., 'Composite TiN–Ni Thin Films Deposited by Reactive Magnetron Sputter Ion Plating', *Surface and Coatings Technology*, **110**, 1998, 168–172.
Mitterer, C., Losbichler, P., Hofer, F., Warbichler, P., Gibson, P.N. and Gissler, W., 'Nanocrystalline Hard Coatings Within the Quasi-Binary System TiN–TiB_2', *Vacuum*, **50**, 1998, 313–318.
Mitterer, C., Mayrhofer, P.H., Beschliesser, M., Losbichler, P., Warbichler, P., Hofer, F., Gibson, P.N., Gissler, W., Hruby, H., Musil, J. and Vlcek, J., 'Microstructure and Properties of Nanocomposite Ti–B–N and Ti–B–C Coatings', *Surface and Coatings Technology*, **120/121**, 1999, 405–411.
Murakami, Y., Sakae, C., Ichimaru, K. and Morita, T., 'Experimental and Fracture Mechanics Study of the Pit Formation Mechanism Under Repeated Lubricated Rolling–Sliding Contact: Effects of Reversal of Rotation and Change of the Driving Roller', *Journal of Tribology, Transactions ASME*, **119**, 1997, 788–796.
Musil, J., Hard and superhard nanocomposite coatings, *Surface and Coatings Technology*, **125**, 2000, 322–330.
Musil, J. and Regent, F., 'Formation of Nanocrystalline NiCr–N Films by Reactive dc Magnetron Sputtering', *Journal of Vacuum Science and Technology*, **A16**(6), 1998, 3301–3304.
Musil, J., Zeman, P., Hruby, H. and Mayrhofer, P.H., 'ZrN/Cu Nanocomposite Film – A Novel Superhard Material', *Surface and Coatings Technology*, **120/121**, 1999a, 179–183.
Musil, J., Leipner, I. and Kolega, M., 'Nanocrystalline and Nanocomposite CrCu and CrCu–N Films Prepared by Magnetron Sputtering', *Surface and Coatings Technology*, **115**, 1999b, 32–37.
Nastasi, M., Kodali, P., Walter, K., Embury, J., Raj, R. and Nakamura, Y., Fracture toughness of diamondlike carbon coatings, *Journal of Materials Research*, **14**(5), 1999, 2173–2180.
Neuville, S. and Matthews, A., 'Hard Carbon Coatings: The Way Forward', *MRS Bulletin*, **22**, 1997, 22–26.
Nie, X., Leyland, A., Song, H.W., Yerokhin, A.L., Dowey, S.J. and Matthews, A., 'Low Temperature Deposition of Cr(N)/TiO_2 Coatings Using a Duplex Process of Unbalanced Magnetron Sputtering and Micro-Arc Oxidation', *Surface and Coatings Technology*, **133**, 2000a, 331–337.
Nie, X., Wilson, A.D., Leyland, A. and Matthews, A., 'Deposition of Duplex Al_2O_3/DLC Coatings on Al Alloys for Tribological Applications Using a Combined Micro-Arc Oxidation and Plasma Immersion Ion Implantation Technique', *Surface and Coatings Technology*, **131**, 2000b, 506–513.
Nie, X., Tsotsos, C., Wilson, A., Yerokhin, A. L., Leyland, A. and Matthews, A., 'Characteristics of a Plasma Electrolytic Nitrocarburizing Treatment for Stainless Steels', *Surface and Coatings Technology*, **139**, 2001, 135–142.
Oberle, T.L., 'Properties Influencing Wear of Metals', *Journal of Metals*, **3**, 1951, 438–439.
Oliveira, S.A.G. and Bower, A.F., 'An Analysis of Fracture and Delamination in Thin Coatings Subject to Contact Loading', *Wear*, **198**, 1996, 15–32.
Olsson, U., Sjöström, H. and Sjödin, U., 'Increased Wear Resistance of Roller Bearings Using Me-C:H Coated Rollers', *Journal of Tribology, Transactions ASME*, **122**, 2000, 682–688.

Ortmann, S., Savan, A., Gerbig, Y. and Haefke, H., 'In-Process Structuring of CrN Coatings, and Its Influence on Friction in Dry and Lubricated Sliding', *Wear*, **254**, 2003, 1099–1105.

Page, T.F. and Knight, J.C., 'Factors Affecting the Tribological Behaviour of Thin Hard TiN and TiC Coatings', *Surface and Coatings Technology*, **39/40**, 1989, 339–354.

Perry, M. and Harrison, J., 'Molecular Dynamics of the Effects of Debris Molecules on the Friction and Wear of Diamond', *Thin Solid Films*, **290–291**, 1996, 211–215.

Polonsky, I. and Keer, L., 'Numerical Analysis of the Effect of Microstructure on Three-Dimensional Crack Propagation in the Coating Under Rolling Contact Fatigue Conditions', *Journal of Tribology, Transactions ASME*, **124**, 2002, 14–19.

Polonsky, I.A., Chang, T.P., Keer, L.M. and Sproul, W.D., 'An Analysis of the Effect of Hard Coatings on Near-Surface Rolling contact Fatigue Initiation Induced by Surface Roughness', *Wear*, **208**, 1997, 204–219.

Rabinowicz, E., 'Variation of Friction and Wear of Solid Lubricant Films with Film Thickness', *ASLE Transactions*, **10**, 1967, 1–9.

Raehle, B., 'Design and Development of a Tester to Evaluate the Impact Resistance of Coatings', MSc thesis, University of Hull, 1988.

Ramalingam, S. and Zheng, L., 'Film-Substrate Interface Stresses and Their Role in the Tribological Performance of Surface Coatings', *Tribology International*, **28**(3), 1995, 145–161.

Rebholz, C., Schneider, J., Ziegele, H., Rähle, B., Leyland, A. and Matthews, A., 'Deposition and Characterisation of Carbon-Containing Tungsten Coatings Prepared by Reactive Magnetron Sputtering', *Vacuum*, **49**(4), 1998a, 265.

Rebholz. C., Ziegele, H., Leyland, A. and Matthews, A., 'Structure, Mechanical and Tribological Properties of Ti–B–N and Ti–Al–B–N Thin Films Produced by Electron Beam Evaporation', *Journal of Vacuum Science and Technology A*, **16**(5), 1998b, 2851–2857.

Rebholz, C., Zielgele, H., Leyland, A. and Matthews, A., 'Structure, Mechanical and Tribological Properties of Nitrogen-Containing Chromium Coatings Prepared by Reactive Magnetron Sputtering', *Surface and Coatings Technology*, **115**, 1999a, 222–229.

Rebholz, C., Schneider, J.M., Leyland, A. and Matthews, A., 'Wear Behaviour of Carbon-Containing Tungsten Coatings Prepared by Reactive Magnetron Sputtering', *Surface and Coatings Technology*, **112**, 1999b, 85–90.

Rebholz, C., Leyland, A., Schneider, J.M., Voevodin, A.A. and Matthews, A., 'Structure, Hardness and Mechanical Properties of Magnetron-Sputtered Titanium–Aluminium Boride Films', *Surface and Coatings Technology*, **120/121**, 1999c, 412–417.

Rebholz. C., Schneider, J.M., Voevodin, A.A., Steinebrunner, J., Charitidis, C., Logothetidis, C., Leyland, A. and Matthews, A., 'Structure, Mechanical and Tribological Properties of Sputtered TiAIBN Thin Films', *Surface and Coatings Technology*, **113**, 1999d, 126–133.

Rebholz, C., Leyland A. and Matthews, A., 'Deposition and Characterization of TiAIBN Coatings Produced by Direct Electron-Beam Evaporation of Ti and Ti–Al–B–N Materials from a Twin Crucible Source', *Thin Solid Films*, **343–344**, 1999e, 242–245.

Rebholz, C., Leyland, A., Larour, P., Charitidis, C., Logothetidis, S. and Matthews, A., 'The Effect of Boron Additions on the Tribological Behaviour of TiN Coatings Produced by Electron-Beam Evaporative PVD', *Surface and Coatings Technology*, **116–119**, 1999f, 648–653.

Rebouta, L., Vaz, F., Andritschy, M. and Da Silva, M.F., 'Oxidation Resistance of (Ti,Al,Zr,Si)N Coatings in Air', *Surface and Coatings Technology*, **76–77**, 1995, 70–74.

Rickerby, D. and Matthews, A. (eds), '*Advanced Surface Coatings: A Handbook of Surface Engineering*', Blackie, London, 1991.

Robbins, M. and Krim, J., 'Energy Dissipation in Interfacial Friction', *MRS Bulletin*, June, 1998, 23–26.

Roberts, E.W., 'Ultralow Friction Films of MoS_2 for Space Applications', *Thin Solid Films*, **181**, 1989, 461–473.

Roberts, E.W., 'The Advantages and Limitations of Sputtered Molybdenum Disulphide as a Space Lubricant', *Proceedings of 4th European Symposium on "Space Mechanisms and Tribology"*, Cannes, France, 20–22 September 1989 (ESA SP-299, March 1990a), 1990, 59–65.

Ronkainen, H., 'Tribological Properties of Hydrogenated and Hydrogen-Free Diamond-Like Carbon Coatings', *VTT Publications*, No. 434, VTT Technical Research Centre of Finland, 2001.

Ronkainen, H., Nieminen, I., Holmberg, K., Leyland, A., Fancey, K.S., Matthews, A., Matthes, A. and Broszeit, E., 'Evaluation of Some New Titanium-Based Ceramic Coatings in Tribological Model Wear and Metal Cutting Tests', *Materials Science Engineering A*, **140**, 1991, 602–608.
Ronkainen, H., Nieminen, I., Holmberg, K., Leyland, A., Matthews, A., Matthes, B. and Broszeit, E., 'Evaluation of Some Titanium-Based Ceramic Coatings on High Speed Steel Cutting Tools', *Surface and Coatings Technology*, **49**, 1992, 468.
Ronkainen, H., Varjus, S. and Holmberg, K., 'Friction and Wear Properties in Dry, Water- and Oil-Lubricated DLC Against Alumina and DLV Against Steel Contacts', *Wear*, **222**, 1998a, 120–128.
Ronkainen, H., Koskinen, J., Varjus, S. and Holmberg, K., 'Estimations of Load Carrying Capacity in Diamond-Like Carbon Coated Systems', *Proceedings of COST 516 Tribology Symposium*, Espoo, Finland, 14–15 April, 1998b, 233–242.
Ronkainen, H., Varjus, S. and Holmberg, K., 'Friction and Wear Properties in Dry, Water- and Oil-Lubricated DLC Against Alumina and DLC Against Steel Contacts', *Wear*, **222**, 1998c, 120–128.
Ronkainen, H., Koskinen, J., Varjus, S. and Holmberg, K., 'Load-Carrying Capacity Evaluation of Coating/Substrate Systems for Hydrogen-Free and Hydrogenated Diamond-Like Carbon Films', *Tribology Letters*, **6**, 1999, 63–73.
Rosado, L., Jain, V. and Trivedi, H., 'The Effect of Diamond-Like Carbon Coatings on the Rolling Fatigue and Wear of M50 Steel', *Wear*, **212**, 1997, 1–6.
Roth, T., Kloos, K.H. and Broszeit, E., 'Structure, Internal Stresses, Adhesion and Wear Resistance of Sputtered Alumina Coatings', *Thin Solid Films*, **153**, 1987, 123–133.
Ruff, A.W., Shin, H. and Evans, C.J., 'Damage Processes in Ceramics Resulting from Diamond Tool Indentation and Scratching in Various Environments', *Wear*, **181–183**, 1995, 551–562.
Schaefer, L., Gaebler, J., Mulcahy, S., Brand, J., Hieke, A. and Wiltorf, R., 'Tribological Applications of Amorphous Carbon and Crystalline Diamond Coatings', in *Proceedings of 43rd Annual Technical Conference of the Society of Vacuum Coaters*, Denver, USA, SVC, Albuquerque, 15–20 April, 2000.
Scharf, T. and Singer, I., 'Monitoring Transfer Films and Friction Instabilities with In Situ Raman Tribometry', *Tribology Letters*, **14**(1), 2003a, 3–8.
Scharf, T. and Singer, I., 'Thickness of Diamond-Like Carbon Coatings Quantified with Raman Spectroscopy', *Thin Solid Films*, **440**, 2003b, 138–144.
Scharf, T. and Singer, I., 'Quantification of the Thickness of Carbon Transfer Films Using Raman Tribometry', *Tribology Letters*, **14**(2), 2003c, 137–145.
Schuelke, T., 'Overview of Wear Coating Technologies', in *Gorham Workshop*, Atlanta, USA, 13 November, 2000.
Shepard, S.R. and Suh, N.P., 'The Effects of Ion Implantation on Friction and Wear of Metals', *Transactions of ASME, Journal of Lubrication Technology*, **104**, 1982, 29–38.
Sherbiney, M.A. and Halling, J., 'Friction and Wear of Ion-Plated Soft Metallic Films', *Wear*, **45**, 1977, 211–220.
Sin, H., Saka, N. and Suh, N.P., 'Abrasive Wear Mechanisms and the Grit Size Effect', *Wear*, **55**, 1979, 163–190.
Singer, I., 'How Third-Body Processes Affect Friction and Wear', *MRS Bulletin*, June, 1998, 37–40.
Singer, I., Dvorak, S., Wahl, K. and Schafer, T., 'Role of Three Bodies in Friction and Wear of Protective Coatings', *Journal of Vacuum and Science Technology*, **A21**(5), 2003, 9.
Singer, I.L., Fayeulle, S. and Ehni, P.D., 'Wear Behaviour of Triode-Sputtered MoS_2 Coatings in Dry Sliding Contact with Steel and Ceramics', *Wear*, **195**, 1996a, 7–20.
Singer, I.L., Le Mogne, T., Donnet, C. and Martin, J.M., 'In Situ Analysis of the Tribological Films Formed by SiC Sliding Against Mo in Partial Pressures of SO_2, O_2 and H_2S Gases', *Journal of Vacuum Science and Technology*, **A14**(1), 1996b, 38–45.
Sjöström, H. and Wicksröm, V., 'Diamond-Like Carbon Coatings in Rolling Contacts', *Proceedings of the Institution of Mechanical Engineers, Part: J*, **215**, 2001, 545–561.
Spalvins, T. and Sliney, H.E., 'Frictional Behavior and Adhesion of Ag and Au Films Applied to Aluminium Oxide by Oxygen-Ion Assisted Screen Cage Ion Plating', *Surface and Coatings Technology*, **68/69**, 1994, 482–488.
Springer, R.W. and Catlett, D.S., 'Structure and Mechanical Properties of Al/Al_xO_y Vacuum Deposited Laminates', *Thin Solid Films*, **54**, 1978, 197–205.

Springer, R.W. and Harford, C.D., 'Characterization of Aluminium–Aluminium Nitride Coatings Sputter-Deposited Using the Pulsed Gas Process', *Journal of Vacuum Science and Technology*, **20**, 1982, 462–464.

Sproul, W.D., 'New Routes in the Preparation of Mechanically Hard Films', *Science*, **273**, 1996, 889.

Stupp, B.C., 'Synergistic Effects of Metal Co-Sputtered with MoS_2', *Thin Solid Films*, **84**, 1981, 257–266.

Sue, J.A. and Troue, H.H., 'Friction and Wear Properties of Titanium Nitride Coating in Sliding Contact with AISI 01 Steel', *Surface and Coatings Technology*, **43/44**, 1990, 709–720.

Suh, N.P., '*Tribophysics*', Prentice-Hall, Englewood Cliffs, New Jersey, USA, 1986, p. 489.

Svahn, F., Kassman-Rudolphi, Å. and Wallen, E., 'The Influence of Surface Roughness on Friction and Wear of Machine Element Coatings', *Wear*, **254**, 2003, 1092–1098.

Takagi, R. and Liu, T., 'The Lubrication of Steel by Electroplated Gold', *ASLE Transactions*, **10**, 1967, 115–123.

Tangena, A.G., 'Tribology of Thin Film Systems', Doctoral thesis, Eindhoven Technical University, 1987, p. 130.

Teer, D., 'New Solid Lubricant Coatings', *Wear*, **251**, 2001, 1068–1074.

Tervo, J., 'Wear Properties of High Nitrogen Austenitic Stainless Steels', *Acta Polytechnica Scandinavica*, Mechanical Engineering Series No. 128, Espoo, Finland, 1998, p. 88.

Theunissen, G.S.A.M., 'Wear Coatings for Magnetic Thin Film Magnetic Recording Heads', *Tribology International*, **31**(9), 1998, 519–523.

Thompson, P.A. and Robbins, M.O., 'Simulations of Contact-Line Motion: Slip and the Dynamic Contact Angle', The American Physical Society, *Physical Review Letters*, **63**(7), 1989, 766–769.

Tricoteaux, A., Jouan, P., Guerin, J., Martinez, J. and Djouadi, A., 'Fretting Wear Properties of CrN and Cr_2N Coatings', *Surface and Coatings Technology*, **174–175**, 2003, 440–443.

Tsui, T., Pharr, G., Oliver, W., Bhatia, C., White, R., Anders, S., Anders, A. and Brown, I., 'Nanoindentation and Nanoscratching of Hard Carbon Coatings for Magnetic Discs', *Material Research Society Symposium Proceedings*, Material Research Society, **383**, 1995, 447–452.

Tsuya, Y. and Takagi, R., 'Lubricating Properties of Lead Films on Copper', *Wear*, **7**, 1964, 131–143.

Tuszynski, W., Szczerek, M. and Michalczewski, R., 'Investigation on Anti-Wear Coatings Deposited by the PVD Process', *Tribotest Journal*, **10**, 2003, 3–18.

Van der Kolk, G.J., Hurkmans, T., Trinh, T. and Fleischer, W., 'Coating Evaluations Of Decorative PVD Finishes', in *Proceedings of 41st Annual Technical Conference of the Society of Vacuum Coaters*, SVC, Albuquerque, Boston, USA, 18–23 April, 1998.

Veprek, S., 'Conventional and New Approaches Towards the Design of Novel Superhard Materials', *Surface and Coatings Technology*, **97**, 1997, 15–22.

Veprek, S., 'The Search for Novel, Superhard Materials', *Journal of Vacuum Science and Technology*, **A17**(5), 1999, 2401–2420.

Veprek, S., Nesladek, P., Niederhofer, A., Glatz, F., Jilek, M. and Sima, M., 'Recent Progress in the Superhard Nanocrystalline Composites: Towards Their Industrialization and Understanding of the Origin of the Superhardness', *Surface and Coatings Technology*, **108/109**, 1998, 138–147.

Vercammen, K., Meneve, J., Dekempeneer, E. and Smeets, J., 'Study of RF PACVD Diamond-Like Carbon Coatings for Space Mechanisms Applications', International Conference on Metallurgical Coatings and Thin Films, San Diego, 12–15 April, 1999.

Vincent, L., Bertier, Y. and Godet, M., 'Testing Methods in Fretting Fatigue: A Critical Appraisal', in *Standardization of Fretting Fatigue Test Methods and Equipment* (eds M. Attia and R. Waterhouse), ASTM SP 1159, 1992, 33–48.

Vingsbo, O. and Söderberg, S., 'On Fretting Maps', *Wear*, **126**, 1988, 131–147.

Voevodin, A. and Zabinski, J., 'Supertough Wear-Resistant Coatings with "Chameleon" Surface Adaption', *Thin Solid Films*, **370**, 2000, 223–231.

Voevodin, A.A., Bantle, R. and Matthews, A., 'Dynamic Impact Wear of TiC_xN_y and Ti–DLC Composite Coatings', *Wear*, **185**, 1995a, 151–157.

Voevodin, A.A., Rebholz, C. and Matthews, A., 'Comparative Tribology Studies of Hard Ceramic and Composite Metal-DLC Coatings in Sliding Friction Conditions', *Tribology Transactions*, **38**, 1995b, 829–836.

Voevodin, A.A., Schneider, J.M. Rebholz, C. and Matthews, A., 'Multilayer Composite Ceramic-Metal-DLC Coatings for Sliding Wear Applications', *Tribology International*, **29**(7), 1996, 559–570.

Voevodin, A.A., O'Neill, J.P. and Zabinski, J.S., Nanocomposite Tribological Coatings for Aerospace Applications', *Surface and Coatings Technology*, **116/119**, 1999a, 36–45.

Voevodin, A.A., O'Neill, J.P. and Zabinski, J.S., 'Tribological Performance and Tribochemistry of Nanocrystalline WC/Amorphous Diamond-Like Carbon Composites, *Thin Solid Films*, **342**, 1999b, 194–200.

Wahl, K.J. and Unertl, W.N., 'Formation of Nanometer-Scale Contacts to Viscoelastic Materials', in *Tribology Issues and Opportunities in MEMS* (ed B. Bhushan), Kluwer Academic Publishers, London, 1998, pp. 261–271.

Wahl, K.J., Dunn, D.N. and Singer, I.L., 'Wear Behavior of Pb-Mo-S Solid Lubricating Coatings, *Wear*, **230**, 1999, 175–183.

Wang, D. and Kato, K., 'Coating Thickness Effects on Initial Wear of Nitrogen-Doped Amorphous Carbon in Nano-Scale Sliding Contact: Part I: In Situ Examination', *Tribology International*, **36**, 2003a, 649–658.

Wang, D. and Kato, K., 'Coating Thickness Effects on Initial Wear of Nitrogen-Doped Amorphous Carbon in Nano-Scale Sliding Contact: Part II: Theoretical Modelling', *Tribology International*, **36**, 2003b, 659–665.

Waterhouse, R., '*Fretting Fatigue*', Applied Science, Elsevier, London, 1981.

Wiklund, U., 'Mechanics and Tribology of Micro- and Nanolayered PVD Coatings', Ph.D. thesis, Acta Universitatis Upsaliensis, *Comprehensive Summaries of Uppsala Dissertations from the Faculty of Science and Technology* 428, Uppsala, Sweden, 1999, p. 47.

Wilson, S. and Alpas, A.T., 'TiN Coating Wear Mechanisms in Dry Sliding Contact Against High Speed Steel', *Surface and Coatings Technology*, **108–109**, 1998, 369–376.

Yamada, Y., Tanaka, K. and Saito, K., 'Friction and Damage of Coatings Formed by Sputtering Polytetrafluoroethylene and Polyimide', *Surface and Coatings Technology*, **43/44**, 1990, 618–628.

Yang, S.H., Kong, H., Yoon, E.S. and Kim, D.E., 'An Experimental Study of the Rolling Resistance of Bearings Coated by Pure Silver', *Wear*, **225–229**, 1999, 119–126.

Ye, N. and Komvopoulos, K., 'Effect of Residual Stress in Surface Layer on Contact Deformation of Elastic–Plastic Layered Media', *Journal of Tribology, Transactions of ASME*, **125**, 2003a, 692–699.

Ye, N. and Komvopoulos, K., 'Indentation Analysis of Elastic–Plastic Homogenous and Layered Media: Criteria for Determining the Real Material Hardness', *Journal of Tribology, Transactions of ASME*, **125**, 2003b, 685–691.

Yerokhin, A.L., Nie, X., Leyland, A., Matthews, A. and Dowey, S.J., 'Plasma Electrolysis for Surface Engineering', *Surface and Coatings Technology*, **122**, 1999, 73–93.

Yerokhin, A.L., Nie, X., Leyland, A. and Matthews, A., 'Characterisation of Oxide Films Produced by Plasma Electrolytic Oxidation of Ti–6Al–4V Alloy', *Surface and Coatings Technology*, **130**, 2000, 195–206.

Yuan, F. and Hayashi, K., 'Influence of the Grain Size of the Aluminium Coating on Crack Initiation in Indentation', *Wear*, **225–229**, 1999, 83–89.

Zabinski, J.S. and Donley, M.S., 'Lubricant Coatings', US Patent, 5, 282, 985, 1994.

Zabinski, J.S. and Prasad, S.V., 'Advanced Solid Lubricant Coatings for Aerospace Systems', *Proceedings of the AGARD Conference 'Tribology of Aerospace Systems'*, Sesimbra, Portugal, AGARD-CP-589, 1996.

Zabinski, J.S., Donley, M.S., Dyhous, V.J. and McDevitt, N.T., 'Chemical and Tribological Characterization of PbO–MoS_2 Films Grown by Pulsed Laser Deposition', *Thin Solid Films*, **214**, 1992, 156–163.

Zabinski, J.S., Donley, M.S., Walck, S.D., Schneider, T.T., and McDevitt, N.T., 'The Effects of Dopants on the Chemistry and Tribology of Sputter-Deposited MoS_2Films', *Tribology Transactions*, **38**, 1995, 894–904.

Zabinski, J.S., Prasad, S.V. and McDevitt, N.T., 'Advanced Solid Lubricant Coatings for Aerospace Systems', in *Proceedings of the AGARD Conference on Tribology of Space Systems*, Sesimbra, Portugal, AGARD, CP-589, 31.3.12, 6–7 May, 1996.

Zhang, L. and Tanaka, H., 'Towards a Deeper Understanding of Wear and Friction on the Atomic Scale – A Molecular Dynamics Analysis', *Wear*, **211**, 1997, 44–53.
Zhang, L. and Tanaka, H., 'Atomic Scale Deformation in Silicone Monocrystals Induced by Two-Body and Three-Body Contact Sliding', *Tribology International*, **31**(8), 1998, 425–433.
Zheng, L. and Ramalingam, S., 'Stresses in a Coated Solid Due to Shear and Normal Boundary Tractions', *Journal of Vacuum Science and Technology*, **A13**(5), 1995, 2390–2398.
Zum Gahr, K.-H., 'Wear by Hard Particles', *Tribology International*, **31**(10), 1998, 587–596.

8

Wear of Ceramics: Wear Transitions and Tribochemical Reactions

S. Jahanmir

Abstract

Structural ceramics are used in diverse tribological applications due to their unique properties that include resistance to abrasion and erosion, resistance to corrosive wear, wear resistance at elevated temperatures, low density, and unique electrical, thermal, and magnetic properties. Applications include precision instrument bearings, cutting tool inserts, prosthetic articulating joints, and engine components. Following a brief overview of the processing methods applied to alumina, silicon nitride, and silicon carbide, the wear behavior of these ceramics is reviewed. The wear behavior changes as the load or the coefficient of friction is increased. This change is associated with a drastic increase in wear at a critical load that depends on the material and test conditions. This transition from mild to severe wear occurs through a microfracture process at the sliding contact. Contact mechanics is used to develop a simple model to determine whether contact failure occurs by a classical brittle process (i.e. cone cracking) or a quasi-plastic behavior (i.e. small microcracks distributed below the contact surface). The model provides estimates of the transition load within a factor of 2 of the experimentally obtained values and indicates that the brittle behavior dominates the observed mild to severe wear transition in the three ceramics evaluated in this study. The limitations of the model and its use in design of tribological components are discussed. The role of lubrication and tribochemical reactions between ceramic surfaces and atmospheric moisture on the wear behavior is discussed.

Wear – Materials, Mechanisms and Practice Edited by G. Stachowiak

8.1 Introduction

Ceramics are generally defined as inorganic nonmetallic solid materials. This definition includes not only materials such as pottery, porcelains, refractories, cements, abrasives, and glass, but also nonmetallic magnetic materials, ferroelectrics, and a variety of other new materials. Renewed interest in ceramics is rooted in unique materials classified as advanced structural ceramics, and electronic and optical ceramics [1]. Structural ceramics are presently used in diverse applications as tribomaterials [2, 3] due to their unique properties that include resistance to wear and corrosion at elevated temperatures, low density, and unique electrical, thermal, and magnetic properties. Applications include precision instrument bearings, cutting tool inserts, prosthetic articulating joints, and engine components.

During the past 20 years, a large number of studies have been carried out on the tribological properties of structural ceramics. This chapter is not intended as a comprehensive, all-inclusive review of the subject, due to space limitations. The chapter focuses on two issues that determine performance of structural ceramics in tribological applications: the transition from mild to severe wear and tribochemical reactions. The influence of lubrication on wear and particularly on wear transition is also discussed. The transition from mild to severe wear that occurs through a microfracture process at the sliding contact is modeled to determine whether contact failure occurs by a classical brittle process (i.e. cone cracking) or a quasi-plastic behavior (i.e. small microcracks distributed below the contact surface). The limitations of the proposed model and its use in design of tribological components are discussed.

8.2 Structure and Properties of Ceramics

Structural ceramics, by definition, are high-strength ceramics used in load-bearing applications. In this section, the structure, processing methods, and properties of three commonly used structural ceramics are reviewed. More detailed information of the wear behavior of some of these materials is discussed in the next section.

8.2.1 Alumina Ceramics

Ceramics based on aluminum oxide or alumina have been used in commercial applications for many years because of their availability and low cost. Alumina ceramics are often classified either as high aluminas having mass fractions more than 80% aluminum oxide or as porcelains with less than 80% aluminum oxide. High aluminas are used in many mechanical devices and in electronics.

Pure aluminum oxide, Al_2O_3, has one thermodynamically stable phase at room temperature, designated as alpha phase. Often the term corundum is used for alpha alumina. Commercial aluminas are densified by pressureless sintering, using MgO sintering aid for high-purity (>95%) aluminas or silicates for less-expensive, low-purity grades. Fracture strength improves as the percentage of alumina is increased. However, cost increases because of processing difficulty. The thermal shock resistance of aluminas is generally inferior to other structural ceramics (e.g. silicon nitride) because of the highly anisotropic properties that depend on the crystallographic orientation of alumina grains.

The tribological behavior of alumina depends on the composition and microstructure of the material, as well as on the test conditions. The tribological behavior of different materials can be compared only if the tests are conducted under the same conditions. However, analysis of published data shows certain trends [4]. Although the variability in published results on the effects of contact conditions on wear is large, the wear rate of alumina generally increases as either sliding speed or load is increased.

8.2.2 Silicon Nitride Ceramics

Silicon nitride, one of the strongest structural ceramics, has emerged as an important tribological material, especially in rolling applications. It has low elastic modulus, high strength, and outstanding fatigue resistance, excellent oxidation resistance due to a protective surface oxide layer, and very good thermal shock resistance because of its low thermal expansion coefficient. In oxidizing environments, silicon nitride is stable only at very low partial pressures of oxygen; in air, it rapidly forms a silicon oxide surface layer. This layer protects against further oxidation; if it is removed, for example by wear, oxidation occurs rapidly.

Pure silicon nitride, Si_3N_4, exists in two crystallographic forms designated as alpha and beta. Since the beta phase is thermodynamically more stable, silicon nitride materials are primarily in the beta phase; but starting powder is usually in the alpha phase. Commercial silicon nitride materials are processed with various oxide sintering aids.

Silicon nitride materials are classified according to processing techniques: sintered, hot-pressed, reaction-bonded (or reaction-sintered), sintered-reaction-bonded, and hot-isostatically pressed. Variations in composition, microstructure, and properties depend on the processing route used in fabrication.

Silicon nitride powder compacts are sintered to full density using combinations of rare earth oxides and alumina sintering aids. However, mechanical properties (i.e. strength, hardness, and fracture toughness) of commercially available sintered silicon nitrides are inferior to those processed by hot-pressing, usually containing MgO or Y_2O_3 sintering aids. Application of pressure during sintering leads to materials with nearly full density and very good properties. Shapes that can be formed by hot-pressing hot-isostatic-pressing (HIP) and gas-pressure sintering (GPS), however, are limited and processing cost is relatively high.

Reaction-bonded silicon nitride is made by pressing silicon powder and reacting the preform with nitrogen at high temperatures. While they are much less expensive than hot-pressed or sintered materials, their porosity, which is greater than 10%, results in inferior mechanical properties. Adding oxide sintering aids to the starting silicon powder and a subsequent hot-pressing, or HIP, reduces this porosity and improves the properties. The principal advantage of reaction-bonded silicon nitride is its lower cost of starting powder.

The tribological performance and mechanical properties of silicon nitride depend on composition and microstructure, as well as on the processing procedure and types of starting powders. The published [4] friction and wear data for different types of silicon nitrides show a wide scatter due to different test conditions used in various studies, making it difficult to make a clear distinction between performance of different types of silicon nitrides. The data, however, show that replacing the silicon nitride counterface with steel or alumina slightly increases friction, and a zirconia counterface can decrease the friction coefficient. The wear rate increases as either the load or the temperature is increased, especially above 800°C.

8.2.3 Silicon Carbide Ceramics

Silicon carbide ceramics are widely used in applications requiring wear resistance, high hardness, retention of mechanical properties at elevated temperatures, and resistance to corrosion and oxidation. The oxidation resistance is due to a protective silicon oxide layer, as with silicon nitride. The thermal shock resistance is somewhat lower than that of silicon nitride. Silicon carbide ceramics are grouped into four types depending on processing methods: reaction-bonded (or reaction-sintered), hot-pressed, sintered, and CVD (chemical vapor deposition).

In the reaction bonding process, a mixture of silicon carbide powder, graphite, and a plasticizer is pressed in a mold to prepare a preform or "green" compact. After the plasticizer is burned off to produce a porous product, silicon metal is infiltrated into the pores as a liquid or vapor. The reaction between silicon and carbon to form SiC is not complete, leaving some residual Si and C; usually excess Si is used to fill the pores. The finished material has little porosity and contains a mixture of Si, C, and reaction-formed SiC in between the original SiC particles. Mechanical properties depend on the amount of free Si and C. Since the densification process does not produce shrinkage, tight dimensional tolerances are more easily achieved than with sintering. The primary advantage of reaction-bonded silicon carbide is its relatively low cost. Because components can be made to near net-shape, little machining is required after densification.

Hot-pressing can also produce high-strength silicon carbide of nearly full density. In this process, boron and carbon, and sometimes alumina, are used as sintering aids for processing of both alpha and beta silicon carbide components. Although hot-pressed silicon carbide exhibits very good mechanical properties, its use is limited by the high cost of finished components, due to difficulty in machining after densification.

Silicon carbide components are also produced by sintering without the application of pressure, using carbon and boron sintering aids. The major advantage of this process is that most of the machining can be easily done on the green compact. The densified component is then machined by diamond grinding and polishing.

CVD is used to produce a relatively pure and dense silicon carbide. This material is highly anisotropic, due to the columnar structure developed during the deposition process. In addition to anisotropy, high cost and residual stresses are major drawbacks against widespread use. Nevertheless, CVD SiC is an excellent coating material where resistance to wear, erosion, or oxidation is required.

The tribological performance and properties of silicon carbides are sensitive to the processing conditions and microstructure of the material. Analysis of published data [4] suggests that, in general, the friction coefficient of reaction-bonded silicon carbide is lower than that of other types because the excess carbon acts as a solid lubricant. However, this reduction depends on the specific microstructure and amount of free carbon, as well as on test conditions. The friction coefficient for the sintered materials decreases with increasing temperature and speed. Both friction and wear for hot-pressed silicon carbides are lower in sliding against alumina and zirconia than sliding against silicon nitride and silicon carbide.

8.3 Wear Transitions

Various studies on tribological behavior of ceramics have shown that the wear mechanisms depend on contact conditions during laboratory testing. A common feature in most structural ceramics is that wear occurs through a small-scale fracture (i.e. microfracture) process if the contact load is higher than a threshold value specific to each material [5]. A second common feature associated with many ceramics is their high reactivity with water vapor in the air environment forming oxide and hydroxide surface layers [6]. Such reactions are accelerated during wear testing, and thus are referred to as tribochemical reactions. This section reviews the wear mechanisms associated with a high-purity alumina, a hot-isostatically pressed silicon nitride, and a sintered silicon carbide under conditions of low sliding speeds. The specific properties of the three ceramics are listed in Table 8.1. Particular attention is given to the transition from mild to severe wear and to tribochemical reactions with water. The terms "mild" and "severe" are used to distinguish the two extreme behaviors. Mild wear refers to a wear coefficient lower than 10^{-4} and severe wear refers to unacceptably high wear above this value. The wear coefficient is a dimensionless parameter that is obtained by multiplying the measured wear volume by hardness of the ceramic and dividing this product by the sliding distance and contact load [7]. This normalization implies that the wear volume is proportional to the applied load and inversely proportional to the hardness of the material. Such linear relationships do not generally hold for most ceramics under a wide range of contact conditions. Nevertheless, wear coefficient serves as a universal parameter for comparison of wear data obtained under different test conditions.

8.3.1 Alumina

The tribological behavior of a high-purity alumina (99.8%) sliding in air in self-mated, ball-on-flat, unlubricated tests is displayed in Figure 8.1 [8] as a function of normal load and temperature. The wear coefficients, K, and the coefficients of friction, f, are included in the figure. The wear transition diagram (Figure 8.1) delineates the effects of load and temperature and identifies the boundaries between regions dominated by different mechanisms under low sliding speeds. The wear transition diagram can be used to predict the tribological behavior and to determine the useable range of conditions (i.e. load and temperature) for a given material. The tribological characteristics of this alumina under low speed sliding are divided into four distinct regions on the basis of the fundamental mechanisms involved in the wear process. At low temperatures in Region I (T <200°C), tribochemical reactions between alumina and water vapor in the environment control the behavior; friction coefficient and wear coefficient are relatively low at 0.40 and 10^{-6}, respectively. Formation of a thin hydroxide film has been confirmed with Fourier transform infrared microscopy (FTIR) [9] and is usually associated with formation of cylindrical wear debris, shown in the micrograph in Figure 8.2(a) on the wear surface

Table 8.1 Mechanical properties of selected ceramics

Mechanical properties	Alumina (99.8%)	Alumina (99.5%)	Silicon nitride (HIP)	Silicon carbide (sintered)
Vickers hardness (GPa)	15.0	14.7	19.6	36
Elastic modulus (GPa)	345	372	310	406
Poisson's ratio	0.22	0.22	0.28	0.14
Fracture toughness (MPa $m^{1/2}$)	3.2	3.0	3.9–5.4	3.7
Flexural strength (MPa)	331	262	1000	400

The values listed in the table were obtained from several references [8, 14, 15, 19].

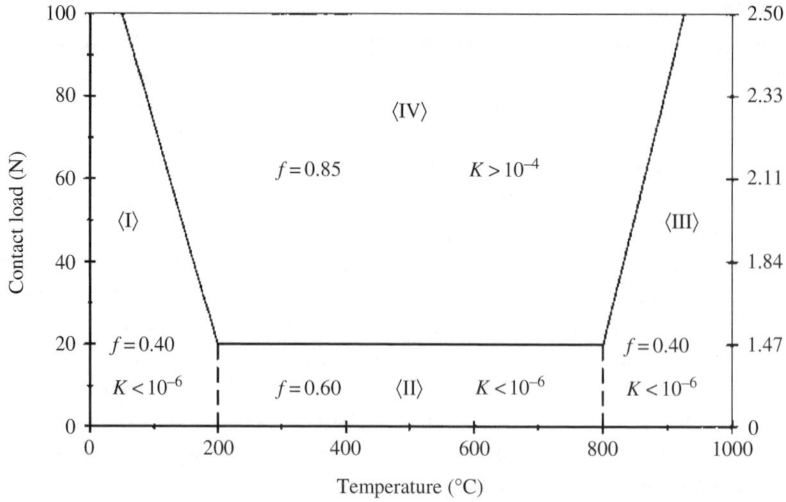

Figure 8.1 Wear transition diagram for a 99.8% alumina [8]. Self-mated, reciprocating, ball-on-flat tests in air; sliding speed: 1.4 mm/s, ball diameter: 12.7 mm, polished surfaces, R_a: 0.1–0.2 μm

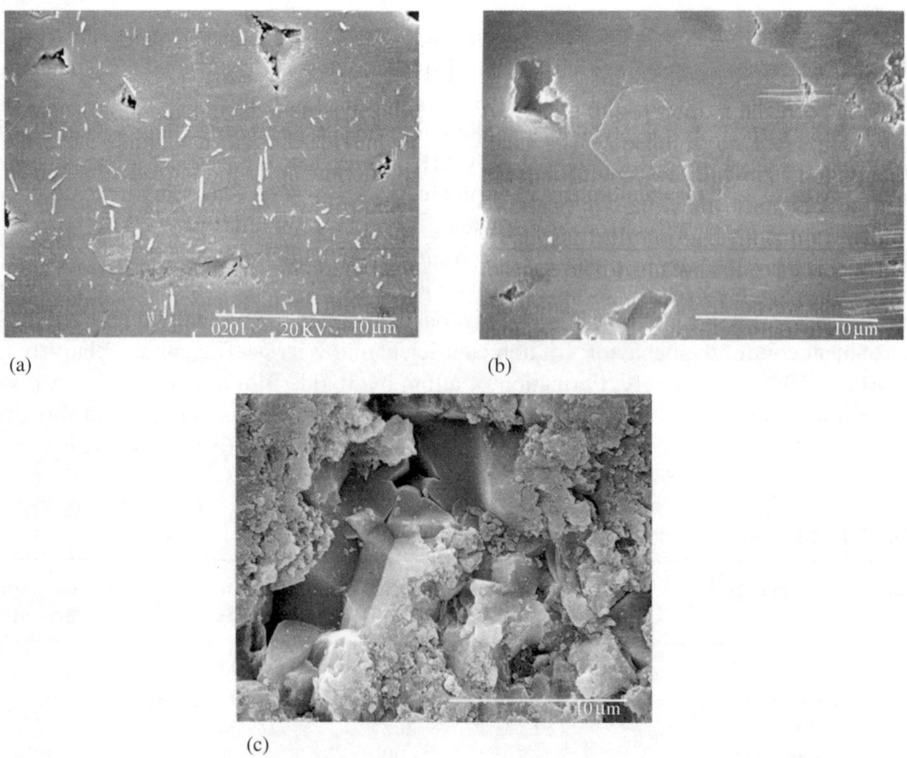

Figure 8.2 Micrographs of alumina wear tracks: (a) Region I, reaction with water; (b) Region II, plowing and onset of microfracture; and (c) Region IV, microfracture [8]

of alumina. At intermediate temperatures (200°C < T < 800°C), wear behavior depends on contact load. At low loads in Region II, wear occurs by plastic flow and plowing (Figure 8.2(b)); friction and wear are low, similar to those observed at the lower temperatures. At loads higher than 20 N, in Region IV, severe wear by intergranular fracture (Figure 8.2(c)) results in a friction coefficient of 0.85 and a wear coefficient larger than 10^{-4}. A fracture mechanics model [10] has confirmed that the observed mild to severe wear transition at the intermediate temperatures is controlled by propagation of microcracks from pre-existing near surface flaws. At temperatures above 800°C (in Region III), both the coefficient of friction and wear coefficient are low because of the formation of a thin surface layer from diffusion and viscous flow of the silica glass grain boundary phase. The transition from mild to severe wear is sensitive to the sliding speed [11, 12]. For example, for a 99.7% purity alumina tested under a constant load of 10 N, the wear increases by more than two orders of magnitude as the speed is increased from 0.003 to 0.3 m/s [12]. Considering the high wear associated with the wear transition, it is clear that the use of alumina ceramics should be avoided under contact conditions that promote microfracture. The wear rate and the wear transition are both influenced by the grain size [13] and the glass content [11] due to the effect of these parameters on the fracture toughness of aluminas.

The transition from mild to severe wear is also observed in lubricated tests with hydrocarbon fluids. Figure 8.3 shows typical data obtained for a 99.5% alumina [14], indicating abrupt increases in wear as the load reaches a critical value that depends on the sliding speed. Observation of the wear scars before and after this transition confirmed that severe wear had occurred through a microfracture process similar to the unlubricated tests. The results in Figure 8.3 were obtained using a step-loading four-ball test configuration with pure paraffin oil as the lubricant. The transition load is higher for the low-speed lubricated contacts compared with the unlubricated tests in Figure 8.1. The transition load in Figure 8.3 decreases from 60 to 20 N as the rotational speed of the upper ball is increased from 8 rpm (0.0031 m/s) to 1000 rpm (0.38 m/s). The friction coefficient prior to transition was about 0.10, independent of speed, and therefore an influence from elastohydrodynamic lubrication

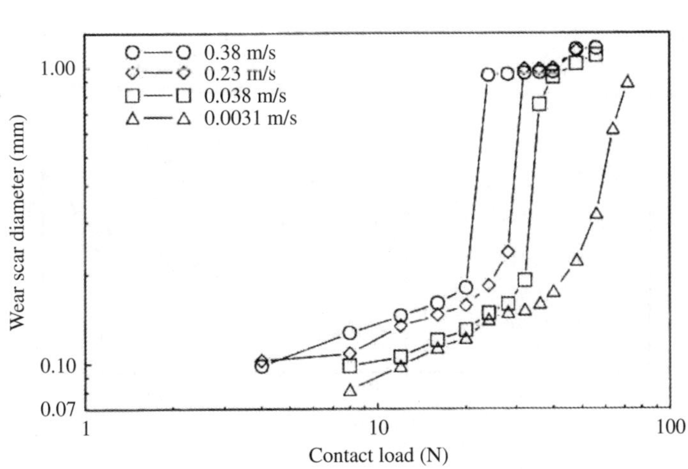

Figure 8.3 Wear transition in a 99.5% alumina under lubrication [14]. Self-mated, four-ball tests with a light paraffin oil, viscosity: 27.4 cSt, ball diameter: 12.7 mm, R_a: 0.1–0.2 μm, room temperature

can be ruled out. The reduction in the transition load is possibly due to thermoelastic stresses as the contact temperature increases at higher speeds.

8.3.2 Silicon Nitride

The wear transition diagram for a hot-isostatically pressed silicon nitride in low-speed, self-mated, ball-on-flat, unlubricated tests in air comprises five regions, Figure 8.4 [15]. The boundaries (the cross-hatched area) between the five regions are broader than those in alumina. At low loads and relatively low temperatures ($T <$ 400°C) in Region I, the tribological behavior is controlled by formation of silicon hydroxide on the wear track [16], with a friction coefficient of 0.30 and a wear coefficient of approximately 10^{-4}. Formation of hydrated silicon oxide has been confirmed by FTIR, electron loss spectroscopy, X-ray diffraction, and Auger spectroscopy [16]. Cylindrical wear debris formed from the hydroxide film is shown in Figure 8.5(a). In Region II (400°C$< T <$ 700°C), selective oxidation of WC inclusions controls the wear behavior. Formation of crystalline precipitates from amorphous magnesium silicate grain boundary phase controls the wear process in Region III (700°C $< T <$ 900°C), where both the friction coefficient and wear coefficient increase. Oxidation of silicon nitride controls the wear behavior in Region IV ($T >$ 900°C); the friction coefficient is approximately 0.70 and the wear coefficient increases to 10^{-2}. In Region V, similar to alumina, microfracture is the primary wear mechanism, Figure 8.5(b). The friction coefficient in this region is approximately 0.80 and the wear coefficient is very high at 10^{-2}. The critical load for onset of microfracture decreases from about 14 to 3 N as the temperature increases to 900°C. This transition is associated with an increase in wear by two orders of magnitude. Unlike alumina, the transition load in silicon nitride is less sensitive to an increase in speed [17]. The high loads in Region V must be avoided because of large wear coefficients when

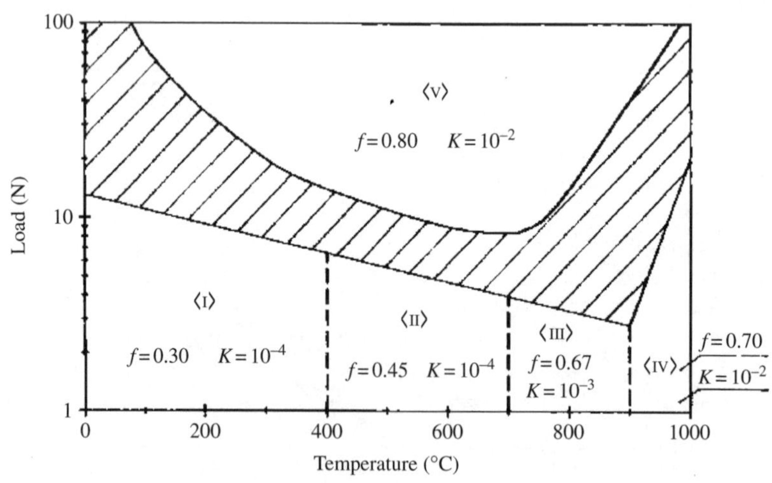

Figure 8.4 Wear transition diagram for a hot-isostatically pressed silicon nitride [15]. Self-mated, reciprocating, ball-on-flat tests in air; sliding speed: 1.4 mm/s, ball diameter: 12.7 mm, polished surfaces, R_a: 0.1–0.2 μm

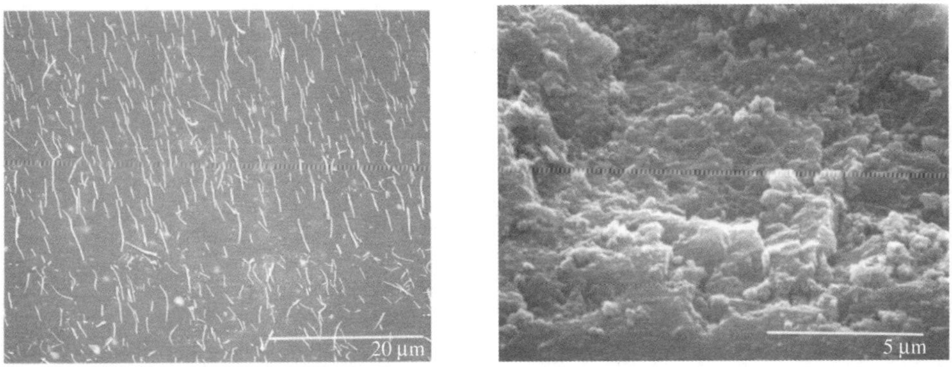

(a) (b)

Figure 8.5 Micrographs of silicon nitride wear tracks: (a) Region I, reaction with water; and (b) Region V, microfracture [15]

wear occurs by microfracture. The wear coefficient of the HIP silicon nitride (Figure 8.4) is too high even under low load conditions. Lubrication is, therefore, necessary to control wear of this silicon nitride.

Pure hydrocarbon fluids (e.g. hexadecane) and boundary lubricant additives (e.g. stearic acid) are effective in reducing the coefficient of friction (to 0.10 or lower) and reducing the wear coefficient (to 10^{-5} or lower) by preventing the microfracture wear mode [16]. In another study [18] where step-loading four-ball tests were used, a pure paraffin oil increased the transition load of an HIP silicon nitride to 100 N and above depending on the rotational speed.

8.3.3 Silicon Carbide

The wear transition diagram for a sintered silicon carbide in low-speed, self-mated, ball-on-flat, unlubricated tests in Figure 8.6 shows four regions [19]. At room temperature, high loads, and high relative humidities, friction coefficient is approximately 0.23 and the wear coefficient is 10^{-3}. In Region I, tribochemical reaction between water vapor and silicon carbide controls the tribological behavior. The tribochemical reactions with water form a continuous thin film on the wear track, Figure 8.7(a). This film fractures during examination in the high vacuum of SEM forming microcrack networks that are similar to "mud cracks." As the humidity decreases from about 70 to 30%, the friction coefficient increases to 0.70. In Region II, wear occurs by plowing, Figure 8.7(b), and the friction coefficient is 0.63, irrespective of humidity. Both friction and wear decrease in Region III due to oxidation of silicon carbide and formation of large cylindrical debris on the wear track. At high loads in Region IV, wear occurs by microfracture, Figure 8.7(c), resulting in a high coefficient of friction and wear coefficient. There is only a slight difference in the behavior of wear debris in the two parts of Region IV. At low temperatures, the wear debris appear loosely attached to the surface, whereas at higher temperatures, some of the debris form compacted regions. The mild to severe wear transition and associated change in the wear mechanisms to microfracture occurs at a load of 7 N, irrespective of temperature. The wear coefficient of this silicon carbide increases only by one order of magnitude. Lubrication with paraffin

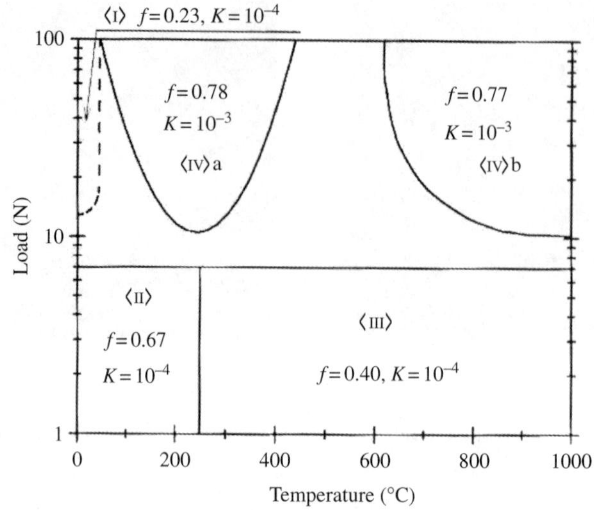

Figure 8.6 Wear transition diagram for a sintered silicon carbide [19]. Self-mated, reciprocating, ball-on-flat tests in air; sliding speed: 1.4 mm/s, ball diameter: 12.7 mm, polished surfaces, R_a: 0.1–0.2 μm

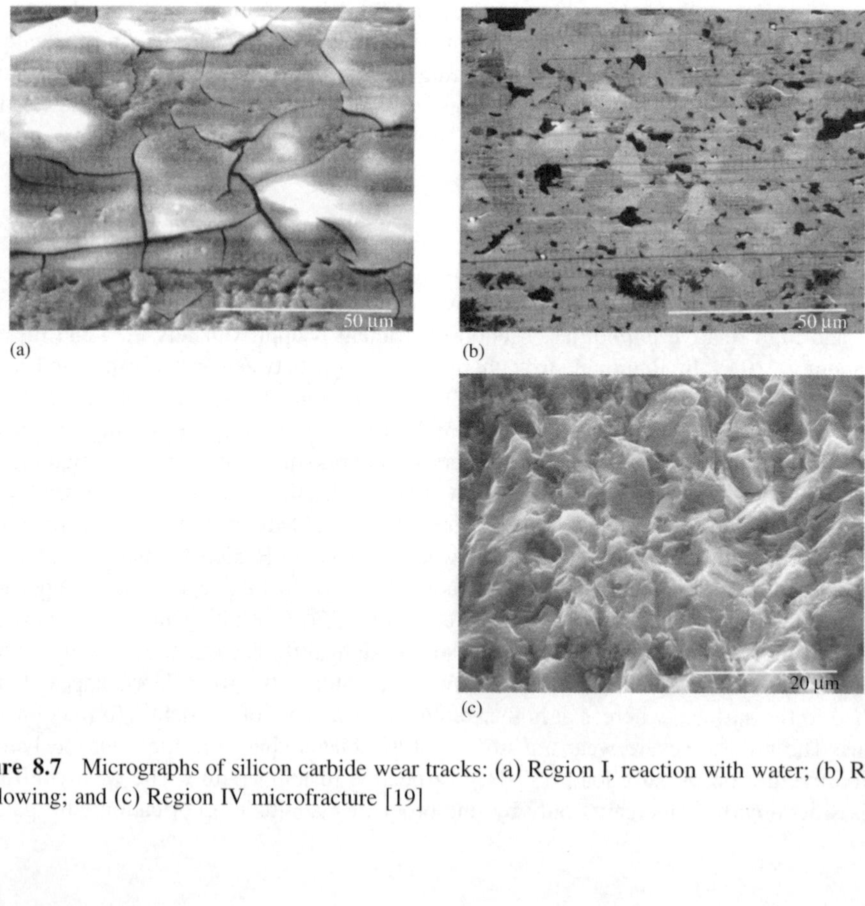

Figure 8.7 Micrographs of silicon carbide wear tracks: (a) Region I, reaction with water; (b) Region II, plowing; and (c) Region IV microfracture [19]

oil increases the transition load to 130–270 N depending on the rotational speed used in step-loading four-ball tests [20].

8.4 Damage Formation in Hertzian Contacts

To ensure reliable performance, tribological contacts must be designed to operate in the mild wear region. Therefore, it is necessary to predict the transition load during the design process. Various models have been suggested in the literature for the prediction of ceramic wear rates [13, 21–23]. Most of these models relate wear to the hardness, elastic modulus, and fracture toughness of the ceramics and therefore model the wear process in the microfracture region. Although these models are useful as they delineate the importance of microstructure and mechanical properties, after transition to microfracture occurs, the wear rate is unacceptably high, and there is no longer a need to predict the wear rate. It is, however, crucial to predict the critical load for the onset of contact failure, i.e. transition from mild to severe wear.

The process of damage formation leading to microfracture in a sliding contact is similar to the types of damage formed in Hertzian contacts, i.e. normal indentation with a spherical body on a flat surface. Recent studies have confirmed that above a critical indentation load, ceramics can exhibit either brittle or quasi-plastic behavior, depending on the microstructure and indenter radius [24–29].

8.4.1 Brittle Behavior

When a spherical solid is pressed against an opposing planar surface, the stress field is initially elastic, and classical Hertzian stress calculations can be used to analyze the stress distribution within the stressed half-plane. Beyond a critical load, the material undergoes irreversible deformation and/or fracture. In brittle materials such as glasses and fine-grain polycrystalline ceramics, a ring crack is initiated in the tensile region just outside the circle of contact produced by the indenting sphere [30]. This crack propagates downward and outward into the material (Figure 8.8) producing a surface-truncated cone-shaped crack.

The critical load, P_C, for the formation of cone cracks, i.e. brittle behavior, is determined by assuming that the condition for crack initiation/extension is related to the stress intensity factor associated with the tensile stresses at the contact circle [29]:

$$P_C/r' = AT^2/E' \tag{1}$$

where r' is the radius of the ball in contact with a flat surface, T is the fracture toughness (i.e. K_{IC} assuming single-valued toughness), and E' is the modulus of elasticity. For a contact between two dissimilar materials and two curved surfaces: $1/E' = 1/E_1 + 1/E_2$ and $1/r' = 1/r_1 + 1/r_2$. In equation (1), A is related to the Poisson's ratio, ν, and can be obtained by experimental calibration. The critical loads for contact fracture in several ceramics that exhibit brittle behavior are shown in Figure 8.9 [29]. The solid line in the figure is the best fit to equation (1) corresponding to a calculated value $A = 8.6 \times 10^3$.

8.4.2 Quasi-Plastic Behavior

Hertzian indentations produced in coarse-grain polycrystalline alumina [24, 25], coarse-grain silicon nitride [26], and mica-containing glass ceramics [27] depart radically from the

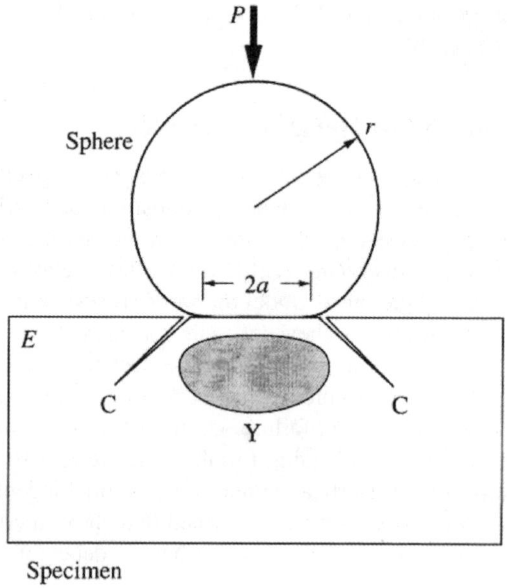

Figure 8.8 Types of damage formed in Hertzian contacts; C: cone cracks (brittle behavior) and Y: plastic yield (quasi-plastic behavior) [29] (Figure reproduced from original publication with authors' permission)

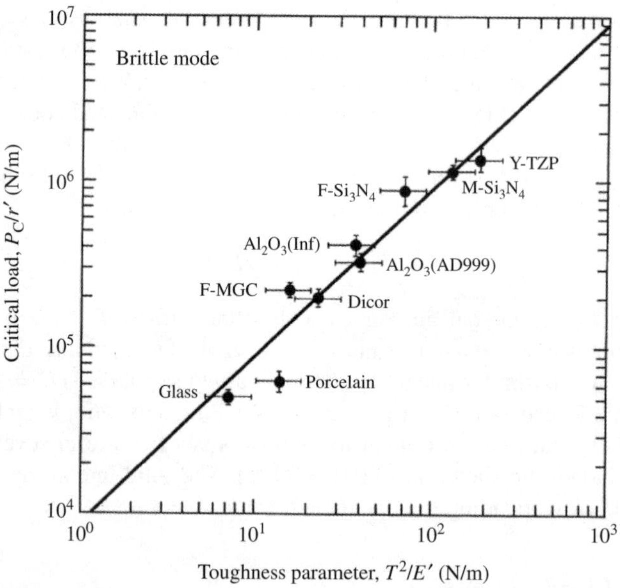

Figure 8.9 Critical load for brittle fracture determined for various ceramics [29] (Figure reproduced from original publication with authors' permission)

classical brittle fracture pattern. In these ceramics, the cone crack is suppressed in favor of distributed damage below the contact. The subsurface damage, which occurs in regions corresponding to high compression/shear expected for plastic deformation (Figure 8.8), consists of intragrain twin/slip bands and grain-localized intergranular microfracture. Microcracks are nucleated in crystalline materials by three principal mechanisms in a plastically deforming field: dislocation pile-ups, twin intersections, and strain incompatibility. The microcrack nucleation mechanism in a purely elastic field has not been clearly identified. It is possible that, even in an elastically deforming region, localized microplastic deformation leads to microcrack nucleation by one of the plastic nucleation processes.

For the quasi-plastic mode, the critical load, P_Y, for onset of failure is calculated assuming that damage formation is driven by the shear component of the Hertzian stress field and that hardness can be used as a predictor for plastic yield [29]:

$$P_Y/r'^2 = DH(H/E')^2 \tag{2}$$

where H is the hardness. The dimensionless constant D is related to the Poisson's ratio and can be determined experimentally. Note that P_Y depends on r'^2 while P_C depends linearly on r'. The critical load for onset of quasi-plasticity for several ceramics as well as for mild steel and polycarbonate is shown in Figure 8.10 [29]. The solid line is best fit to equation (2) corresponding to $D = 0.85$. (The data for silicon nitrides in Figure 8.10 were excluded from the best-fit analysis because the softer tungsten carbide balls used in the experiments were flattened during indentation resulting in higher indentation loads for quasi-plasticity.)

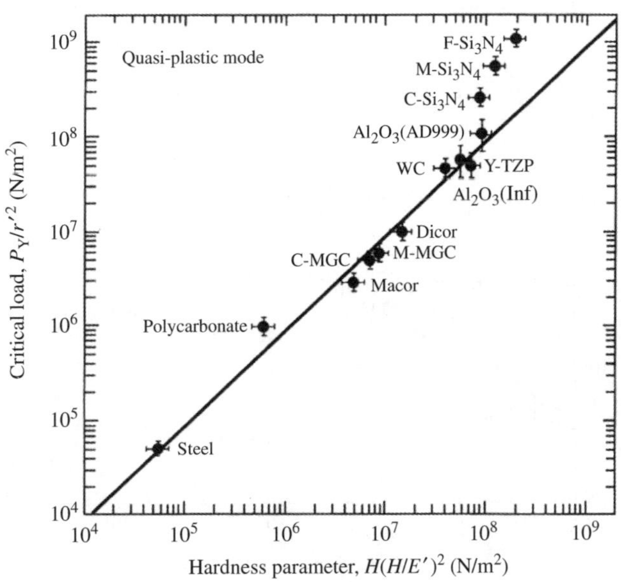

Figure 8.10 Critical load for onset of quasi-plasticity determined for various materials [29] (Figure reproduced from original publication with authors' permission)

8.4.3 Brittleness Index

The transition from traditional brittle behavior (i.e. formation of cone cracks) to a quasi-plastic behavior (i.e. distributed microfractures) depends on the microstructure of the material being stressed; for example in alumina, the transition occurs above a threshold grain size [25]. The competition between brittle fracture and quasi-plasticity can be expressed as a brittleness index: the ratio of critical loads for quasi-plastic yield and initiation of cone cracks, from equations (1) and (2) [29]:

$$P_Y/P_C = (D/A)(H/E')(H/T)^2 r' \tag{3}$$

For a given ball radius, the response is brittle if $P_Y/P_C > 1$ and quasi-plastic if $P_Y/P_C < 1$. The ratio P_Y/P_C is displayed in Figure 8.11 versus the brittleness index for a series of materials tested using a ball radius of 3.18 mm [29]. Note a strong size effect manifested in the value of r'. The predicted line for $P_Y/P_C = 1$ translates to left or right for smaller or larger values of r. Smaller ball radius increases the Hertzian contact stress and promotes plasticity. This is an important phenomenon as in the contact between rough surfaces the asperity radius of curvature may promote plasticity rather than contact fracture at the asperity scale, a fact supported by numerous experimental observations.

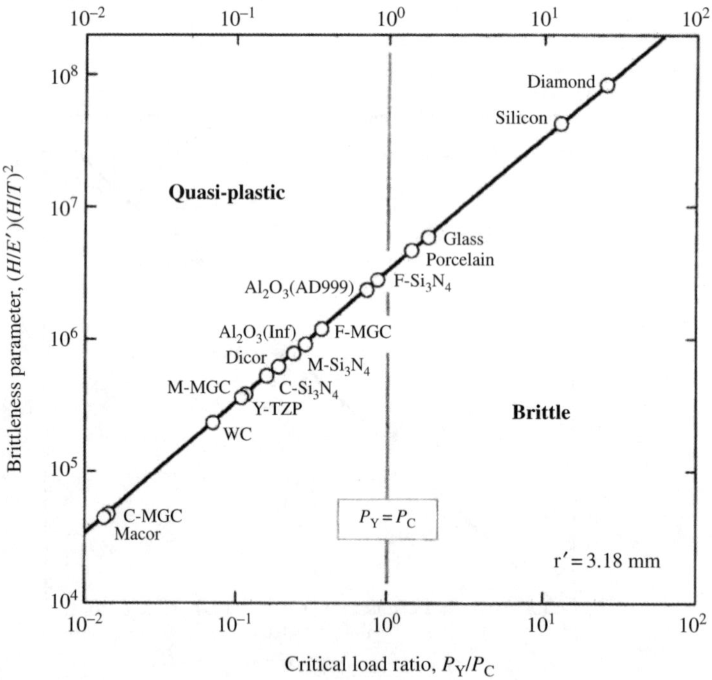

Figure 8.11 Brittle and quasi-plastic behavior determined from critical load ratio and brittleness parameter [29] (Figure reproduced from original publication with authors' permission)

8.5 Transition Loads in Sliding Contacts

The previous analysis for the prediction of transition load for the two contact damage modes should be modified for sliding contacts to account for the addition of friction. In the following theoretical analysis [10], the state of stress under a sliding contact with friction is determined, and the conditions for the onset of subsurface plastic flow (leading to microcrack initiation, i.e. quasi-plastic behavior) are analyzed. Linear elastic fracture mechanics is then used to analyze the process of contact fracture by propagation of microcracks from pre-existing flaws located in the tensile zone behind a moving contact with friction.

Analysis of stresses in a sliding contact should consider both the stress distribution associated with the apparent contact and the asperity contacts. Microcontact models [31–34] have shown that pressure at the real area of contact is related to the surface roughness. Rough surfaces produce microcontact pressures that are much larger than the calculated Hertzian pressure. However, the deviation of microcontact pressure from the Hertzian pressure calculation is small for polished smooth surfaces [34]. Therefore, the following analysis, while applicable to contacts between polished surfaces, should be used with caution if applied to rough surfaces.

8.5.1 Quasi-Plastic Behavior

Explicit equations developed by Hamilton and Goodman [35] for the stress field due to a circular contact region under a hemispherical normal pressure and tangential traction were recently refined and simplified by Hamilton [36]. These equations are not repeated here due to their length. However, it is instructive to examine certain important features of the stress state relevant to the wear process.

For a purely normal loading contact between a spherical body and a semi-infinite plane, the maximum tensile stress occurs on the circumference of the contact circle. Sliding with friction adds a compressive stress to the front edge and intensifies the tensile stress at the trailing edge of the contact. The influence of tensile stresses on the formation of cone cracks was described in Section 8.4 for normal indentation and it will be discussed in the following section for sliding contacts with friction.

The necessary condition for initiation of plastic deformation is analyzed by using the stress distribution to calculate the second stress invariant J_2. The condition for plastic deformation, as well as the location for the initiation of plastic flow, is then determined by invoking the von Mises yield criterion. To analyze the conditions for initiation of plastic flow, the second stress invariant, J_2, is first calculated for a sliding contact, in this case between a ball and a flat surface. The elastic constants, E and ν, are assumed to be independent of temperature to simplify the calculations. It is also assumed that the yield stress can be approximated by a value equivalent to one-third of the indentation hardness. This assumption is based on perfectly plastic slip-line field analysis, and there is ample evidence in its support for metallic alloys [37]. It is used here as a first approximation for ceramics, in the absence of any other reliable criteria.

It can be shown that the maximum yield parameter normalized with respect to the maximum contact pressure $(J_2^{1/2}/P_0)$ raised to the power of 3 is similar in form to equation (2). Therefore, the transition load for quasi-plastic deformation with contact friction can be expressed as

$$P_Y/r'^2 = DH(H/E')^2 h(f) \tag{4}$$

The constant D depends on the value of Poisson's ratio and decreases from 0.89 to 0.76 as the Poisson's ratio is increased from 0.2 to 0.3. The experimental results of Rhee *et al.* [29] on several ceramics suggested that $D = 0.85$, equivalent to using an average Poisson's ratio of 0.22.

In a sliding contact between a sphere and a flat plane, the location of flow initiation depends on the magnitude of the coefficient of friction [36]. For $f = 0$, plastic flow initiates below the contact surface and moves toward the surface as f increases. At $f \geq 0.33$, plastic flow initiates on the surface at the trailing edge of the contact circle, i.e. at the same location as the maximum tensile stress. To a first approximation $h(f)$ in equation (4) can be written as

$$h(f) = 1 \qquad \text{for } 0 \leq f \leq 0.33 \tag{5a}$$

$$h(f) = (3f)^{-3} \qquad \text{for } f \geq 0.33 \tag{5b}$$

These equations imply that plastic flow is not sensitive to the coefficient of friction when plastic yield occurs below the surface at $0 \leq f \leq 0.33$. However, when the location of first yield occurs on the surface at $f \geq 0.3$, the critical load for onset of quasi-plasticity decreases with friction.

It is important to note that the critical load for onset of quasi-plasticity depends strongly on hardness (i.e. hardness raised to the power of 3). The measured values of hardness are sensitive to the specific test methods used for their determination and can decrease by 20 to 30% as the indentation load is increased [38]. Therefore, published hardness values should be used with caution, and comparisons between different ceramics should be made only if the hardness has been measured using a consistent and/or standard test method.

The calculated values of the critical loads for onset of quasi-plasticity both for a Hertzian contact, $P_Y{}^0$, and a sliding contact, $P_Y{}^f$, are listed in Table 8.2. This table also includes the experimentally determined values of the transition loads and the coefficients of friction prior to transition for alumina (unlubricated and lubricated), silicon nitride, and silicon carbide, based on the data in Figures 8.1, 8.3, 8.4, and 8.6. The calculations assumed self-mated contacts between a ball of 6.35 mm radius with a smooth flat surface (for unlubricated tests)

Table 8.2 Transition loads for brittle fracture and quasi-plastic behavior at room temperature

	Transition load Expt. (N)	f	$P_Y{}^0$(N)	$P_Y{}^f$(N)	$P_C{}^0$(N)	$P_C{}^f$(N)	$P_Y{}^0/P_C{}^0$	$P_Y{}^f/P_C{}^f$
Alumina (99.8%)	20	0.60	920	156	1500	8	0.6	19
Alumina (99.5%)	60	0.10	310	310	600	85	0.5	4
Silicon nitride (HIP)	10	0.30	2570	2570	2300	23	1.1	111
Silicon carbide (sintered)	7	0.67	9240	924	1700	11	5.4	84

f: average coefficient of friction from experiments
$P_Y{}^0$: critical load for quasi-plasticity assuming zero friction
$P_Y{}^f$: critical load for quasi-plasticity using the experimental values of f
$P_C{}^0$: critical load for brittle fracture assuming zero friction
$P_C{}^f$: critical load for brittle fracture using the experimental values of f
$P_Y{}^0/P_C{}^0$: brittleness index for Hertzian contacts assuming zero friction
$P_Y{}^f/P_C{}^f$: brittleness index for sliding contacts using the experimental values of f

and with a ball of the same radius (for lubricated tests) using the physical properties for each material listed in Table 8.1. The hardness values used in the analysis were determined using the same indentation loads. The calculated critical loads for quasi-plasticity, P_Y^0, in the absence of friction range from 310 to 9240 N due to the differences in physical properties of the selected ceramics and the contact radius. The addition of friction reduces the critical yield loads, P_Y^f, when $f \geq 0.33$. However, the calculated values are larger than the experimentally obtained transition loads by at least one order of magnitude. This discrepancy indicates that the onset of quasi-plasticity cannot explain the mild to severe wear transition for the selected ceramics.

8.5.2 Brittle Behavior

Chiang and Evans [39] used linear elastic fracture mechanics to analyze the propagation or extension of pre-existing microcracks by the tensile component of stress. Their analysis applies to a homogeneous, isotropic, and linear elastic semi-infinite plane subjected to a combination of normal and tangential surface forces, similar to the stress state solution of Hamilton and Goodman [35]. The analysis procedure consists of prescribing the state of stress and assuming that a pre-existing surface flaw with a semi-circular crack plane is located at the trailing edge of the contact circle. The crack is assumed to be located on a plane normal to the surface. To predict the condition for crack extension, first the stress intensity factor, K, for such a crack is determined. It is assumed that the crack propagates when K becomes equal to or exceeds the critical stress intensity factor, K_c. The stress intensity factor is calculated by integrating the S_x component of prior stress over the crack. (Prior stress is defined as the stress at the crack location, in the absence of the crack, from the Hamilton and Goodman solution.) Allowing $K = K_c$ for crack extension, an equation is obtained for the critical normal load for the occurrence of contact fracture. This equation is based on maximum flaw size for unstable crack growth and calculates a lower bound to the load necessary for fracture. The calculated load is a linear function of ball radius r and inversely related to the elastic modulus E of the contacting materials. The critical load P_c for the onset of brittle fracture is similar in form to equation (1).

$$P_c/r' = A(K_c^2/E')g(f) \qquad (6)$$

The constant A is approximately 7.5×10^3 which is very close to the value determined by Rhee *et al.* [29] in equation (1). An important feature of this equation is the dependence of contact fracture load on the coefficient of friction. Although an explicit equation was not provided by Chiang and Evans [39], the form of this relationship based on the dependence of maximum tensile stress on friction [36] can be written as

$$g(f) = (1 + Bf)^{-3} \qquad (7)$$

where B depends on the value of Poisson's ratio. It can vary from 8 to 12 as the Poisson's ratio increases from 0.2 to 0.3. The cubic relationship in equation (7) demonstrates a strong influence of friction and Poisson's ratio on the critical load for brittle fracture.

The critical load for onset of brittle fracture is sensitive to the fracture toughness (i.e. toughness raised to the power of 2). The value of fracture toughness is highly influenced by the measurement technique and the initial crack size used for toughness measurement [40–42]. Therefore, published values of toughness should be used with great caution.

It is well established [30] that many polycrystalline ceramics exhibit an R-curve behavior (i.e. dependence of toughness on the initial crack size). Experimental data clearly show that while homogeneous materials such as glasses exhibit a single-value toughness, microstructural inhomogeneities such as coarse grains and second phases produce resistance toward an extending crack and result in an increase in toughness [43]. In some cases, the toughness is four times higher in the long-crack regime [44]. Since wear in most structural ceramics occurs by crack propagation at a scale equivalent to only a few grain diameters, it is clear that the short-crack toughness should be used in the analysis of wear transition.

The calculated values of the critical loads for brittle fracture for Hertzian contacts, $P_C{}^0$, and for sliding contacts, $P_C{}^f$, are listed in Table 8.2. The addition of friction drastically decreases the calculated loads for onset of brittle fracture. The predicted values range from 8 to 85 N compared with the experimental transition loads of 7 to 60 N. The difference between the predicted critical loads and the experimental values by a factor of about 2 is reasonable considering the simplifications made in the analysis.

8.5.3 *Transition from Brittle Fracture to Quasi-Plasticity*

The competition between brittle fracture and quasi-plasticity for sliding contacts can be expressed as a brittleness index, i.e. the ratio of equations (4) and (6), similar to the approach used for Hertzian contacts [29].

$$P_Y/P_C = (D/A)(H/E')(H/T)^2 r' k(f) \tag{8}$$

The response is brittle if $P_Y/P_C > 1$ and quasi-plastic if $P_Y/P_C < 1$. The influence of friction is denoted by $k(f) = h(f)/g(f)$, or

$$k(f) = (1 + Bf)^3 \quad \text{for } 0 \leq f \leq 0.33 \tag{9a}$$

$$k(f) = \{(1/f + B)/3\}^3 \quad \text{for } f \geq 0.33 \tag{9b}$$

Note that brittle behavior is promoted for a larger r, smaller toughness, lower elastic modulus, and higher hardness. Equations (9a) and (9b) suggest that quasi-plastic behavior is promoted as the coefficient of friction or the constant B (i.e. the Poisson's ratio) is increased when $f \leq 0.33$. However, as the coefficient of friction increases, its influence diminishes relative to the effect of Poisson's ratio.

The brittleness indexes for Hertzian contacts and sliding contacts with friction are compared in Table 8.2. The brittleness indexes for Hertzian contacts are near unity for alumina and silicon nitride, indicating a possible mixed damage mode consisting of both brittle fracture and quasi-plasticity. However, addition of friction increases the brittleness index by one to two orders of magnitude, suggesting a brittle behavior for the sliding contacts. A brittle behavior is predicted for the silicon carbide, irrespective of friction. The results of this analysis are consistent with the experimental observations confirming that the transition from mild to severe wear for the selected ceramics occurs by a brittle fracture process that initiates under the influence of tensile stresses at the trailing edge of contact circle. Therefore, the mechanics analysis presented here provides a simple means for predicting contact failure in low speed sliding.

8.6 Ceramics in Tribological Applications

Implementation of ceramics in engineering applications requires experience in designing with brittle materials, design data, field performance data, and performance and failure prediction methods. Selection of materials for tribological applications should be based not only on friction and wear behavior, but also on other application requirements such as strength, fatigue resistance, corrosion resistance, dimensional stability, thermal properties, reliability, ease of fabrication, and cost.

A common feature for most ceramics in tribological applications is their excellent wear resistance – often three to four orders of magnitude better than that of metals. This advantage, however, is lost at higher contact loads because of severe wear and larger coefficients of friction. The transition from mild to severe wear, for a given material, depends on the sliding velocity and temperature, as well as on the coefficient of friction, which can be modified by tribochemical reactions and lubrication. Therefore, design of ceramic tribological contacts requires data regarding their tribological behavior and models to predict the transition from mild to severe wear.

Reliable data and fundamental knowledge on the mechanisms of friction and wear are essential for developing guidelines for design and material selection. The wear transition diagrams discussed here and the wear maps [45, 46] serve as simple and easy-to-use tools for predicting the performance of different ceramics under service conditions. These diagrams clearly show the varied tribological performance based on different sets of fundamental mechanisms, and alert the designer to the conditions that could promote catastrophic failures.

The simple mechanics analysis outlined here can be used to predict the onset of severe wear in ceramics under low speed sliding. It is, however, important to recognize the limitations of such analysis. The contact model is based on the assumption of "smooth" contacts, i.e. the presence of asperities is not included. The analysis should be modified for rougher surfaces. The contact stress calculations are based on the assumption of a homogeneous elastic material; it may not apply to composites and layered materials (e.g. films and coatings). The model should be modified to include the effect of thin layers (e.g. tribochemical film) on stress distributions below the contact. The analysis does not account for thermoelastic stresses that may develop at high sliding speeds, nor does it consider the influence of residual stresses either due to processing or developed during sliding. It is also important to note that transition to severe wear by microfracture could occur by extension of several smaller cracks or by interactions between surface and subsurface microcracks rather than by extension of a single crack, thus making the precise prediction of contact fracture more complex. Another important issue that should be considered is the potential effect of machining damage [47] on wear of ceramics [48]. Other issues such as the possibility for subsurface damage accumulation due to cyclic nature of sliding wear [49] and its effect on contact failure [50–52] are important factors that limit our ability to make more accurate predictions.

To maintain low friction and avoid severe wear, lubrication is often required. Tribochemical reactions between ceramic surfaces and water vapor provide a natural opportunity for lubrication of ceramics [53]. However, the adsorption tendency of water is reduced as the temperature is increased to about 200°C and the formation of oxides is favored over the hydroxides. The temperature limit may be shifted to lower temperatures at larger loads and higher speeds, due to increased contact temperatures. The coefficient of friction can be also controlled through boundary lubrication due to adsorption of hydrocarbons or chemical reactions of certain chemical compounds with the sliding surface [54–57]. Other schemes used for lubrication of ceramics include solid lubricating coatings [58, 59], supply of reacting lubricants through a gas stream [60], deposition of carbonaceous layers by catalytic decomposition of hydrocarbons [61], and self-lubricating ceramic matrix composites [62, 63].

Water, under certain sliding conditions, is also an excellent lubricant for silicon nitride and silicon carbide. Tomizawa and Fischer [64] in a seminal article presented convincing results that the coefficient of friction for self-mated silicon nitride in water decreases to a value less than 0.002 at sliding speeds higher than 60 mm/s. Since the worn surfaces were extremely smooth, they attributed the low friction to the initiation of hydrodynamic lubrication by a thin water film at the interface. Since no wear debris was observed in the water, they suggested that wear of silicon nitride occurred through tribochemical reactions and dissolution of reaction products (i.e. $(SiO_x)\ nH_2O$) in water. The formation of tribochemical reaction products on the wearing surfaces and the role of the hydrated oxides on the tribological properties of silicon nitride have been confirmed through several investigations [15, 16, 65, 66].

The unique low-friction behavior of silicon nitride has been confirmed by Xu *et al.* [67] and Chen *et al.* [68] in a series of studies that included a narrow range of loads, speeds, and temperatures. They reported a coefficient of friction below 0.03 after the run-in period. The low friction was only obtained at higher speeds and lower loads. The coefficient of friction remained high at about 0.5 when the water temperature was raised above 50°C [69]. The low friction was attributed to a combination of hydrodynamic lubrication due to the smooth surfaces and boundary lubrication due to the presence of colloidal silica on the wearing surfaces and was confirmed by introducing an additive to water that accelerated the formation of silica and reduced the run-in period [69].

Data by Kato and Adachi [70] showed that the initial surface roughness must be 10 nm or less for the low-friction behavior to occur. Saito *et al.* [71] have shown that the run-in period necessary for the transition from high- to low-friction behavior for self-mated silicon nitride depends on both the surface roughness and the sliding speed. The run-in period was shorter for a lapped surface ($R_a = 10$ nm) compared to a surface finished by grinding ($R_a = 130$ nm). They also showed that the friction remained high (about 0.8) at speeds above 130 mm/s for both surface finishes. Jahanmir *et al.* [72] recently showed that low-friction behavior of silicon nitride in water, however, can be unstable and occasional high-friction spikes can occur under certain sliding conditions. More recently, Heshmat and Jahanmir [73] confirmed that low coefficient of friction can be achieved also for silicon nitride sliding against silicon carbide in steam and showed that the coefficient of friction increases to a high value (about 0.6) as the speed is raised above 1000 mm/s.

The fundamental understanding of the wear mechanisms discussed in this chapter also can be used to improve the performance of ceramics by making changes in the composition and/or microstructure. Modifications that increase the fracture toughness should result in increasing the transition load. Approaches to improve fracture toughness include transformation toughening and control of microstructure and the grain boundary phase [30]. Since the short-crack toughness controls the microfracture process at sliding contacts, only the approaches that increase toughness in the short-crack range are beneficial. Changes in the composition of the grain boundary phase can result in improved short-crack toughness and may result in lubrication if the grain boundary phase can form a thin surface film under certain operating conditions. In alumina, diffusion and viscous flow of the glassy grain boundary phase at high temperatures are beneficial in reducing the coefficient of friction and avoiding the transition to severe wear. In silicon nitride and silicon carbide, however, oxidation reactions associated with the impurities and grain boundary phases tend to increase the wear rate. Since oxidation reactions and diffusion processes are thermally activated, an increase in the sliding speed or the contact load may accelerate these processes. However, thermoelastic stresses generated at higher speeds and increased wear at higher loads may override the beneficial effects. Because of the pronounced effects of such minor chemical compositions on tribological performance, it is important to select ceramics carefully for each application.

Acknowledgments

This material has been based upon an article that first appeared in the *Journal of Engineering Tribology* – Proceedings Part J, 2002, Vol. 216, No. J6, ISSN 1350–6501, published by Professional Engineering Publishing. Permission is granted by the Institution of Mechanical Engineers.

References

1. Richerson, D.W., '*Modern Ceramic Engineering*', Marcel Dekker, New York, NY, 1982.
2. Schwartz, M., '*Handbook of Structural Ceramics*', McGraw Hill, New York, NY, 1992.
3. Jahanmir, S., '*Friction and Wear of Advanced Ceramics*', Marcel Dekker, New York, NY, 1993.
4. Jahanmir, S. and Fischer, T.E., 'Friction and Wear of Ceramics', *Handbook of Lubrication and Tribology* (ed E.R. Booser), Vol. 3, CRC Press, Boca Raton, FL, 1994, pp. 103–120.
5. Kato, K., 'Tribology of Ceramics', *Wear*, **136**, 1990, 117–133.
6. Fischer, T.E. and Tomizawa, H., 'Interaction of Tribochemistry and Microfracture in the Friction and Wear of Silicon Nitride', *Wear*, **105**, 1985, 29–42.
7. Archard, J.F., 'Contact of Rubbing Surfaces', *Journal of Applied Physics*, **24**, 1953, 981–988.
8. Dong, X., Jahanmir, S. and Hsu, S.M., 'Tribological Characteristics of Alumina at Elevated Temperatures', *Journal of the American Ceramic Society*, **74**, 1991, 1036–1044.
9. Jahanmir, S. and Dong, X., 'Wear Mechanisms of Aluminum Oxide Ceramics', *Friction and Wear of Advanced Ceramics* (ed S. Jahanmir), Marcel Dekker, New York, NY, 1993.
10. Jahanmir, S. and Dong, X., 'Mechanism of Mild to Severe Wear Transition in Alumina', *Journal of Tribology*, **114**, 1992, 403–411.
11. Chen, Y.M., Rigaut, B. and Armanet, F., 'Friction and Wear of Alumina Ceramics at High Sliding Speed', *Lubrication Engineering*, **47**, 1991, 531–537.
12. Woydt, M. and Habig, K.H., 'High Temperature Tribology of Ceramics', *Tribology International*, **89**, 1989, 75–88.
13. Cho, S.J., Hockey, B.J., Lawn, B.R. and Bennison, S.J., 'Grain-Size and R-Curve Effects in the Abrasive Wear of Alumina', *Journal of the American Ceramic Society*, **72**, 1989, 1249–1252.
14. Deckman, D.E. and Jahanmir, S., 'Wear Mechanisms of Alpha Alumina Lubricated with a Paraffin Oil', *Wear*, **149**, 1991, 155–168.
15. Dong, X. and Jahanmir, S., 'Wear Transition Diagram for Silicon Nitride', *Wear*, **165**, 1993, 169–180.
16. Jahanmir, S. and Fischer, T.E., 'Friction and Wear of Silicon Nitride Lubricated by Humid Air, Water, Hexadecane, and Hexadecane +0.5 Percent Stearic Acid', *ASLE Transactions*, **31**, 1988, 32–43.
17. Skopp, A., Woydt, M. and Habig, K.H., 'Lubricated Sliding Friction and Wear of Various Silicon Nitride Pairs Between 22 and 1000°C', *Tribology International*, **23**, 1990, 189–199.
18. Jahanmir, S. and Deckman, D.E., 'Wear Test Methodology for Lubricated Ceramics', DOE-ECUT Tribology Program, US Department of Energy, Report No. TRIB-ECUT 88-4, 1988, pp. 137–143.
19. Dong, X. and Jahanmir, S., 'Wear Transition Diagram for Silicon Carbide', *Tribology International*, **28**, 1995, 559–572.
20. Jahanmir, S. and Deckman, D.E., 'Wear Test Methodology for Lubricated Ceramics', DOE-ECUT Tribology Program, US Department of Energy, Report No. TRIB-ECUT 88-1, 1988, pp. 163–172.
21. Evans, A.G. and Marshall, D.B., 'Wear Mechanisms in Ceramics', in *Fundamentals of Friction and Wear of Materials* (ed D.A. Rigney), American Society for Metals, Metals Park, OH, 1980, pp. 439–452.

22. Wang, Y. and Hsu, S.M., 'Wear and Wear Transition Modeling of Ceramics', *Wear*, **195**, 1996, 35–46.
23. Kim, S.S., Kato, K., Hokkirigawa, K. and Abe, H., 'Wear Mechanism of Ceramic Materials in Dry Rolling Friction', *Journal of Tribology*, **108**, 1986, 522–528.
24. Guiberteau, F., Padture, N.P., Cai, H. and Lawn, B.R., 'Indentation Fatigue: A Simple Cyclic Hertzian Test for Measuring Damage Accumulation in Polycrystalline Ceramics', *Philosophical Magazine Series A*, **68**, 1993, 1003–1016.
25. Guiberteau, F., Padture, N.P. and Lawn, B.R., 'Effect of Grain Size on Hertzian Contact in Alumina', *Journal of the American Ceramic Society*, **77**, 1994, 1825–1831.
26. Xu, H.H.K., Wei, L., Padture, N.P., Lawn, B.R. and Yeckley, R.L., 'Effect of Microstructural Coarsening on Hertzian Contact Damage in Silicon Nitride', *Journal of Materials Science*, **30**, 1995, 869–878.
27. Cai, H., Kalceff, S. and Lawn, B.R., 'Deformation and Fracture of Mica-Containing Glass Ceramics in Hertzian Contacts', *Journal of Materials Research*, **9**, 1994, 762–770.
28. Lawn, B.R., Lee, S.K., Peterson, I.M. and Wuttiphan, S., 'Model of Strength Degradation from Hertzian Contact Damage in Tough Ceramics', *Journal of the American Ceramic Society*, **81**, 1998, 1509–1520.
29. Rhee, Y.W., Kim, H.W., Deng, Y. and Lawn, B.R., 'Brittle Fracture Versus Quasi Plasticity in Ceramics: A simple Predictive Index', *Journal of the American Ceramic Society*, **84**, 2001, 561–565.
30. Lawn, B.R., '*Fracture of Brittle Solids*', Cambridge University Press, Cambridge, UK, 1993.
31. Ioannides, E. and Kuijpers, J.C., 'Elastic Stresses Below Asperities in Lubricated Contacts', *Journal of Tribology*, **108**, 1986, 394–402.
32. McCool, J.I., 'Comparison of Models for the Contact of Rough Surfaces', *Wear*, **107**, 1986, 37–60.
33. Chang, W.R., Etsion, I. and Bogy, D.B., 'An Elastic–Plastic Model for the Contact of Rough Surface', *Journal of Tribology*, **109**, 1987, 257–263.
34. Merriman, T. and Kannel, J., 'Analyses of the Role of Surface Roughness on Contact Stresses Between Elastic Cylinders With and Without Soft Surface Coating', *Journal of Tribology*, **111**, 1989, 87–89.
35. Hamilton, G.M. and Goodman, L.E., 'The Stress Field Created by a Circular Sliding Contact', *Journal of Applied Mechanics*, **88**, 1966, 371–376.
36. Hamilton, G.M., 'Explicit Equations for the Stress Beneath a Sliding Spherical Contact', *Proceedings of the Institution of Mechanical Engineers*, **197C**, 1983, 53–59.
37. Tabor, D., 'Indentation Hardness and Its Measurement', *Microindentation Techniques in Materials Science and Engineering* (eds P.J. Blau and B.R. Lawn), ASTM-STP 889, American Society for Testing and Materials, Philadelphia, PA, 1985, pp. 129–159.
38. Quinn, G.D., 'Indentation Hardness Testing of Ceramics', *Materials Testing and Evaluation*, Vol. 8, Mechanical Testing, ASM International, Metals Park, OH, 2001, pp. 24–255.
39. Chiang, S. and Evans, A.G., 'Influence of a Tangential Force on the Fracture of Two Contacting Elastic Bodies', *Journal of the American Ceramic Society*, **66**, 1983, 4–10.
40. Quinn, G.D., Salem, J.A., Bar-On, I. and Jenkins, M.G., 'The New ASTM Fracture Toughness of Ceramics Standard: PS 070-97', *Ceramic Engineering and Science Proceedings*, **19**, 1998, 565–579.
41. Quinn, G.D., Xu, K., Gettings, R.J., Salem, J.A. and Swab, J.J., 'Does Anyone Know the Real Fracture Toughness? SRM 2100: the World's First Ceramic Fracture Toughness Reference Material', *Fracture Resistance Testing of Monolithic and Composite Brittle Materials* (eds J.A. Salem, G.D. Quinn and M.G. Jenkins), ASTM STP 1409, ASTM, West Conshohocken, PA, 2001.
42. Quinn, G.D., Gettings, R.J. and Kübler, J.J., 'Fracture Toughness of Ceramics by the Surface Crack in Flexure (SCF) Method', *Fracture Mechanics of Ceramics* (eds R.C. Bradt, D.P.H. Hasselman, D. Munz, M. Sakai and V. Yashevchenko), Vol. 11, Plenum, New York, 1996, pp. 203–218.
43. Lee, S.K., Wuttiphan, S. and Lawn, B.R., 'Role of Microstructure in Hertzian Contact Damage in Silicon Nitride: Mechanical Characterization', *Journal of the American Ceramic Society*, **80**, 1997, 2367–2381.
44. Xu, H.H.K., Jahanmir, S., Ives, L.K., Job, L.S. and Ritchie, K., 'Short-Crack Toughness and Abrasive Machining of Silicon Nitride', *Journal of the American Ceramic Society*, **79**, 1996, 3055–3064.
45. Kong, H. and Ashby, M.F., 'Wear Mechanisms in Brittle Solids', *Acta Metallurgical Materials*, **40**, 1992, 2907–2920.
46. Hsu, S.M., Lim, D.S., Wang, Y.S. and Munro, R.G., 'Ceramics Wear Maps: Concept and Method Development', *Lubrication Engineering*, **47**, 1991, 49–55.
47. Jahanmir, S., Xu, H.H.K. and Ives, L.K., 'Mechanisms of Material Removal in Abrasive Machining of Ceramics', *Machining of Ceramics and Composites* (eds S. Jahanmir, M. Ramulu and P. Koshy), Marcel Dekker, New York, 1999, pp. 11–84.

48. Woydt, M. and Effner, U., 'Influence of Machining on Friction and Wear of Lubricated Ceramics in Slip-Rolling Contacts', *Machining Science and Technology*, **1**, 1997, 275–287.
49. Jahanmir, S., 'On Mechanics and Mechanisms of Laminar Wear Particle Formation', *Advances in the Mechanics and Physics of Surfaces* (eds R.M. Latanision and T.E. Fischer), Vol. 3, Harwood Academic, New York, NY, 1986, pp. 261–331.
50. Lee, S.K. and Lawn, B.R., 'Contact Fatigue in Silicon Nitride', *Journal of the American Ceramic Society*, **82**, 1999, 1281–1288.
51. Kim, D.K., Jung, Y.G., Peterson, I.M. and Lawn, B.R., 'Cyclic Fatigue of Intrinsically Brittle Ceramics in Contact With Spheres', *Acta Materialia*, **47**, 1999, 4711–4725.
52. Lee, K.S., Jung, Y.G., Peterson, I.R., Lawn, B.R., Kim, D.K. and Lee, S.K., 'Model for Cyclic Fatigue of Quasi Plastic Ceramics in Contact with Spheres', *Journal of the American Ceramic Society*, **83**, 2000, 2255–2262.
53. Fischer, T.E., Liang, H. and Mullins, W.M., 'Tribochemical Lubricious Oxides on Silicon Nitride', *Materials Research Society Symposium Proceedings*, **140**, 1989, 339–345.
54. Fischer, T.E., Anderson, M.P., Salher, R. and Jahanmir, S., 'Friction and Wear of Tough and Brittle Zirconia in Nitrogen, Air, Water, Hexadecane, and Stearic Acid', *Wear*, **122**, 1988, 133–148.
55. Gates, R.S. and Hsu, S.M., 'Effect of Selected Chemical Compounds on Lubrication of Silicon Nitride', *Tribology Transactions*, **34**, 1991, 417–425.
56. Gates, R.S. and Hsu, S.M., 'Silicon Nitride Boundary Lubrication: Effect of Phosphorus-Containing Organic Compounds', *Tribology Transactions*, **39**, 1996, 795–802.
57. Gates, R.S. and Hsu, S.M., 'Silicon Nitride Boundary Lubrication: Effect of Sulfonate, Phenate, and Salicylate Compounds', *Tribology Transactions*, **43**, 2000, 269–274.
58. Erdemir, A., Busch, D.E., Erck, R.A., Fenske, G.R. and Lee, R., 'Ion-Beam-Assisted Deposition of Silver Films on Zirconia Ceramics for Improved Tribological Behavior', *Lubrication Engineering*, **47**, 1991, 863–872.
59. Erdemir, A., 'Lubrication of Ceramics with Thin Solid Films', *Friction and Wear of Ceramics* (ed S. Jahanmir), Marcel Dekker, New York, 1993, pp. 119–162.
60. Klaus, E.E., Jeng, G.S. and Dudda, J.L., 'A Study of Tricresyl Phosphate as Vapor Delivered Lubricant', *Lubrication Engineering*, **45**, 1989, 717–723.
61. Lauer, J.L. and Bunting, B.G., 'High Temperature Solid Lubrication by Catalytically Generated Carbon', *Tribology Transactions*, **31**, 1988, 338–349.
62. Gangopadhyay, A. and Jahanmir, S., 'Friction and Wear Characteristics of Silicon Nitride–Graphite and Alumina–Graphite Composites', *Tribology Transactions*, **34**, 1991, 257–265.
63. Gangopadhyay, A. and Jahanmir, S., 'Self-Lubricating Ceramic Matrix Composites', *Friction and Wear of Ceramics* (ed S. Jahanmir), Marcel Dekker, New York, 1993, pp. 163–198.
64. Tomizawa, H. and Fischer, T.E., 'Friction and Wear of Silicon Nitride and Silicon Carbide in Water: Hydrodynamic Lubrication at Low Sliding Speed Obtained by Tribochemical Wear', *ASLE Transactions*, **30**, 1987, 41–46.
65. Woydt, M. and Schwenzien, J., 'Dry and Water-Lubricated Slip-Rolling of Si_3N_4 and SiC-Based Ceramics', *Tribology International*, **26**, 1993, 165–174.
66. Woydt, M. and Effner, U., 'Influence of Machining on Friction and Wear of Lubricated Ceramics in Slip-Rolling Contacts', *Machining Science and Technology*, **1**, 1997, 275–287.
67. Xu, J., Kato, K. and Hirayama, T., 'The Transition of Wear Mode During the Running in Process of Silicon Nitride Sliding in Water', *Wear*, **230**, 1997, 55–63.
68. Chen, M., Kato, K., and Adachi, K., 'The Comparisons of Sliding Speed and Normal Load Effect on Friction Coefficients of Self-mated Si_3N_4 and SiC Under Water Lubrication',*Tribology International*, **35**, 2002, 129–135.
69. Xu, J. and Kato, K., 'Formation of Tribochemical Layer of Ceramics Sliding in Water and Its role for Low Friction', *Wear*, **245**, 2000, 61–75.
70. Kato, K. and Adachi, K., 'Wear of Advanced Ceramics', *Wear*, **253**, 2002, 1097–1104.
71. Saito, T., Hosoe, T. and Honda, F., 'Chemical Wear of Sintered Si_3N_4, hBN and Si_3N_4–hBN Composites by Water Lubrication', *Wear*, **247**, 2001, 223–230.
72. Jahanmir, S., Ozmen, Y. and Ives, L.K., 'Water Lubrication of Silicon Nitride in Sliding', *Tribology Letters*, **17**, 2004, 409–417.
73. Heshmat, H. and Jahanmir, S., 'Tribological Behavior of Ceramics Sliding at High Speeds in Steam', *Tribology Letters*, **17**, 2004, 359–366.

9

Tribology of Diamond and Diamond-Like Carbon Films: An Overview

A. Erdemir and Ch. Donnet

Abstract

Detailed tribological studies on diamond and diamond-like carbon (DLC) films have confirmed that they are inherently self-lubricating and resistant to erosion and abrasive and adhesive wear. Because of their high chemical inertness, they are also very resistant to corrosion and oxidation (even at fairly elevated temperatures). The combination of such exceptional qualities makes these films ideal for a wide range of demanding tribological applications (microelectromechanical systems, cutting tools, mechanical seals, magnetic hard disks, invasive and implantable medical devices, etc.). These films, available for more than three decades, have been used extensively for tooling and magnetic-hard-disk applications. Their potential in other application areas is currently being explored around the world. With the development of new and more robust deposition methods in recent years, it is envisioned that the production of high-quality diamond and DLC films will become very cost-effective and reliable for large-scale applications in the medical, transportation, and manufacturing sectors.

In this chapter, sliding friction and wear mechanisms of diamond and DLC films will be presented. Specifically, it will be shown that, in general, the wear of these films is extremely low (mainly because of their exceptional hardness and low-friction characteristics). Specific test conditions established during each sliding test, however, may dramatically affect the wear performance of certain diamond and DLC films. One of the dominant wear mechanisms relates to a phase transformation that is primarily the result of very high mechanical and thermal loadings of sliding contact interfaces. The transformation products

Wear – Materials, Mechanisms and Practice Edited by G. Stachowiak

(such as disordered graphite) trapped at the sliding interface may transfer to the mating surface and significantly affect their friction and wear behavior. This chapter describes, in terms of structural and fundamental tribological knowledge, the ideal film microstructures and chemistry, as well as operational conditions under which diamond and DLC films perform best and provide superlow friction and wear properties in sliding tribological applications. Present and future potential applications for these films are also highlighted.

9.1 General Overview

Diamond and diamond-like carbon (DLC) films offer numerous outstanding properties that make them ideal candidates for a wide range of industrial applications. In particular, their extreme hardness (diamond is harder than any other known material), chemical inertness (they resist attack by almost all known acids and bases), and inherently low friction and wear coefficients make them very desirable for a wide range of demanding applications that involve sliding, rolling, and rotating contacts between moving mechanical surfaces. Because of their high hardness, these films hardly wear at all under normal operating conditions (i.e. open air and moderate temperatures and sliding velocities). However, under extreme conditions (i.e. high temperatures, vacuum, and high sliding velocities), they may experience high friction and wear losses.

Diamond films can be grown on certain ceramic and metallic substrates such as Si and W and their carbides by chemical vapor deposition (CVD) methods. DLC coatings can be produced on virtually all kinds of metallic and ceramic substrates by several CVD and physical vapor deposition (PVD) methods, such as plasma-enhanced CVD, sputtering, ion plating, and laser ablation [1–3]. Gas discharge plasmas used in these processes consist mainly of hydrocarbon gases (as the carbon source) and hydrogen or argon. Solid carbon materials (such as graphite) can also be used for the extraction of carbon for film growth. The deposition parameters are controlled very carefully to dissociate the hydrogen/hydrocarbon mixtures into ionic or highly energetic species that are essential for the initial nucleation and subsequent growth of solid diamond and/or DLC films. The use of hydrogen and/or argon is critically important for the initial nucleation and subsequent growth of micro- or nanocrystalline diamond films, whereas the energetic carbonaceous species are considered key to the synthesis of DLC films. Figure 9.1 illustrates the available types of carbon films, with respect to their sp^2 and sp^3 bond characteristics and hydrogen content [4].

The synthesis of high-quality diamond films requires rather high temperatures (i.e. 600–950°C), whereas DLC films can be deposited at room or even at subzero temperatures. Most diamond films consist of large columnar grains, with tetrahedrally bonded carbon atoms. Such a structural anisotropy may cause some weaknesses along the large columnar grains, but the grains themselves exhibit most of the desired properties of natural diamonds. Unlike diamond, DLC coatings are structurally amorphous. Depending on the carbon sources (such as methane, acetylene, ethylene, and cyclohexane) they may contain considerable amounts of hydrogen. Films derived from solid carbon sources are essentially hydrogen-free. They can be diamond- or graphite-like, depending on the deposition method or conditions.

Since the initial syntheses of diamond and DLC films in the early seventies, researchers have made great strides in the deposition, characterization, and diverse utilization of these films [5–9]. Although some applications are still in the exploratory stages (such as

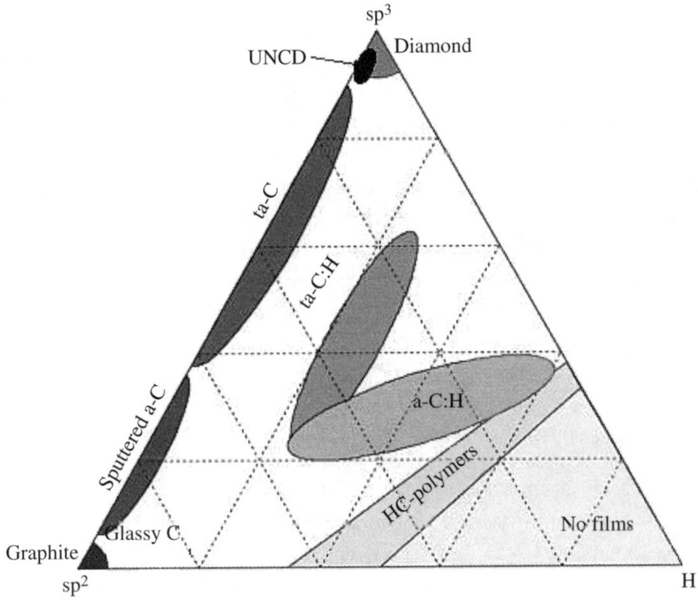

Figure 9.1 Ternary phase diagram for carbon films (after Ferrari and Robertson [4]). UNCD refers to ultrananocrystalline diamond

microelectromechanical systems, or MEMS), others have been fully developed and are offered on a commercial scale (such as diamond-coated cutting tools, DLC-coated razor blades, automotive parts and components, mechanical seals, surgical tools, magnetic hard disks, biomedical implants, optical windows, heat spreaders, sensors, acoustic-wave filters, flat-panel displays, photomultiplier and microwave power tubes, night vision devices, and various sensors). Because it exhibits high thermal conductivity and electrical insulation, diamond is used as a heat sink in X-ray windows, circuit packaging, and high-power electronic devices. The high chemical stability and inertness of diamond and DLC make them ideal for use in corrosive environments and prosthetic devices that require biocompatibility [9–12]. Recently, our fundamental understanding of the chemistry and tribology of these films has increased tremendously. This understanding has been used to design and customize better coatings that are capable of meeting the multifunctional property needs of increasingly demanding tribological applications [1, 13]. Besides phase-pure diamond and DLC films, metallic and nonmetallic species have been introduced into diamond and DLC structures to achieve better electrical, mechanical, and tribological properties [14, 15].

The goal of this chapter is to provide an overview of recent developments in friction and wear studies of diamond and DLC films that are used for a wide range of engineering applications. The major emphasis is on the sliding friction and wear mechanisms of these films in various test environments and at high temperatures. Current and future applications for both films are also presented. In terms of structural and fundamental tribological knowledge, the chapter also describes deposition conditions that lead to the synthesis of highly optimized films that have superlow friction and wear in various industrial applications.

9.2 Diamond Films

About 30 years ago, researchers discovered that diamond can be grown as a thin overcoat on various substrates at fairly low deposition pressures (10^2–10^3 Pa) from hydrocarbon/hydrogen gas mixtures by CVD methods [16]. Despite such a long time having passed, interest in these films has remained very high. With its super hardness, stiffness, chemical inertness, and thermal conductivity, diamond represents an ideal material for demanding multifunctional applications. Besides their exceptional mechanical, chemical, and thermal properties, sliding diamond surfaces provide some of the lowest friction and wear coefficients. The combination of these qualities in a material is very rare but extremely desirable for demanding engineering applications. In fact, if diamond were cheap and abundant, it would undoubtedly be the material of choice for many applications.

9.2.1 Deposition and Film Microstructure

As mentioned above, diamond can be grown as thin films at fairly low deposition pressures (10^2–10^3 Pa) using essentially hydrocarbon/hydrogen gas mixtures by CVD methods, including plasma-enhanced CVD, hot-filament CVD, microwave CVD, and DC-arc jets [1, 5–9]. Plasma conditions in these methods are tailored to dissociate hydrogen/hydrocarbon mixtures into ionic or highly energetic species that are essential for the initial nucleation as well as subsequent growth of the diamond films. These films can be deposited on certain metallic and ceramic substrates with excellent bonding. Typical substrate temperatures for diamond deposition are 600–950°C. Deposition of diamond at temperatures below 600°C has also been demonstrated, but growth rates were reduced substantially and the amount of nondiamond phases became substantially higher. The large amounts of atomic hydrogen (i.e. up to 99%) used in these methods during film growth caused etching of the smaller diamond grains and prevented continuous renucleation. During the etching and renucleation, only the grains that had attained a certain critical ratio of bulk-to-surface atoms survived. Consequently, these films consisted of large grains with a rough surface finish. When such films were used in a sliding application, they caused severe abrasive wear on the mating surfaces; thus, their uses were generally limited to some tooling applications [17].

During the nineties, researchers developed effective polishing methods (e.g. laser polishing) to achieve a smoother surface finish on diamond films [18, 19]. Such smooth surfaces were less damaging to mating surfaces; hence, they were very desirable for sliding wear applications. However, the polishing processes were rather slow and costly, and, in the case of complex geometries, they were highly impractical. More recently, researchers have developed new procedures for the deposition of very smooth (hence, less abrasive) nanocrystalline diamond films. In one method, they used a higher-than-normal carbon-to-hydrogen ratio in a microwave plasma. In another, they applied a DC bias to the substrates to increase their nucleation density [20, 21]. Although growth rates were somewhat reduced, the surface finish of the resultant diamond films was much smoother (i.e. 25 nm, centerline average (CLA)), and their friction coefficients were 0.05–0.1, depending on test environment. In some cases, structural analysis of the films showed the existence of significant amounts of nondiamond phases (such as amorphous carbon or sp^2-bonded disordered graphite). By using proper seeding techniques and plasmas based on Ar–C_{60} and Ar–CH_4 gas mixtures in a modified microwave CVD system, diamond films have

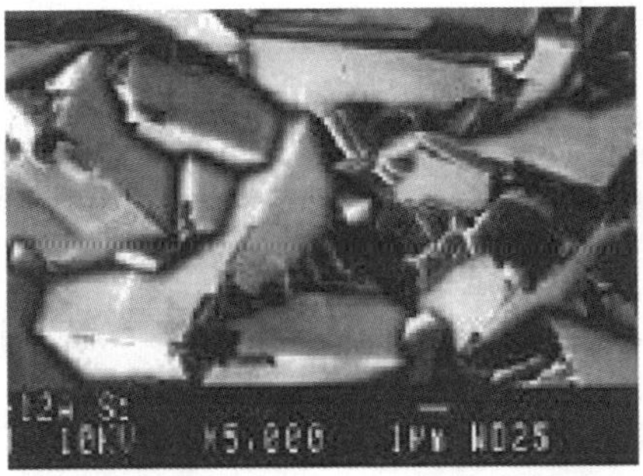

(a)

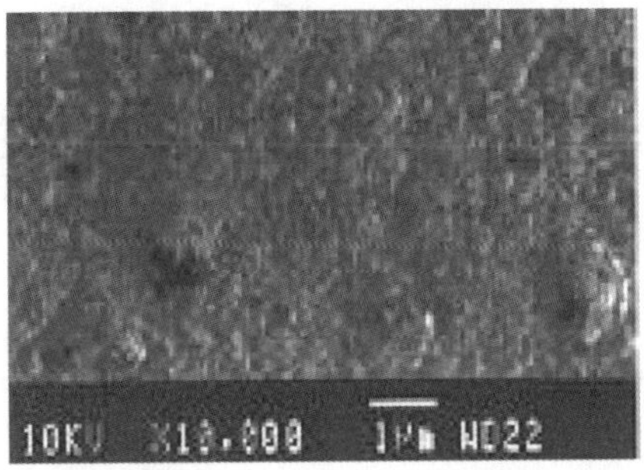

(b)

Figure 9.2 Surface morphology of (a) microcrystalline (rough) and (b) smooth nanocrystalline diamond films

been deposited with ultrananocrystalline grain morphology [22, 23]. The grain size of the films is very small (5–10 nm), and their surface finish is much smoother (root mean square surface roughness of 10–20 nm). Figure 9.2 shows the surface morphology of a diamond film with a microcrystalline and a nanocrystalline grain structure. The images in Figure 9.3 are the corresponding surface roughness of these films as measured by atomic force microscopy (AFM).

9.2.2 *Tribology of Diamond Films*

Much effort has been spent in the past to study the tribological behavior of synthetic diamonds with macro- (or poly-), micro-, and nanocrystalline grain structures. In general, it was found that under most sliding situations, diamond films suffer very little wear. This finding is thought to be due to their extreme hardness, which minimizes abrasive wear, and their high chemical inertness, which reduces adhesive and corrosive wear. However, depending on grain size, orientation, and surface morphology, diamond films can inflict severe wear damage on mating surfaces [24–26].

Researchers also found that the chemistry of the test environment (relative humidity, vacuum or inert gases, partial pressures of hydrogen and oxygen, etc.) may play significant roles in the friction and wear performance of diamond films. The low friction and wear

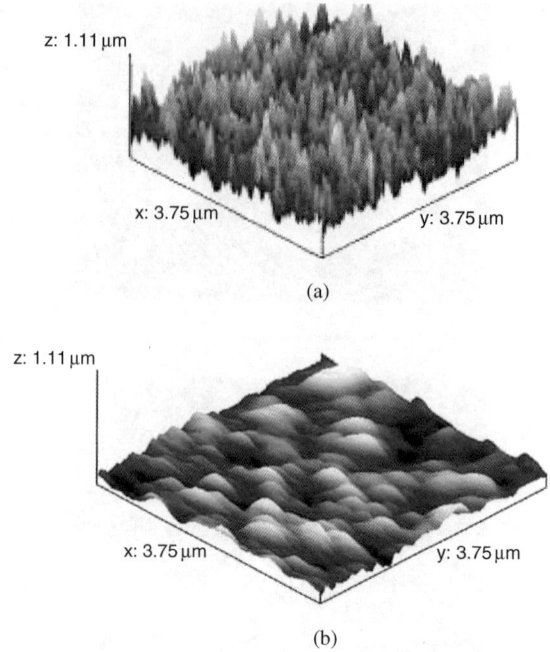

Figure 9.3 AFM images of (a) rough and (b) smooth diamond films

coefficients with diamond films are achieved only in air and relatively dry inert gases, and at lower ambient temperatures. Whereas in vacuum and at high temperatures, friction and wear are much higher [27, 28]. To achieve low friction and wear in vacuum, researchers have performed additional surface engineering (such as ion implantation) and/or used appropriate counterface materials (such as cubic boron nitride) [29].

Mechanistically, the self-lubricating nature of natural diamond and/or smooth diamond films has long been attributed to the high chemical passivity of their sliding surface [27–32]. Figure 9.4 illustrates hydrogen-terminated diamond surfaces sliding against one another. Hypothetically, gaseous adsorbates such as hydrogen, oxygen, or water vapor can effectively passivate the free σ-bonds of diamond on its surface. When these bonds become highly passive, the adhesion component of friction is dramatically reduced, and hence low friction is achieved on such smooth diamond surfaces.

Under certain sliding conditions, i.e. high thermal and mechanical loading, rubbing surfaces of diamond films may undergo phase transformation. The transformation product is often graphitic and thus may exert a dominant effect on friction and wear. Specifically, in inert gases or vacuum, the presence of graphitic phase at the sliding interface may cause intermittent or periodic friction spikes, while in moist air, graphitic phase may act as a self-lubricating interfacial film between sliding surfaces. The extent of ambient temperature can have a dramatic effect on the friction and wear behavior of diamond films. Specifically, at high temperatures, the sliding surfaces of diamond films may oxidize or gradually transform into a graphitic phase which can dominate the friction and wear of such surfaces.

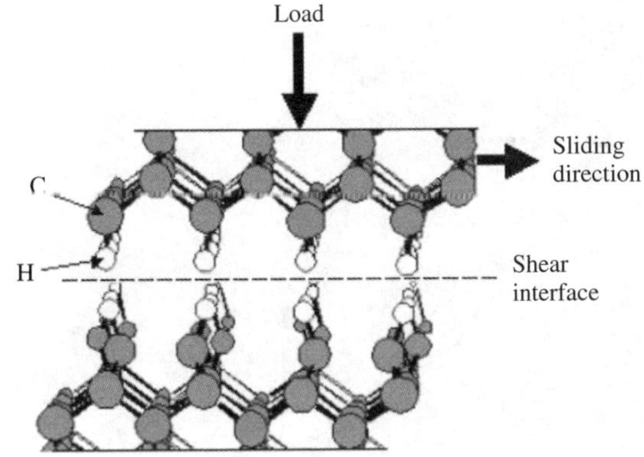

Figure 9.4 Lubrication mechanism of H-terminated diamond surfaces

9.2.2.1 Effect of Surface Roughness on Friction and Wear

Most synthetic diamond films are highly faceted (and thus rough) and may contain some nondiamond carbon phases such as graphite and amorphous carbon in their microstructures (in particular, at grain boundaries). Surface roughness could be very detrimental to the friction and wear properties of these films. Because of their extreme hardness, diamond films hardly wear but can severely abrade much softer counterface materials during sliding contacts. Figure 9.5 shows the surface conditions of two Si_3N_4 balls that were rubbed against an uncoated and a rough diamond-coated SiC substrate. The CLA surface roughness of this diamond film was 0.25 μm, and its crystallographic growth orientation was <111>. As is clear, the rough diamond film caused severe abrasive wear damage on the softer Si_3N_4 ball. The bulk of the wear debris particles consisted of Si_3N_4, which was mostly smeared on the rough diamond surface and had effectively filled in the valleys between tall surface asperities (Figure 9.6). Figure 9.7 shows the relationship between friction coefficient and surface roughness of diamond films. In general, the smoother the surface finish of diamond films, the lower the friction and wear coefficients [26, 27]. Ultrananocrystalline diamond films with a smooth surface finish, as well as laser-polished microcrystalline diamond films, can provide very low friction and wear coefficients in sliding contacts [33–35].

The grain morphology and/or growth orientation of diamond films can be controlled to give a smooth surface finish. For example, the growth orientation of diamond films can be changed from <111> to <100>, which gives a very smooth surface finish. During sliding tests, these films cause very little wear on mating surfaces and thus afford very low friction and wear to counterface materials during sliding, as shown in Figure 9.8. The films with a <111> orientation are rather abrasive and can cause severe wear damage, as was shown in Figure 9.5. Again, the high friction and wear coefficients of rough diamond films are largely attributed to the severe abrasive cutting and plowing effects of sharp surface asperities on the sliding surfaces of softer counterfaces. Progressive polishing of rough diamond films to a very smooth surface finish or use of fine-grained or ultrananocrystalline diamond films leads to low friction and wear, as shown in Figure 9.9 [34–37].

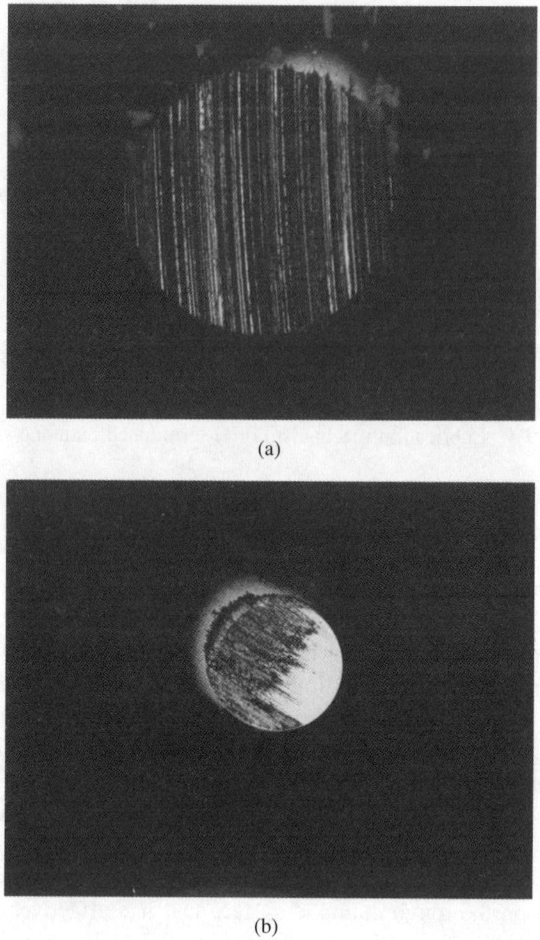

Figure 9.5 Wear scar of Si_3N_4 formed during sliding against (a) rough diamond and (b) uncoated SiC substrate

9.2.2.2 Effect of Environment on Friction and Wear of Diamond Films

Apart from the effects of physical roughness, surface chemical and/or structural changes that occur on sliding contact surfaces of diamond films can also play a major role in their friction and wear performance [36–39]. The extent of these changes may be dominated by the environmental species present in the surrounding air or by the ambient temperature [38, 39]. In open air, liquid water, or oils, smooth or polished diamond films can provide very low friction and wear to mating surfaces (Figure 9.10(a)). However, in inert gas or vacuum environments, the friction and wear performance of diamond films may be rather poor (Figure 9.10(b)). Increased σ-bond interactions and/or phase transformation can occur at sliding interfaces and dominate friction and wear.

In most sliding contact situations, the actual contact occurs between sharp asperity tips of the sliding surfaces. Under static conditions, the severity of contact pressure over these

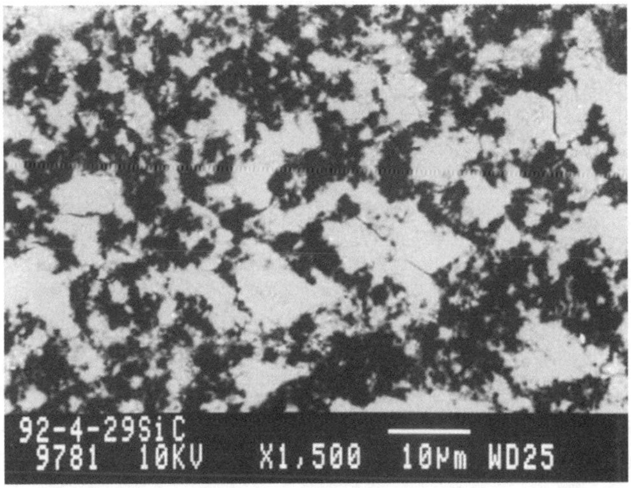

Figure 9.6 Smeared or transferred wear debris particles on rough diamond surfaces

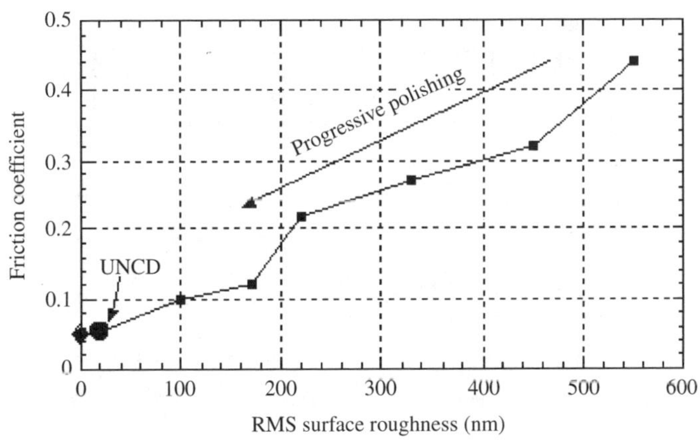

Figure 9.7 Relationship between surface roughness and friction coefficients of diamond films [34] (UNCD = ultrananocrystalline diamond)

asperity tips could be enormous. In the case of diamond films, some of the asperities may either fracture or undergo irreversible phase transformations. However, under dynamic sliding conditions, frictional heating of asperity tips can also occur and accelerate the kinetics of phase transformation [36, 40, 41].

It is possible that the asperity tips of diamond film may first be chipped away without phase transformation, become trapped at the sliding interface, and then undergo further wear and phase transformation under the influence of repeated sliding passes. Diamond, which is thermodynamically unstable, when excited thermally or by ion bombardment, can be transformed to a more stable carbon form, such as graphite [42, 43]. Apparently, the bond

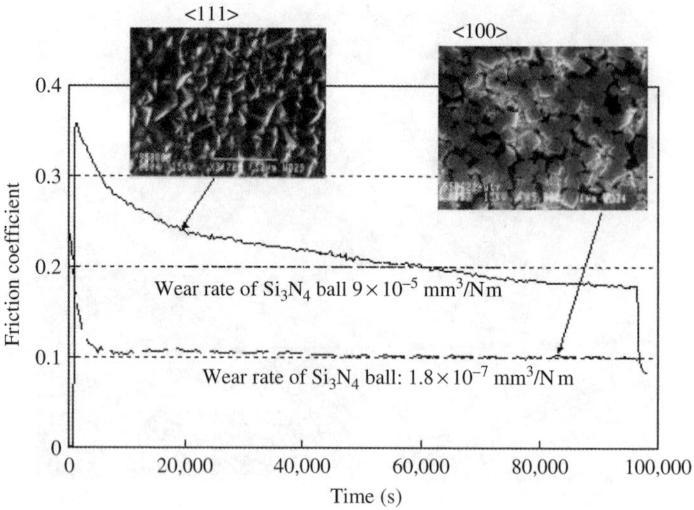

Figure 9.8 Effect of grain orientation on friction and wear of diamond films

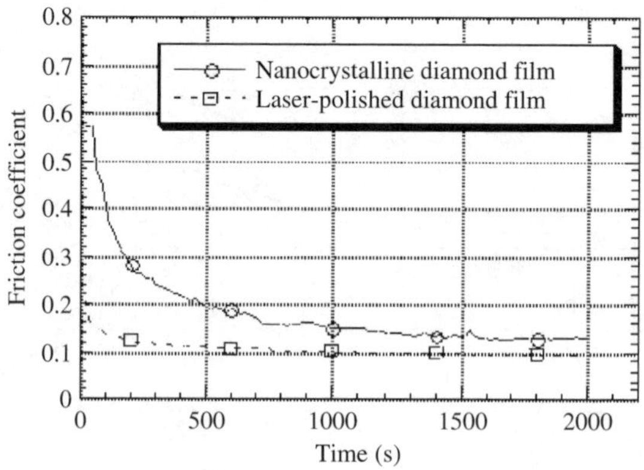

Figure 9.9 Frictional performance of nanocrystalline and laser-polished diamond film

characteristics of these debris particles and transfer layers are graphitic. Figure 9.11 shows the Raman spectra of debris particles and transfer layers formed on and around the wear scar of a Si_3N_{4th}ball that was rubbed against a nanocrystalline diamond film in dry N_2. The graphitic debris particles accumulated at the sliding contact interface begin to dominate the long-term friction and wear performance of diamond films when sliding in dry and inert gas environments. Because graphite is a poor solid lubricant in the absence of water molecules or other chemical adsorbates, the friction coefficients of such sliding interfaces become high. Figure 9.12 shows the long-duration frictional performance of a nanocrystalline diamond film in dry nitrogen. Friction spikes are thought to be the result of the presence of the graphitic debris particles (Figure 9.11) that are generated at the sliding contact interfaces [36, 40]. When moist laboratory air is introduced into the test chamber, the trace of the

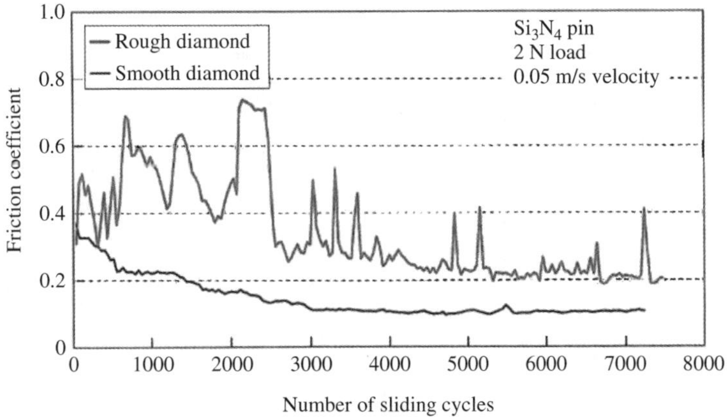

(a)

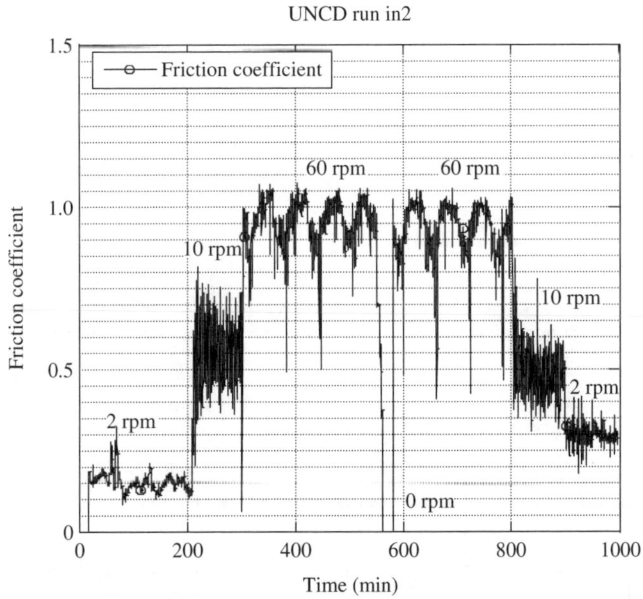

(b)

Figure 9.10 Frictional behavior of smooth diamond films in (a) open air and (b) high vacuum. Initially low friction in (b) is thought to be due to the presence of some adsorbed layers on sliding surfaces

friction coefficient becomes very smooth, and the large fluctuations in the traces disappear. Erdemir and co-workers have conducted electron energy loss spectroscopy, transmission electron microscopy (TEM), and electron diffraction of the sliding diamond surfaces, as well as the debris particles found in and around the wear tracks and scars. The combined results provided more convincing evidence for the transformation of sliding diamond surfaces to a disordered graphitic state [36, 40].

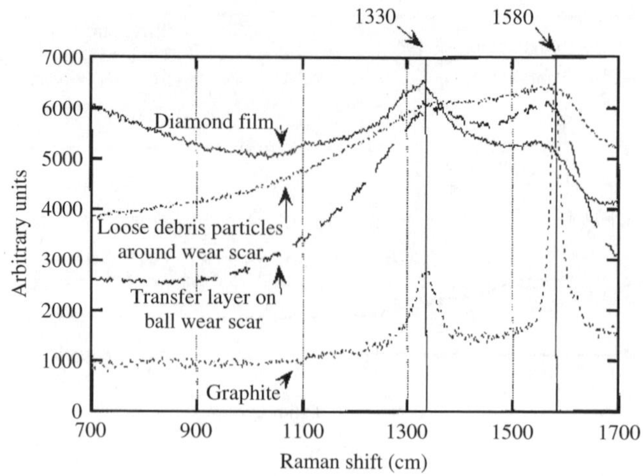

Figure 9.11 Raman spectra of debris particles and transfer layers formed during sliding against ball surfaces

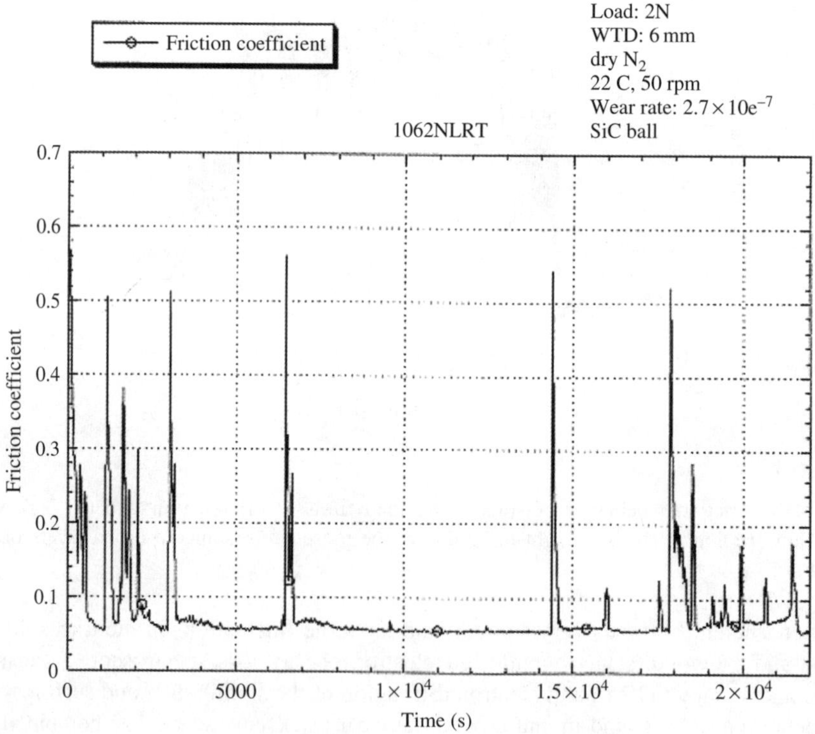

Figure 9.12 Long-duration friction performance of nanocrystalline diamond film in dry N_2

In water or oil environments, sliding diamond surfaces experience little or no wear. Recent testing of ultrananocrystalline diamond-coated face seals led to very little wear of the sliding surfaces. The wear was limited to the asperity tips. Beyond the heights of the asperities, there was no measurable wear even after 21 days of testing at 3500 rpm (Figure 9.13) [44]. The friction and wear performance of smooth diamond films under oil-lubricated sliding

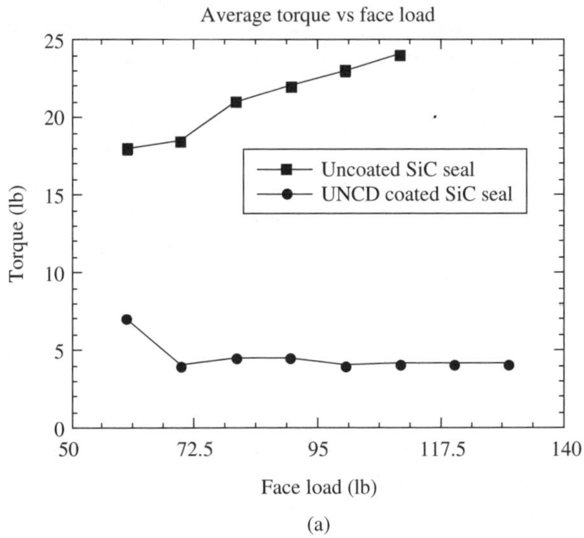

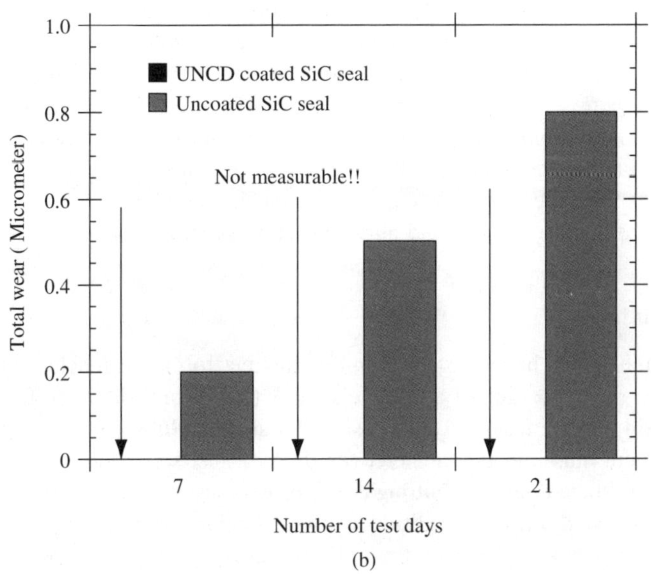

Figure 9.13 (a) Friction (torque) and (b) wear performance of nanocrystalline diamond-coated SiC seal in water

conditions are also exceptional. Beyond some run-in-type initial wear, these films experience essentially no wear. Such impressive friction and wear performance in water or oils are considered the result of the highly passive nature of sliding diamond surfaces. Specifically, no adhesive or abrasive interactions occur between highly passive surfaces of diamond and mating surfaces.

At elevated temperatures, the friction and wear performance of diamond films degrade considerably. Surface adsorbates (both chemisorbed and physisorbed) begin to desorb and gradually expose the strong σ-bonds for adhesive interactions across the sliding interface. Such strong interactions can cause high friction and wear of sliding diamond surfaces. Graphitization of sliding diamond surfaces becomes feasible at elevated test temperatures and can cause high wear losses [39].

Previous studies have shown that counterface materials used in the sliding tests with diamond films can have some effect on friction and wear of these films [45, 46]. When sliding against steel surfaces, high friction and severe wear losses can occur on the steel side [47], perhaps due to the relatively high solubility of carbon in iron. This is one of the factors preventing the use of diamond-coated tools from machining of iron and iron-based alloys. Against most ceramic counterfaces, sliding diamond surfaces provide very low friction and wear.

9.2.3 Practical Applications

Despite the great interest in the tribological and industrial applications of diamond films in recent years, more fundamental studies still remain to be pursued to better understand and tailor the tribological properties of these films for specific tribological applications. Until now, only cemented carbide tool inserts have been successfully coated with diamond and made available commercially. Although diamond film prototypes of other tribological systems have also been demonstrated, their large-scale utilization has not yet been realized. Such delays in commercialization of diamond films are thought to be due to the very high deposition temperatures, inadequate adhesion, oxidation when used at elevated temperatures, and, most important, high fabrication cost for large-scale applications. Other potential applications for diamond coatings besides cutting tools are mechanical seals, sliding bearings, MEMS, wire-drawing dies, high-precision microdrills, surgical blades, and various wear parts that are used against erosion and abrasion (such as jet nozzles).

9.2.3.1 Machining

Currently, diamond films have been used in two distinct forms for machining purposes. One form is based on the deposition of a relatively thick (0.5-1 mm) diamond film on a Si wafer from which small bits are laser-cut into desired bit shapes. Subsequently, the Si substrate is etched out, and cut diamond bits are recovered. As a last step, the free-standing diamond bits are bonded or brazed onto the cutting edges of tool inserts or end mills. This technology currently competes well with the traditional products prepared by high-pressure sintering of synthetic diamond powders produced by high-pressure, high-temperature methods.

The second form involves direct deposition of diamond on cutting tools by various methods (hot filament, microwave CVD, DC-arc jet, etc.). These methods are robust and capable of producing high-quality diamond films on carbide-based cutting tool inserts and other suitable parts and components. Diamond coatings are mostly applied to WC-Co-based cemented carbide tool inserts because these tools exhibit excellent toughness, hardness, and high-temperature durability. Co is mainly used as a binder and it also controls the toughness of the tool material. In principle, lower Co content ($<5\%$) in cemented carbide is highly desirable for diamond deposition because Co has a detrimental effect on the adhesion of diamond coatings to carbide inserts. Owing to its high mobility and vapor pressure at high temperatures, Co tends to interfere with the nucleation and growth processes during deposition. It can also increase the difference in thermal expansion coefficient between diamond and substrate; hence, during cooling, very high compressive stresses may build up at the interface region. When these stresses are combined with the stresses associated with cutting action, premature adhesive failure occurs at or near the cutting edges.

Cutting tool manufacturers and commercial coaters are now offering a variety of diamond-coated tools at a reasonable cost. These coated tools provide much improved performance during metal-cutting operations by allowing high-speed machining at increased feed rates. Some of the advantages that diamond provides in these applications are extreme hardness and wear resistance; good fatigue strength; high chemical inertness; excellent resistance to abrasion, erosion, and corrosion; high thermal conductivity; low friction; and excellent environmental compatibility. Despite these exceptional qualities, the market share of diamond-coated tools has remained rather small. It would increase substantially if the cost of producing diamond on tools were further reduced.

Currently, diamond-coated tools are primarily used in the machining of nonferrous metals, alloys, and composite materials that are inherently very difficult to cut or machine. Materials that are suitable for machining by diamond-coated tools include aluminum and its alloys (in particular, high-silicon aluminum alloys); magnesium and its alloys; copper, lead, and manganese alloys; graphite; carbon; plastics; fiberglass composites; carbon–carbon composites; metal–matrix composites; epoxy resins; green ceramic; fiber-reinforced plastics; concrete; and tools for various mining and rock-drilling operations. However, diamond-coated tools cannot be used for machining of ferrous alloys and the alloys of groups IV^a, V^a, VI^a, VII^a, and VIII^a in the periodic table. Diamond can chemically react and/or dissolve and rapidly wears out in these materials at the high temperatures that are generated during machining.

Diamond-coated tools are used in machining of high-silicon-containing aluminum alloys for automotive applications. In particular, hypereutectic aluminum–silicon alloys are very hard to machine with base carbide tools or tools coated with other types of hard materials. These alloys are increasingly being used in the making of main engine blocks in all kinds of transportation vehicles. Their lighter weights can reduce the overall weight of the vehicles and thus improve fuel economy. The increased use of lightweight aluminum alloys and other materials (such as magnesium) in the automotive sector is likely to continue; hence, the need for diamond-coated tools will also increase. Woodworking tools represent another major market segment for diamond coatings. Tungsten carbide blades already comprise a large part of the woodworking tool market. Such blades can easily be coated with diamond and will find a market if the cost of the diamond coating process can be further decreased.

9.2.3.2 Mechanical Seals

In addition to cutting tools, seals represent the next best application possibility for CVD diamond films. The worldwide mechanical seals market is very large. Over three million high-end pumps rely on the tribological performance of SiC seals, which are ideal substrates for diamond film growth. In fact, because of its extreme hardness and low coefficient of friction, diamond is an ideal candidate for sealing surfaces of rotating SiC shafts. Improper use or leakage in mechanical pump systems can be very costly and cause environmental problems. Total energy losses due to high friction and wear are estimated to be on the order of 10^{10} kWh/year, which translates into roughly \$1 billion/year. Also, it is estimated that 10–15% of total hazardous emissions are due to leakage or catastrophic failure of mechanical seals. If applied properly, diamond coatings may substantially extend the useful life of these components and also reduce energy losses because of the very low friction they give to the sliding surfaces of seals. Early studies by Hogmark *et al.* [19] demonstrated that smooth diamond films can lead to substantial reductions in frictional torque and wear rates of shaft seals.

Rough diamond coatings, mentioned above, cannot be used in seal applications. One of the sealing faces (inserts) is mostly made out of a soft carbon–graphite composite, which wears out rather quickly when rubbed against a rough and superhard diamond film. Hard SiC inserts may also be used as a mating surface, but they are much more expensive. When rubbed against rough diamond, they can also suffer major wear losses. Polished or smooth coatings of nanocrystalline diamond will be much more desirable for seal applications. Coated seals would be especially attractive in applications that require chemical inertness (especially in environments that involve concentrated acids or bases), the pumping of grit slurries that cause rapid erosion in conventional seals, and pumping without a liquid lubricant. Figure 9.13 shows smooth diamond-coated mechanical seals and their friction and wear performance in water [44].

9.2.3.3 Other Tribological Applications

MEMS can be used for a wide range of high-precision applications, including sensors, actuators, and micro-scale positioning devices. MEMS devices such as gear assemblies and micromotors have been largely fabricated by Si micromachining. However, Si exhibits very poor mechanical and tribological properties. When these microdevices are used in a very high-speed application, they always suffer from unacceptably high friction and wear losses. Consequently, they are unsuitable for high-speed applications at realistic loads. Researchers have developed alternative fabrication methods that would allow production of MEMS devices from SiO_2-, Si_3N_4-, or SiC-type materials. However, the tribological performance of these materials is also very poor; hence, they may not function well in a dynamic MEMS application.

Researchers have recently been contemplating the use of diamond coatings for MEMS applications. However, some difficulties have been encountered in fabricating diamond coatings and/or components with a morphology suitable for MEMS. The nanocrystalline diamond films discussed earlier appear to be better suited because their grain size is very small (10–30 nm). With these films, various free-standing MEMS structures can be produced using special micro-fabrication methods. Alternatively, such films may be produced on the surfaces of Si-based MEMS as a thin protective layer; it is possible to attain film coverage at very small thicknesses (as thin as 50 nm) and thus preserve the fine structural features of the base Si MEMS [48]. Figure 9.14 shows some of the MEMS structures fabricated from nanocrystalline diamond films.

Figure 9.14 Examples of various MEMS structures and microfeatures fabricated from nanocrystalline diamond [48]

9.3 Diamond-like Carbon Films

9.3.1 Structure and Composition

DLC films are amorphous and may combine many of the mechanical and chemical properties of crystalline diamonds. Mechanically, they are quite resilient and can exhibit Vickers hardness values as high as 9000 kg/mm^2 [1]. Chemically, DLC films can be made very stable and resistant to acidic and saline media. They are electronically insulating and optically transparent to visible and ultraviolet light [49]. Unlike diamond, they can be deposited at temperatures that range from subzero to 300°C and at deposition rates that are fairly high by various methods, including ion-beam deposition, DC and RF sputtering, arc-plasma, deposition plasma-enhanced CVD, and laser ablation [50–54]. As illustrated in Figure 9.1, depending on the deposition method and carbon sources, these films may contain large amounts of hydrogen (10–50 at.%). Their sp^2-to-sp^3 bond characteristics vary significantly [55]. Hydrogen-free DLC films are largely deposited by sputtering, laser ablation, and/or arc-plasma deposition methods that use solid carbon targets. Structurally, most DLC films are amorphous, consisting of very short-range-ordered diamond (characterized by sp^3-type tetrahedral bonds) and graphitic phases (characterized by sp^2-type trigonal bonds). Hence, these films can be regarded as degenerate forms of bulk diamond and/or graphite, as illustrated in Figure 9.15.

Various C-based materials, similar in structure to a (amorphous)-CH or ta (tetrahedral amorphous)-C, include elements such as B, Si, F, and N, and metals such as Cr, Ti, and W. The introduction of dopants or alloying elements generally decreases compressive stress due to fewer interconnections in the random carbonaceous network. Grischke *et al.* [56] have shown that the specific chemical composition of the modified films strongly influences the surface energy, depending on the nature of the dopant, and may modify various physical properties, making some doped DLCs suitable for various applications, e.g. to improve field emission characteristics. As highlighted in the next section, the introduction of alloying elements into DLC affects many properties, including tribological behavior.

From the point of view of optimizing the routine production of DLC films, the primary need is for effective process control, primarily the tailoring of adhesion properties. Indeed, the adhesion of DLC films varies widely, depending on the nature of the substrate. To be used as tribological coatings, DLC films must adhere well to the substrate, and the adhesive forces must overcome the high internal stresses that otherwise would cause film delamination.

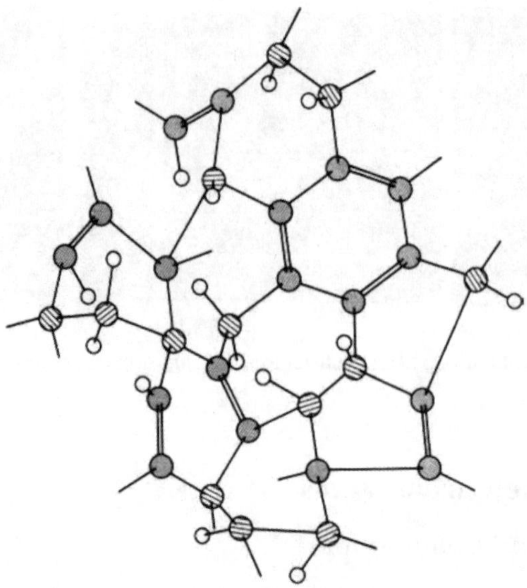

(a)

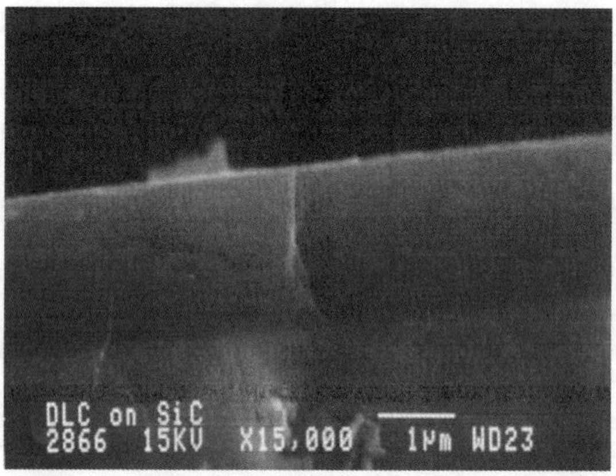

(b)

Figure 9.15 (a) Illustration of molecular structure and (b) cross-sectional scanning electron microscopy image of DLC film, revealing a featureless microstructure

Adhesion depends on the deposition method, in combination with the nature of the substrate. Good adhesion of DLC films is observed on carbide- and silicide-forming substrates. The adhesion of DLC coatings to silicide-forming metals can be improved by depositing a 2–4 nm thick interfacial layer of amorphous Si between the metal and the C film, thus forming an interfacial silicide layer promoted by the plasma, even at relatively low substrate temperatures. Titanium-based, functionally gradient films intercalated between the substrate and the DLC top layer have also been studied [57, 58]. The deposition of adhesion and interface layers is ideally achieved in the same reactor as the deposition of the DLC films, i.e. by multiplex processes that minimize the introduction of defects and impurities between the superimposed layers and allow precise control of the entire deposition methodology.

9.3.2 Tribology of DLC Films

The tribological properties of DLC films have been extensively investigated by many scientists during the past decade. In general, these films were shown to provide some of the lowest friction and wear coefficients under certain sliding conditions. In most cases, friction and wear are dominated by test conditions, environment, temperature, type of film, and counterface material [54–72]. Also, a strong correlation seems to exist between friction coefficients and wear rates of most DLC films. In general, the higher the friction coefficient, the shorter the wear lives. For example, the friction coefficients of hydrogen-free or -poor DLC films are very high in an inert gas or vacuum, and these films wear out rather quickly in such environments. Conversely, the hydrogenated DLC films provide very low friction and hence low wear rates in such environments [55, 68–70, 73]. Figure 9.16 shows the close relationship between friction and wear coefficients of various DLC films in dry nitrogen. Under conditions where friction is extremely low, the DLC films may attain a wearless

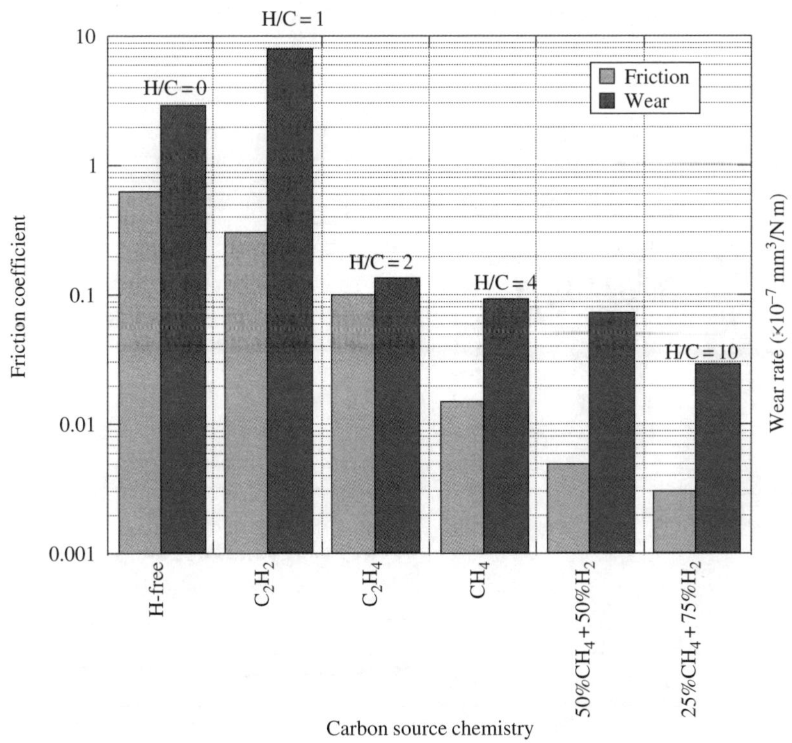

Figure 9.16 Friction and wear coefficients in dry nitrogen of DLC films derived from different carbon sources with various hydrogen-to-carbon (H/C) ratios

sliding regime in which sliding seems to take place at the contact interface without causing much wear [73].

9.3.2.1 Effects of Test Environments on Friction and Wear of DLC Films

The composition of the test environment can have a dramatic effect on the friction and wear behavior of both hydrogenated and hydrogen-free DLC films. For example, in dry air and inert gases, the friction coefficient of a highly hydrogenated DLC film is as low as 0.003 when sliding against itself [74]. However, in humid air, the friction coefficient of a DLC film may increase to values as high as 0.15 while sliding against itself or metallic and ceramic counterface materials [74, 75]. Depending on the partial pressures of various reactive gaseous species (such as oxygen and water), the friction coefficients of such DLC films may also go up in vacuum environments [76, 77]. The hydrogen-free DLC films seem to work best in humid test environments [74, 75, 77]. Figure 9.17 shows the effect of humidity on the friction coefficient of a hydrogen-free and a highly hydrogenated DLC film at atmospheric pressure as well as in vacuum [74, 77].

Addition of small amounts of elemental species (N, F, Si, Ti, etc.) to DLC films may reduce their tribological sensitivity to humidity [72, 78]. Paulmier *et al.* [79] found that atomic hydrogen exerts the most beneficial effect on the friction coefficients of DLC films; they reported a friction coefficient of 0.06 for a DLC/steel pair in the presence of atomic hydrogen. In molecular hydrogen, the friction coefficient increased to 0.08, and in air, it became 0.1.

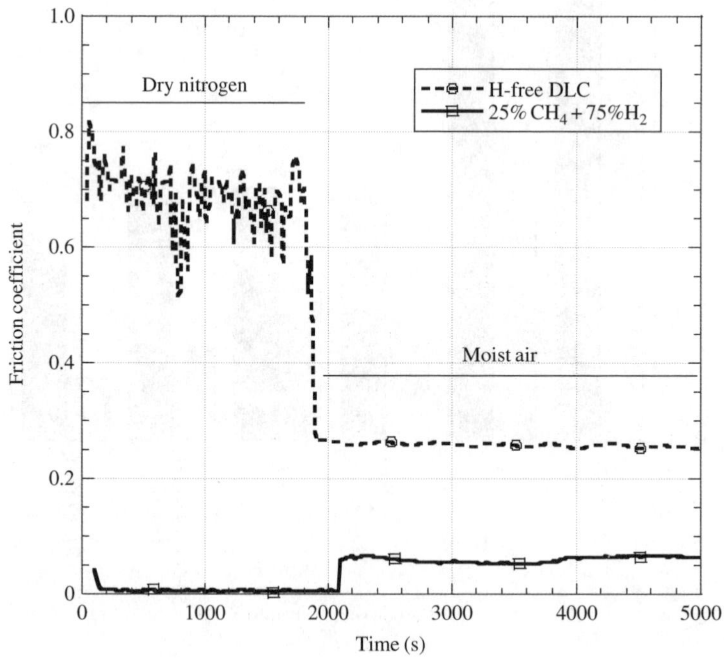

Figure 9.17 Effect of moist air on friction coefficient of hydrogen-free and hydrogenated DLC films

Tribochemical and structural changes that occur at the sliding contact interfaces of DLC films may also have a strong influence on the friction and wear performance of DLC films. Olsen *et al.* [80] reported that, in addition to humidity, oxygen can significantly increase the friction coefficients of the DLC films produced by a plasma-activated CVD process. In the presence of oxygen, sliding contact areas of most DLC films reacted with oxygen. Such a reaction reduced the wear life and generated high friction. However, when the tests were run in open air or dry N_2, the friction coefficients were much lower, and the wear lives were much longer. The higher friction and wear in oxygen-containing environments were attributed to tribo-oxidation of the rubbing surfaces. Similar observations were reported by Strom *et al.* [81] for tests performed in oxygen-rich test environments. Again, tribo-oxidation was suggested as the main reason for generally higher friction and wear of the DLC films. Ramirez *et al.* [82] further confirmed that oxygen and water molecules are detrimental to the tribological performance of DLC films.

In ultrahigh vacuum, hydrogenated DLC films performed quite well. Donnet and Grill [55] reported friction coefficients below 0.01 for a hydrogenated DLC in vacuum, whereas the friction coefficient in open air was 0.17. For films with a hydrogen content <34 at.%, the friction coefficients were significantly higher in vacuum, but similar in open air.

Several researchers have noticed the formation of a transfer layer on surfaces sliding against a DLC film [59, 60, 62–73, 75, 83–85]. They have claimed that these layers were in a graphitic form. When compared with crystalline graphite, DLC films are structurally amorphous and thermodynamically unstable. When subjected to thermal and/or mechanical loading in sliding contact, they can transform to the more stable graphitic forms. In most sliding interfaces, frictional heating occurs in the real areas of contact, in which transient temperatures can reach very high values. Repeated occurrence of such heating at the real contact spots can convert sp^3-bonded carbon to the sp^2-bonded form.

Erdemir and co-workers have shown that the wear debris particles that were smeared onto, or accumulated around, the edges of the sliding surfaces of DLC had a disordered graphitic structure [75, 83–85]. Figure 9.18 shows the transfer layer formed on the wear scar of a steel ball. It also shows the debris particles accumulated on the trailing edge of the sliding ball surface. These studies used a combination of TEM, electron diffraction, bright-field/dark-field imaging, Raman, and Fourier transform infrared spectroscopy. Raman spectra taken from the sliding ball surfaces and wear debris particles after long-duration wear tests reveal Raman bands indicative of micrographitization. Specifically, as shown in Figure 9.19, the microlaser Raman spectra of black transfer layers indicate that their structural chemistry is quite different from that of the original DLC film, but similar to that of the crystalline graphite (provided as a standard). Two broad Raman bands (at 1336 and 1594/cm) match the D and G bands of crystalline graphite. The peaks of the carbonaceous transfer layer are not as sharp as those of the crystalline graphite, possibly because the debris particles are very small, and because phonon damping occurs. The Raman spectrum also indicates that the degree of structural disorder in the carbonaceous transfer layer is very high. Furthermore, bright- and dark-field images (Figure 9.20), together with the electron diffraction pattern of the wear debris particles collected from the rubbing surfaces, further enhance the Raman findings by revealing an increasingly graphitic structure [85]. Therefore, the carbonaceous transfer layers or debris particles found at sliding interfaces are graphitic, but in a disordered state. The friction coefficients of sliding contact interfaces that contain such debris particles may partially be influenced by the shear rheology of these particles. TEM studies by Ramirez

Figure 9.18 Carbon transfer layer and debris particles on and around wear scar formed on steel ball slid against DLC film

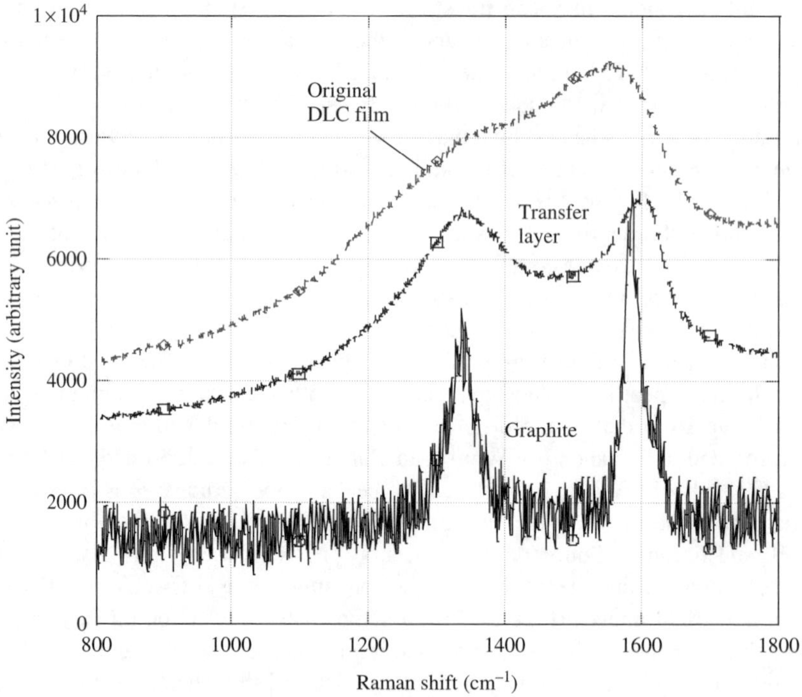

Figure 9.19 Raman spectra of original DLC film and transfer layer formed on ball side. Raman of crystalline graphite is provided as reference

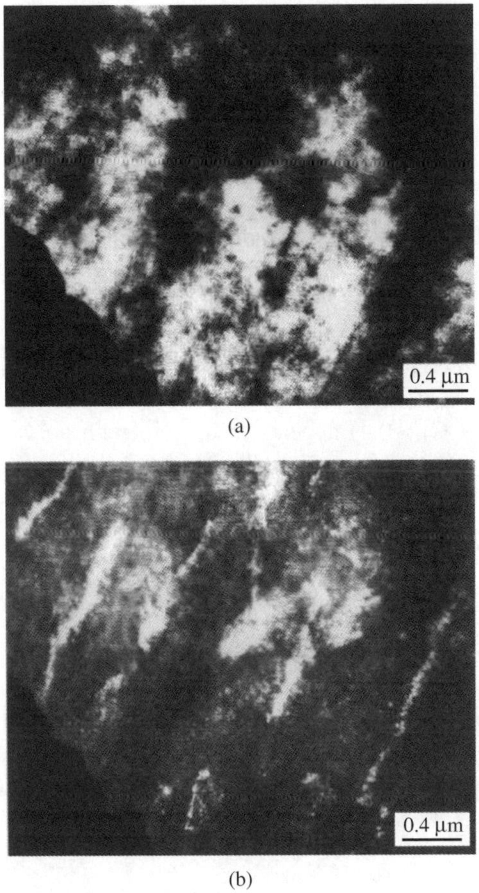

Figure 9.20 (a) Bright- and (b) dark-field images of transfer layers recovered from sliding ball surfaces, suggesting micrographitization

et al. [82] and Liu *et al.* [85] also verified that the sliding contact areas of DLC films were transformed into a graphitic structure.

In recent years, researchers have investigated the triboemission of electrons from amorphous carbon films during sliding in dry air and vacuum [86, 87]. They confirmed the emission of electrons during sliding contact. The intensity of the electron emission increased sharply with an increase in the hydrogen content of the source gases from which the carbon films were derived. In general, the specific relationship between triboemission and tribological properties of DLC films is not exactly known, but it was reported by Nakayama and co-workers that higher levels of triboemission correspond to higher wear in carbon films [86, 87].

9.3.2.2 Effect of Dopant or Alloying Elements on Friction and Wear of DLC Films

A compilation discussing the tribology of doped DLC and C alloy coatings has been prepared by Donnet [78]. Silicon incorporation into the DLC structure is shown to affect most film properties (including a decrease in surface energy and internal stress) and its tribological behavior. Friction appears to be significantly reduced (below 0.1), when compared with that of conventional undoped DLC in ambient humid air, and the wear resistance is improved. However, this tribological behavior seems to be observed only when the contact pressure remains below 1 GPa. At higher contact pressures, conventional ta-C and a-CH films cannot be surpassed. Consequently, ta-CHSi films may be used in applications that require both low friction (<0.1) and high wear resistance ($<10^{-7}$ $mm^3/N\,m$) under moderate mechanical conditions. Such applications include the protection of low-stress aerospace or automotive components, precision ball bearings and gears, sliding bearings, and magnetic recording media.

The incorporation of Si and F into the DLC structure affects the surface properties. The reduction of stress, when compared with conventional DLC, is in the same range as that in a-CHSi. However, the reduction of the surface energy is higher with F than with Si. Highly fluorinated DLC ($F/[F+C] > 0.4$) appears to be soft, with no wear resistance. Moderate fluorination ($F/[F+C] < 0.2$) can be controlled by the deposition conditions to obtain films with comparable wear resistance and friction levels of conventional a-CH films, but with lower stress and surface energy. When compared with conventional DLC and other doped DLC, less work has been performed on the tribological investigation of N-containing DLC films because of their recent discovery. The tribological behavior of C nitride films is summarized elsewhere in this chapter.

The range of composition and structures attainable with metal-containing DLC coatings appears to be enormous. One should keep in mind that the optimization of the material combination and deposition parameters is a challenging subject for each element or combination of elements. When optimization is achieved, metal-containing DLC films may exhibit promising tribological properties in terms of steady-state friction levels and wear rates, for various applications. Friction coefficients in the range of 0.02–0.04 have been observed in ambient air when a significant amount of Ta or W is incorporated in a-CH films [88].

9.3.2.3 Effects of Other Parameters on Friction and Wear of DLC Films

The friction coefficients of DLC films are generally lower under heavier loads and reach low steady-state friction coefficients faster at higher sliding velocities [65]. At elevated temperatures, the DLC films may transform to a graphitic phase, and their friction and wear coefficients increase substantially. In particular, the hydrogenated DLC films undergo rapid phase transformation/oxidation and wear out quickly at elevated temperatures [39]. Figure 9.21 shows the wear rates of a hydrogenated DLC film with increasing temperature. The hydrogen-free DLC films resist oxidation or phase transformation at much higher temperatures (up to 500°C), but their friction coefficients become very high. This is thought to be caused by most of the surface adsorbates being removed from the surfaces; hence, the amount of covalent bond interaction (which can cause high friction) becomes significant.

The tribological behavior of a surface can be affected by its precise patterning at the micro-scale. This rather new concept in tribology will probably become better understood in the near future. Plasma, chemical, or laser techniques as widely performed by the microelectronic industry for many years can be used to achieve a specific geometrical design on a surface. Laser treatment has advantages related to its versatility and fast adaptability. Excimer lasers

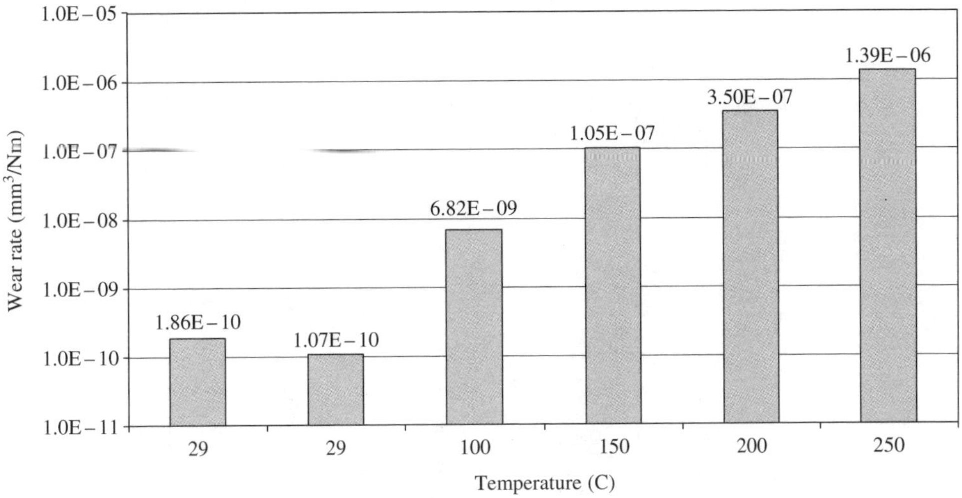

Figure 9.21 Effect of temperature on wear performance of hydrogenated DLC film

have been used to induce patterns on DLC films [89], but emerging femtosecond laser facilities will probably introduce new perspectives in that field. Indeed, the ultrashort laser pulses (in the 100-fs range, with power density as high as 10^{13} W/cm^2) can ablate all kinds of materials without any collateral effects (negligible thermal-affected zone), while nanosecond pulses cannot [90]. Lifetime increases up to a factor of 10 have been achieved by patterning of TiN [91] and TiCN [92] films with a femtosecond laser. Recently, laser patterning of wear-resistant DLC has been studied for two configurations, coating already patterned substrates (indirect processing) and direct laser processing of deposited DLC films [93]. Pore depths that yield positive tribological improvements (>10 μm) are larger than film thickness (>5 μm). The authors indicate that debris particles were trapped in the surface pore obtained by indirect laser processing, thus preventing the breakdown of the tribological system. Coating patterning will probably be a major avenue of development in coating technology in the next decade.

9.3.3 Synthesis of Carbon Films with Superlow-Friction and -Wear Properties

Recent studies indicated that source gases used in the deposition of DLC films dramatically affect the friction and wear performance of these films (especially in inert gas or vacuum). As shown in Figure 9.22, the friction and wear coefficients decrease with increasing hydrogen content in the source gases. In general, the higher the amount of hydrogen, the lower the friction coefficients [69, 70, 73]. The lowest friction coefficients (Figure 9.23) were measured on DLC films derived from superhydrogenated source gases (75–90% H_2 and 10–25% CH_4 or C_2H_2) [70]. Fundamental tribological and surface analytical studies indicated that, in such nearly frictionless and wearless DLC films, all of the surface σ-bonds are terminated by hydrogen. This hypothesis has been confirmed by Fontaine *et al.* [94], who

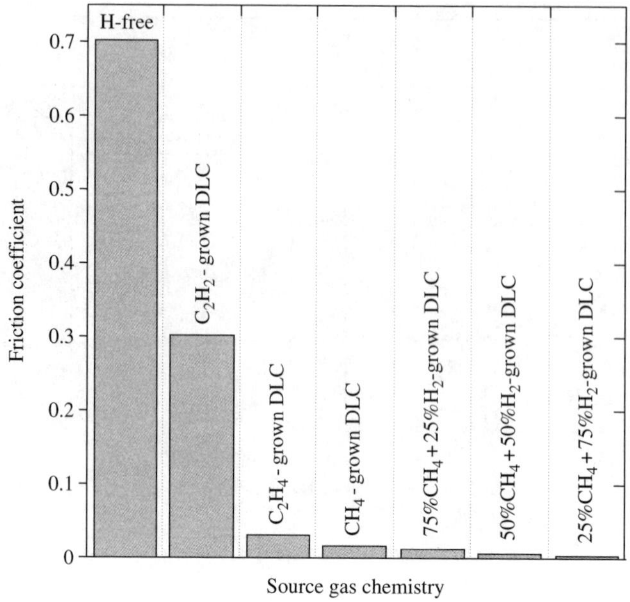

Figure 9.22 Effect of carbon source on friction coefficient of DLC films in dry nitrogen

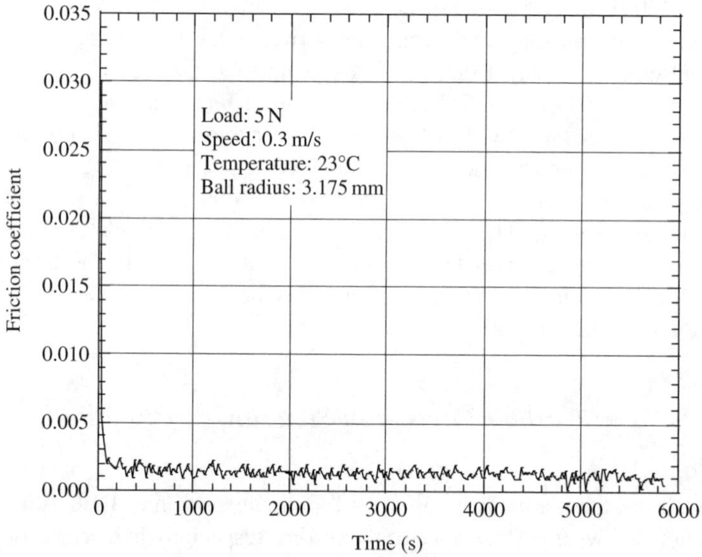

Figure 9.23 Friction coefficient of DLC film grown in 25% CH_4 + 75% H_2

investigated the friction behavior of DLC films with various hydrogen contents, tested in vacuum or in an atmosphere of pure molecular hydrogen. Furthermore, perhaps some of the surface carbon atoms are di-hydrated (two hydrogens bonded to one carbon). Such high levels of surface hydrogen concentration provide a high degree of surface passivation,

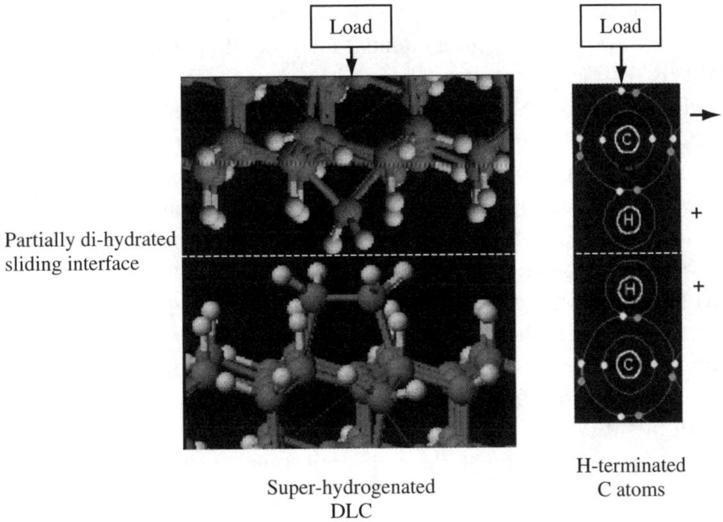

Figure 9.24 Schematic illustration of superlow-friction mechanism of superhydrogenated DLC films.

and hence superlow friction, as schematically illustrated in Figure 9.24 [69, 70, 73, 74]. Nevertheless, the exact role of hydrogen in DLC films, including its capacity to passivate the dangling bonds, is not trivial, as recently pointed out by Robertson [95], and is a subject of fundamental investigations. Recent fundamental studies involving the use of molecular dynamic simulations have indicated that DLC films with high levels of hydrogen can indeed provide very low friction to sliding contact interfaces [96, 97].

9.3.4 Practical Applications

The unique properties of DLC films and their modifications, together with the possibility of adjusting the properties by choosing the right deposition parameters, make them suitable for various applications. The exploited properties so far include high wear resistance and low friction coefficients, chemical inertness, infrared transparency, and high electrical resistivity. They potentially include field emission and low dielectric constants. The announcement of the latest MACH3 razor blades by Gillette underscores the use of DLC as a coating to improve the quality and performance of the blades. Such a popular application plays a key role to promote DLC films widely.

One of the major applications of DLC films is probably in the automotive industry [98]. Metal-containing DLC (Me-C:H) is a DLC type of coating which has been tested on a large number of automotive components during the last decade. Today, parts coated in serial production are mainly to be found in diesel injection systems. Diesel injection systems are constantly and quickly moving toward higher pressures. In many injection systems today, the pressure is too high for ordinary steel substrates, which would wear rapidly. Both erosive and adhesive wear are present in these systems. Me-C:H deposition has been the best alternative to retain tolerances. For several years, parts in the diesel injection system have been the largest application for Me-C:H coatings. However, the use of Me-C:H coatings in other parts of the motors is increasing rapidly. Nearly 30 million coated parts are delivered to the automotive industry every year, with an annual increase of approximately 50%.

Because DLC is IR-transparent, it can be used as an antireflective, scratch-resistant, wear-protective coating for IR optics (at wavelengths of 8–13 μm) made of Ge, ZnS, and ZnSe [99]. The low deposition temperature of DLC allows its use as a wear-protective layer on products made of plastic; therefore, it is used to protect polycarbonate sunglass lenses from abrasion. Other optical applications include the protection of aluminum mirrors and the photothermal conversion of solar energy [100]. Another widespread use of DLC films is wear and corrosion protection of magnetic storage media. Nanosmooth and very thin (even <5 nm) DLC films are now used as corrosion and wear-protective coatings for both magnetic disks and magnetic heads [101]. Video recording or magnetic-data storage tapes, in which ferromagnetic metal is the recording medium, and metallic capstans are in contact with the tapes, are also being protected with DLC coatings to reduce wear and friction and thus extend both the life of the tapes and their reliability.

DLC films appear to be biocompatible, and applications are being developed for their uses in biological environments [102, 103]. Because they are chemically inert and impermeable to liquids, DLC coatings could protect biomedical implants against corrosion and serve as diffusion barriers. DLC is being considered for coating metallic and polymer-based invasive and implantable biomedical devices. It is thought that such coatings can improve the biocompatibility of polyurethane, polycarbonate, and polyethylene with body tissues. DLC, deposited on stainless steel, Co(Cr) alloys, or Ti alloys, which are components of artificial heart valves and stents, have been found to be capable of satisfying the mechanical and biological requirements and improving the performance of these components [104]. The same properties may make DLC useful as a protective coating for hip and knee joint prosthetics. Carbon/carbon composite prosthetics have also been improved by DLC coatings. Yet, the short-term data obtained so far by *in vitro* and *in vivo* experiments indicate that DLC coatings are biocompatible. The investigations of their long-term biological performance are further required before being able to use them in practical applications at large scale.

Currently, DLC and its modifications are being considered as low-dielectric (low k) materials in ultra-large-scale integrated circuits. The objective is to reach lower dielectric constants than attainable with SiO_2 in order to reduce the capacitance of the insulator. Besides low k, these materials must satisfy various criteria. They must be thermally stable to 400°C; have adequate rigidity, low mechanical stress, low dissipation, low leakage, and good adhesion; and be processible using acceptable means. Various carbon-based materials have been tried, such as a-C:H, fluorinated amorphous carbon (a-C:F), carbon nitrides, and methyl-substituted siloxanes, in addition to a wide range of organosilane derivatives [105]. SiCOH films applied by plasma-enhanced CVD from a proper choice of precursor and optimization of the deposition conditions exhibit dielectric constant values in the 2.2–2.8 range [106]. Another potential application of carbon-based films is field emission displays. The advantages of a carbon cathode are its chemical and physical stability compared to the Si and Mo used in usual tips. The carbon is more resistant to poisoning by elements such as sulfur. Carbon has a low sputter yield when bombarded by residual ions in the device. In addition, DLC devices could be very low-cost as DLC can be deposited at room temperature. The need is for any low-field-emitting material, so it does not really matter if it is diamond, DLC, or carbon nanotubes (CNT). All three must be compared. The advantage of DLC over diamond or CNTs is its low-temperature deposition. This has led to the intensive study of field emission in DLC. But it is now realized that field emission from DLC has serious problems because their emission site density is too low, and their emission is somehow extrinsic and not fully controllable. Nanotubes are now viewed as the best carbon system for field emission [105].

A recent interest for DLC films is in MEMS or nanoelectromechanical systems (NEMS). The recent development of ta-C films with high-modulus, hydrophobic properties and a low ability to stiction seems to be promising. However, the intrinsic stress of ta-C films significantly restricts their use. Indeed, the stress limits the maximum thickness of adherent film, and MEMS (NEMS) require a zero stress film as the device must be free-standing. If the film remains stable up to about 600°C, thermal annealing at this temperature range leads to a complete stress relaxation. A promising way seems to use DLC multilayers of alternately hard and soft a-C [105].

9.4 Summary and Future Direction

Diamond and DLC films are unique and have much to offer for a wide range of tribological applications. Over the years, scientists have made great strides in understanding the growth mechanisms, surface chemical/physical states, and tribological behaviors of diamond and DLC films. This understanding has been used to design and customize new coatings, such as ultrananocrystalline diamond and near-frictionless carbon, which can meet the multifunctional needs of future tribological applications. Unlike most other solids, diamond and DLC films enjoy a combination of low friction and high wear resistance under a wide range of sliding contact conditions. Detailed tribological studies show that test conditions and environmental species can dramatically affect the friction and wear performance of these films. Depending on the tribological and environmental constraints, tribo-oxidation, graphitization, and gaseous adsorption/desorption can occur at their sliding contact interfaces and control their friction and wear behavior. Overall, with the manipulation of process parameters and proper structural engineering, diamond and DLC films may live up to their promise. DLC films produced in highly hydrogenated plasmas can provide superlow-friction and -wear properties to sliding tribological interfaces. However, for applications that involve high ambient temperatures or sliding velocities, these films may lose their effectiveness as protective layers. The development of doped and alloyed films, together with multilayered and nanostructured carbon-based coatings, is in progress and will probably extend the potential of DLC films in a wide range of applications in the near future.

Acknowledgments

This work is supported by the US Department of Energy, under Contract W-31-109-Eng-38. The authors thank their colleagues and students who participated in the preparation, testing, and characterization of numerous diamond- and DLC-coated samples.

This material has been based upon an article that first appeared in the *Journal of Engineering Tribology* – Proceedings Part J, 2002, Vol. 216, No. J6, ISSN 1350–6501, published by Professional Engineering Publishing. Permission is granted by the Institution of Mechanical Engineers.

References

1. Erdemir, A. and Donnet, C., 'Tribology of Diamond, Diamondlike Carbon and Related Films', *Modern Tribology Handbook* (ed B. Bhushan), CRC Press, Boca Raton, FL, 2000, p. 871.
2. Cuomo, J.J., Doyle, J.P., Bruley, J. and Liu, J.C., *Applied Physics Letters*, **58**, 1991, 1.
3. Holmberg, K., *Tribologia*, **12**, 1998, 33–62.
4. Ferrari, A.C. and Robertson, J., *Physical Review B*, **61**, 2000, 14095.
5. Williams, B.E. and Glass, J.T., *Journal of Materials Research*, **4**, 1989, 373.
6. Tamor, M.A. and Wu, C.H., *Journal of Applied Physics*, **67**, 1990, 1007.
7. Bar-Yam, Y. and Moustakes, T.D., *Nature*, **342**, 1989, 786.
8. Zhu, W., Badzian, A.R. and Messier, R., *Journal of Materials Research*, **4**, 1989, 659.
9. Seal, M., 'The Current Status of CVD Diamond Applications and Prospects for the Future', in *Applications of Diamond Films and Related Materials* (eds A. Feldman *et al.*), 3rd International Conference, NIST Special Publication 885, 1995, pp. 3–12.
10. Bull, S.J. and Mathews, A., *Diamond and Related Materials*, **1**, 1992, 1049.
11. Ravi, K.V., 'Technological Applications of CVD Diamond', Chapter 14, in *Synthetic Diamond: Emerging CVD Science and Technology* (eds K.E. Spear and J.P. Dismukes), Electrochem. Soc. Monograph, John Wiley and Sons, New York, 1994, pp. 419–504.
12. Matthews, A. and Eskildsen, S.S., *Diamond and Related Materials*, **3**, 1994, 902.
13. Holmberg, K., Ronkainen, H. and Matthews, A., Proceedings of the World Tribology Congress, New Directions in Tribology, Mechanical Engineering Publishers Limited for Institute of Mechanical Engineers, London, UK, 1997, pp. 251–268.
14. Harris, S.J., Weiner, A.M. and Meng, W.J., *Wear*, **211**, 1997, 208.
15. Voevodin, A.A., Capano, M.A., Laube, S.J.P., Donley, M.S. and Zabinski, J.S., *Thin Solid Films*, **298**, 1997, 107.
16. Angus, J., in *Synthetic Diamond: Emerging CVD Science and Technology* (eds K.E. Spear and J.P. Dismukes), John Wiley & Sons, Inc., New York, 1994, p. 21.
17. Bhat, D.G., Johnson, D.G., Malshe, A.P., Naseem, H., Brown, W.D., Schaper, L.W. and Shen, C.-H., *Diamond and Related Materials*, **4**, 1995, 921.
18. Ramesham, R. and Rose, M.F., *Thin Solid Films*, **320**, 1998, 223–227.
19. Hogmark, S., Hollman, O., Alahelisten, A. and Hedenqvist, P., *Wear*, **200**, 1996, 235.
20. Choi, S.K., Jung, D.Y., Kweon, S.Y. and Jung, S.K., *Thin Solid Films*, **279**, 1996, 110.
21. Hollman, R.P., Wanstrand, O. and Hogmark, S., *Diamond and Related Materials*, **7**, 1998, 1471–1477.
22. Gruen, D.M., *MRS Bulletin*, **23**(9), 1998, 32–35.
23. Zuiker, C., Krauss, A.R., Gruen, D.M., Pan, X., Li, J.C., Csencsits, R., Erdemir, A., Bindal, C. and Fenske, G., *Thin Solid Films*, **270**, 1995, 154.
24. Grillo, S.E., Field, J.E. and van Bouwelen, F.M., *Journal of Physics D*, **33**, 2000, 985.
25. Schmitt, M., Paulmier, D. and Huu, L., *Surface and Coatings Technology*, **120**, 1999, 585.
26. Hayward, I.P., Singer, I.L. and Seitzman, L.E., *Wear*, **157**, 1992, 215.
27. Miyoshi, K., Wu, R.L.C., Garscadden, A., Barnes, P.N. and Jackson, H.E., *Journal of Applied Physics*, **74**, 1993, 4446.
28. Chandrasekar, S. and Bhushan, B., *Wear*, **153**, 1992, 79.
29. Miyoshi, K., NASA Technical Memorandum, 1999, n208905.
30. Bowden, F.P. and Young, J.E., *Proceedings of the Royal Society, London*, **208**, 1951, 444.
31. Bowden F.P. and Hanwell, A.E., *Proceedings of the Royal Society, London*, **A295**, 1966, 233.
32. Pepper, S.V., *Journal of Vacuum Science and Technology*, **20**, 1982, 643.
33. Buhshan, B., Subramanian, V.V., Malshe, A., Gupta, B.K. and Ruan, J., *Journal of Applied Physics*, **74**, 1993, 41.
34. Gupta, B.K., Malshe, A., Bhushan, B. and Subramaniam, V.V., *Journal of Tribology*, **116**, 1994, 445.
35. Erdemir, A., Halter, M., Fenske, G.R., Krauss, A., Gruen, D.M., Pimenov, S.M. and Konov, V.I., *Surface and Coatings Technology*, **94–96**, 1997, 537.
36. Erdemir, A., Bindal, C., Fenske, G.R., Zuiker, C., Csencsits, R., Krauss, A.R. and Gruen, D.M., *Diamond Films Technology*, **5**, 1996, 923.
37. Gardos, M.N. and Ravi, K.V., *Diamond Films Technology*, **4**, 1994, 139.

38. Erdemir, A., Fenske, G.R. and Wilbur, P., in *Proceedings of NATO Advanced Research Workshop on Protective Coatings and Thin Films: Synthesis, Characterization, and Applications* (eds Y. Pauleau and P.B. Barna), Kluwer Academic Publishers, Netherlands, 1996, pp. 169–184.
39. Erdemir, A. and Fenske, G.R., *Tribology Transactions*, **39**, 1996, 787.
40. Erdemir, A., Halter, M., Fenske, G.R., Csencsits, R., Krauss, A.R. and Gruen, D.M., *Tribology Transactions*, **40**, 1997, 667.
41. Gardos, M.N. and Soriano, B.L., *Journal of Materials Research*, **5**, 1990, 2599.
42. Lee, E.H., Hembree, D.M., Rao, G.R. and Mansur, L.K., *Physical Review*, **48**, 1993, 15540.
43. Lee, E.H., Lewis, M.B., Blau, P.J. and Mansur, L.K., *Journal of Materials Research*, **6**, 1991, 610.
44. Sumant, A.V., Krauss, A.R., Gruen, D.M., Auciello, O., Erdemir, A., Williams, M., Artiles, T. and Adams, W., *Tribology Transactions*, **48**, 2005, 24.
45. Zeiler, E., Klaffke, D., Hiltner, K., Groegler, T., Rosiwal, T., Rosiwal, S.M. and Singer, R.F., *Surface and Coatings Technology*, **116**, 1999, 599–608.
46. Skopp, A. and Klaffke, D., *TriboTest*, **4**, 1997, 3.
47. Gangopadhyay, A.K. and Tamor, M.A., *Wear*, **169**, 1993, 221.
48. Krauss, A., Auciello, O., Gruen, D., Jayatissa, A., Sumant, A., Tucek, J., Macini, D., Molodvan, M., Erdemir, A., Ersoy, D., Gardos, D., Busmann, H., Meyer, E. and Ding, M., *Diamond and Related Materials*, **10**, 2001, 1952.
49. Grill, A., *Thin Solid Films*, **355**, 1999, 189.
50. Tsai, H. and Bogy, D.B., *Journal of Vacuum Science and Technology*, **A5**, 1987, 3287.
51. Enke, K., *Thin Solid Films*, **80**, 1981, 277.
52. Wei, R., Wilbur, P.J., Erdemir, A. and Kustas, F.M., *Surface and Coatings Technology*, **51**, 1992, 139.
53. Weissmantel, C., Bewilogua, K., Breuer, K., Dietrich, D., Ebersbach, U., Erler, H.-J., Rau, B. and Reisse, G., *Thin Solid Films*, **96**, 1982, 31.
54. Miyoshi, K., Pouch, J.J. and Alterovitz, S.A., *Materials Science Forum*, **52/53**, 1989, 645.
55. Donnet, C. and Grill, A., *Surface and Coatings Technology*, **94–95**, 1997, 456.
56. Grischke, M., Bewilogua, K., Trojan, K. and Dimigen, H., *Surface and Coatings Technology*, **74/75**, 1995, 739–745.
57. Voevodin, A.A., Capano, M.A., Laube, S.J.P., Donley, M.S. and Zabinski, J.S., *Thin Solid Films*, **298**, 1997, 107–115.
58. Donnet, C., Fontaine, J., Le Mogne, T., Belin, M., Heau, C., Terrat, J.P., Vaux, F. and Pont, G., *Surface and Coatings Technology*, **120/121**, 1999, 548–554.
59. Miyoshi, K., Wu, R.L.C. and Garscadden, A., *Surface and Coatings Technology*, **54/55**, 1992, 428.
60. Erdemir, A., Switala, M., Wei, R. and Wilbur, P., *Surface and Coatings Technology*, **50**, 1991, 17.
61. Hollman, P., Hedenqvist, P., Hogmark, S., Stenberg, G. and Boman, M., *Surface and Coatings Technology*, **96**, 1997, 230.
62. Grill, A., Patel, V. and Meyerson, B., *Surface and Coatings Technology*, **49**, 1991, 530–536.
63. Ronkainen, H., Koskinen, J., Likonen, J., Varjus, S. and Vihersalo, J., *Diamond and Related Materials*, **3**, 1994, 1329.
64. Erdemir, A., Nichols, F.A., Pan, X., Wei, R. and Wilbur, P., *Diamond and Related Materials*, **3**, 1993, 119.
65. Erdemir, A., Bindal, C., Fenske, G.R. and Wilbur, P., *Tribology Transactions*,**39**, 1996, 735.
66. Donnet, C., Belin, M., Martin, J.C., Martin, J.M., Grill, A. and Patel, V., *Surface and Coatings Technology*, **68–69**, 1994, 626.
67. Kim, D.S., Fischer, T.E. and Gallois, B., *Surface and Coatings Technology*, **49**, 1991, 537.
68. Holmberg, K., Koskinen, J., Ronkainen, H., Vihersalo, J., Hirvonen, J.P. and Likonen, J., *Diamond Films Technology*, **4**, 1994, 113.
69. Erdemir, A., Eryilmaz, O.L., Nilufer, I.B. and Fenske, G.R., *Diamond and Related Materials*, **9**, 2000, 632.
70. Erdemir, A., Eryilmaz, O.L. and Fenske, G.R., *Journal of Vacuum Science and Technology*, **18A**, 2000, 1987.
71. Heimberg, J.A., Wahl, K.J., Singer, I.L. and Erdemir, A., *Applied Physics Letters*, **78**, 2001, 2449.
72. Gangopadhyay, A.K., Wilermet, P.A., Tamor, M.A. and Wassell, W.C., *Tribology International*, **30**, 1997, 9.
73. Erdemir, A., Erylmaz, O.L., Nilufer, I.B. and Fenske, G.R., *Surface and Coatings Technology*, **133–134**, 2000, 448–454.
74. Erdemir, A., *Surface and Coatings Technology*, **146–147**, 2001, 292.
75. Sanchez-Lopez, J.C., Erdemir, A., Donnet, C. and Rojas, T.C., *Surface and Coatings Technology*, **163–164**, 2003, 444.
76. Andersson, J., Erck, R.A. and Erdemir, A., *Surface and Coatings Technology*, **163–164**, 2003, 535.

77. Andersson, J., Erck, R.A. and Erdemir, A., *Wear*, **254**, 2003, 1070.
78. Donnet, C., *Surface and Coatings Technology*, **100–101**, 1998, 180.
79. Paulmier, D., Zaidi, H., Nery, H., Lee Huu, T. and Mathia, T., *Surface and Coatings Technology*, **62**, 1993, 570.
80. Olsen, J.E., Fischer, T.E. and Gallois, B., *Wear*, **200**, 1996, 233.
81. Strom, D.B., Bogy, D.B., Bhatia, C.S. and Bhushan, B., ASME Paper No. 90-Trib-69, American Society of Mechanical Engineers, 1990.
82. Ramirez, A.G., Kelly, M.A., Strom, D.B. and Walmsley, R.G., *Tribology Transactions*, **39**, 1996, 710.
83. Erdemir, A., Bindal, C., Pagan, J. and Wilbur, P., *Surface and Coatings Technology*, **76–77**, 1995, 559.
84. Erdemir, A., Bindal, C.,Fenske, G.R., Zuiker, C. and Wilbur, P., *Surface and Coatings Technology*, **86/87**, 1996, 692.
85. Liu, Y., Erdemir, A. and Meletis, E.I., *Surface and Coatings Technology*, **94–95**, 1997, 463.
86. Nakayama, K. and Nevshupa, R.A., *Journal of Physics D: Applied Physics*, **35**, 2002, L53.
87. Nakayama, K. and Nevshupa, R.A., *Journal of Tribology*, **125**, 2003, 780.
88. Dimigen, H. and Klages, C.P., *Surface and Coatings Technology*, **49**, 1991, 543–547.
89. Agreev, V.P., Glushko, T.N., Dorfman, V.F., Kuzmichev, A.V. and Pypkin, B.N., *SPIE Proceedings*, **1503**, 1991, 453.
90. Le Harzic, R., Huot, N., Audouard, E., Jonin, C., Laporte, P., Valette, S., Fraczkiewicz, A. and Fortunier, R., *Applied Physics Letters*, **80**(21), 2002, 3886.
91. Kononenko, T.M., Garnov, S.V., Pimenov, S.M., Konov, V.I., Romano, V. and Borsos, B., *Applied Physics*, **A71**, 2000, 627.
92. Dumitru, G., Romano, V., Weber, H.P., Haefke, H. and Gerbig, Y., Proceedings WLT Laser 2001, 2001, 351.
93. Dumitru, G., Romano, V., Weber, H.P., Pimenov, S.M., Kononenko, T.M., Hermann, J., Bruneau, S., Gerbig, Y. and Shupegin, M., *Diamond and Related Materials*, **12**, 2003, 1034.
94. Fontaine, J., Donnet, C., Grill, A. and Lo Mogne, T., *Surface and Coatings Technology*, **146/147**, 2001, 286–291.
95. Robertson, J., *Diamond and Related Materials*, **12**, 2003, 79–84.
96. Harrison, J., private communication, 2003.
97. Ciraci, S., private communication, 2003.
98. Gahlin, R., Larsson, M. and Hedenqvist, P., *Wear*, **249**, 2001, 302–309.
99. Grill, A., *Thin Solid Films*, **355/356**, 1999, 189–193.
100. Lettington, A.H., *Carbon*, **36**(5/6), 1998, 555–560.
101. Bhushan, B., 'Macro- and Microtribology of Magnetic Storage Devices', in *Modern Tribology Handbook* (ed B. Bhushan), CRC Press, Boca Raton, FL, 2000, p. 1413.
102. Grill, A., *Diamond and Related Materials*, **12**, 2003, 166–170.
103. Hauert, R., *Diamond and Related Materials*, **12**, 2003, 583–589.
104. Gutensohn, K., Beythien, C., Bau, J., Fenner, T., Grewe, P., Koester, R., Padamanaban, K. and Kuehn, P., *Thrombosis Research*, **99**, 2000, 577–585.
105. Robertson, J., *Materials Science and Engineering*, **37**(4/6), 2002, 129–281.
106. Grill, A., *Thin Solid Films*, **398**, 2001, 527–532.

10

Tribology of Polymeric Solids and Their Composites*

B.J. Briscoe and S.K. Sinha

Abstract

This chapter reviews recent progresses made, as well as the historical precedents, in the area of tribology (mainly wear and friction) of polymers and their composites. The now classical theories of friction are summarized briefly. Wear mechanisms are classified under three broad approaches that reflect primarily the way this subject has been historically studied. It is demonstrated here that the wear of polymers is influenced by at least the contact conditions, bulk mechanical properties of the polymer and the properties of the 'third body' which generally appears in the form of transfer film or degraded polymer particles between two sliding surfaces. Further, this chapter establishes a link between the different contact and material parameters and shows how they are important in elucidating wear mechanisms for polymers. The influences of fillers such as soft and hard phases upon the improvement, or not, in the wear resistance of polymers are presented. The effects of environment and lubrication on polymer wear are briefly explained in terms of the chemical interactions between the liquid phase and the polymer and the potential lubricating action. Capabilities and limitations of current predictive friction, damage and wear models for polymers are also highlighted. A case study on the wear of hip and knee prosthesis materials has been added at the end of the chapter to provide a detailed analysis of wear in a well-examined tribological application.

* This chapter is an extended version of a review paper published by the same authors in the special issue of IMechE Proceedings Part J: J of Engineering Tribology, 2002, Vol. 216, pp. 401–413. While the main structure of the current chapter is same as that in the paper, several new sections and figures have been included.

Wear – Materials, Mechanisms and Practice Edited by G. Stachowiak

List of symbols and abbreviations

A	Real area of contact
F	Friction force
K	Proportionality constant also known as wear factor
S	Fracture stress/Ultimate tensile stress
d	Sliding distance
e	% elongation to fracture
E	Elastic modulus
H	Hardness
p	Pressure
R	Radius of spherical slider
R_a	Surface roughness measured as centre-line average
t	Time
T	Duration of sliding
T_g	Glass transition temperature
v	Sliding speed
V	Wear volume
W	Normal load
Δw	Weight loss
α	Coefficient
β	Fraction of energy lost due to frictional hysteresis
γ	Surface energy
μ	Coefficient of friction
ν	Poisson's ratio
ϕ	Elastic work done in deforming the polymer per unit sliding distance
τ	Interfacial shear stress
τ_0	Interfacial shear stress at zero imposed pressure
ψ	Semi-included angle of conical indenter/slider
GF	Glass fibre
CF	Carbon fibre
AF	Aramid fibre
PBI	Poly(benzimidazole)
PMMA	Poly(methylemethacrylate)
PEEK	Poly(etheretherketone)
PE	Poly(ethylene)
UHMWPE	Ultrahigh-molecular-weight poly(ethylene)
PS	Poly(styrene)
PC	Poly(carbonate)
POM	Poly(oxymethylene)

10.1 Introduction

Tribological usages of polymers and their composites have been increasing in engineering applications for several technical and economic reasons. The advantages of using polymers over other materials include self-lubricity, light weight, resistance to corrosion or general oxidation, non-toxic nature and potential ease of near-net shape form manufacture. Despite several potential advantages, there are still many challenges which remain for the effective and economic use of polymers for specific tribological applications. For example, the wear rate can be extremely high if the contact pressure is slightly raised, and low friction is not necessarily associated with low wear rate. Therefore, it is extremely important to make a proper selection of the polymer materials for any tribological usage. It is also possible to improve and optimize their tribological performance by certain changes in the bulk constituents or the surface conditions of the materials.

The science and engineering of polymer tribology is still naturally in the process of development, but a large amount of literature may be found in relevant technical journals. Although some convergence can be found in the conclusions of many researchers who have studied the friction and wear properties of polymers, some conflicting views can also be frequently encountered in the interpretation of the data. The main problem, which is also true with any other materials, is that the tribological performance depends upon a large number of factors including those which are system related. Therefore, great care must be taken in the interpretation of the data, or the explanation of the associated friction and wear mechanisms in a tribological process involving polymers.

The purpose of this chapter is to rationalize and highlight some important established phenomena in polymer tribology. Explanations are presented, where appropriate, on the mechanisms of friction (Section 10.2) and wear (Section 10.3) for polymers and their composites (Section 10.4) which are also supported by some data obtained from the literature. A separate but brief section has been provided on the effects of lubricants and environments (Section 10.5). A specific example encountered in the application of hip and knee joint replacement is presented as a case study to exemplify the implementation of knowledge in a related mature application (Section 10.6). Finally, the chapter ends with some general concluding remarks (Section 10.7).

10.2 The Mechanisms of Polymer Friction

The origin of the friction between two sliding surfaces has been described variously; however, from the modelling view-point, it is much more useful to utilize the description of frictional energy dissipation at the interface zone and relate frictional force to the energy lost at the interface. This certainly applies to sliding friction, both 'smooth' and discontinuous, but may also be usefully adopted for the so-called 'static' friction. To a first order it is often now assumed that there are two major processes by which the frictional energy is dissipated. The first process is the ploughing actions of the asperities of a harder surface and the second is the dissipation in overcoming the adhesive actions between the two surfaces in the absence of sharp asperities. This scheme of describing interfacial friction is also known as the two-term non-interacting model [1]. Figure 10.1 is a suitable description of this model [2]. The question as to which component in the two-term model would be dominant in a particular frictional event will be largely decided by the surface roughness of the hard counterface and by the shear properties of the polymer.

There are many comprehensive accounts of these processes[2].

10.2.1 The Ploughing Term – Brief Summary

The ploughing component of frictional work arises from the sub-surface deformation (grooving) which involves plastic flow and fracture depending upon the polymer's

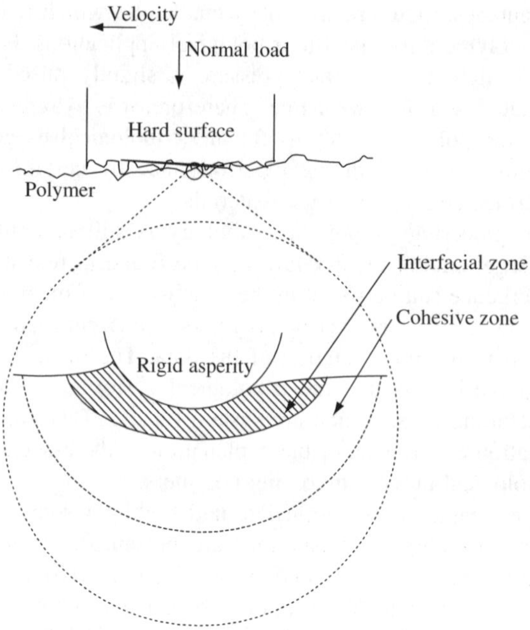

Figure 10.1 Two-term model of wear process. The distinction between interfacial and cohesive wear processes arises from the extent of deformation in the softer material (usually polymer) by rigid asperity of the counterface. For interfacial wear the frictional energy is dissipated mainly by adhesive interaction, while for cohesive wear the energy is dissipated by adhesive and abrasive (sub-surface) interactions [3]

mechanical response under the imposed contact conditions (contact severity, attack angle of the asperity, speed, temperature, etc.); the adhesion component is suppressed by efficient lubrication or by rolling. The friction force for ploughing has been modelled for elastomers considering only the elastic work done and that fraction which is dissipated. Here the friction force is expressed as a function of the elastic work done (basically, the friction force is equated to the elastic energy lost per unit of the sliding distance),

$$F = \beta\phi \tag{1}$$

where β is the fraction of the energy lost in frictional hysteresis work and ϕ is the elastic work done in deforming the polymer per unit sliding distance. The equations for ϕ for spherical- and conical-shaped sliders have been proposed as

$$\phi = 0.17\ W^{4/3}\ R^{-2/3}(1-\nu^2)^{1/3}\ E^{-1/3} \tag{2}$$

for spherical slider of radius R [4] and

$$\phi = (W/\pi)\cot\psi \tag{3}$$

for conical sliders with semi-included angle ψ.

Thus, by substituting equations (2) or (3) in equation (1), it is rather trivial to compute the friction force for ploughing. The above analytical model has application in scratch deformation work where a conical (or spherical) indenter is used to model asperity interaction and the computation of scratch hardness. The main problem with the analysis using only elastic energy loss is that it significantly underestimates the actual ploughing friction force. It has been reported that a more rigorous theoretical analysis of experimental data would indicate that the friction force is equal to $3\alpha\phi$ and not $\alpha\phi$ per unit sliding distance. The reasons for this have been explained earlier by one of the present authors [2]. This process has practical significance in the grip of automobile tyres on wet roadways [5].

The case for the friction processes which are accompanied by extensive ductile (and brittle) fractures has been examined in some detail and was originally recorded by Bowden and Tabor [1].

10.2.2 The Adhesion Term – Brief Summary

The adhesion term in the two-term model of friction is conceptually less easy to understand and less tractable in terms of frictional modelling. The physical explanation [6] of the nature and strength of the adhesion forces between two bodies if brought in close contact is well recorded; electrostatic and van der Waals forces, and, in the presence of polar atoms, due to the dipole interactions and hydrogen bonding and all contribute. The common feature is that they are all, if of significant magnitude, short-range and thus operate at contact areas. The static friction and the dynamic friction, due to adhesion, are related to what happens at the interface in terms of shear and deformation of a very thin layer of the polymer which is directly in contact with a counter surface. Depending upon the polymer behaviour (elastomeric or non-elastomeric), we can explain frictional work dissipation in two ways. For elastomeric polymers, the frictional work is sometimes done in moving waves of attachment on the surface of the polymer. This type of behaviour involves contact adhesion and dhesion (energy loss) as the wave (also known as Schallamach waves) moves forward and it resembles the passage of a 'ruck' through a carpet [7]. The movement of the wave from one end to the other brings incremental movement of the elastomer and this type of movement may not be termed as true sliding. The main reason for the above-mentioned behaviour shown by elastomers is their visco-elasticity and their ability to extend giving large local recoverable strains. The ratio of the modulus of elastomers to the surface adhesion force is low. However, as this ratio exceeds a certain value for harder elastomers or if the elastomer is kept at a temperature below T_g, the 'rucks' or the waves become finer. For non-elastomeric polymers, the waves of attachment are either non-existent or they occur at molecular scale and are not detected in a normal optically visualized tribological test. If the strength of the adhesive forces is weaker in comparison to the strength of the polymer, true sliding appears to take place. The main mechanism of frictional energy dissipation in this case will be shear of a very thin zone of the polymer at the interface. The thickness of this shearing layer may be only a few nanometres. A layer of the polymer film may be deposited on the smooth counterface as a transfer layer and subsequently the frictional energy dissipation mainly takes place either in the transfer film or in a very thin layer of the polymer which is in direct contact with the counter surface.

It is possible to measure the strength of this shearing layer by careful design of the experiment and it has been shown previously that the shear strength at constant velocity and temperature is related to the contact pressure, p, by an expression of the form [2]:

$$\tau = \tau_0 + \alpha p \quad (4)$$

It should be mentioned here that equation (4) resembles that for the shear bulk of the material under hydrostatic constraint. However, the value of τ_0 is considerably smaller (about one-tenth typically) than the bulk value probably because the material at the interface is weaker due to molecular alignment. The frictional force is a product of the shear strength of the interface and the real area of contact. The coefficient of friction for the adhesive component then may be written as

$$\mu_a = \tau_0/p + \alpha \quad (5)$$

Equation (5) has been shown to hold good experimentally with reasonable confidence [8–10]. There are some notably interesting properties of this equation. Firstly, as the pressure increases for high-load conditions, the first term on the right-hand side effectively becomes very small, and in that case equation (5) may be approximated as

$$\mu_a \sim \alpha \quad (6)$$

Secondly, the effect of sliding speed, at least in the low-speed range, on τ is small compared to the effect of temperature. Thirdly, above T_g the influence of temperature is much more striking than below T_g. And, finally, the shear of the interfacial film during experiment does not always take place in a continuous fashion but rather in a discontinuous 'saw-toothed type profile' for the shear stress which is also known as stick-slip phenomenon. Here, we may add that the stick-slip profile measured, in any tribological test, is not only a result of the true interfacial events occurring between the two surfaces but also a reflection of the stiffness of the friction force sensing device; it is a combined response.

For completeness, it is worth noting that the coulombic, or ratchet, friction is quite common in polymeric systems such as fibres [11] and powders. Here the restraint and its release arise from special geometric asperity engagements.

The mechanisms of friction for polymers against hard counterfaces, as explained above, have direct implications on the mechanism and extent of damage and subsequent wear processes. Therefore, in subsequent sections, we will further delve into the two-term model as a means to rationalize and explain the damage and wear of polymers.

10.3 Wear

10.3.1 Semantics and Rationalizations

The process of 'wear' may be variously defined but most generally it is measured in terms of the mass, or volume, loss from a moving or eroding contact. The sequence of events is as follows. Mechanical forces, frictional work, impact forces, cavitation forces and so on induce damage in the contact members. Eventually, or may be immediately, the surfaces lose mechanical cohesion and debris is produced. Chemical wear has a similar character but at a smaller scale which may involve extensive chemical reactions. Subsequently, but perhaps immediately, this debris is expelled from the contact zone and the process of wear is observed. The pathway is convoluted not least because the damage and wear processes produce, by their action, significant modifications to the properties of the materials in the contact zone; the 'third body' evolves [12]. For polymeric systems, there is also the very significant prospect of marked environmental influences which will include the role of lubricants as well as the often dominant consequences of frictional heating.

There are at least three ways in which the subject, in the context of wear, may be rationalized by a review such as this (Figure 10.2). The choice of approach depends upon the audience; here we offer the three methodologies together as each has a unique value.

First, we may take the generic scaling approach which emerges from the accepted value of the two-term non-interacting model of friction [2]. Friction is, as a first approximation, of two kinds: interfacial and bulk (ploughing); the model has been popular but actually not well understood. It has major limitations for multiphase systems (discussed later). As frictional work causes damage and subsequent wear, this is a potentially useful means to classify the wear as 'cohesive' and 'interfacial' in nature. This scheme will be explored and discussed in Section 10.3.2. A more pragmatic approach would follow a more classical line and list 'wear' processes by some perceived judgement of the origins and consequences. Thus, we may have 'abrasion', 'transfer wear', 'fretting', 'chemical wear', 'erosion', 'fatigue wear', 'delamination wear' and so on; we may describe this as a phenomenological classification. This method of analysis has value for those who practice extensive electron microscopic examination (Section 10.3.3), which is actually a very useful practical approach. The third approach recognizes the extremely wide diversity of response of polymeric systems and focuses upon the material response upon contact deformation. Thus, we will then subdivide wear according to material response and produce a scheme where each polymer 'class' is dealt with in relative isolation from the rest. In this type of division, we distinguish

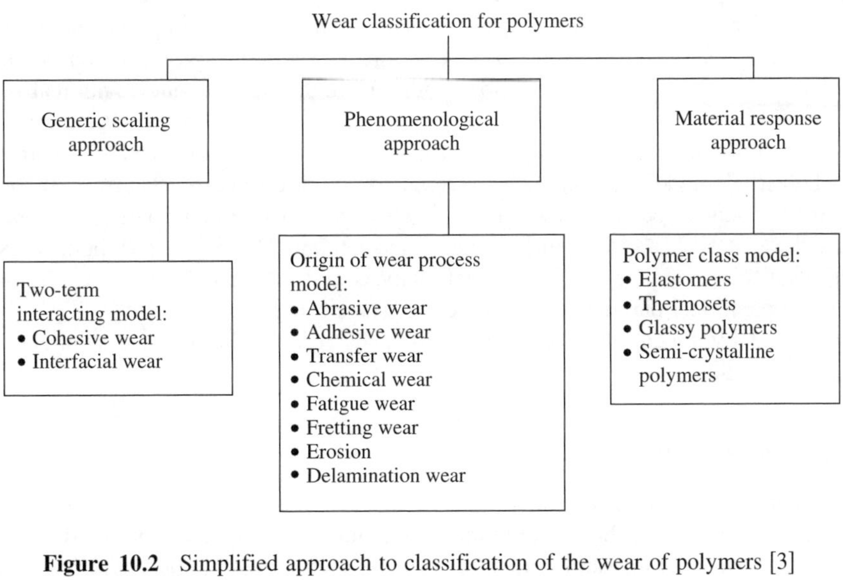

Figure 10.2 Simplified approach to classification of the wear of polymers [3]

elastomers, thermosets, glassy polymers and semi-crystalline polymers as having unique attributes in the context of their wear behaviour. This has been historically the way the subject has evolved; this bias has been practiced for many reasons. Section 10.3.4 follows this theme.

The subsequent sections will deal with these three methods of division: generic scaling response (Section 10.3.2), the phenomenological classification (Section 10.3.3) and the material-based responses (Section 10.3.4).

10.3.2 Wear Classification Based on Generic Scaling Responses

10.3.2.1 Cohesive Wear

The classification of cohesive (bulk, ploughing, sub-surface) wear arises, in our mind at least, from the success, or popularity, of the two-term non-interacting model of friction (Figure 10.1). If there are seen to be, as a first-order approximation, two modes of friction energy dissipation, then there ought to be two general modes of damage and the consequent wear. Clearly the same argument will apply to the interfacial component. This is a crude division and has major flaws in its application in certain aspects of process engineering and in the detailed investigation of such areas as the scratch damage of polymers in lubricated environments. The two-term division has merits in that it makes a clear distinction between the relatively mild deformations in cohesive wears and the more intense interfacial wears.

Another definition of cohesive wear modes, as opposed to interfacial modes, might be that, in some way to be defined, wear rate might be related to an accessible bulk mechanical failure property. For example, the Ratner–Lancaster correlation provides an interrelationship between abrasive wear (single pass) and a crude measure of tensile toughness [13, 14]. In the abrasion of elastomers there is evidence of correspondence between fatigue life and wear life [15, 16].

The models thus seek to define the contact stress environment and then relate this to the endurance of the polymer. A work (frictional) and damage rate is first predicted and then, by scaling, a wear rate law is predicted. The models may then easily incorporate environmental influences. The central thesis is that a bulk endurance or failure property is accessible by some other means even if the deformation conditions in the contact zone do not realize those which are addressed in a 'conventional' mechanical testing; the pressures, strains, strain rates and loading transients are not comparable for example. Lately, some modifications of the Ratner–Lancaster model have also been presented which do take into account the contact load and the counterface roughness effects [17]. Other notable contributions in modelling wear are those of Kar and Bahadur [18] and Viswanath and Bellow [19]. In both cases, they have applied the principles of dimensional analysis and included factors such as material properties, operational parameters and the counterface roughness.

Section 10.3.4 will deal with these aspects of cohesive wears in the contexts of elastomers and semi-crystalline polymers.

10.3.2.2 Interfacial Wear

The argument for distinguishing this class of damage and wear was mentioned above; it comprises the other part of the two-term model and, more importantly, those modes which do not readily, and more importantly directly, correlate with accessible bulk failure properties.

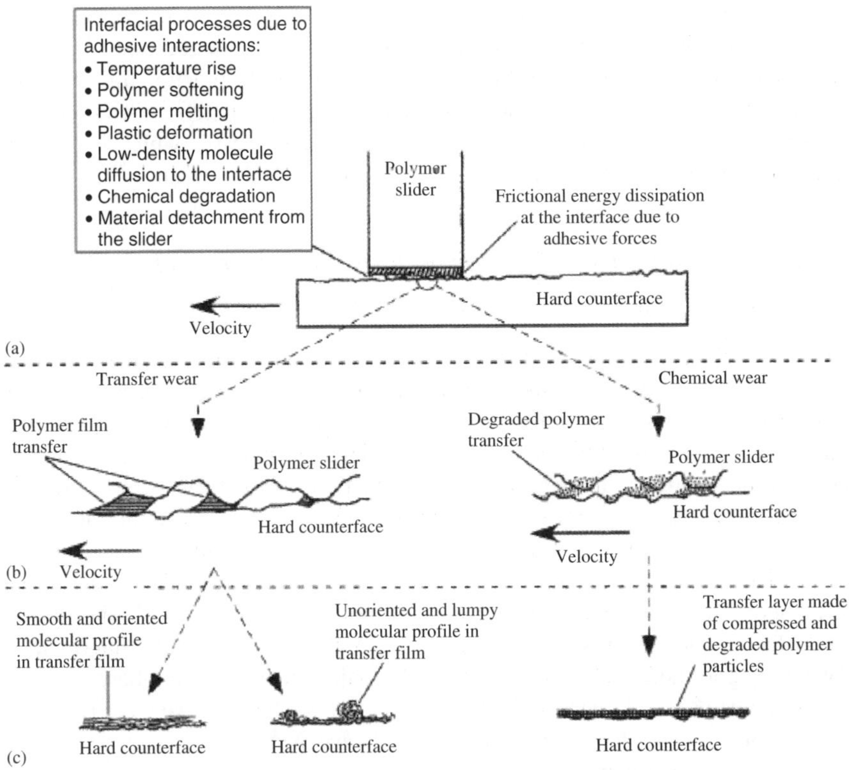

Figure 10.3 Interfacial wear processes for neat polymers [3]: (a) initial contact of the two surfaces; (b) running-in process where the soft polymer molecules are gradually transferred to the hard counterface as third body; and (c) steady-state wear process where the wear and friction phenomena are influenced mainly by the shear and adhesive properties of the transferred film

It is simplest to deal with the so-called transfer damage modes as the first subdivision. The first distinction is that of whether the failure, via transfer, is isothermal or not. Then, there is the question of the consequence of quasi-adiabatic interfacial shear behaviour. A scheme is shown in Figure 10.3. Some 'special' polymers appear to accommodate near isothermal interface shear induced softening; basically the interface zone, as opposed to the actual interface, reorders and as a consequence a highly oriented film, and polymeric contact, is produced. The frictions are then often very low and the rate of transfer, and also subsequent wear, may be prolific. Poly(tetrafluoroethylene) (PTFE) is the classical example in the unfilled forms [20]. Much has been published on this topic for a small but uniquely important polymer group of semi-crystalline materials: the 'smooth molecular profile' polymers [21–24].

The 'adiabatic' cases are of two types: the glassy polymers which are uncross-linked and the cross-linked systems which include the elastomers and the thermoset systems. The thermosets have much importance in brakes [25] and traction couples, whilst the elastomers have much application in automobile tyre and belt drive applications [26]. In either case, during damage, there is the prospect of significant chemical degradation and extensive chain scission. The transfer layers, if they occur, are not immediately related to chemical or physical composition of the 'host' polymer. The transfer material does not correspond to what may be described as the 'whole' polymer but to a degraded form. In practice, this case does not naturally lead to interface shear induced softening, and a reduction in friction. However, the transfer layers will naturally impede the heat transfer processes which is of significant consequence in brake applications and the phenomenon of brake 'fading' [27].

10.3.3 Phenomenological Classification of Wear Damages

10.3.3.1 Abrasive Wear

Abrasive wear is the common name for a cohesive wear as was explained earlier on the basis of the two-term model. Based on the type of the interfaces, there can be at least two types of abrasive wear: two-body and three-body abrasive wear. As the names suggest, two-body wear takes place when only the two surfaces (the polymer and the mating counterface) are interacting, whereas three-body wear involves interactions of hard debris or foreign particles trapped between the two surfaces that may further increase or decrease the rate of wear by several orders. The third body can also drastically change the nature of the interface if the loose debris particles can form strong and tenacious thin films on either of the two mating surfaces. This is the main mechanism of wear in polymer composites sliding against hard metallic surfaces. The steady-state wear rate is normally very low if the interfacial film is strongly adhering to the metallic surface; this is a major practical technology.

The classical works by Ratner and Lancaster [13, 14] for the abrasive wear relate the wear rate property of polymers to some quasi-static bulk mechanical properties. The relation is given by the following equation [13]:

$$V = \frac{(K\mu Wv)}{(HSe)} \tag{7}$$

where V is the wear volume, K is a proportionality constant, W is the normal load, v is the sliding velocity, H is the hardness of the polymer, S is the ultimate tensile stress and e is the elongation to break. A plot of V as a function of $1/Se$ would furnish a straight line which has been shown by Ratner and Lancaster and many other researchers [14, 15, 28]. The correlation has resemblance with the classical Archard's wear law [29]. Figure 10.4 presents a plot of wear volume as a function of $1/Se$ for a number of polymers using the data from References [15, 28]. Despite the fact that the data were obtained at different sliding speeds, a reasonable linearity is noticeable in the plot which mainly confirms some correlation between abrasive wear and the bulk mechanical properties. Equation (7) also predicts that wear volume is inversely proportional to the hardness of the materials. This means the harder the materials, the less will be the wear. While this is true for metals, polymers show very complex correlation of wear with hardness and in some cases even reverse is the effect, that is, increasing hardness effectively increases the wear volume for abrasive wear [28]; the case for elastomers is even more problematic. Budinski [30] has concluded that for the abrasive wear of polymers and elastomers, the wear volume may be correlated better with a product of μ, S and e (where μ is the coefficient of friction of the abrasive material on the polymer surface; S and e have been defined earlier). No specific correlation may be found in the work published by Budinski [30] for the wear volume with hardnesses such as Shore

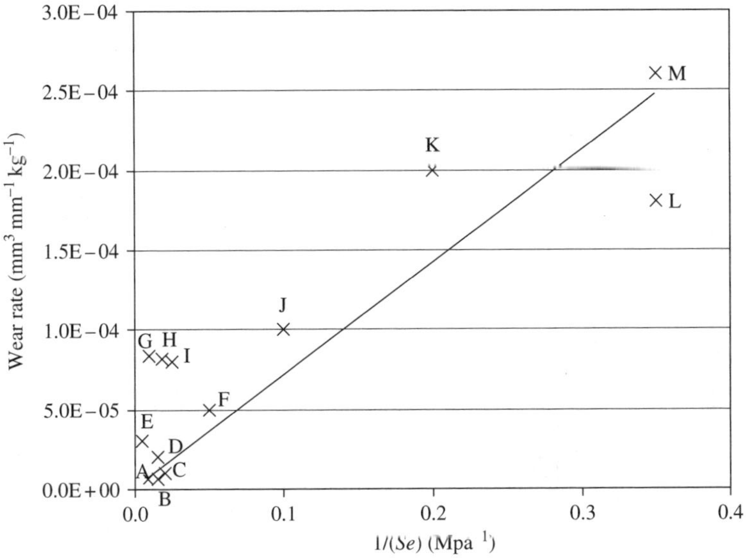

Figure 10.4 A plot of wear rate ($mm^3\,mm^{-1}\,kg^{-1}$) as a function of the reciprocal of the product of ultimate tensile stress and elongation to fracture. The data are taken from literature. A – poly(ethylene) [15]; B – nylon 66 [15]; C – PTFE [15]; D – poly(propene) [15]; E – high-density poly(ethylene) [28]; F – acetal; G – poly(carbonate) [15]; H – poly(propylene) [28]; I – poly(ethyleneterephthalate glycol) [28]; J – poly(vinyl chloride) [28]; K – PMMA [28]; L – poly(styrene) [12]; M – PMMA [28]

D hardness [31] and scratch hardness for polymers and elastomers [32]. It may be worth mentioning here that scratch hardness of elastomers, obtained by the common method of measuring the scratch width, can introduce large errors in the measurement due to the highly elastic recovery nature of the material over large strains. The width measurement method for the scratch deformation extent can only be used for materials which are inherently ductile and irrecoverably yield at a relatively low value of strain (such as thermoplastics and thermosets). Considering this, a plot of scratch hardness versus the wear rate for only semi-crystalline polymers was plotted for the data from Reference [30]; however, no particular trend was observed. Hence, it appears that further work is needed to resolve the issue of the effect of a hardness upon the abrasive wear of polymers. Budinski [30] has suggested that the role of the coefficient of friction in abrasive wear of polymer is important especially for third-body abrasion. He argues that a low friction between the abrasive particle and the polymer will cause the particle to dig into the material giving high wear; however, there is tendency for the particle to easily roll if the coefficient of friction is high, which gives low wear rate. This is the reason why some of the elastomers, such as polyurethane, have a high wear resistance. Figure 10.5 gives some data for the abrasive wear for different polymers as has been recently reported in the literature. It is interesting to observe that the specific wear rates for the abrasive wears do not correlate with the specific wear rates for adhesive and fatigue-related wear. Data on the adhesive types of wear will be presented in subsequent sections.

It is notable that the abrasive wear of polymers has also been extensively studied by model scratch experiments; the Schallamach [7] precedent has been mentioned.

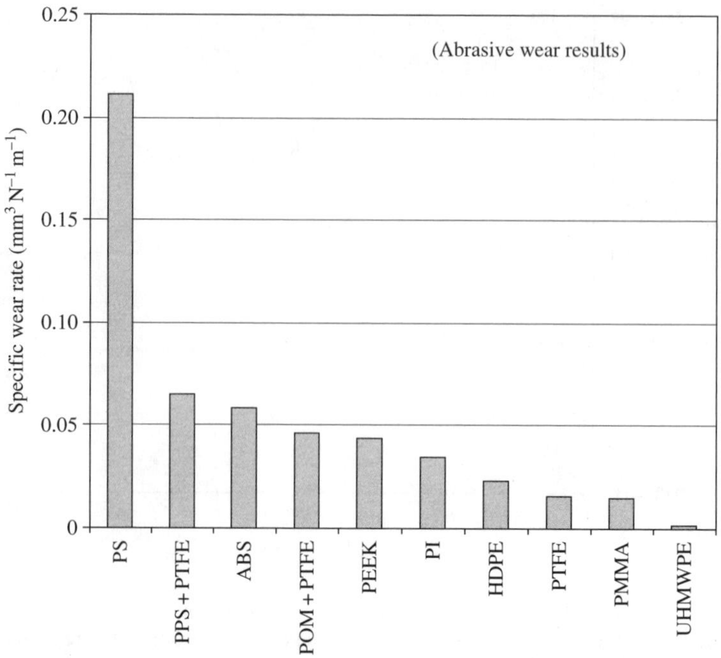

Figure 10.5 Abrasive wear results for a number of polymers [30]. The tests were in a dry sand rubber wheel tester that uses a 228-mm-diameter chlorobutyl rubber wheel (60Shore A) as an abrader. Width of the wheel = 12.7 mm, rotational speed = 20.9 rad s^{-1}, loading force = 140 N, abrasive: 215–300 μm silica sand

Adams *et al.* [33] and later Wong *et al.* [34] have found that wear can be extremely high when scratches intersect each other compared to when scratching is carried out on a single track [35]. Hence, modelling wear by such techniques must employ the consideration of intersecting scratches which closely resembles any real and random abrasive wear process. A number of theoretical and empirical works on creating wear models have developed equations for predicting wear of polymers in predominantly abrasive situations; the Ratner–Lancaster correlation is the classical example. Not enough comparative data are available at this point of time to judge which model is the better. Therefore, a simple list of such wear models is presented in Table 10.1.

10.3.3.2 Adhesive Wear

Adhesive wear is a name for interfacial wear as was mentioned previously in the generic response classification of wear. Arguably, the most important phenomenon in adhesive wear is the prospect of the formation of an interfacial transfer film on the hard counterface. The friction coefficient can be high or low, and the wear rate then predominantly depends upon the adhesive and rheological properties of the interfacial layer or film. An adhering and tenacious film would, in certain cases, reduce the wear rate by several orders in the steady-state stage when a polymer is slid against a hard but smooth thermally conducting counterface. Hence,

Table 10.1 Models for abrasive wear of polymers as proposed by many researchers

Archard's wear law [29]	
$V = K\frac{Wd}{H}$	where K is the wear factor, W is the normal load, d is the sliding distance and H is the hardness of the softer materials
Lewis model [36]	
$V = KWvT$	where T is the duration of sliding
Ratner and Lancaster correlation [14]	
$V = \frac{K\mu Wv}{HSe}$	
Rhee's model [37]	
$\Delta w = KW^{a}v^{b}t^{c}$	where Δw is the weight loss of the polymers, K is the wear factor, W is the normal load, v is the sliding speed, t is the time, and, a, b and c are material-dependent set of parameters for a given system
Kar and Bahadur [18]	
$V = \frac{1.5K\gamma^{1.775}p^{1.47}d^{1.25}}{E^{3.225}}$	where γ is the surface energy, p is the contact pressure and d is the sliding distance (for POM system)
Wang *et al.* [17]	
$V\alpha\frac{W^{3/2}R_{a}^{\ 3/2}}{S^{3/2}e}$	Applicable to microscopic abrasive wear of polymer

the friction and wear properties, to a large extent, depend upon the properties of the film (third body), though the adjacent bulk property is also important in deciding the friction and wear. It is important that the strength of the bulk of the polymer is higher compared to the shear strength of the film. A weaker bulk in comparison to the interfacial film will eventually result in a high wear rate as the interfacial shear stress will cause premature failure of the bulk. Here, interfacial temperature also plays a strong role in controlling the properties of both the interface layer and the part of the bulk which is close to the interface. Several polymers have been studied for their adhesive wear properties using smooth metallic counterface. The low-friction (self-lubricating) and low-wear properties of many polymeric materials are specially attractive for many industrial applications where metals and ceramics cannot be used due to their inappropriate mechanical and chemical properties, the prospect of contamination or for economic reasons. Amongst many polymers (and their composites which will be discussed in a later section), two polymers are worth mentioning here because of their low friction and mainly adhesive wear properties. These polymers are PTFE and the linear (low and high density) poly(ethylene)s (PE)s. Early frictional studies on PTFE was inspired by the possibility of a very low coefficient of friction of PTFE against hard, such as steel, counterfaces. Works by Steijn [24] and Pooley and Tabor [20] investigated the cause of the low friction and the molecular architectural changes at the interface of PTFE and in the transfer film due to frictional stresses. Both studies indicated that for PTFE there is an extensive molecular reorientation in the direction of sliding. Such reorientation of the molecules occurs for the surface of the PTFE in contact with the counterface as well as for the transfer film that is deposited on the counterface. The measured dynamic friction coefficient is directly related to the molecular structure of the polymer. PTFE, which is composed of rigid rod-like molecules with a 'smooth molecular' profile, shows a low coefficient of friction because of an easy slip between the aligned molecule chains, basically a 'smooth molecular topography' at the interface. Whereas, for low-density PE which has molecules with many side chains ('rough molecular topography') tends to give a higher friction. High-density PE has very few side chains and hence the coefficient of friction is lower compared to that for the low-density PE. It is also important to note here that transfer film and consequently the frictional properties are greatly influenced by the prevailing interfacial temperature and also the kinematics of the relative motion between the two bodies. Smurugov *et al.* [38] have mentioned that the self-lubricating nature of PTFE originates from the low activation energy for the motion of the PTFE molecular bonds. In the temperature range of 30–90°C, which is easily achieved in any frictional event, PTFE fails along weak intermolecular bonds due to even minor temperature rise which involves slippage of crystalline formations of the bond structure. The wear rate is generally very high which can further increase at temperatures higher than 90°C. Wear data on adhesive wear for many polymers will be presented in Section 10.3.4.4 where we will specifically deal with the friction and wear behaviour of semi-crystalline polymers. The topic of non-linear motion and its influences on friction and wear of certain polymers is discussed later.

10.3.3.3 Chemical Wear

It has been suggested that some chemical reaction is essentially present in almost all tribological interactions; this is fair. However, chemical wear is defined as the process of significant material removal as a result of extensive chemical reactions between the mating materials and the environment. There are four main reasons to promote a tribo-chemical reaction in polymer–metal interaction. The first reason is the elevated interfacial temperature. The flash temperature can potentially be extremely high (certainly in filled systems) but the mean steady-state interface temperature could be found in the range of 200–400°C depending upon the system, but the upper level will be defined by the T_g or the melting temperature of the polymer where appropriate. At these temperatures, polymers can melt, ions can be pulled out or some chemical reaction (such as oxidation) can take place. The second reason for chemical reaction is the catalytic actions of the exposed clean metal surface; again likely for composite systems. The third reason is the actions of the fillers in either catalysing the reaction or in actually taking part in making reaction products. Finally, the fourth reason is the contribution from the mechanical straining of the materials which can result in the enhanced chain scission of the molecules. A detailed knowledge of these contributors to the chemical wear is lacking; however, several studies have focused on the bulk decomposition and oxidation of the interface. The subject is still in its infancy.

At less severe situations of pressure (load and friction work) and sliding speed, the interfacial conditions (PV factor, a product of the magnitudes of pressure and sliding speed – see later) seem to help in the process of forming strong adhering transfer film which may reduce the rate of wear. Brainhard and Buckley [39] studied the interactions of PTFE with several clean and oxidized metal surfaces in UHV conditions. On the basis of the augur emission spectroscopy study of the slid surfaces, they concluded that there is possibility of the direct metal–carbon bonding reaction leading to the formation of very strongly attached thin transfer film. In addition to this kind of chemical reaction, van der Waals forces and some other physical means of attachment of the polymer molecules to the metal surface have also been suggested. Briscoe *et al.* [40] and Pratt [41] have suggested a reaction between PTFE and Pb (lead) in the presence of bronze or CuO which results in the formation of strong adhering PTFE film in porous bronze bearings. Several oxides (PbO, Pb_3O_4, Cu_2O, Fe_3O_4) can catalyse or themselves react with PTFE forming hydrogen or chemical bonds with ferrous substrates [42]. In a, now classical, differential scanning calorimetric study, Pocock and Cadman [43] found that PTFE reacts with tin at a temperature of 450 °C after a prolonged thermal soak. Though a temperature of 450 °C may be unrealistically high for a normal tribological polymeric contact of this type, the presence of strain-induced mechanical actions and available oxygen can lower this reaction temperature, which means this reaction is common in dry bearing sliding situations. Conversely, in severe sliding situations chemical attack which involves mainly the oxidation of the polymer can result in an extremely high wear rate [44]. In the case of nylon 6,6, it has been shown that the presence of particulate CuS promotes the formation of $FeSO_4$ when nylon 6,6 is slid against smooth steel counterface. The electronegativity in sulfur and oxygen atoms promotes the formation of hydrogen bond with the nylon [45].

10.3.3.4 Fretting Wear

Fretting is a very common engineering problem often involving polymers and metals and possibly polymers and ceramics. Polymers have often been used as a bulk or coating to counter fretting problems in metal components. Fretting involves movements of two adjacent surfaces with a small amplitude of the relative displacement and often at high frequencies. There is invariably a strong influence of the action of the third body, that is the debris particles, in modifying the fretting damage of the polymer surfaces and the counterface. Fretting damage is often a combination of two competing phenomena: wear debris formation on the one hand, and the crack nucleation and growth on the other. While crack nucleation and growth is concerned with the loss in the fatigue life of the component, wear debris generation is primarily a wear problem. In order to rationalize different material responses in fretting, Vincent *et al.* [46] have introduced the concept of the fretting map. Such a map is reproduced in Figure 10.6. One conclusion that can be drawn from this map is that the tendency for the contact to produce wear debris is higher as the displacement amplitude is increased or as the normal force is decreased. At higher contact amplitude, there is more interfacial slip between the two surfaces leading to more apparent wear. Higher loads influence the formation and the propagation of cracks and the loss of fatigue life and thus reduce the wear damage. Most of the fretting process, between a polymer surface and a mating steel ball, at low normal load situations shows scratches parallel to the direction of motion of the steel ball within the first few strokes. These scratches will subsequently grow into microcracks and deepen and join together to detach themselves forming debris particles. Loose polymer debris become compacted between the two contacting surfaces and then essentially take the cylindrical form and roll structures as third body between the two surfaces; a similar situation is noted with the elastomers (erasers) used to remove pencil marks from paper. At some very high number of cycles of the fretting process, the slip between the surfaces is actually reduced or accommodated by the process of shear and roll

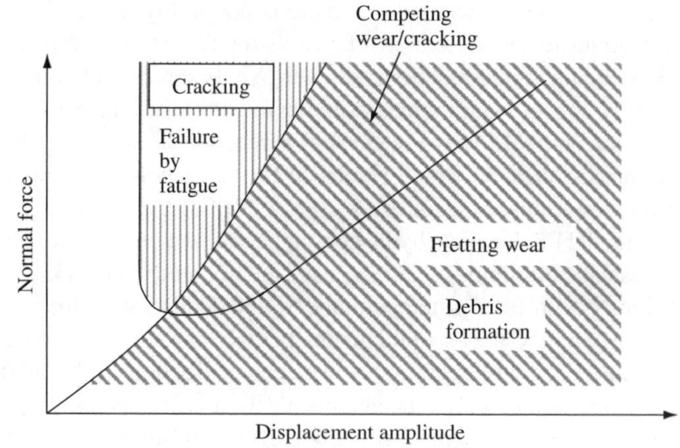

Figure 10.6 Fretting map for polymers as introduced by Vincent *et al.* [46]. Increasing normal load induces crack formation, whereas increasing displacement amplitude generates debris and wear due to larger slip between the two surfaces

relaxation of the third body and the wear is practically arrested. Briscoe *et al.* [47, 48] have noted that the mechanical properties (elastic modulus and hardness) of the compacted and rolled debris can be comparable or even higher than those of the host polymer. Also, here it may be worth mentioning that the fretting wear can be high if there is a combination of linear and torsional motion at the contact [47]. Increasing the torsional component increases wear. Cracks form at the interface at a stress which is much below the nominal ultimate tensile stress mainly because of the repetitive fatigue process [47]. Fretting wear scars always develop in a specimen at its periphery because the slip is at the maximum in the periphery region compared to the centre of the fretting motion [48]. A metal ball surface can also oxidize during the fretting operation and this tendency is much higher at higher normal loads. In fretting tests involving poly(carbonate) (PC) and poly(methylemethacrylate) (PMMA) against a steel ball, there is a large amount of the oxide that is formed at the contact which essentially protects the polymer surface from further wear [49].

Thus, it is apparent that the third body plays a vital role in reducing or controlling the process of fretting wear. Crack formation and wear debris production are the two competing processes during fretting and the tendency for each to prevail will depend upon the contact pressure and the relative displacement amplitude as is shown in the typical fretting map [46]. The environment (presence of humidity, for example) has stronger effect upon the fretting life of the majority of polymeric contacts.

10.3.3.5 Fatigue Wear/Rolling Contact

Fatigue is normally associated with a simple internal failure mechanism for the bulk of materials. Fatigue failure occurs as a result of cyclic or random changes in the stress levels below the fracture strength of the materials. A similar phenomenon of fatigue failure is also observed in the case of cyclic sub-surface stress variation and this type of failure is known as fatigue wear. The surface stress variation induces sub-surface cracks which finally propagate to the surface producing a chip or flake of the wear debris. Fatigue wear is very common in rolling contacts in metallic gear applications where the contact stress varies during the operation. Fatigue also naturally leads to the so-called delamination wear processes as are observed and reported in the case of metals [50].

Traditionally, polymers have not been widely used for rolling contact or for high load bearing gear applications; however, such applications have naturally increased in recent years. This increase in the use of polymers has arisen, first, because of the commercial manufacture of many high-temperature engineering polymers that have a combination of excellent strength and toughness properties; the other reason is the advantages of polymers over other materials such as 'self-lubricity', non-contaminant and low unit product cost. Although the low load applications of polymer gears and rolling contacts have been widespread in electronics industries, many new formulations of polymers (such as PEEK) and their composites show excellent high load applications with performances even exceeding those of metals [51] for more substantial engineering applications. Very few examples of the fatigue wear/rolling contact results can be found in the literature for polymers. Major contributions have been by Stolarski [52, 53], Eiss and Potter [54] and by some more recent works [55]. The energy dissipation processes involved during the rolling of elastic and visco-elastic materials is quite well understood. The energy is dissipated mainly due to the visco-elastic plastic deformation of the materials in the beginning of rolling process and due to elastic hysteresis losses within the contacting surfaces. However, for polymers, there is an added effect of the visco-elastic behaviour of the materials which means the material does not relax instantaneously as the contact loci pass through a point. This produces a stress inhomogeneity close to the surface of the polymer. Also, the visco-elastic deformation starts at a much lower stress than the plastic yielding in a Hookean elastic body which means permanent deformation can be introduced in a visco-elastic material even at low stresses for every cycle of stress leading to a loss of fatigue life. Elastic hysteresis loss, which can be comparatively large (say 5–20%) in elastomeric and other polymeric bodies, leads to sharp temperature rise and the chemical degradation or softening of the polymer; the 'blistering' of automobile tyres is a familiar example. For filled polymers, cracks can be introduced on the entire contacting surface due to matrix–fibre debonding as a result of localized matrix deformation and the associated temperature rise. Though the process of fatigue for polymers and polymer composites is somewhat understood, creating useful models has been very challenging especially for the composite systems.

Limited experimental data are available on the fatigue wear in rolling contacts for polymers. Stolarski has worked on the acetal, nylon 6,6 [56] and PEEK [52]. Results on a three-ball rolling contact tester (with a hemispherical polymer pin in contact with the three balls) show that PEEK is superior in fatigue performance while acetal is better than nylon 6,6. Nylon 6,6, which wore most under the same given condition, showed formation of flaky debris indicative of the fatigue wear. Acetal showed a 'hair-cuticle' type of crack formation without much visible evidence of wear. In contrast to nylon 6,6 and acetal, PEEK showed no observable wear even at slightly elevated load compared to the other two polymers. The surface of PEEK had some deformation mark characteristics of plastic flow rather than that of brittle fatigue. Although PEEK showed better wear performance when tested in water emulsion as lubrication, the same authors have found that PEEK performs poorly in other lubricants containing linear carboxylic acids due to the occurrence of significant bulk plasticization. Nylon 6,6 and acetal gave improved fatigue wear resistance when tested in a commercial lubricant (Shell Vitrea Oil 100) probably due to the development of hydrodynamic lubrication conditions [52]. PEEK, in both virgin and composite forms, has been used by several researchers and, currently, commercial gear products made of PEEK are in extensive use. Kurokawa *et al.* [55] found that the introduction of a pitch-type carbon fibre provided better wear properties for gear application. Limited data on fatigue wear in the rolling contacts of polymers show that PEEK is superior to many other polymers in fatigue probably due to the polymer's good high temperature and better strength properties. Composites of PEEK with carbon or glass fibres will give even better performance if properly optimized in terms of fibre wt.% and fibre orientation.

10.3.4 Wear Classification Based on Polymeric Responses

10.3.4.1 Elastomers

The subject of the 'wear of elastomers' has evolved rather separately from that of the wear of other polymers where the main focus has been devoted to the understanding of the behaviour of semi-crystalline polymers. Thermosets and glassy polymeric systems have not been studied as extensively as these two classes.

To review this area, it is necessary to recognize the seminal work of Schallamach [7], although the subject has usefully evolved since then [57, 58]. Schallamach provided very useful precedents, not least the attempt to relate 'single deformation of events' (he termed these as isolated stress intensity events) to the prevailing overall wear and contact damage processes.

The work published by Schallamach and his collaborators was of two types: the isolated stress single asperity works and the generic study of wear (abrasive) processes and the subsequent appearance of the worn surfaces (Figure 10.7). The former sets useful precedents for the now fashionable studies of the scratch deformation of polymers, particularly glassy polymers (discussed later).

The single contact deformation studies provided a good picture of the nature of the damage events. The use of sharp needles was a useful illustration of the importance of surface strain relaxation [59], while the use of 'blunt' indenters provided results comparable to those observed with brittle solids, basically the formation of traction cracks. These studies emphasized the tensile nature of the damage process and also the geometric anisotropy of the consequent damage.

Schallamach, his colleagues and indeed others [60] complemented these studies with an extensive series of investigations of wear processes both in the laboratory and on the road; obviously the major driving force of these studies was to elucidate the wear processes of automobile tyres. Two types of work were reported – unidirectional sliding and random directional sliding. The counterfaces were invariably abrasive bodies or, in some cases, gauzes; Ratner innovated the use of gauzes to suppress the natural 'clogging' of rough counterfaces, caused by wear debris [61].

The unidirectional sliding produced the familiar 'abrasion patterns' often notable on the surfaces of misaligned automobile tyres. The normal section of the worn surfaces has the cuticle structure seen in animal hair which resembles the tiling on roofs. Several authors have sought to simulate and model this process often using metal blades contacting rotating rubber discs. Roberts and colleagues [62] related wear to tensile and fatigue properties and Gent and Pulford [63] noted that the high radical concentration, generated by chain scission,

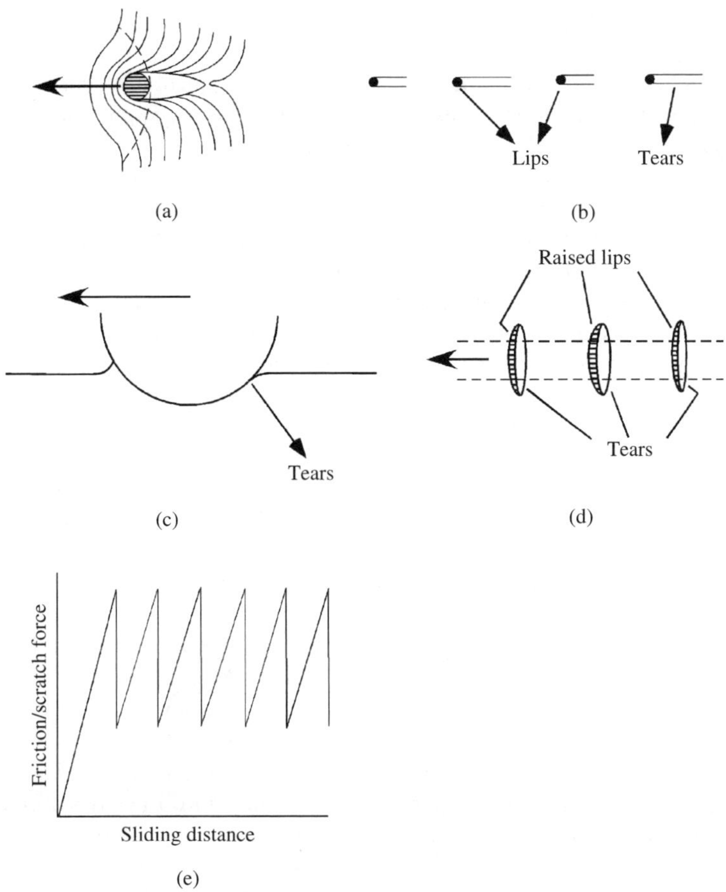

Figure 10.7 Damage created on the surface of an elastomer by isolated stress concentration [3]. (a) Surface deformation pattern when a sharp needle or conical indentor with acute angle is slid on the surface of an elastomer. The elastomer surface is pulled in the direction of motion and fails in tension behind the contact at $\pi/2$ to the tensile field. (b) After the needle jumps forward the surface relaxes and tensile tears are evident on the surface but are now in the direction of motion. (c) Tearing of an elastomer due to tractive stress with a large unlubricated indentor. The tear is generated at the rear of the contact region and is almost at right angles to the sliding motion. (d) A raised tip of elastomer is formed but no material is actually removed. (e) A typical friction/scratching force profile when a slider is passed over an elastomer

could erode metal (steel) blades. The picture is fairly well resolved; the wear is a result of low- or high-cycle fatigue in tensile fields.

The case of random directional abrasion is less interesting and less studied. The wear rates are generally significantly higher than the unidirectional case under comparable conditions. The lip or cuticular structure is vulnerable to disruption from stresses imposed from non-orthogonal directions. Data on the abrasive wear of elastomers can be found in a few published works [30, 64]; however, they have been obtained under very different test conditions and hence it would be difficult to bring them into a single plot or rationalization in some meaningful way.

A few others have since noted other effects. Aharoni [65] mentions scroll formations and Russian work [66] notes the importance of radical scavenges to suppress the propagation of chain scission during low-cycle fatigue-induced abrasive wear.

There was much literature published on the topic of wear, and tribology, of elastomers in the period up to 1980. The book by Moore [67] was a useful survey and the recent books by Zhang [68] and Denton and Keshavan [69] are very comprehensive. There has also been recent and innovative work produced by Uchiyama and co-workers [70, 71] on the fundamental origins of wear processes. We might reflect, however, that the overall picture has not greatly changed since the early and seminal work published by Schallamach. The additional parts focus upon fatigue processes and the chemical component of the wear process.

The friction for elastomers is different from most of the other materials in that for elastomers in addition to the interfacial and cohesive term there is one more mode of energy dissipation which is by hysteresis. Some facets of this subject were introduced in Sections 10.2.1 and 10.2.2. The hysteresis loss is a characteristic of all visco-elastic materials and this comes from the visco-elastic 'flow' of the materials over the asperities of the hard counterface. For smooth and dry contacts, the interfacial component can be very large in comparison to the hysteresis component; however, for rough surfaces and for smooth and wet surfaces, hysteresis loss contributes substantially to the overall frictional force. An excellent and exhaustive review on the theoretical models of hysteresis loss in the tribology of elastomers have been presented by Moore and Geyer [72]. Friction of elastomers can be very difficult to model because of the effects of load and velocity that produce considerable interfacial heat. However, Grosch [60] has shown that if true interfacial temperature is measured, the friction data for different load and velocity can be brought to a single master curve by using time–temperature transformation such as the Williams, Landel and Ferry (WLF) equation. Thus, friction can be plotted as a function of velocity at a fixed temperature [30, 60].

Additionally, for elastomers there is also the unique phenomenon of the formation and the propagation of the Schallamach waves [7] (Section 10.2.2).

10.3.4.2 Thermosets

The main practical area here is associated with the formulation of friction couples such as dry clutches and brake components; there has also been interest in bearing systems. Like the case of elastomers, but more so, this subject has been an academic Cinderella amongst the various polymer classes. Surprisingly, there has been little published in this area. Rhee has produced a range of articles [73, 74] and there are some new contributions [25]. Even here, the focus has been mainly on frictional stability, or 'fade' (traction) suppression, rather than on wear *per se*. The importance of chemical degradation and degradative transfer is mentioned mainly in the context of thermal stability and heat balance. The concoctions used in practice do make a fundamental study impracticable.

The subject did take on a renaissance when there was a move to replace asbestos by 'Kevlar' (aramid fibres) as the reasons for the use of certain fibrous fillers had to be revisited. This area touches upon the utilization of fibrous composites as bearing materials (see also Section 10.4). There is interesting literature on this [75–78]. The fibre orientation in the matrix is apparently important.

There is useful and extensive literature on the wear behaviours of short- and long-fibre reinforced thermosets and the influences of fibre content and orientation [79, 80]. Friedrich [81, 82] has surveyed the area and has provided recent research seminars. Naturally, there must be much speculation as to the origins of the wear behaviour of such complex systems. However, some interesting trends emerge. The influence of the principal fibre orientation direction, with respect to the sliding direction, is very interesting if not quite predictable. There is, of course, the nature of the transformed interface (third body) to be considered but the data do fit into the basic geometric models of how contact mechanical stress may propagate and subsequently disrupt the interfacial bonding between the matrix and the reinforcing fibres. The published data note that the strength of the fibre/matrix interface is a crucial factor governing the wear life of the composite in some circumstances. The tribology of polymer composites is further briefly elaborated in Section 10.4.

10.3.4.3 Glassy Polymers

The common glassy polymers, such as PMMA, PC and poly(styrene) (PS), have not been used as bearings but rather as optical windows. The Achilles' heel of polymeric materials, in the context of bearing applications, is their inability to dissipate frictional heat. The rather abrupt thermal softening of glassy polymers leads to uncontrolled thermally induced failure akin to scuffing in metals; a similar technical problem exists with some other non-cross-linked systems such as the PEEK [83, 84].

Uniquely, some polymeric bearing contacts have the propensity to 'scuff'; this is a very notable feature of the behaviour of the non-cross-linked systems where molecular mobility and flow are not constrained by the cross-linking process. When a polymeric contact reaches a temperature near to the glass transition or crystalline melting temperature, there is naturally a pronounced decrease in the Young's modulus and the hardness. The process resembles that seen in ductile metals [85, 86].

It is a simple matter to examine the consequences of such an event by considering the rate of energy dissipation in the contact (refer Section 10.2). The basic premise is that there will be, at the transition, a loss of 'asperity persistence' [87]. The model [1] is simple and predicts that the friction is

$$F = \tau A \tag{8}$$

where τ is the interfacial shear stress and A, which is roughly inversely proportional to the hardness, is the contact area; the material properties are decreasing functions of temperature. When the frictional heating is localized at the interface, τ decreases but the hardness, and hence the contact area, remains largely unchanged. As the thermal front propagates into the polymer, the hardness decreases and the contact area correspondingly increases. The frictional work increases as a result and at some stage a catastrophic failure mode is instituted; there is a complete loss of asperity persistence and the contact area approaches the apparent contact area. The process has many features which are similar to the 'Junction Growth' models developed for metals in early literature [1]. The overall consequence is a massive and highly damaging contact failure. The phenomenon is well exemplified by the behaviour of virgin PEEK against metal counterfaces in certain lubricated media [84]. Certain additives in conventional lubricating fluids have the capacity to induce plasticization, at elevated temperatures, in such engineering polymers [88]. The combination of frictional heating and ambient plasticization leads to high contact areas and scuffing failures (a Rubenstein [89] effect) . The same problem is not seen with modified materials (PEEK combined with PTFE [90] for example), as the self-lubrication additive is capable, because of third-body formation, of producing a low-friction counterface.

'Self lubricating' polymers such as PTFE and linear PEs have the capacity to form oriented layers which are 'weak' in shear and hence do not generate the same degree of adiabatic shear heating at the interface (see next section) [20].

Other parts of this chapter will deal with the subtle effects induced by special contact displacement modes; linear motion combined with load axis spin for example. One of the interesting features of the overall wear response is the influence of the contact configuration. The process of fretting is a good example of the importance of debris accumulation and expulsion from a contact zone. Again, the problem resolves around the imposed complexity of the sliding motions and the details of the specific actions of the interface shear fields. Briscoe and Chateauminois have recently encountered this problem in the context of a glassy polymer, PMMA [47, 48]. The imposed sliding kinematics has a major influence upon the overall rate of fretting wear.

It is unlikely that polymers of this class would be generally used for advanced tribological service. More likely is the prospect, because of their cost and utility, they may have a tribological function as part of an overall design requirement. The most obvious examples currently are the use of poly(acrylate) in 'bathware' and 'domestic' surfaces and the various usages of PCs in optical windows. A noticeable interest in the marring or scratching of such polymers is apparent in recent literature [91, 92]. This would not, by conventional definition, be regarded as 'wear'; the parallel with the Schallamach precedents is of interest and better wear models may accrue because of this focus upon the origins of the perceived surface damage.

The scratch work, mainly for PMMA, leads to several general conclusions. The damage evolves through a range of severity as the contact strain is increased: visco-elastic smoothening or 'ironing', plastic or visco-plastic grooving, extensive plastic flow and tearing, pronounced fracture or tearing and finally cutting or chip formation. The contact strain may be loosely defined as being proportional to the tangent of the included semi-angle of the indenter.

There is an expanding literature on the topic of the scratching of polymers which provides a refreshing and exciting addition to the understanding of damage processes [48, 91–95]. A typical damage map is shown in Figure 10.8 [95]. A combination of frictional force measurements, AFM and topographic studies, SEM images and modelling has shed much light on the processes involved, all be it for only a few polymers, in particular PMMA.

10.3.4.4 Semi-Crystalline Polymers

For a very large variety of reasons, these systems have been very extensively studied not least because of the historically practical and academic interest in the wear of the poly(tetrafluoroethylene)s and the sister polymers such as the linear PEs, in particular the high-molecular-weight PEs; the latter has found great favour in orthopaedic replacement components (see later in Section 10.6). The class is also one which sees a wide range of applications: PTFE and the derived high-temperature polymer composites for bearings and

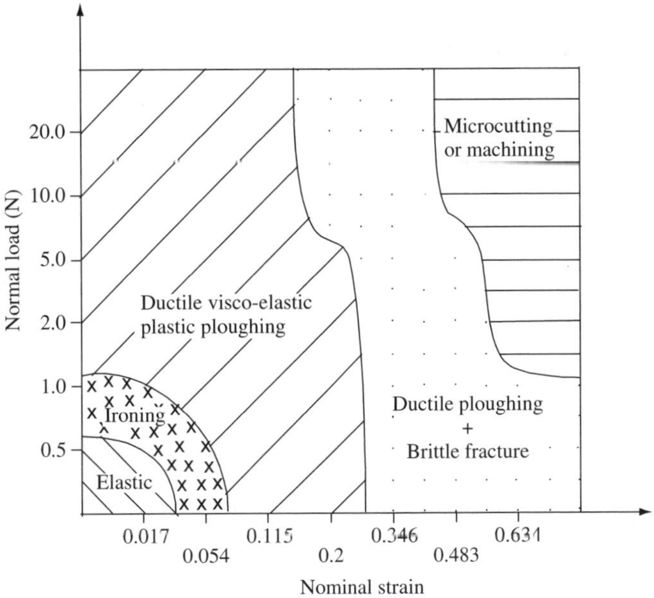

Figure 10.8 Scratching mode map for PMMA [3]. The diagram shows the dependence of the observed scratching deformation mode upon the nominal contact strain and the applied load. The nominal contact strain was calculated as $0.2 \times \tan \psi$, 2ψ being the included angle of the indenter. The scratches were produced at a constant scratching velocity of 0.004 mm s^{-1}, at ambient temperature of 20°C and under unlubricated contact conditions

automobile piston seals [96], nylons as common automobile bearings [97], PEs in marine environments and poly(vinylidenefluoride) and PEEK in wire line electrical shielding.

The intriguing facet of the semi-crystalline polymers is the variety of their 'transfer' wear or damage responses. There are probably three types of behaviour; the classification is an old one and many would cite the work of Tabor and colleagues [20, 98] as being seminal here – Figure 10.9 illustrates. A distinction is first made between isothermal and adiabatic processes although this is made more on a judgement of extent rather than upon kind. This apart, for the isothermal cases, three situations are regarded as being encountered (Figure 10.10). First, there is transfer or not; poly(propylene)s are regarded as not forming thin and ordered transferred layers, but the situation is unclear. What is clear, however, is that the PTFEs and linear PE do transfer under a wide range of contact conditions to many counterfaces [99–101]. There is evidence that ambient temperature and sliding velocity are important variables [98]. This apart, the isothermal condition which gives transfer does so in two ways. There is the case of the so-called 'smooth molecular profile, or special, polymers' which produce, once linear sliding has commenced, thin highly oriented and weakly adherent layers; this is typical of the PTFE and UHMWPE systems. The static friction is high but the dynamic friction is low and the resulting wear (see later) may be very high. Surface roughness and rotation coupled with sliding can suppress thin film transfer formation and orientation leading to a reduction in the wear [102, 103]. Technically, the prospect of low

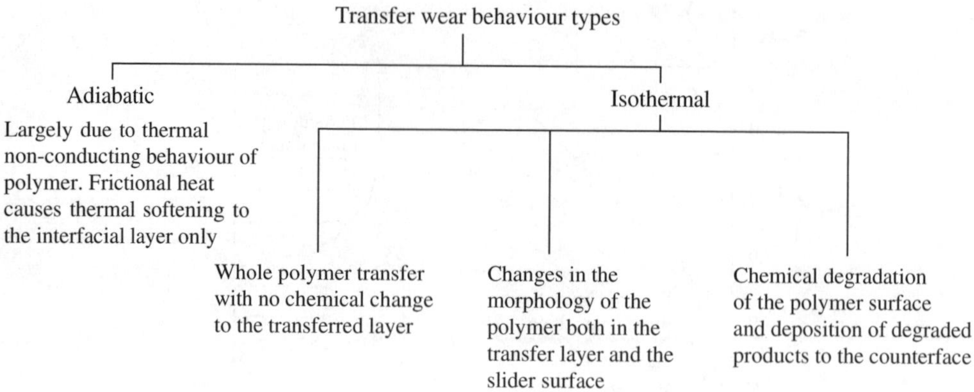

Figure 10.9 Generic types of transfer wear behaviour when semi-crystalline polymers are slid onto a hard smooth surface. In every case there is a formation of transfer layer on the counterface though the shear and adhesive properties of the transfer films will vary depending upon the polymer's mechanical properties and the surface topography of the counterface [3]

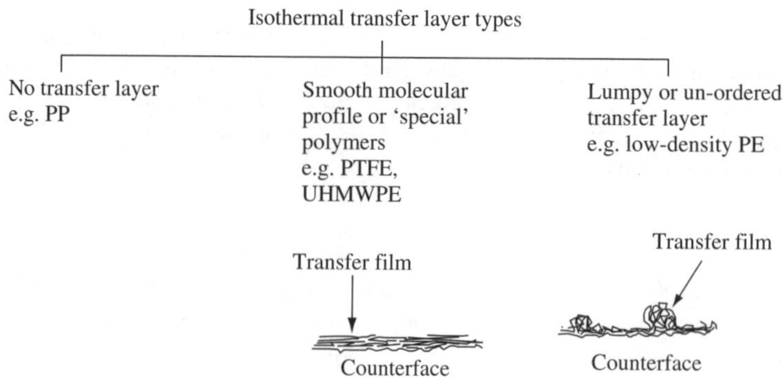

Figure 10.10 Types of transfer layers for semi-crystalline polymers when slid against a hard smooth surface [3]

friction motivated the great interest in these materials. Figure 10.11 summarizes wear rate data for many different classes of polymers as found in the literature [104]. These data were obtained for different conditions of load, velocity and the counterface surface roughness and hence should be used with caution. However, relative wear performances of the different polymers are evident.

The seminal work on transfer film was probably that by Pooley and Tabor [20] but Tanaka has made useful contributions [70, 99] (also see reviews by Rhee and Ludema [112], Bahadur [45] and by Biswas and Vijayan [113]). The basic idea was that certain polymers, under the action of interfacial shear gradients, would produce reoriented interface zones and indeed transferred layers [45]. Stolarski and co-workers [102, 103] demonstrated this point nicely by undertaking studies of friction in wear of a range of polymers, including the 'special'

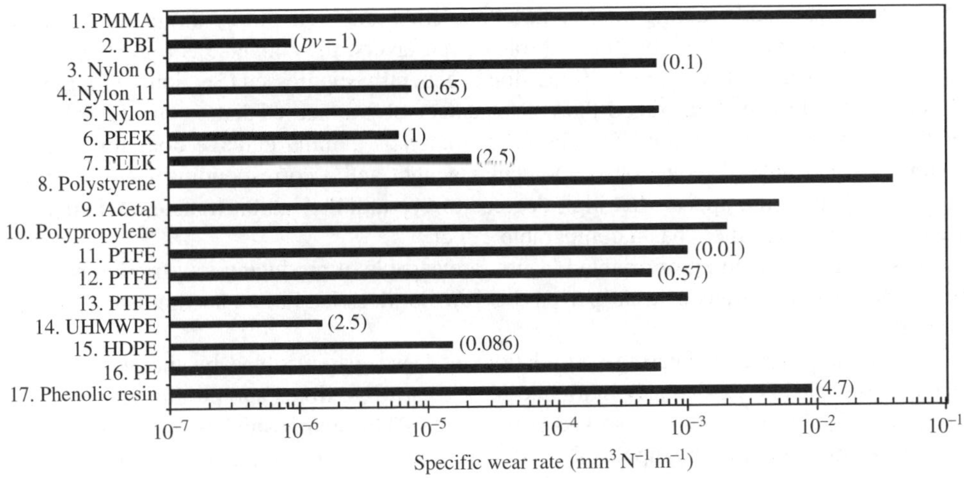

Specimen No.	Material	Counterface roughness, R_a(μm)	Sliding speed, v (m s^{-1})	1/Se	Normal pressure, p (MPa)	Temperature (°C)	References
1.	PMMA	1.2	–	0.09	–	–	[105]
2.	PBI	–	1	–	1	20	[106]
3.	Nylon 6	–	5×10^{-3}	–	20	–	[107]
4.	Nylon 11	0.11	1	–	0.65	–	[108]
5.	Nylon	1.2	–	0.1	–	–	[105]
6.	PEEK	–	1	–	1	20	[106]
7.	PEEK	0.05	0.5	–	5	–	[109]
8.	Polystyrene	1.2	–	5	–	–	[105]
9.	Acetal	1.2	–	0.5	–	–	[105]
10.	Polypropelene	1.2	–	0.1	–	–	[105]
11.	PTFE	–	0.2	–	0.05	–	[110]
12.	PTFE	–	0.1	–	5.66	29	[71]
13.	PTFE	1.2	–	0.2	–	–	[105]
14.	UHMWPE	0.05	0.5	–	5	–	[109]
15.	HDPE	0.9	0.03	–	2.8	–	[111]
16.	Polyethylene	1.2	–	0.09	–	–	[105]
17.	Phenolic resin	0.05	5.6	–	0.84	–	[78]

Figure 10.11 Specific wear rate for a number of polymers in adhesive wear modes [4]. Specimen legends and test conditions are given below [reprinted from Sinha (2002) with permission from ASM International] [104]

polymers under the action of combined linear and rotational sliding. Linear sliding produces oriented interfaces with low friction and high wear. The addition of the rotation disrupts the natural orientation process. As a result, the frictional forces increase towards the static (un-oriented) values. Wang *et al.* have recently revisited this problem in the context of the wear of UHMW PEs [114] . The evidence is persuasive; rotation when combined with linear motion disrupts the process of highly oriented transfer. The fact that the consequent dry wear is reduced is explained by an argument that high friction and high energy dissipation promote the formation of more strongly attached transferred layers; the converse is true for synovial fluid lubricated contacts where film formation is generally suppressed (Section 10.6). Earlier, Briscoe *et al.* [115] had examined the consequences of gamma irradiation for the tribology of these systems. Interestingly, for the PTFE materials gamma damage does not produce chain scission; rather a reduction in molecular weight and a corresponding increase in the crystallinity. It would appear that high degrees of crystallinity inhibit transfer and suppress wear. The action of fillers has a comparable effect.

These 'highly structured or oriented layers' are capable of producing low friction but high wear; the volume of transfer wear per pass is low but does lead to an accumulated high net wear rate.

The problem here, for bearing formulations at least, was resolved by the addition of a range of organic and particularly inorganic fillers: particles, fibres and porous supports. The use of a, say, 10% by weight of a glass particle in a PTFE composite has often been shown to reduce the wear rate by up to at least one to two orders of magnitude [116]. Many authors [110, 117, 118] have speculated upon why this occurs. Polymer blends are another example for enhanced wear resistance of polymers [18, 119, 120]. Certain transition metal oxide fillers have also been reported to induce mild polymer degradation at the interface and thus create strong valence bonds between the transferred layer and the counterface [40, 41]. Similar tribo-chemical effects leading to stronger bonding of the transferred layer to the counterface have also been tested for a polyphenylene sulphide composite filled with 30–35% CuS powder when slid against a steel counterface [121]. Various models are available but basically the fillers are thought to provide a thin, coherent, securely attached transfer layer of the polymer species on the adjacent counterface. A stable transfer layer promotes low friction and low wear. Other factors such as the suppression of transfer and the generation of local stress events at the interface may be important [122]. The wear morphology of fibre-filled polymers has been shown to exhibit crack nucleation at the polymer–fibre interface and crack propagation parallel to the surface and crack shearing to the surface resulting in flakes of polymer debris [123]. Environmental influences, particularly water, may further disrupt this interface and thus cause higher rates of wear [23]. The friction and wear of polymer composites is considered further in the next section.

Another important parameter that controls the wear of semi-crystalline polymers is the initial and steady-state counterface roughnesses. This aspect has been dealt with at length in an earlier review [15]. However, it may be appropriate to mention here that for both pure and filled polymers, there seems to be an optimum initial counterface roughness that produces strong adhering transfer layer and consequently low wear [124–126]. For example, PTFE filled with 10% carbon shows a minimum in the wear rate at about initial counterface roughness of 0.4 μm (R_a value) [127]. The optimal roughness value for medical grades of UHMWPE has been extensively studied by Dowson and co-workers [124].

Much less is known and recorded regarding the other two cases: no transfer, if it occurs in this class, and the 'lumpy' and un-ordered transfer which is seen with such polymers as low-density (chain-branched) PE. Pooley and Tabor [20], for example, reported high frictions and relatively thick but un-oriented transfer for this case. At the initiation of sliding in linear motion sliding with load axis spin 'lumpy' transfer is seen with PTFE and the linear PE. Steward and co-workers [127, 128] observed similar behaviours for carbon-filled PTFEs. The general result for those cases is that high friction produces thick un-oriented layers and also relatively low rates of wear.

Central to all of these observations is that nature and adhesive tenacity of the transferred layers is apparently a dominant factor. The abrasion resistance of the class, by and large, may be rationalized using the Ratner–Lancaster correlation for the virgin polymers at least. At a fundamental level, there are now issues to be considered regarding the scale of the deformation and the corresponding scale of the homogeneity of the system; for composite materials this is an issue of interest for practitioners but one which is not resolved by those who wish to produce generic models. Friedrich [106] and one of the current authors [15] have commented upon this problem.

10.4 Tribology of Polymer Composites

Although unfilled (or virgin) polymers have shown excellent tribological properties in some very uneconomical cases, they are rarely used as monolithic one-phase materials; not least for economic reasons as fillers are often relatively less expensive than the polymer. An addition of second phase in the form of particulate, short- and long-fibre fillers, making composites has many structural advantages and the potential and actual tribological advantages of these composites will be discussed in this section.

From the material deformational standpoint, we can simplify the classification of the formulation of composites by dividing them into two types. One is where 'soft and lubricating phases' are added to a matrix of a relatively continuous 'hard phase' and the second is the type where 'hard and strong phase' is dispersed in a 'soft phase' matrix [129]. The polymer may actually be the matrix of the dispersed phase and a wide spectrum of such cases exists in practice. This classification will help in the understanding of the tribological behaviours of polymer composites because the deformational characteristics of the second phase in composites play a very important role in modifying the interface. It is now well understood that the tribological responses of polymer composites when slid against hard (and 'smooth') metal surfaces are dependent more on the interface properties and less on the cohesive or sub-surface properties. However, the latter controls the sub-surface fracture properties in the high wear rate regimes and is a significant fraction in gross abrasion and rolling contact fatigue.

10.4.1 'Soft and Lubricating' Phases in a Harder Matrix

This is a type of composite where the intention of adding soft phase is to provide a low shearing or self-lubricating material in a hard matrix and thus a lubricating third body. Use of PTFE or graphite in harder polymers such as PEEK or nylon 66 is a good example of this kind of widely utilized composite. The second phase does not contribute towards the improvement in any other mechanical properties such as the elastic modulus or fracture stress. In fact, there can be drastic reduction in the bulk mechanical properties as a result of the inclusion of the soft phase in the matrix. The main advantage of using such a soft lubricating material such as PTFE in the matrix is the reduction in the coefficient of friction and wear up to an optimum wt.% of the soft phase and a corresponding increase in the PV limit; a volume fraction inclusion of 5–20% is typical. The use of a very large wt.% of soft phase will probably be counterproductive by reducing the elastic modulus and strength and thus finally increasing the wear rate. Several experimental data on, specially, the use of PTFE are available in the literature. Figure 10.12 summarizes some of the results that in fact

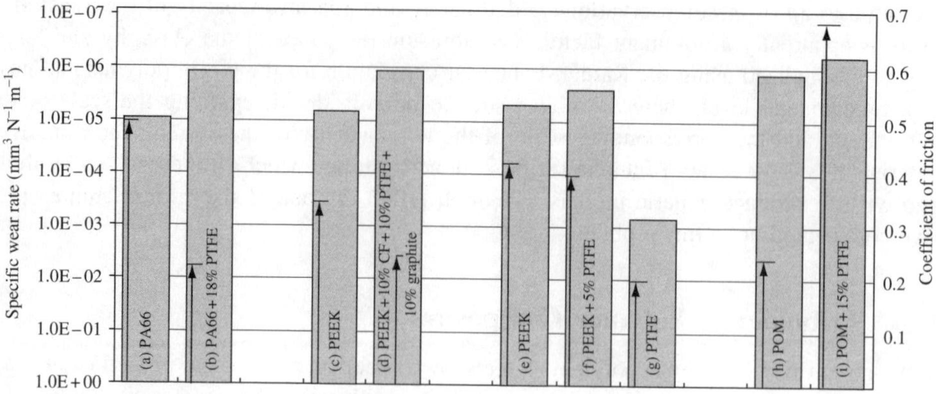

Figure 10.12 The effect of PTFE addition to hard polymer matrices. The rectangular bar chart indicates specific wear rate (units on the left of the graph) and vertical arrows indicate the coefficient of friction (units on the right of the graph). Test conditions are as follows: (a) pin-on-disc apparatus, steel counterface roughness $R_a = 2.8$ μm, $p = 3.14$ MPa, $v = 0.5$ m s^{-1} [130]; (b) test conditions same as for (a) [130]; (c) counterface – steel disc grade DINX20Cr13, counterface roughness $R_a = 0.03$ μm, $p = 0.5$ MPa, $v = 0.25$ m s^{-1} [131]; (d) counterface – steel disc grade DINX20Cr13, counterface roughness $R_a = 0.03$ μm, $p = 0.5$ MPa, $v = 1.2$ m s^{-1} [131]; (e) block-on-ring apparatus, $p = 1$ MPa, $v = 1$ m s^{-1} [132]; (f) test conditions same as for (e) [132]; (g) test conditions same as for (e) [132]; (h) counterface roughness $R_a = 0.1$ μm, $p = 2.5$ MPa, $v = 0.3$ m s^{-1} [133]; and (i) test conditions same as for (h) [133]

lead to the same conclusion. Use of PTFE up to a moderate 18 wt.% can help in reducing the coefficient of friction by up to half and the specific wear rate by an order of magnitude; the PV limits will also be greatly improved. There will also be a marked increase in the scuffing load (PV limit) as the heat generation will also be very greatly reduced. The presence of PTFE basically helps in the formation of a continuous PTFE-enriched interfacial transfer film on the counterface. Flow by shearing action is facilitated by the molecular structure of PTFE and thus the coefficient of friction drops. The wear rate as a function of wt.% PTFE initially decreases with the increase in the percentage of PTFE; however, it will reach a minimum, and a further increase in the percentage of PTFE will increase the wear rate. Other soft and lubricating phases that have been used for polymer composites include MoS_2, CuS and CaF_2 [110], aliphatic amides [134] and silicone resins and fluids [135]. MoS_2 is solid lubricant, whereas compounds of copper decompose to copper, or its oxides, during the wear process. It is believed that copper acts as a catalyst in the formation of strong adhering interfacial film on steel counterface [40].

10.4.2 'Hard and Strong' Phases in a 'Soft' Matrix

The main advantage of adding hard and strong phases to soft matrix is in increasing the stiffness of the matrix; there may be a corresponding loss of toughness or fatigue life. While the friction is controlled by the self-lubricating property of the matrix, wear resistance is improved because the load-carrying capacity of the composite is basically increased due to the presence of hard and strong second phases. Very good examples of this kind of polymer composites would be those with glass or carbon fibres in thermoplastic or thermoset matrices. Several other kinds of hard phases such as ceramic oxides are also added to polymer matrix with varying effects.

PTFE, polyamides (nylon), PEEK or phenolic resin filled with glass fibre (GF), carbon fibre (CF), aramid fibre (AF) or other fillers are very common examples of the composites where the addition of second phase essentially raises the strength of the matrix. The wear rates for these composites are generally lower than that for the neat polymers; however, the magnitude and the relative effectiveness of each filler will depend upon a number of factors such as the environment and whether the fibres are used as long or short fibres. Figure 10.13 compares the wear rates of neat and filled polymers. It is important to note that although there is often an increase in the coefficient of friction as a result of the fibre addition, the wear rates are significantly reduced. For most of the fibres, there is an optimum wt.% inclusion after which the wear rates tend to increase. The optimum level of the fibre in polymer matrix has been experimentally found to be in the range of 15–30 wt.% [106, 108, 137], though this value may vary drastically from system to system. Approximately 20 wt.% is considered to be optimal by many researchers. One important aspect of the wear mechanism here is that of the formation of a strong and coherent adhering transfer film upon

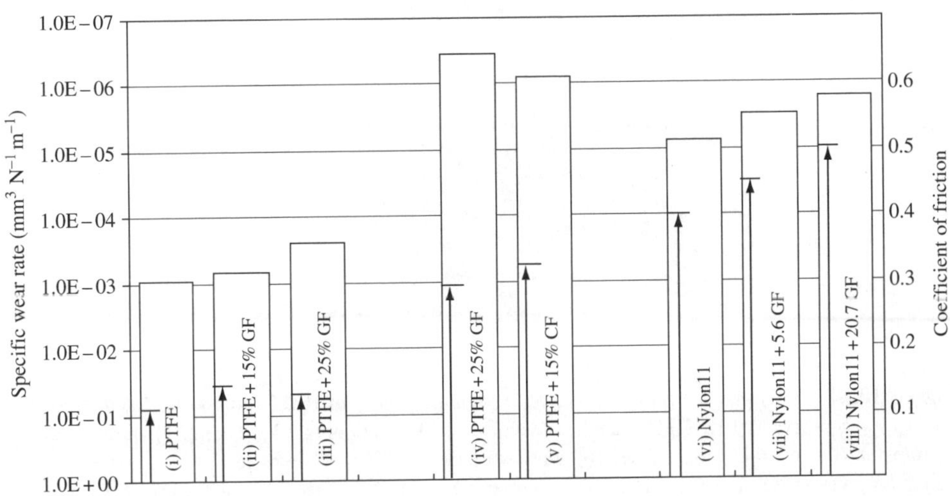

Figure 10.13 The effect of fibre addition on the specific wear rates of a few polymers. The rectangular bar chart indicates specific wear rate (units on the left of the graph) and vertical arrows indicate the coefficient of friction (units on the right of the graph). Test conditions are as follows: (i) 440 C steel ball (diameter = 9 mm) sliding on polymer specimen, normal load = 5 N, $v = 0.1$ m s^{-1}, roughness of polymer surface $R_a = 400$ nm, 30% humidity [136]; (ii) test conditions same as for (i) [136]; (iii) test conditions same as for (i) [136]; (iv) reciprocating pin–steel plate apparatus, counterface roughness $R_a = 0.051$ μm, N_2 environment [96]; (v) test conditions same as for (iv) [96]; (vi) pin-on-steel (AISI02 quench hardened) disc apparatus, counterface roughness $R_a = 0.11$ μm, $p = 0.66$ MPa, $v = 1$ m s^{-1} [108]; (vii) test conditions same as for (vi) [108]; and (viii) test conditions same as for (vi) [108]

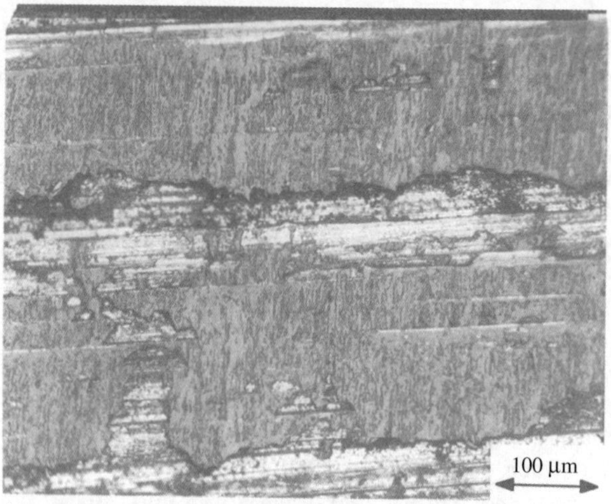

(a)

(b)

Figure 10.14 (a) High speed–low pressure reciprocating wear of PEEK composite (containing 10% CF, 10% graphite and 10% PTFE) against stainless steel (containing 13% Cr) [131] shows existence of transfer film on the counterface; (b) worn surface of PEEK composite showing extensive fibre debonding and some abrasive marks. The sample was collected after 70 h of testing at a velocity = 1.2 m s^{-1} pressure = 0.5 MPa [reprinted from Schelling and Kausch (1993) with permission from Elsevier] [131]

the counterface (normally smooth ferrous surfaces). The presence of a stable thin film appears to prevent further removal and deposition of the matrix. Figure 10.14 gives an example of such a transfer film on a steel surface. The wear of the composite is further reduced because the load-carrying capacity of the composite is much higher than that of the matrix and thus any sub-surface fracture and yielding of the material are minimized due to the presence of the hard and strong second phases. The effects of several different types of fibres on the friction and wear properties of polymers have been studied very extensively. A summary of some of these results can be usefully presented in a friction and wear map format as has been shown by Friedrich *et al.* [106]. A very general conclusion on the inclusion of fibres in polymer matrix may be made here which is as follows. At ambient room temperatures, the addition of glass fibre sometimes increases the coefficient of friction by 0.2–0.3 units, whereas addition of carbon fibre will either often have no effect upon the coefficient of friction or it will reduce by 0.1–0.2 units. At higher temperature (in the range of 150°C), the friction largely remains unaffected or drops slightly by 0.1–0.2 units. Both carbon fibre and glass fibre are equally effective in reducing the extent of the wear rate. Wear rate can by reduced by one to two orders of magnitude, or better, for the optimum levels of fibre wt.%.

One disadvantage of using hard phases in soft matrix is the possibility of transfer film removal or even the aggressive abrasion of the hard counterface if the hard phases (particles or fibres) are released to the interface as third body due to loosening of the bond between the filler and the polymer matrix [138]. Perhaps to overcome the abrasive effects of hard phases and to reduce the coefficient of friction the majority of composites that are used for tribological applications today are in fact hybrid in nature. Therefore, the following section will present some of the various tribological trends of polymeric hybrid composites.

10.4.3 Hybrid Polymer Composites

The optimum levels of friction and wear can often be obtained for a polymer matrix if both soft (and lubricating) and hard (and strong load bearing) phases are selectively added to enhance the self-lubricating and load-carrying properties of the matrix. The potential and actual advantages of using hybrid polymer composites for tribological applications have been shown by many researchers and manufacturers. The addition of a strong fibre such as glass fibre (GF), carbon fibre (CF) or aramid fibre (AF) coupled with some wt.% of soft and lubricating phases such as PTFE, MoS_2 or CuS_2 could provide excellent wear resistance and low friction. Figure 10.15 gives an example of friction and wear behaviour when CF and PTFE are added to poly(amide) matrix.

The selection of the second phase for enhancing the tribological performances of polymeric matrices will depend upon the normal pressure, sliding velocity (the PV factor) and the surface conditions of the counterface as well as on the ambient environment and contact kinematics. A high PV factor requirement will necessitate the presence of fibres to increase the strength of the materials, and the presence of soft lubricating phases will keep the coefficient of friction, and thus the interfacial temperature, low. Polymers with high T_g, the so-called engineering polymers, are generally preferred for heavy-duty applications: PEEK, PES, PPS, for example. Counterface roughness also has profound influence on the wear resistance of such composites. For high counterface roughness, where abrasive wear is predominant, the presence of second phase may not have desired effects. Many studies have suggested that the presence of a second phase in polymer matrix has deleterious effect on the abrasive wear resistance [139]. As always, the thermal conduction and heat capacity of the system will be a controlling factor at high values of the PV factor.

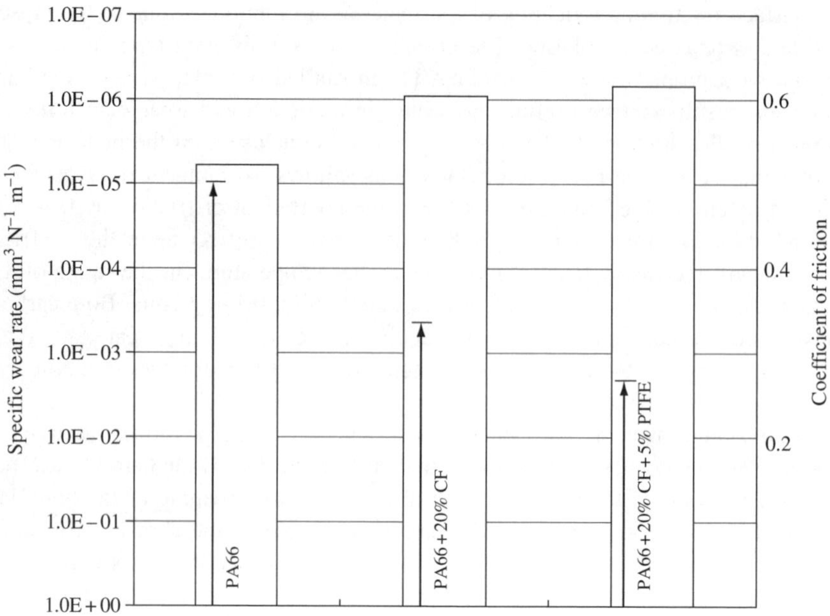

Figure 10.15 The effect of fibre and soft phase addition on the specific wear rates of PTFE. The rectangular bar chart indicates specific wear rate (units on the left of the graph) and vertical arrows indicate the coefficient of friction (units on the right of the graph). Test conditions: pin-on-steel disc apparatus, counterface roughness $R_a = 2.8$ μm, $p = 3.14$ MPa, $v = 0.5$ m s^{-1}, $T = 23$°C [130].

10.5 Environmental and Lubrication Effects

Polymers, in general, are not used in the presence of lubricants. Nevertheless, there has been interest from polymer tribologists in this area; recall that a major virtue of organic polymers is their intrinsic 'self-lubrication' capacity. One obvious reason is that polymers, intentionally or un-intentionally, do get subjected to lubricants or humidity or contaminants present in the operating environment. Examples are industrial bearings subjected to leaked lubricants or water, household plastics subjected to humidity and polymer human hip/knee joint replacements working in the environment of proteinaceous synovial fluid. One example of elastomers in lubricating environment is that of hydraulic seals, a major area of application. Here, the presence of a thin layer of lubricant is essential for the reduction of friction and wear of the seal material. Working fluid film thickness for viscous oils present in seals has often been predicted from elasto-hydrodynamic theory [140, 141]. This film generally protects the elastomer surface from coming into direct contact with the solid surface. In the case of prolonged stationary condition or for very high contact stresses, the lubricating film tends to get squeezed out of the contact region, leading to high start-up friction and dry sustained contact [142–144]. Such conditions eventually lead to high wear and seal failure. The wear mechanism for such dry condition can be treated as similar to the case where there is no presence of lubricant.

For the non-elastomeric polymers, depending upon the polymer and the lubricant polarity, the presence of lubrication has various effects such as polymer plasticization, the reduction of the contact area due to fluid film formation at the interface, modification of the shear properties of the interfacial junctions, changes in the adhesion properties of the transfer film and anti-plasticization effects. In many cases, a polymer/metal contact will exhibit the classical Stribeck–Hershy curve; PEEK/steel in hydrodynamic fluids is such an example (the system used in the manufacture of Mc Pherson stuts [145, 146]). Briscoe and Kreminitzer [147] have shown that for poly(ethyleneterephthalate) fibre/isopropanal–water mixture systems, there are two situations of lubrication depending upon the normal load. At low loads, the lubricating efficiency of the fluid depends upon the surface tension and the wettability of the fluid. Low fluid surface tensions generally lower the coefficient of friction due to the formation of a uniform lubricating layer at the interface. On the contrary, at high loads the elevated contact stresses tend to extrude the interposed fluid out of the interface zone leading to direct solid–solid contact and, as a result, a high friction situation is generated. The effect of humidity on high-density PE, PTFE and nylon 6,6 has also been reported [23, 148–150]. In general, these polymers allow a limited sorption of water molecules into the amorphous regions of the bulk of the polymer contributing to a decrease in the mechanical properties such as hardness, elastic modulus and the shear strength of the polymer. Such sorption is accelerated in the presence of high hydrostatic pressures that are very often present at the polymer/counterface contact regions. Modest surface plasticization of polymers such as nylons often has the effect of reducing the interface shear stress and thus the sliding friction: the so-called Rubenstein effect [129]. The effect is comparable to an increase in interfacial temperature induced by, say, an increase in sliding velocity. Excessive and deep plasticization of the surface will cause a disproportionate increase in the contact area. The failure for fracture resembles thermally induced scuffing. The wear rates have been found to increase by two- to threefolds under such conditions. The interesting feature of lubricated sliding of transferring polymers is the presence of very little or no transfer film on the counterface. Probably, this is one of the major factors contributing to their high wear as there is continuous removal of the debris from the interface in the absence of a securely attached transfer film. The effects of water on the tribology of polymer composites have also been studied [151]. The wear rate generally is higher for all composites (PTFE reinforced with either of the fillers – GF, CF, graphite, bronze, MoS_2) studied; however, the reduction of the wear rate is much higher (two to three times) for GF reinforced composite than for others. For lubricants with larger molecules (fatty acid and paraffins), the plasticization of the bulk polymer is minimized and thus the friction and wear depend upon the ability of the polymer surface to generate a coherent layer of lubricant at the interface. A good lubrication is possible if the contact stress conditions are not too severe. In the case of PE used for artificial hip replacement, oxygen-plasma treatment has been used to enhance the adsorption of protein molecules present in the human synovial fluid to the polymer surface due to increased hydrophilicity [152]. Such polymer surface modification can lead to considerable reduction (50%) in friction; however, its effect on wear is still not known. To counter the damaging effects of mixed lubrication and the consequent wear of UHMWPE in artificial hip joint prosthesis [153], different joint designs consisting of a compliant layer such as water-swollen hydrogel polymer sliding over hard metal has also been reported to have desirable friction and wear properties [154, 155]. Unlike most metals, lubrication of polymers, in general, greatly depends upon how the lubricant molecules attach themselves at the polar sites, if any of consequence, of the polymer surface and often less upon the viscosity of the lubricant [156]. Further consideration of the tribology of hip–knee joints will be presented in the case study in Section 10.6.

The effects of humidity on the fretting wear characteristics of some polymers have also been reported. A poly(imide), with carbonyl linkages in the molecule, in the form of coatings of about 17-μm solvent cast on steel and aluminium surfaces shows a decrease in the fretting life by twofold for an increase in the relative humidity from 10 to 40% [157]. The reason for this has been suggested that water molecules attach to the carboxyl sites by hydrogen bonding. At temperatures lower than the T_g of the polymer, the bonded water molecules behave as anti-plasticizer and suppress the secondary relaxation process and flexibility in the polymers; effectively the crystallinity is considered to increase. Less flexible polymers (higher crystallinity) tend to fracture rather than deform plastically and hence the wear rate is higher for the less elastically flexible case. This has also been proved to be the case in linear sliding experiments [158].

The available literature on the lubrication and environmental effects on the wear of polymer clearly suggest that polymers differ from metals and to some extent ceramics in at least two aspects. The first is the ability of the fluid phase to penetrate into the bulk of the polymer and thus change, often in rather subtle ways, the mechanical properties of the polymer. The second effect of the lubricant or the environment is on the change in the adhesive and shear properties of the transfer film. Despite some progress in the understanding of the lubricated polymer wear, further research is needed to tap the great potential of using polymers for specific applications in the presence of lubricants. Molecular modification of the polymer surface to effectively accommodate lubricants without affecting the mechanical properties of the bulk of the polymer appears to have some potential.

10.6 A Case Study: Polymers in Hip and Knee Prosthetic Applications – Ultrahigh-Molecular-Weight Poly(ethylene) (UHMWPE)

UHMWPE is frequently used as one of the articulating materials for prosthesis implants in total hip and knee joint replacements (Figure 10.16) [159]. Total joint replacements (TJR) require two articulating components that are made to slide against each other during service. Historically, the choice of the pair of materials for TJR has been metal-on-metal, metal-on-ceramics, ceramics-on-ceramics, metal-on-polymer and ceramic-on-polymer. Amongst polymers, PTFE [160, 161], UHMWPE [17, 162–164] and POM (commercial name Delrin™ [165]), and a few of their composites, have been used in clinical trials. However, a vast number of studies, both in the laboratory and in actual implants, have clearly shown that UHMWPE is by far the most useful material for the TJR application for reasons such as excellent biocompatibility, low friction and wear, good strength and toughness and chemical stability in the body physiological environment. However, UHMWPE suffers from the problem of wear particle generation due to the abrasive or transfer film actions of asperities of the counterface or hard third body entrapped at the interface and due to fatigue. These wear debris are about 10 μm in size or less and they are not biocompatible, meaning they can trigger a macrophage reaction in the body leading to osteolysis in total hip arthroplasty [166, 167]. It is believed that the presence of wear debris is the main cause of implant failure. Therefore, a great deal of research has been carried out, first, to elucidate the mechanisms of wear debris generation when UHMWPE is slid against smooth metallic or ceramic surfaces in dry or lubricated environment, and, second, to engineer the surface and bulk of this polymer in order to reduce the rate of wear (or particle generation).

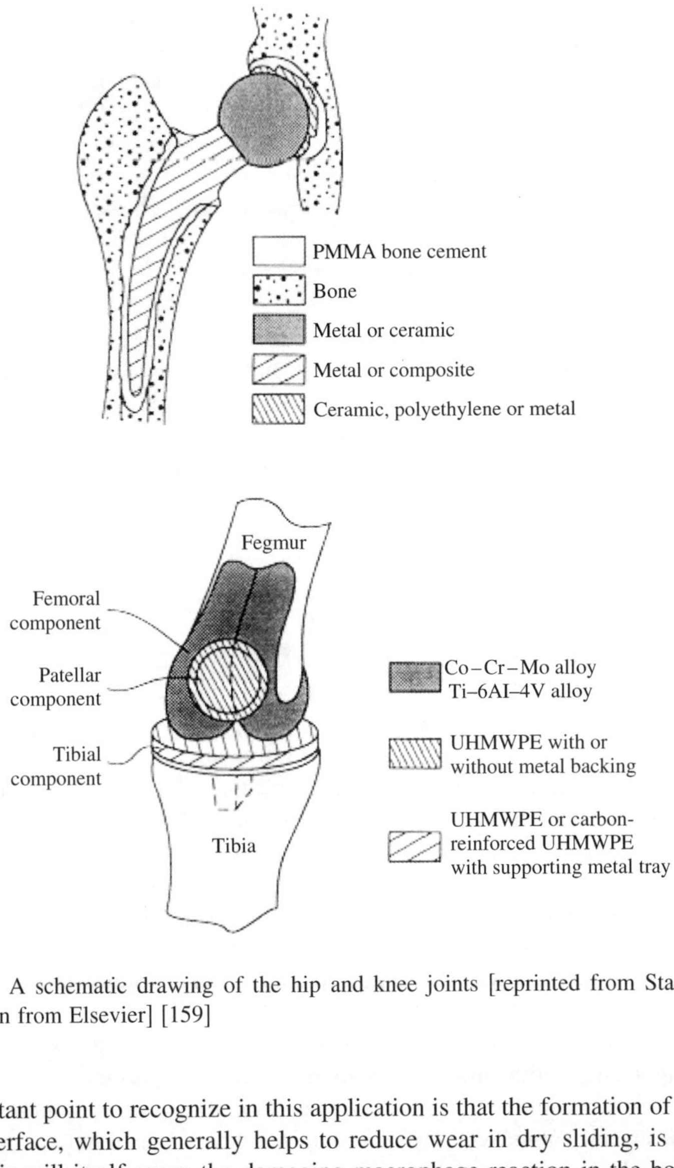

Figure 10.16 A schematic drawing of the hip and knee joints [reprinted from Stachowiak (1993) with permission from Elsevier] [159]

One important point to recognize in this application is that the formation of a transfer film on the counterface, which generally helps to reduce wear in dry sliding, is not desired as the film debris will itself cause the damaging macrophage reaction in the body. Therefore, a variety of lubricants are used mainly during simulated experiments to study the wear behaviour of UHMWPE. Early researchers used distilled or deionized water as a lubricant; however, UHMWPE wears heavily in water and this thus does not represent the actual wear behaviour in service. It is now believed that protein biomolecules (such as phospholipids) present in the serum help in the boundary lubrication of the joints and thus reducing wear [168–170]. The wear mechanism of UHMWPE involves plastic deformation, abrasion and delamination due to fatigue (in part rolling contact)-related phenomenon. Normally, hip joints encounter lower stresses than the knee joints [171] and therefore the wear mechanism can vary for these two joints. In the case of knee joint, the surface stress generated can be higher (~45 MPa) than the yield strength of the UHMWPE (~20–25 MPa) and hence may be described by a low-cycle fatigue model such as the Coffin–Manson equation [17]. Wear for such a kind of process occurs by a process often termed as delamination where the cracks initiate in the sub-surface. For the hip joint, the contact stress (<10 MPa) is much below the yield stress of the polymer and hence wear takes place by microscopic abrasion or transfer and can be described by a model similar to the one presented by Ratner–Lancaster. Wang *et al.* [17] derived a model for the microscopic abrasive wear process and gave a relation as

$$\Delta V \alpha W^{3/2} R_a^{3/2} (S^{3/2} e)^{-1} \tag{9}$$

where W is the applied load and R_a is the centre-line average surface roughness of the counterface (femoral head for hip joint). Equation (9), as does the Ratner–Lancaster correlation presented earlier in this chapter, suggests that the wear can be reduced by increasing the tensile fracture stress (strength) and the elongation to fracture (toughness) of UHMWPE. The model also shows that the wear volume is a non-linearly increasing function with the counterface roughness. However, many experimental results have shown that this may not be true for very low (below a roughness threshold value) roughness values. The wear volume initially decreases as the roughness is increased from a very low value and reaches a minimum before increasing [124, 163]. This is also true for other polymeric systems when sliding against metal counterface [15]. It has been noted by many researchers that the main cause of abrasive wear of UHMWPE in hip joints is the loose hard particles that are released from the PMMA bone cement with time and by the scratches that are made on femoral head by these hard particles (or by the surgeon). Therefore, in order to reduce the wear of the polymer, it is absolutely important to avoid the loose particles coming into the interface and thus making scratches upon the articulating surface which further increase the rate of wear. One solution to this problem is to make the metal part as hard as possible while retaining the toughness so that the metal part will not suffer damage [172]. Ceramic counterfaces have advantages here.

The specific wear rate of UHMWPE from retrieved UHMWPE actabular cups has been found to be in the range of 0.9×10^{-6} to 7.2×10^{-6} $mm^3 N^{-1} m^{-1}$ [173]. This study was conducted over a period of 16 years with 25 explanted actabular cups. One of the important objectives of many simulated laboratory tests for TJR is to provide a wear phenomenon *in vitro* that matches the *in vivo* clinical trials. It is extremely difficult to conduct any material selection or improvement tests unless the laboratory test exactly simulates, with confidence, the real human body conditions. Major successes have been achieved in this regard and today there are several commercial hip and knee simulated wear testers as well as ASTM standards for the friction and wear tests for total joint prostheses [174]. Obviously, the laboratory test must simulate the lubrication system, stress cycles and orientations, presence of third body, physiological environment and the human body temperature. Because of the difficulty in matching all of these parameters in a laboratory test, often pin-on-disc type of test apparatus is used with a very common lubricant, bovine serum, for the initial material screening purpose only. Despite several successes in simulating the wear of hip and knee joints, major inconsistencies in the laboratory measured values of wear are common [175].

Several studies have been carried out in order to improve the current materials for reducing wear and debris generation. Saikko [176] found that wear rate of UHMWPE is lower when slid against Al_2O_3 and ZrO_2 by two orders of magnitude than when slid against Co–Cr stainless steel. Si_3N_4 gives wear at an order lower when compared to Co–Cr stainless steel counterfaces. The tests were conducted in deionized water on a pin-on-disc type simulator. Average wear factors ($mm^3\,N^{-1}\,m^{-1}$) for UHMWPE in Saikko's tests were 1×10^{-7}, 3.3×10^{-9}, 2.6×10^{-9} and 2.5×10^{-8} for Co–Cr stainless steel, Al_2O_3, ZrO_2 and Si_3N_4 counterfaces, respectively, whereas the measured coefficients of friction were in the ranges of 0.1, 0.10–0.12, 0.12–0.15 and 0.09–0.18, respectively. It was concluded from this study that ceramics or ceramic coatings on metals are viable alternatives for the femoral head component. Derbyshire *et al.* [177] have found that the hardening of the femoral head can help in the reduction of wear of UHMWPE as long as the surface roughness, R_a, is kept in the range of 0.01–0.02 μm.

There have been some (probably ill-conceived) attempts to introduce polymer composites for better wear resistance of the acetabular cup. The main problem with polymer composites in lubricated sliding contact is the weakening of the matrix–fibre bond due to the presence of synovial fluid. Further, glass fibre or even carbon fibre is generally not recommended due to their abrasive nature if fragments of glass fibre or carbon fibre are released as debris material [138]. Davim and Marques [109] found that under water-lubricated environment and fixed conditions of contact stress (5 MPa) and sliding velocity ($0.5\,m\,s^{-1}$ at 30-mm radius of the disk) on a pin-on-disc type apparatus, PEEK/30% carbon fibre composite gave wear rate which was lower by about two orders of magnitude compared to UHMWPE. Virgin PEEK had higher wear by roughly about one order when compared to that of UHMWPE. The measured specific wear rates ($mm^3\,N^{-1}\,m^{-1}$) for UHMWPE, neat PEEK and PEEK/30% carbon fibre composites were 1.5×10^{-6}, 2.2×10^{-5} and 6.8×10^{-8} (wear units), respectively. Suh *et al.* [119] have found that a fabric or fibre reinforced homocomposite of UHMWPE has much lower wear rate compared to normal UHMWPE in dry sliding and slightly lower in bovine serum lubricated sliding test when all other parameters are kept constant.

The other facet of the work on UHMWPE for TJR is on the improvement of the lubrication system for the wear simulation and their actual use in patients. Widmer *et al.* [152] have claimed that an increase in the hydrophilicity of UHMWPE surface would enhance the adsorption of the proteins present in human serum albumin (HSA), thus making a denser boundary layer bringing beneficial effect on the wear. These authors have shown that O_2 plasma treatment, which increases the hydrophilicity of UHMWPE surface, can decrease the coefficient of friction of UHMWPE sliding against Al_2O_3 discs in a pin-on-disc apparatus by about half when the experiments are carried out in Ringer's solution containing HSA. Saikko and Ahlroos [178] have studied the behaviour of phospholipids as boundary lubricants. The final results are still inconclusive; however, the work does show that increasing phospholipids with further addition of cholesterols (liquid crystals [179]) may provide an improved boundary lubricant for both experimental and actual joint application purposes. There is the question as to what continues to work when these surface modifications are worn out.

The above essay provides a very brief summary of the current research in the area of the application of UHMWPE for total hip and knee joint replacements. So far, no alternative organic polymeric material to the UHMWPE has been found; however, a great deal of research is continuing on the improvement of the UHMWPE's tribological properties and the overall joint lubrication system. Another major hurdle in this kind of research is the long time that takes to perform any clinical trial even when a tribologically improved material or lubricant system has been found. An up-to-date compendium of papers on this topic has recently appeared in Wear (Vol. 259, 2005, various papers from page 882–1011).

10.7 Concluding Remarks

This chapter has surveyed only briefly a rather complex subject of the tribology of polymers and their composites. Based on data available in current literature and their analysis some conclusions can be drawn.

The wear and friction properties of polymers are influenced by three major groups of parameters. The first group includes sliding contact mechanical and geometric conditions such as the surface roughness and the contact kinematics. The second group incorporates the bulk mechanical properties of the polymer and how these mechanical properties change with temperature and environmental conditions. The third parameter group, which is in fact defined by the first two groups, involves the role and properties of the 'third body', the transfer film and loose degraded polymer particulates. The wear mechanism and its magnitude are defined by the contact conditions, the mechanical properties of the bulk polymer, and these parameters lead to the subsequent events of transfer film formation of interface modification. The wear classification based on generic scaling and phenomenological and material response approaches do account for the different parameters in discrete ways and thus make the study of polymer wear readily described.

Wear mechanism of polymers under known contact conditions and the mechanical properties of the polymers are now fairly understood within the general framework of mechanics and material science. However, the theoretical *a priori* prediction of polymer friction, let alone wear, is still far from being resolved. On a limited scale, adhesive interactions (interfacial) can be modelled by the Bowden and Tabor approach, while abrasive (cohesive) interaction is correlated using the Ratner–Lancaster relationship and some other relationships available in the literature. Major problems in the wear prediction are the undefined roles and the mechanical properties of the third body and the fact that polymers show a variety of mechanical responses under a slight variation in the thermal or mechanical stress conditions. Besides, there is always an overlap of different wear mechanisms in any particular wear process. Therefore, accounting for every wear mechanism without proper knowledge of the extent to which these mechanisms are active in a wear process adds numerous uncertainties to a wear predictive model. For polymer composites, the problem is further complicated by factors such as the filler–matrix bonding properties and the role of the fillers in the third body. Empirical wear relations and wear maps have been successfully developed and used for the practical purpose of material and machine component designs. Despite limited theoretical predictive capabilities for the wear of polymers, this fascinating class of materials has been industrially employed in a very large number of tribological applications.

Lubricated wear of polymers is a very complex phenomenon and almost no theoretical model is currently available in the literature for predicting wear. Friction may sometimes be predicted using isoviscous elasto-hydrodynamic lubrication models for elastomers. Experiments have shown that polymer wear in the presence of lubricants or any environmental fluid will depend primarily upon the interaction between the fluid phase and the polymer. Polymer plasticization, fluid sorption into the polymer surface, alteration of the adhesion properties of the counterface-transfer film and changes in the polymer bulk and polymer–counterface junction mechanical properties are the important aspects of the lubricated sliding of polymers. Generally, except in cases where there is sorption of the lubricant molecules by the polymer surface, the rate of polymer wear has often been reported to be higher in the presence of an external fluid. For polymer composites, there is possibility of lubricant fluid seeping into the bulk whereby weakening the interfacial bonding between the matrix and the filler. This eventually increases wear of the polymer composites in these media.

Acknowledgements

This material has been based upon an article that first appeared in the *Journal of Engineering Tribology* – Proceedings Part J, 2002, Vol. 216, No. J6, ISSN 1350–6501, published by Professional Engineering Publishing. Permission is granted by the Institution of Mechanical Engineers.

References

1. Bowden, F.P. and Tabor, D., '*The Friction and Lubrication of Solids, Part II*', Clarendon Press, Oxford, 1964.
2. Briscoe, B.J. and Tabor, D., 'Friction and Wear of Polymers', Chapter 1, in *Polymer Surfaces* (eds D.T. Clark and J. Feast), John Wiley and Sons, New York 1978, pp. 1–23.
3. Briscoe, B.J. and Sinha, S.K., 'Wear of Polymers', *Proceedings of the Institution of Mechanical Engineers, Part J: Journal of Engineering Tribology*, **216**, 2002, 401–413.
4. Greenwood, J.A. and Tabor, D. 'The Friction of Hard Sliders on Lubricated Rubber: The Importance of Deformation Losses', *Proceedings of the Physical Society*, **71**, 1958, 989–1001.
5. Briscoe, B.J., 'Slippery Customers; Sticky Problems', in *Imperial College Inaugural Lecture in 'Material Science and Material Engineering*' (ed D.W. Pashley), Imperial College Press, London 2001.
6. Briscoe, B.J., 'Interface Friction of Solids', in *Indo-UK Forum on Solid-Solid Interactions* (eds B.J. Briscoe, M.J. Adams and S.K. Biswas), Imperial College Press, London, 1996.
7. Schallamach, A., 'How Does Rubber Slide?', *Wear*, **17**, 1971, 301–312.
8. Briscoe, B.J. and Tabor, D., 'Shear Properties of Thin Organic Films', *Symposium on Lubricant Properties, American Chemical Society*, **21**, 1976, 10–25.
9. Briscoe, B.J. and Tabor, D., 'Rheology of Thin Organic Films', *ASLE Transactions*, **17**, 1974, 158–165.
10. Kraghelsky, I.V. and Sabelnikov, V.P., Experimental Check of Elementary Laws of Friction', Paper 7, Proceedings of the Institution of Mechanical Engineers, Conference on Lubrication and Wear, London, 1957, pp. 247–251.
11. Adams, M.J., Briscoe, B.J. and Wee, T.K., 'The Differential Friction Effect of Keratin Fibres', *Journal of Physics D Applied Physics*, **23**, 1990, 406–428.
12. Godet, M., 'The Third-Body Approach: A Mechanical View of Wear', *Wear*, **100**, 1984, 437–452.
13. Lancaster, J.K., 'Relationship Between the Wear of Polymers and Their Mechanical Properties', Tribology Conv. 1969, Institution of Mechanical Engineers, London, 1969, pp. 100–108.
14. Ratner, S.N., Farberoua, I.I., Radyukeuich, O.V. and Lure, E.G., 'Correlation Between Wear Resistance of Plastics and Other Mechanical Properties', *Soviet Plastics*, **7**, 1964, 37–45.
15. Briscoe, B.J., 'Wear of Polymers: An Essay on Fundamental Aspects', *Tribology International*, August, 1981, 231–243.
16. Lancaster, J.K., 'Friction and Wear', Chapter 14, *Polymer Science, A Materials Science Handbook* (ed A.D. Jenkins), North-Holland Publishing Company, Amsterdam, 1972.
17. Wang, A., Sun, D.C., Stark, C. and Dumbleton, J.H., 'Wear Mechanism of UHMWPE in Total Joint Replacement', *Wear*, **181–183**, 1995, 241–249.
18. Kar, M.K. and Bahadur, S., 'The Wear Equation for Unfilled and Filled Polyoxymethylene', *Wear*, **30**, 1974, 337–348.
19. Viswanath, N. and Bellow, D.G., 'Development of an Equation for the Wear of Polymers', *Wear*, **181–183**, 1995, 42–49.
20. Pooley, C.M. and Tabor, D., 'Friction and Molecular Structure: The Behaviour of Some Thermoplastics', *Proceedings of the Royal Society of London*, **A 329**, 1972, 251–274.
21. Tanaka, K. and Miyata, T., 'Studies on the Friction And Transfer Of Semi-Crystalline Polymers', *Wear*, **41**, 1977, 383.
22. Hooke, C.J., Kukureka, S.N., Liao, P., Rao, M. and Chen, Y.K., 'The Friction and Wear of Polymers in Non-Conformal Contacts', *Wear*, **200**, 1996, 83–94.

23. McNicol, A., Dowson, D. and Davies, M., 'The Effect of Humidity and Electrical Fields upon the Wear of High Density Polyethyle and Polytetrafluoroethyle', *Wear*, **181–183**(2), 1995, 603–612.
24. Steijn, R.P., 'The Sliding Surface of Polytetrafluoroethylene: An Investigation with the Electron Microscope', *Wear*, **12**, 1968, 193–212.
25. Eriksson, M. and Jacobson, S., 'Tribological Surfaces of Organic Brake Pads', *Tribology International*, **33**(12), 2000, 817–827.
26. Zhang, S.W., 'State of the Art of Polymer Tribology', *New Directions in Tribology* (ed I. Hutchings), Mechanical Engineering Publications Ltd., London, 1997, pp. 469–481.
27. Briscoe, B.J. and Tweedale, P.J., 'Aramid Fibre Friction: A Replacement for Asbestos in High Friction Materials', Proceedings of the American Society of Materials Conference on 'Tribology of Composite Materials', Oak Ridge, Tennessee, May 1990.
28. Shipway, P.H. and Ngao, N.K., 'Microscale Abrasive Wear Of Polymeric Materials', *Wear*, **255**, 2003, 742–750.
29. Archard, J.F., 'Contact and Rubbing of Flat Surfaces', *Journal of Applied Physics*, **24**, 1953, 981.
30. Budinski, K.G., 'Resistance to Particle Abrasion of Selected Plastics', *Wear*, **203–204**, 1997, 302–309.
31. ASTM D2240-00 Standard Test Method for Rubber Property – Durometer Hardness.
32. Briscoe, B.J. and Sinha, S.K., 'Scratch Resistance and Localized Damage Characteristics of Polymer Surfaces – A Review', *Materialwissenschaft und Werkstofftechnik*, **34**(10), 2003, 989–1002.
33. Adams, M.J., Allan, A., Briscoe, B.J., Doyle, P.J., Gorman, D.M. and Johnson, S.A., 'An Experimental Study of the Nano-Scratch Behaviour of Poly(Methyl Methacrylate)', *Wear*, **251**, 2001, 1579–1583.
34. Wong, B.K.P., Sinha, S.K., Tan, J.P.Y. and Zeng K.Y., 'Nano-Wear Mechanism for Ultra-High Molecular Weight Polyethylene (UHMWPE) Sliding Against a Model Hard Asperity', *Tribology Letters*, **17**(3), 2004, 613–622.
35. Evans P., Abrasion of Polymers in Fluids Ph.D. thesis, Imperial College, London, UK, 1987.
36. Lewis, R.B., *Mechanical Engineering*, **86**, 1964, 33 (Ref. [78] in Steijn, R.P., *Friction and Wear in Failure of Plastics* (ed W. Brostow and R.D Cornelivssen), Hanser Publishers, Munich 1986, pp. 357–392).
37. Rhee, S.K., 'Wear Equation for Polymers Sliding Against Metal Surfaces', *Wear*, **16**, 1970, 431–445.
38. Smurugov, V.A., Senatrev, A.I., Savkin, V.G., Biran, V.V. and Sviridyonok, A.I., 'On PTFE Transfer and Thermoactivation Mechanism of Wear', *Wear*, **158**, 1992, 61–69.
39. Brainhard, W.A. and Buckley, D.H., 'Adhesion and friction of PTFE in contact with Metals as Studies by Augur Spectroscopy, Field Ion and Scanning Electronic Microscopic', *Wear*, **26**, 1975, 75–93.
40. Briscoe, B.J., Pogosian, A.K. and Tabor, D., 'The Friction and Wear of High Density Polyethylene: The Action of Lead Oxide and Copper Oxide Fillers', *Wear*, **27**, 1974, 19–34.
41. Pratt G.C., 'Recent Developments in Polytetrafluoroethylene–Based Dry Bearing Materials and Treatments', *Transactions of the Plastics Institution*, **32**, 1964, 255–271.
42. Jiantang, G., 'Tribochemical Effects in Formation of Polymer Transfer Film', *Wear*, **245**, 2000, 100–106.
43. Pocock, G. and Cadman, P., 'The Application of Differential Scanning Calorimetry and Electron Spectroscopy of PTFE – Metals Reactions of Interest in Dry Bearing Technology', *Wear*, **37**, 1976, 129–141.
44. Pepper, S.V., 'Auger Analysis of Films Formed on Metals in Sliding Contact with Halogenated Polymers', *Journal of Applied Physics*, **45**(7), 1976, 2949–2956.
45. Bahadur, S., 'The Development of Transfer Layers and Their Role in Polymer Tribology', *Wear*, **245**(1–2), 2000, 92–99.
46. Vincent, L., Berthier, Y., Dobourg, M.C. and Godet, M., 'Mechanics and Materials in Fretting', *Wear*, **153**, 1992, 135–148.
47. Briscoe, B.J., Chateauminois, A., Lindley, T.C. and Parsonage, D., 'Fretting Wear Behaviour of Polymethylmethacrylate Under Linear Motions and Torsional Contact Conditions', *Tribology International*, **31**(11), 1998, 701–711.
48. Briscoe, B.J., Chateauminois, A., Lindley, T.C. and Parsonage, D., 'Contact Damage of Poly(Methylmethacrylate) During Complex Microdisplacements', *Wear*, **240**, 2000, 27–39.
49. Dahmani, N., Vincent, L., Vannes, B., Berthier, Y. and Godet, M., 'Velocity Accommodation in Polymer Fretting', *Wear*, **158**, 1992, 15–28.
50. Suh, N.P., 'The Delamination Theory of Wear', *Wear*, **25**, 1973, 111–124.
51. Tsukamoto, N., Taki, T. and Nishida, N., 'A Study on Strength Design Methods of Plastic Gears', *Transactions of the Japanese Society of Mechanical Engineers*, **53**(486C), 1987, 432–435.
52. Stolarski, T.A., 'Rolling Contact Fatigue of Polymers and Polymer Composites', *Advances in Composite Tribology* (ed K. Friedrich), Elsevier Science Publishers B.V., Amsterdam, 1993, pp. 629–667.
53. Stolarski, T.A., 'Tribology of Polyetheretherketone', *Wear*, **158**, 1992, 71–78.

54. Eiss Jr, N.S. and Potter, J.R., 'Fatigue Wear of Polymers. Polymer Wear and Its Control', *ACS Symposium Series*, **287**, 1985, 59.
55. Kurokawa, M., Uchiyama, Y. and Nagai, S., 'Performance of Plastic Gear Made of Carbon Fiber Reinforced Poly-ether-ether-ketone: Part 2', *Tribology International*, **33**, 2000, 715–721.
56. Lawrence, C.C. and Stolarski, T.A., 'Rolling Contact Wear of Polymers: A Preliminary Study', *Wear*, **132**, 1989, 183–191.
57. Barquins, M., 'Adherence, Friction and Wear of Rubber-Like Materials', *Wear*, **158**, 1992, 87–117.
58. Arnold, J.C. and Hutchings, I.M., 'The Mechanisms of Erosion of Unfilled Elastomers by Solid Particle Impact', *Wear*, **138**, 1990, 33–46.
59. Schallamach, A., 'Abrasion of Rubber by a Needle', *Journal of Polymer Science*, **9**(5), 1952, 385–404.
60. Grosch, K.A., 'The Relation Between the Friction and Visco-Elastic Properties of Rubber', *Proceedings of the Royal Society of London*, **A274**, 1963, 21–39.
61. Ratner, S.B. and Lure, E.G., 'Abrasion of Polymers as a Kinetic Thermo-Activation Process', *Dokl. Akad. Nauk SSSR*, **166**(4), 1966, 909.
62. Barquins, M. and Roberts, A.D., 'Rubber Friction Variation with Rate and Temperature: Some New Observations', *Journal of Physics D: Applied Physics*, **19**, 1986, 547–563.
63. Gent, A.N. and Pulford, C.T.R., 'Wear of Steel by Rubber', *Wear*, **49**, 1978, 135–139.
64. Uchiyama, Y. and Ishino, Y., 'Pattern Abrasion Mechanism of Rubber', *Wear*, **158**, 1992, 141–155.
65. Aharoni, S.M., 'Wear of Polymers by Roll Formation', *Wear*, **25**, 1973, 309–327.
66. Kraghelskii, I.V. and Nepomnyashchii, 'Fatigue Wear Under Elastic Contact Conditions', *Wear*, **8**, 1965, 303.
67. Moore, D.F., '*The Friction and Lubrication of Elastomers*', Pergamon Press Ltd., Oxford, 1972.
68. Zhang S.W., '*Wear of Elastomers*', Petroleum University Press, Dongying, Shandong Province, PRC.
69. Denton, R. and Keshavan, M.K. (eds), 'Wear and Friction of Elastomers', ASTM Special Technical Publication 1145, 1992.
70. Tanaka, K., Uchiyama, Y. and Toyooka, S., 'The Mechanism of Wear of Polytetrafluoroethylene', *Wear*, **23**, 1973, 153–172.
71. Uchiyama, Y. and Tanaka, K., 'Wear Law for Polytetrafluoroethylene', *Wear*, **58**, 1980, 223–235.
72. Moore, D.F. and Geyer, W., 'A Review of Hysteresis Theories for Elastomers', *Wear*, **30**, 1974, 1–34.
73. Liu, T. and Rhee, S.K., 'High Temperature Wear of Semi-Metallic Disk Brake Pads', *Wear of Materials 1977*', The ASME, New York, 1977, pp. 552–554.
74. Rhee, S.K. and Thesier, P.A., 'Effects of Surface Roughness of Brake Drums on Coefficient of Friction and Lining Wear', Society of Automotive Engineers (SAE) Paper 720449, 1971.
75. Briscoe, B.J., Ramirez, I. and Tweedle, P.J., Proceedings of the International Conference on Disc Brakes for Commercial Vehicles, London, 1–2 November, Mechanical Engineering Publication Ltd., 1988, p. 5.
76. Sinha, S.K. and Biswas, S.K., 'Friction and Wear Behaviour of Continuous Fibre as Cast Kevlar-Phenolic Resin Composite', *Journal of Materials Science*, **27**, 1992, 3085–3091.
77. Sinha, S.K. and Biswas, S.K., 'Effect of Sliding Speed on Friction and Wear of Uni-Directional Aramid Fibre-Phenolic Resin Composite', *Journal of Materials Science*, **30**, 1995, 2430–2437.
78. Kato, T. and Magario, A., 'The Wear of Aramid Fibre Reinforced Brake Pads: The Role of Aramid Fibres', *STLE Tribology Transactions*, **37**(3), 1994, 559–565.
79. Lancaster, J.K. and Giltrow, J.P., 'The Role of the Counterface Roughness in the Friction and Wear of Carbon Fibre Reinforced Thermosetting Resins', *Wear*, **16**, 1970, 357–374.
80. Lancaster, J.K., 'Polymer Based Bearing Materials: The Role of Fillers and Fibre Reinforcement', *Tribology*, **5**, 1972, 249–255.
81. Freidrich, K., '*Friction and Wear of Polymer Composites*', Elsevier, Amsterdam, 1986.
82. Freidrich, K., '*Advances in Composite Tribology*', Elsevier, Amsterdam, 1993.
83. Briscoe, B.J., Tweedale, P. and Lin H.Y., 'The Ultimate Power Dissipation Capacity of PTFE/PEEK Composite Bearings', Eurotribology, Lyon, Paper 5.17, Elsevier Scientific Publications, Amsterdam, 1985.
84. Briscoe, B.J., Stuart, B.H., Sebastian, S. and Tweedale, P.J., 'The Failure of Poly(ether ether ketone) in High Speed Contacts', *Wear* (Special Issue), **162–164**, 1993, 407–417.
85. Briscoe, B.J., 'Isolated Contact Stress Deformations of Polymers; The Basis for Interpreting Polymer Tribology', World Tribology Congress (ed I.M. Hutchings), Institution of Mechanical Engineers, London, 1997.
86. Bell, J.C. and Willemse, P.J., 'The Development of Scuffing Failures in an Innovative Valve Train System', World Tribology Congress, London, 1997.
87. Briscoe, B.J., 'The Friction of Polymers: A Short Review', in *Friction and Traction* (eds D. Dowson, M. Godet, C.M. Taylor and D. Berthe), Westbury House, IPC Press, Guildford, 1981, pp. 81–93.

88. Briscoe, B.J., Stuart, B.H., Thomas, P.S. and Williams, D.R., 'A Comparison of the Thermal and Solvent Induced Relaxation of Poly(etheretherketone) Using FT Raman Spectroscopy', *Spectrochemica Acta*, **47A**, 1991, 1299–1303.
89. Rubenstein, C., 'Lubrication of Polymers', *Journal of Applied Physics*, **32**, 1961, 1445.
90. Zhang, R., Hager, A.M., Friedrich, K., Song, Q. and Dong, Q., 'Study on Tribological Behaviour of Plasma-Treated PEEK and Its Composites', *Wear*, **181–183**, 1995, 613–623.
91. Briscoe, B.J., Pelillo, E. and Sinha, S.K., 'Characterisation of the Scratch Deformation Mechanisms for Poly(methylmethacrylate) Using Surface Optical Reflectivity', *Polymer International*, **43**, 1997, 359–367.
92. Gauthier, C., Lafaye, S. and Schirrer, R., 'Elastic Recovery of a Scratch in a Polymeric Surface: Experiments and Analysis', *Tribology International*, **34**(7), 2001, 469–479.
93. Kaneko, R., 'Microtribological Applications of Probe Microscopy', *Tribology International*, **28**(3), 1995, 195–202.
94. Jardret, V., Zahouani, H., Loubet, J.L., and Mathia, T.G., 'Understanding and Quantification of Elastic and Plastic Deformation During a Scratch Test', *Wear*, **218**(1), 1998, 8–14.
95. Briscoe, B.J., Evans, P.D., Pelillo, E. and Sinha, S.K., 'Scratching Maps for Polymers', *Wear*, **200**, 1996, 137–147.
96. Bhushan, B. and Wilcock, D.F., 'Wear Behavior of Polymeric Compositions in Dry Reciprocating Sliding', *Wear*, **75**, 1982, 41–70.
97. Neale, M.J. (ed), '*Tribology Handbook*', Butterworth and Co. Ltd., London, 1973, A4.
98. Ludema, K.C. and Tabor, D., 'The Friction and Visco-Elastic Properties of Polymeric Solids', *Wear*, **9**, 1966, 329–348.
99. Tanaka, K., 'Transfer of Semi-Crystalline Polymers Sliding Against Smooth Steel Surfaces', *Wear of Materials 1981* (eds S.K. Rhee, A.W. Ruff and K.C. Ludema), ASME, 1981, pp. 98–106.
100. Czichos, H., 'Friction Regimes and Film Transfer of Polymers Sliding Against Various Materials', *The Wear of Non-Metallic Materials*, Mechanical Engineering Publications Ltd., London, 1978, pp. 285–287.
101. Briscoe, B.J. and Tabor, D., 'Self Lubricating Polythenes', Colloques International du CNRS, No. 233, Polymeres et Lubrification, pp. 425–431.
102. Stolarski, T.A., 'The Influence of Combined Linear and Rotating Motions on the Wear of Polymers', Diploma of Imperial College Thesis, Imperial College, London, 1979.
103. Briscoe, B.J. and Stolarski, T.A., 'The Influence of Combined Motions on the Sliding of Polymers', *Nature*, **281**, 1979, 206–208.
104. Sinha, S.K., 'Wear Failures of Plastics', *ASM Handbook, Vol. 11 Failure Analysis and Prevention*, ASM International, Ohio, 2002, pp. 1019–1027.
105. Evans, D.G. and Lancaster, J.K., 'The Wear of Polymers', *Wear, 13, Materials Science and Technology* (ed D. Scott), Academic Press, New York, 1979, pp. 85–139.
106. Friedrich, K., Lu, Z. and Hager, A.M., 'Recent Advances in tribology', *Wear*, **190**(2), 1995, 139–144.
107. van de Velde, F. and De Baets, P., 'The Friction and Wear Behaviour of Polyimide 6 Sliding Against Steel at Low Velocity Under Very High Contact Pressure', *Wear*, **209**, 1997, 106–114.
108. Bahadur, S. and Polineni, V.K., 'Tribological Studies of Glass Fabric-Reinforced Polyamide Composites Filled with Cuo and PTFE', *Wear*, **200**, 1996, 95–104.
109. Davim, J.P. and Marques, N., 'Evaluation of Tribological Behaviour of Polymeric Materials for Hip Prostheses Application', *Tribology Letters*, **11**(2), 2001, 91–94.
110. Bahadur, S. and Gong, D., 'The Action of Fillers in the Modification of the Tribological Behaviour of Polymers', *Wear*, **158**, 1992, 41–59.
111. de Silva, C.H., Tanaka, D.K. and Sinatora, A., 'The Effect of Load and Relative Humidity on Friction Coefficient Between High Density Polyethylene on Galvanized Steel – Preliminary Study', *Wear*, **225–229**, 1999, 339–342.
112. Rhee, S.K. and Ludema, K.C., 'Transfer Films and Severe Wear of Polymers', *The Wear of Non-Metallic Materials* (eds D. Dowson, M. Godet and C.M. Taylor), Mechanical Engineering Publications Ltd., London, 1978, pp. 11–17.
113. Biswas, S.K. and Vijayan, K., 'Friction and Wear of PTFE – A Review', *Wear*, **158**(1–2), 1992, 193–211.
114. Wang, A., Essner, A., Polineni, V.K., Sun, D.C., Stark, C. and Dumbleton, J.H., 'Lubrication and Wear of Ultra-High Molecular Weight Polyethylene in Total Joint Replacements', *New Directions in Tribology* (ed I. Hutchings), Mechanical Engineering Publications Ltd., London, 1997, pp. 443–458.
115. Briscoe, B.J., Evans, P.D. and Lancaster, J.K., 'Single Point Deformation and Abrasion of γ-Irradiated Poly(tetrafluoroethylene)', *Journal of Physics D: Applied Physics*, **20**, 1987, 346–353.

116. Briscoe, B.J. and Steward, M.D., 'The Effect of Carbon and Glass Fillers on the Transfer Film Behaviour of PTFE Composites', Tribology 1978, Institution of Mechanical Engineers, 1978, pp. 19–23.
117. Blanchet, T.A., Kennedy, F.E. and Jayne, D.T., 'XPS Analysis of the Effect of Fillers on PTFE Transfer Film Development in Sliding Contacts', *Tribology Transactions*, **36**, 1993, 535–544.
118. Liu, C., Ren, L., Arnell, R.D. and Tong, J., 'Abrasive Wear Behaviour of Particle Reinforced Ultrahigh Molecular Weight Polyethylene Composites', *Wear*, **225–229**, 1999, 199–204.
119. Suh, N.P., Arinez, J. and Mosleh, M., 'Tribology of Polyethylene Homocomposites', *Wear*, **214**(2), 1998, 231–236.
120. Hashmi, S.A.R., Neogi, S., Pandey, A. and Chand, N., 'Sliding Wear of PP/UHMWPE Blends: Effect of Blend Composition', *Wear*, **247**(1), 2001, 9–14.
121. Yu, L. and Bahadur, S., 'An Investigation of the Transfer Film Characteristics and the Tribological Behaviours of Polyphenylene Sulfide Composites in Sliding Against Tool Steel', *Wear*, **214**, 1998, 245–251.
122. Briscoe, B.J., 'Tribology of Polymers: State of an Art', *Physicochemical Aspects of Polymers* (ed K.L. Mittal), Plenum Press, NY, 1983, p. 387.
123. Clerico, M. and Patierno, V., 'Sliding Wear of Polymeric Composites', *Wear*, **53**, 1979, 279–301.
124. Dowson, D., Challen, J.M., Holmes, K. and Atkinson, J.R., 'The Influence of Counterface Roughness on the Wear Rate of Polyethylene', *The Wear of Non-Metallic Materials*, Mechanical Engineering Publications Ltd, London, 1978, pp. 99–105.
125. Eiss, N. and Bayraktaroglu, M.M., 'The Effect of Surface Roughness on the Wear of Low Density Polyethylene', *ASLE Transactions*, **23**(3), 1980, 269–278.
126. Ovaert, T.C. and Cheng, H.S., 'Counterface Topographical Effect on the Wear of Polyetheretherketon and a Polyetheretherketon-Carbon Fibre Composite', *Wear*, **150**, 1991, 275–287.
127. Steward, M.P., 'Friction and Wear of PTFE Composites', Ph.D. thesis, University of Cambridge, UK, 1978.
128. Briscoe, B.J., Steward, M.D. and Groszek, A., 'The Effect of Carbon Aspect Ratio on the Friction and Wear of PTFE', *Wear*, **42**, 1977, 99–108.
129. Briscoe, B.J., 'The Tribology of Composites Materials: A Preface', *Advances in Composite Tribology* (ed K. Friedrich), Elsevier Science Publishers B.V., 1993, pp. 3–15.
130. Song, J. and Ehrenstein, G.W., 'Friction and Wear of Self-Reinforced Thermoplastics', *Advances in Composite Tribology* (ed K. Friedrich), Elsevier Science Publishers B.V., Amsterdam, 1993, pp. 19–63.
131. Schelling, A. and Kausch, H.H., 'Reciprocating Dry Friction and Wear of Short Fibre Reinforced Polymer Composites', *Advances in Composite Tribology* (ed K. Friedrich), Elsevier Science Publishers B.V., 1993, pp. 65–105.
132. Hager, A.M. and Davies, M., 'Short-Fibre Reinforced, High-Temperature Resistant Polymers for a Wide Field of Tribological Applications', *Advances in Composite Tribology* (ed K. Friedrich), Elsevier Science Publishers B.V., Amsterdam, 1993, pp. 107–157.
133. Odi-Owei, S. and Schipper, D.J., 'Tribological Behaviour of Unfilled and Composite Polyoxymethylene', *Wear*, **148**, 1991, 363–376.
134. Briscoe, B.J., Mustafaev, V. and Tabor, D., 'Lubrication of Polythene by Olemide and Stearamide', *Wear*, **19**, 1972, 399.
135. Briscoe, B.J., *Friction and Wear of Polymer Composites* (ed K. Friedrich), Chapter 2, Elsevier, Amsterdam, 1986.
136. Kheldkar, J., Negulescu, I. and Meletis, E., 'Sliding Wear Behaviour of PTFE Composites', *Wear*, **252**, 2002, 361–369.
137. Hanmin, Z., Guoren, H. and Guicheng, Y., 'Friction and Wear of Poly(phenylene Sulphide) and Its Carbon Fibre Composites: I Unlubricated', *Wear*, **116**, 1987, 59–68.
138. Friedrich, K., Lu, Z. and Hager, A.M., 'Overview of Polymer Composites for Friction and Wear Applications', *Theoretical and Applied Fracture Mechanics*, **191**, 1993, 1–12.
139. Bijwe, J., John Rajesh, J., Jeyakumar, A., Ghosh, A. and Tewari, U.S., 'Influence of Solid Lubricants and Fibre Reinforcement on Wear Behaviour of Polyethersulphone', *Tribology International*, **33**, 2000, 697–706.
140. Dowson, D. and Swales, P.D., An Elastohydrodynamic Approach to the Problem of the Reciprocating Seal, Proceedings of the 3rd International Conference on Fluid Sealing, Cranfield: BHRA, paper F3.
141. Field, G.J. and Nau, B.S., 'Theoretical Study of EHL of Reciprocating Rubber Seals', *ASLE Transactions*, **18**, 1974, 48–54.
142. Richards, S.C. and Roberts, A.D., 'Boundary Lubrication of Rubber by Aqueous Surfactant', *Journal of Physics D: Applied Physics*, **25**, 1992, A76–A80.

143. Roberts, A.D. and Tabor, D., 'Extrusion of Liquids Between Highly Elastic Solids', *Proceedings of the Royal Society of London*, **A 325**, 1971, 323–345.
144. McClune C.R. and Briscoe, B.J., 'EHL Films Formed Between a Rubber Cylinder and Glass Plate: Comparison of Theory and Experiment', *Journal of Physics D: Applied Physics*, **10**, 1977, 587–598.
145. Briscoe, B.J. and Stolarski, T.A., 'The Frictional Characteristics of Reciprocating Polymer Bushes in Fluid Media', *Wear*, **112**, 1986, 371–394.
146. Briscoe, B.J., Stolarski, T.A. and Davies, G.J., 'Boundary Lubrication of Thermoplastic Polymers in Model Fluids', *Tribology International*, **17**(3), 1984, 129–137.
147. Briscoe, B. and Kreminitzer, S.L. 'A Study of the Friction and Adhesion of Polyethylene-Terephthalate Monofilaments', *Journal of Physics D: Applied Physics*, **12**, 1979, 505–516.
148. Tanaka, K., *Journal of Lubrication Technology*, **102**, 1980, 526–533.
149. Dowson, D., Diab, M.M., Gillis, B.J. and Atkinson, J.R., in Proceedings of the American Chemical Society, 287 – Polymer Wear and Its Control, ACS, New York, pp. 172–187.
150. Cohen, S.C. and Tabor, D., 'The Friction and Lubrication of Polymers', *Proceedings of the Royal Society of London*, **A291**, 1966, 186–207.
151. Watanabe, M., 'Wear Mechanism of PTFE Composites in Aqueous Environments', *Wear*, **158**, 1992, 79–86.
152. Widmer, M.R., Heuberger, M., Vörös, J. and Spencer, N.D., 'Influence of Polymer Surface Chemistry on Frictional Properties Under Protein-Lubrication Conditions: Implications for Hip-Implant Design', *Tribology Letters*, **10**(1–2), 2001, 111–116.
153. Unsworth, A., 'The Effects of Lubrication in Hip Joint Prostheses', *Physics in Medicine and Biology*, **23**, 1978, 253–268.
154. Unsworth, A., Roberts, A. and Thompson, J.C., 'The Application of Soft Layer Lubrication to Hip Prostheses', *The Journal of Bone and Joint Surgery*, **633**, 1981, 297.
155. Caravia, L., Dowson, D., Fisher, J., Corkhill, P.H. and Tighe, B.J., 'Friction of Hydrogel and Polyurethane Elastic Layers when Sliding Against Each Other Under a Mixed Lubrication Regime', *Wear*, **181–183**, 1995, 236–240.
156. Senior, J.M. and West, G.H., 'Interaction Between Lubricants and Plastic Bearing Surfaces', *Wear*, **18**, 1971, 311–323.
157. Kang, C. and Eiss Jr, N.S., 'Fretting Wear of Polysiloxane–Polyimide Copolymer Coatings as a Function of Varying Humidity', *Wear*, **158**, 1992, 29–40.
158. Chitsaz-zadeh, M.R. and Eiss Jr, N.S., 'Friction and Wear of Polyimide Thin Films', *Wear*, **110**, 1986, 359–368.
159. Stachowiak, G.W., 'Friction and Wear of Polymers, Ceramics and Composites in Biomedical Applications', *Advances in Composite Tribology* (ed K. Friedrich), Elsevier Science Publishers B.V., Amsterdam, 1993, pp. 509–557.
160. Charney, J., '*Low Friction Arthroplasty of the Hip*', Springer, Berlin, 1979.
161. Dowson, D. and Wallbridge, N.C., 'Laboratory Wear Tests and Clinical Observations of the Penetration of Femoral Heads into Acetabular Cups in Total Replacement Hip Joints I: Charnley Prostheses with Polytetrafluoroethylene Acetabular Cups', *Wear*, **104**, 1985, 295–215.
162. Dumbleton, J.H., '*Tribology of Natural and Artificial Joints*', Elsevier, Amsterdam, 1981.
163. Dowson, D., 'Tribological Characteristics of Polymers with Particular Reference to Polyethylene', Chapter 19, *Polymer Surfaces* (ed D.T. Clark and J. Feast), John Wiley and Sons, New York 1978, pp. 400–424.
164. McKellop, H., Clarke, I.C., Markolf, K.L. and Amstutz, H.C., 'Wear Characteristics of UHMW Polyethylene: A Method for Accurately Measuring Extremely Low Wear Rates', *Journal of Biomedical Materials Research*, **12**, New York 1978, 895–927.
165. Christiansen, T., *Acta Chirurgica Scandinavica*, **135**, 1969, 43–46.
166. Zhu, Y.H., Chiu, K.Y. and Tang, W.M., 'Review Article: Polyethylene Wear and Osteolysis in Total Hip Arthroplasty', *Journal of Orthopaedic Surgery*, **9**(1), 2001, 91–99.
167. Harris, W.H., 'The Problem is Osteolysis', *Clinical Orthopaedics and Related Research*, **311**, 1995, 46–53.
168. Hills, B.A. and Butler, B.D., 'Surfactants Identified in Synovial Fluid and Their Ability to Act as Boundary Lubricants', *Annals of the Rheumatic Diseases*, **43**, 1984, 641–648.
169. Hills, B.A., Oligolamellar Lubrication of Joints by Surface Active Phospholipids', *The Journal of Rheumatology*, **16**, 1989, 82–91.
170. Williams III, P.F. *et al.*, 'Sliding Friction Analysis of Phosphatidylcholine as a Boundary Lubricant for Articular Cartiledge', *Journal of Engineering in Medicine*, **207**, 1993, 59–66.

171. Bartel, D.L., Bicknell, V.L. and Wright, T.W., *The Journal of Bone and Joint Surgery, American Volume*, **68**(7), 1986, 1041–1051.
172. Peterson, C.D., Hillberry, B.H. and Heck, D.A., 'Component wear of total knee prostheses using Ti-6Al-4V, TiN coated Ti-6Al-4V, and Co-Cr-Mo femoral components', *Journal of Biomedical Materials Research*, **22**, 1988, 887–903.
173. Atkinson, J.R., Dowson, D., Isaac, G.H. and Wroblewski, B.M., 'Laboratory Wear Tests and Clinical Observations of the Penetration of Femoral Heads into Acetabular Cups in Total Replacement Hip Joints III: The Measurement of Internal Volume Changes in Explanted Charnley Sockets After 2–16 Years In Vivo and the Determination of Wear Factors', *Wear*, **104**, 1985, 225–244.
174. ASTM Designation: F 732-82. Standard Practice for Reciprocating Pin-on-Flat Evaluation of Friction and Wear Properties of Polymeric Materials for Use in Total Joint Prostheses', *Annual Book of ASTM Standards*, Vol. 13.01, ASTM, Philadelphia, PA, 1989, pp. 212–217.
175. Wallbridge, N.C. and Dowson, D., 'Distribution of Wear Rate Data and a Statistical Approach to Sliding Wear Theory', *Wear*, **119**, 1987, 295–312.
176. Saikko, V., 'Wear and Friction Properties of Prosthetic Joint Materials Evaluated on a Reciprocating Pin-on-Flat Apparatus', *Wear*, **166**, 1993, 169–178.
177. Derbyshire, B., Fisher, J., Dowson, D., Hardaker, C.S. and Brummitt, K., 'Wear of UHMWPE Sliding Against Untreated, Titanium Nitride-Coated and 'Hardcor'-Treated Stainless Steel Counterfaces', *Wear*, **181–183**, 1995, 258–262.
178. Saikko, V. and Ahlroos, T., 'Phospholipids as Boundary Lubricants in Wear Tests of Prosthetic Joint Materials', *Wear*, **207**, 1997, 86–91.
179. Kupchinov, B.I. *et al.*, 'The Effect of Liquid Crystals on Joint Lubrication', *Wear*, **171**, 1994, 7–12.

11

Wear of Polymer Composites

K. Friedrich, Z. Zhang and P. Klein

Abstract

The use of polymers and polymer composites in various tribological situations has become state-of-the-art. Nevertheless, new developments are still under way to explore new fields of application for these materials and to tailor their properties for more extreme loading conditions. Some of these developments can be followed when searching through the references given at the end of this chapter. The present overview describes, in particular, some of the authors' approaches in designing polymeric composites in order to operate under low friction and low wear against steel counterparts. Special emphasis is focused on special filler (including nanoparticle) reinforced thermoplastics and thermosets. An attempt is made to predict their wear properties and to do systematic parameter studies by the use of artificial neural networks. Further information will be given on the fibre orientation dependence of wear of continuous carbon fibre/polymer matrix composites, and on attempts to predict their load-bearing capacity and related wear mechanisms by the use of finite elements. In addition, some new steps towards the development of functionally graded tribo-materials are illustrated.

11.1 Introduction

Nowadays, there are more and more applications in which friction and wear are critical issues. Polymer composites (Figure 11.1) [1] containing different fillers and/or reinforcements are frequently used for these purposes. However, how these materials must exactly be designed depends on the requirement profile of the particular application. This means the tribological characteristics, i.e. the friction coefficient and the wear resistance, are no real material properties, but depend on the system in which these materials have to function. Sometimes, a high coefficient of friction, coupled with low wear, is required (e.g. for brake pads or clutches). In most of the cases, however, it is of primary concern to develop polymeric

Wear – Materials, Mechanisms and Practice Edited by G. Stachowiak

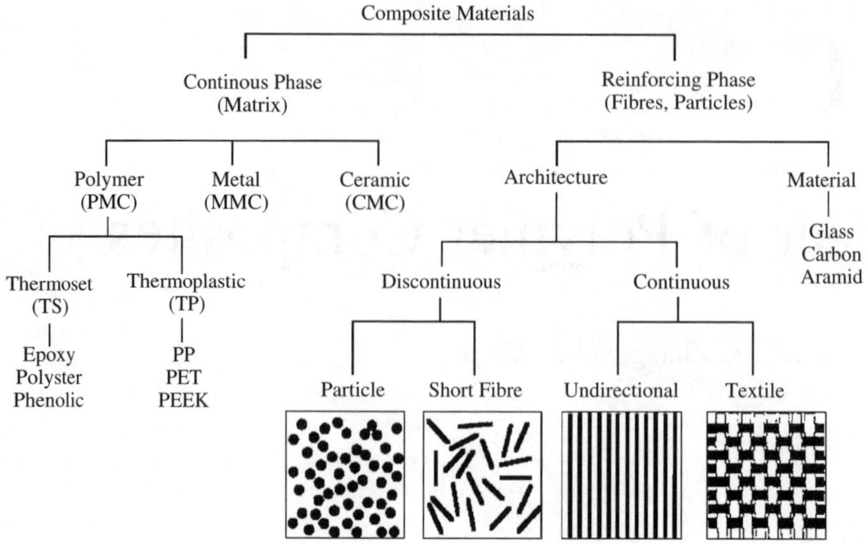

Figure 11.1 Systematic illustration of the structural components of composite materials [1]

composites that possess low friction and low wear properties under dry sliding conditions against smooth metallic counterparts (e.g. as gears or bearings). The following sections illustrate how to design such materials, using both traditional and new concepts.

To characterize the tribological behaviour of polymer materials in the laboratory, standard tests are used. The pin-on-disc test, one of the most frequently used test configurations, is described elaborately [2]. It allows determining the most important tribological property, the specific wear rate W_s of the material to be optimized, by using the equation

$$W_s = \frac{\Delta V}{F_N L} \quad [\mathrm{mm^3/N\,m}] \tag{1}$$

where ΔV is loss in volume, F_N the normal load and L the sliding distance. The inverse of the wear rate is usually referred to as the wear resistance of a material.

It should be noted here that the choice of the type of wear test configuration must be based on the tribo-technical system under consideration. The latter determines the elements of the basic structure of the tribo-system which yields information on the existing wear mechanisms (surface variations), and the loss of material (wear rates). Therefore, tribological testing of the materials under laboratory test configurations can only be considered as a helpful screening tool. The final choice of the right material combination is always dependent on the results of subsequent field tests by the use of the real structural components.

11.2 Sliding Wear of Filler Reinforced Polymer Composites

11.2.1 Short Fibres and Internal Lubricants

One of the traditional concepts for improvement of the friction and wear behaviour of polymeric materials is to reduce their adhesion to the counterpart material and to enhance their hardness, stiffness and compressive strength. This can be achieved quite successfully by using special fillers. To reduce the adhesion, internal lubricants such as polytetrafluoroethylene (PTFE) and graphite flakes are frequently incorporated. One of the mechanisms of the corresponding reduction in the coefficient of friction is the formation of a PTFE-transfer film on the surface of the counterpart [3]. Short aramid fibres (AF), glass fibres (GF) or carbon fibres (CF) are used to increase the creep resistance and the compressive strength of the polymer matrix system used [3–7]. Figure 11.2 shows in a systematic way one of the possible principles how to design the composition of wear-resistant polymer composites. Normally the matrix should possess a high temperature resistance and have a high cohesive strength. However, sometimes it is also advantageous to have a PTFE-based matrix in which a stiffer and more wear-resistant polymer phase along with other fillers provides more optimum conditions for the tribological situation under particular consideration, e.g. its use at cryogenic temperatures [8]. Additional fillers that enhance the thermal conductivity are often of great advantage, especially if effects of temperature enhancement in the contact area must be avoided in order to prevent an increase in the specific wear rate. It should also be noted that not all the fillers are of benefit to the wear performance of composites. The wear resistance is increased when fillers decompose and generate reaction products which enhance the bonding between the transfer film and the counterface [5], whereas other fillers decrease the wear resistance because they generate more discontinuities in the material. It is thus important to understand the growth, bonding and loss of transfer films, which are strongly related to the wear mechanisms. It should also be noted that chemical and mechanical interactions of transfer films are very complicated; therefore, further efforts to understand these relationships in more detail are still a subject of current and future studies [9, 10].

In spite of these discrepancies, more and more polymer composites are now being used as sliding elements that were formerly composed of metallic materials only. As an example, Table 11.1 summarizes the specific wear rate of various filler-modified and/or short fibre reinforced thermoplastic composite systems tested against steel counterparts on a block-on-ring test configuration. They have been developed for special applications in

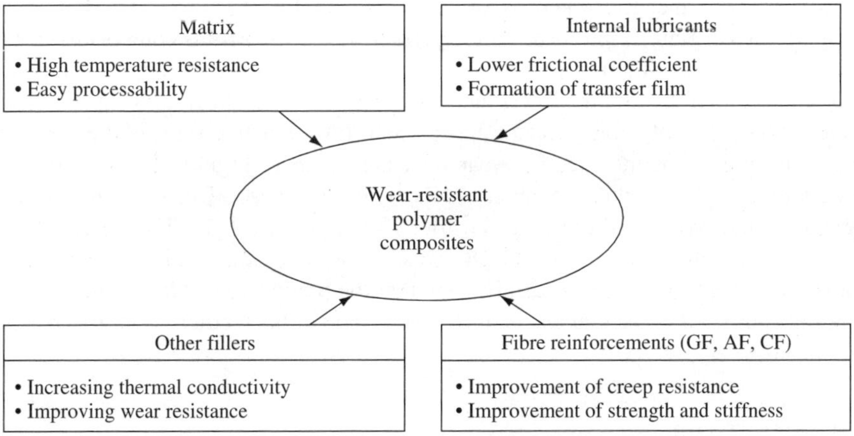

Figure 11.2 Schematic presentation of how to design the composition of wear-resistant polymer composites [23]

Table 11.1 Example of various newly developed, filled polymer systems with excellent tribological properties under various loading conditions

Material compositions (vol.%)							Specific wear rate (10^{-6} mm^3/N m)
PTFE	PPS	PEEK	Graphite	CF	Bronze	Al_2O_3	
51.6	31.8	–	4.8	11.8	–	–	1.53
51	31.6	–	3.9	13.5	–	–	1.40
51.9	32.4	–	2.6	13.1	–	–	1.20
12.4	–	61.8	11.7	14.1	–	–	4.31
9.7	–	49.7	12.5	28.1	–	–	6.33
76.8	–	–	19.8	–	–	3.4	22.40
84.1	–	–	–	12.6	–	3.3	1.25
52.5	28	–	–	19.5	–	–	1.69
78.6	–	–	–	21.4	–	–	1.75
80	–	–	–	10	10	–	0.565

Block-on-ring tests; Testing conditions: $p = 2$ MPa; $v = 1$ m/s; T =RT; $t = 8$ h; counterpart: steel.

which low friction, high wear resistance and good thermal conductivity under sliding wear conditions against smooth steel counterparts were of great importance. It can be found that PTFE + 10 vol.%bronze + 10 vol.%CF exhibits an excellent wear resistance. In particular, the addition of bronze improves significantly the tribological properties of the composite because of its outstanding thermal conductivity. This is, in fact, a good example to prove the effectiveness of this kind of filler for the design of a wear-resistant polymer composite, as it was also illustrated in the schematic presentation shown in Figure 11.2. The microstructure of such a material is shown in Figure 11.3 [11] (in this case PTFE + 39vol.%PPS + 13vol.%GF + 4 vol.%graphite, used in low temperature compressor applications). It should be mentioned that there are certainly many more compositions available on the commercial market, which, due to a lack of space, cannot be listed here [12–20].

The concept is further illustrated by the data generated by Reinicke *et al.* [21, 22] for a set of short carbon fibre reinforced, PTFE particle modified PA4.6 composites at various measuring conditions. Although the set of experimental data is not too large, it demonstrates the joint interaction of lubricating fillers and reinforcing fibres in the improvement of the wear resistance of the relatively high temperature-resistant PA4.6 matrix. Parts of the experimental results, as measured for the specific wear rate, are given in Figure 11.4 as a function of carbon fibre and PTFE volume contents (dots with error bars). Detail of the fretting test configuration and wear conditions can be found in Reference [21]. The plane adjusted to the experimental data points in the 3D plot was predicted by the use of an artificial neural networks (ANN) approach [23]. It can be seen that the predictive results are well acceptable when being compared to the real test results. Details about this predictive approach are given in Section 11.3.

11.2.2 PTFE Matrix Composites

PTFE is commonly used as a filler incorporated into most engineering polymers in order to improve their sliding properties. For special applications, where the unique properties of

Figure 11.3 Reflected light micrograph of a PTFE-based composite material containing a second stiff polymer phase, i.e. 39 vol.% polyphenylene-sulphide (PPS), for improved thermal and creep resistance, short glass fibres (13 vol.%) for a higher creep resistance, and graphite flakes (4 vol.%) for improved frictional behaviour [11]

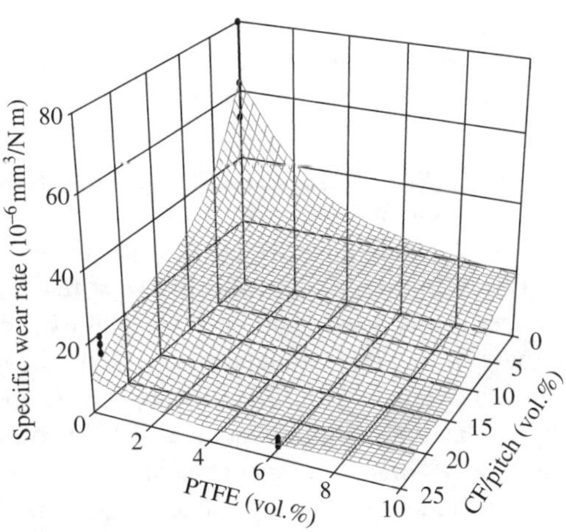

Figure 11.4 Specific wear rate of a polyamide 4.6 (PA4.6) composite as a function of PTFE- and short CF-volume content. The experimental data points and their scatters are given by the black bars, whereas the rest of the 3D plane was calculated by an artificial neural network approach (fretting test configuration, measuring conditions: testing temperature $T = 20°C$; normal force $F = 10$ N; sliding speed $v = 0.04$ m/s) (reprinted with permission from Elsevier) [23]

PTFE provide advantageous use, also pure PTFE and PTFE-based composites are approved. The temperature range in which PTFE is applicable reaches from very low temperatures in cryogenic media up to the maximum continuous use temperature of 260°C. In general, PTFE and PTFE-based composites are a good choice for applications in severe environments where no conventional lubrication is possible.

The automotive industry uses PTFE-based materials for sealing applications, bushings and journal bearing. Increasing temperatures due to the general tendency towards high-performance engines and aggregates and additionally automotive media enhance the use in this field [24]. In cryogenic or vacuum systems where no conventional lubrication is possible PTFE composites show good tribological performance [25–27].

Owing to their molecular structure, PTFE and also other fluoropolymers offer an almost universal chemical resistance and stability also at elevated temperatures. The large fluorine atoms act as a shield protecting the carbon backbone which contributes to the chemical inertness and thermal stability [28]. Low attraction forces between PTFE chains lead to relatively low mechanical strength and stiffness, weak wear resistance and also allow a tendency to cold flow. The very low surface energy of PTFE contributes to a low coefficient of friction and non-stick properties [29]. In addition, these materials can be used at very low temperatures and in cryogenic media because PTFE exhibits less embrittlement towards low temperatures than most other polymers.

In order to improve the rather poor mechanical properties, PTFE is modified by incorporating different fillers in order to increase key properties for tribological applications such as hardness, modulus, compression strength, thermal conductivity as well as resistance to wear and cold flow [30–35]. Commonly used fibre reinforcements are short glass and carbon fibres. They strongly enhance the wear resistance and reduce creep compared to neat PTFE. Particle reinforcements are often done by polymer particles, e.g. polyphenylene-sulphide (PPS), polyimide (PI) and polyetheretherketone (PEEK) and also carbon black and graphite. Metal powder, especially bronze, reinforced PTFE compounds provide a high load-bearing capability, very low wear rates and good frictional coefficients. The higher thermal conductivity of these compounds allows to conduct the frictional heat and thus reduces the contact temperature.

The tribological properties of PTFE have been widely investigated in the past. Among these studies also PTFE-based composites were tested under specific conditions. As an example, Figure 11.5 shows the results of pin-on-disc experiments on the material systems PTFE/PEEK and PTFE/PEEK/CF [36, 37]. The specific wear rate is shown as a function of the blending ratio. The experiments were done on a pin-on-disc wear test apparatus using 100Cr6 steel disc counterparts with a mean roughness of $R_a \approx 0.32\,\mu m$. The speed was set to 1 m/s and an apparent contact pressure of 1 MPa was applied. At these specific conditions, PTFE/PEEK compounds have an optimum ratio at about 10–20 vol.% PTFE where not only good wear rates of around $2 \times 10^{-6} mm^3/N\,m$ can be achieved, but also, which is not shown in the chart, the coefficient of friction reaches low values of about 0.23. To show the effect of a carbon fibre reinforcement, the results of PTFE/PEEK/CF composites tested under similar conditions are added to this chart. These materials reach significantly lower wear rates, being in the range of $8 \times 10^{-7} mm^3/Nm$. A typical worn surface of a PTFE/PEEK/CF composite is shown in Figure 11.6.

PTFE and PTFE composites show a tendency of building up a transfer film during sliding wear against metal surfaces. In case of neat PTFE it has been observed that lamellae are

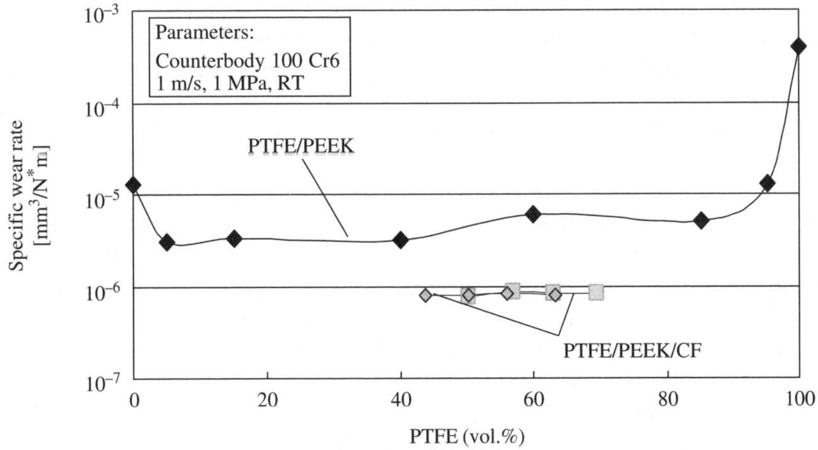

Figure 11.5 Comparison of the specific wear rate of PTFE/PEEK and PTFE/PEEK/CF composites (◇: 22 vol.%CF, □.16 vol.%CF)

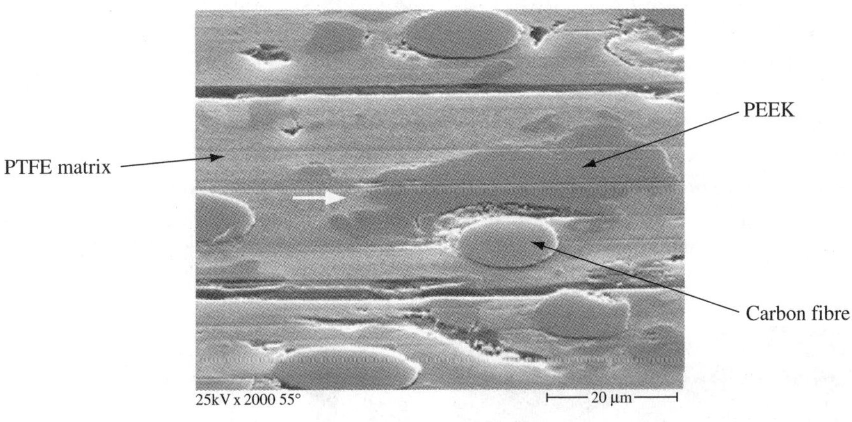

Figure 11.6 Worn sample surface of a PTFE + 13vol.%PEEK + 23.5 vol.%CF composite

deformed and finally torn off from the bulk material to adhere on the counterpart surface. When hard fillers like carbon fibres are used, which are able to cause wear on the steel counterpart, the build-up process of transfer films is time dependent and strongly influences the coefficient of friction [37–39].

11.2.3 Micro- and Nanoparticle Reinforcements

Inorganic particles are well known to enhance the mechanical and tribological properties of polymers; this has been widely investigated in the past decade. Various kinds of microparticle fillers, e.g. copper compounds (CuO, CuS, CuF_2), SiC, TiO_2 and ZrO_2, were selected to reinforce different matrices, e.g. PEEK [40, 41], polyamide (PA) [42, 43] and PPS [44–46]. It has been found that the size of the particles plays an important role to improve the wear resistance. Reducing the particle size to a nano-scale level is assumed to reach a significant efficiency [47, 48].

Polymers reinforced with nanometre-sized fillers are recently under discussion because of some excellent properties they have shown under various testing conditions [49, 50]. Regarding to their wear resistance, not so many efforts have been undertaken so far towards the development of such materials. However, some results were achieved in various studies which give hints that this method is also promising for new processing routes of wear-resistant materials [51–55]. Two studies shall be briefly mentioned in this respect here: (a) the use of TiO_2 nanoparticles in an epoxy resin matrix (Figure 11.7(a) and (b)), and (b) the incorporation of SiC nanoparticles, grafted with another polymer to enhance the adhesion of the particle agglomerates with the surrounding epoxy resin matrix [55]. In both cases it is of high importance that the fine particles are uniformly dispersed rather than being agglomerated in order to yield a good property profile, in general.

It can be seen from Figure 11.7(b) that 5 vol.% nano-TiO_2 reinforcement exhibits a 10 times reduction of the specific wear rate, compared to that of the pure matrix. However, higher filler contents lead to a deterioration in the wear properties which may be due to a tendency of particle agglomeration. This phenomenon also occurs for the other nanoparticle reinforced polymers. Nano-ZrO_2-filled PEEK, for example, exhibits the lowest specific wear rate at 7.5 wt.% [51], and 3 vol.% nano-SiC/PEEK has the best wear property [53]. The size of the nanoparticles significantly influences the wear rate and other mechanical properties of composites in addition. Fine particles seem to contribute better to the property improvement than larger nanoparticles [47, 48], although opposite size effects were found in the field of glass micro-sphere reinforced polyamides under abrasive wear conditions [56]. A similar trend was detected in the case of glass particle reinforced polymethylmethacrylate (PMMA) dental composites when tested against steel counterparts under sliding wear conditions (superimposed by interfacial bond quality effects) [57]. A good particle distribution (as

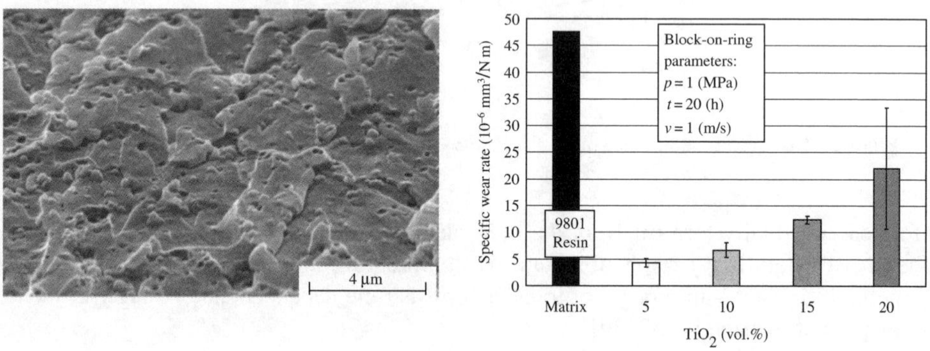

(a) (b)

Figure 11.7 (a) Fracture surface of an epoxy resin filled with 4 vol.%TiO_2 nanoparticles (ϕ300 nm). The good distribution is an important factor in the improvement of the properties. (b) Specific wear rate as a function of TiO_2 content. Higher contents lead to a deterioration in the properties due to a tendency of particle agglomeration shown in Figure 11.7(a)) is also an important factor in the improvement of the properties. It is the subject of further studies in the future to investigate the reasons why the very reduced size of the fillers yields such a significant improvement in the wear properties.

11.2.4 Integration of Traditional Fillers with Inorganic Nanoparticles

In order to fully promote the effect of nanoparticles, further investigations have been preformed on a series of epoxy-based composites blended with different contents of PTFE powders and graphite flakes, additionally reinforced with various amounts of short-CF and nano-TiO_2. The best wear-resistant composition was found as a combination of nano-TiO_2 with conventional reinforcements. As an example, epoxy + 15vol.%graphite + 5 vol.%nano-TiO_2 + 15 vol.%short-CF exhibits a specific wear rate of 3.2×10^{-7} mm^3/N m at 1 MP m/s, which is more than 100 times lower than that of the neat epoxy [58, 59].

Investigations have been carried out with a constant amount of graphite and short-CF, but with a varying content of nano-TiO_2 particles. When the nanoparticle content was either reduced to 2 vol.% or increased to 8 vol.%, the specific wear rates of their composites became higher. Figure 11.8 summarizes the results of the dependence of the specific wear rate on the nano-TiO_2 content with constant 5 or 15 vol.% of both graphite and short-CF, respectively. A comparison to the epoxy, reinforced with various amounts of nanoparticles only [54], is also shown in this figure. It is interesting to note that a content of 4~6 vol.% of nano-TiO_2 exhibited an optimum effect in all the three cases.

To evaluate the wear behaviour of materials under different sliding conditions, i.e. various contact pressure and sliding speed, the time-related depth wear rate, W_t, was introduced:

$$W_t = k^* pv = \frac{\Delta h}{t} \tag{2}$$

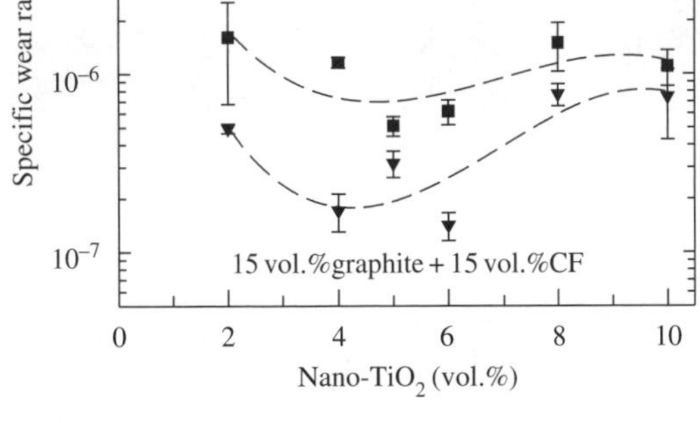

Figure 11.8 Dependence of the specific wear rate on the nano-TiO_2 content of epoxy nanocomposites (reprinted with kind permission from Elsevier) [58, 59]

where k^* is the wear factor (equivalent to the specific wear rate defined in equation (1)), p is the normal pressure, v is the sliding velocity, t is the test time and Δh is the height loss of the specimen. In this equation, the wear factor, k^*, is supposed to be a material parameter, which also can be more or less dependent on the testing conditions, e.g. p, v and T (temperature). To reduce the basic wear factor k^* and to enhance the "limiting *pv*" value are the general targets of the design of wear-resistant polymer composites.

Figure 11.9 compares the time-related depth wear rate of epoxy composites filled with and without nano-TiO_2 as a function of *pv* factor [60, 61]. W_t of the neat epoxy is already very high at a rather low *pv* value (1 MPa m/s) and cannot be drawn within this plot for higher *pv*s (i.e. the material cannot be operated in this condition). By incorporating traditional micro-fillers, i.e. 15 vol.% of short-CF and 5 vol.% of each graphite and PTFE, the wear rate is significantly reduced. The depth wear rate generally increases with an increase in the *pv* factor. For the composite only with these traditional micro-fillers, the wear rate raises to quite a high value when the *pv* factor reaches to 4 MPa m/s. However, the depth wear rate of the composite with additional 5 vol.% of nanoparticles is much lower at this condition, and even at much higher *pv* factors, e.g. 12 MPa and 1 m/s or 4 MPa and 3 m/s. It is, therefore, clear that the wear factor is reduced, and the "limiting *pv*" value is enhanced as well by these nanoparticles. On the basis of further investigations on the worn surfaces, using the scanning electron microscopy (SEM) [59, 62], a positive rolling effect of the nanoparticles between the material pairs was proposed. This rolling effect helped to reduce the frictional coefficient during sliding, accordingly to reduce the shear stress and contact temperature. In addition, this rolling effect protected the short carbon fibres from more severe wear mechanisms, especially at high sliding pressure and speed situations.

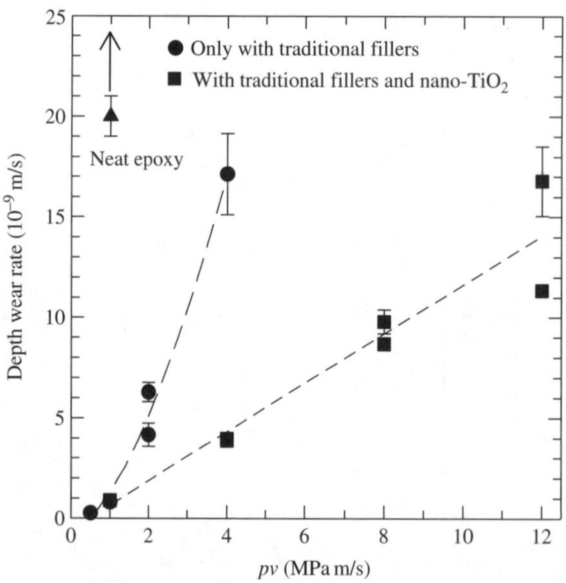

Figure 11.9 Comparisons of the time-related depth wear rate of epoxy composites filled with and without nano-TiO_2 as a function of *pv* factor [60]

11.2.5 Functionally Graded Tribo-Materials

This kind of material can be of technical interest if one wants to produce components that have (a) different properties at different locations of their cross-section, e.g. hardness and abrasive wear resistance on the outside, and toughness and damage tolerance on the inside (e.g. gears), but (b) no sudden property jumps, e.g. as in the case of hard coatings, which can cause stress singularities at the interface, associated with local coating failure due to spalling-off effects. The preparation of functionally gradient materials (FGM) with a thermosetting polymer matrix by centrifugation has been studied from the beginning of the last decade [63–65]. The aim in Reference [66] was to systematically create materials with a gradient in the wear resistance using an epoxy resin matrix with different fillers, including aramid and/or wollastonite powder. This was done by controlling the parameters of centrifugation speed and time, as well as viscosity of the mixture.

Since the possible application was focused on sliding bearings and roller covers with a gradient over the radius, the shape of short tubes was chosen as the sample geometry. The latter was centrifuged in a double wall mould on a lathe creating a filler gradient over the thickness of the wall. Turning speeds up to 1000 rpm were applied to the isotropic mixtures before the resin was cured. Isotropic samples were prepared as a reference. Figure 11.10(a) shows a graded tube filled with wollastonite particles. In the same figure, an SEM overview of the radial cross-section with a gradient of aramid particles is shown. A slightly higher concentration of aramid particles near to the wear surface as opposed to the inner part is

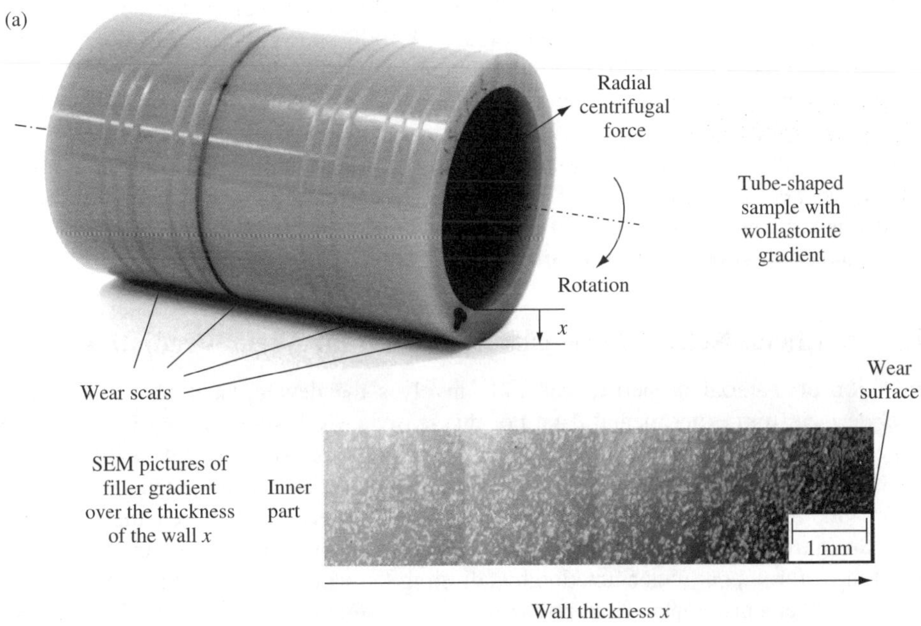

Figure 11.10 (a) Schematic representation of the filler gradient preparation across the wall of a centrifuged cylinder bearing (with micrograph of filler distribution). (b) Variation in specific wear rate and hardness over the wall thickness of a sample

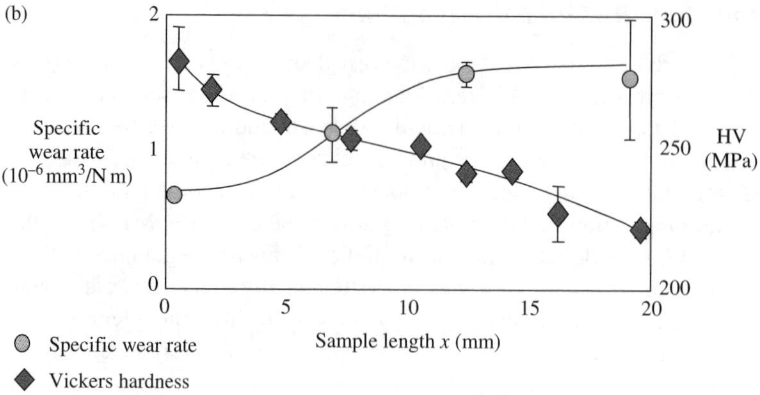

Figure 11.10 (Continued)

visible. The wear scars on the surface of the cylinder result from a wear test, in which ceramic counter-bodies were pressed against the surface of the rotating FGM tube, after curing, to determine the wear characteristics. The wear resistances of the FGM tubes reinforced with all the different filler reinforcements were better than those of the corresponding isotropic samples.

The gradient in the wear rate was also determined by fretting tests. During this test a ceramic ball was oscillated on the sample with a defined normal force, frequency and amplitude over a certain period of time. It was found that the specific wear rate decreased with increasing aramid particle content. Figure 11.10(b) shows the fretting wear rate over the length x of the sample. Also shown in Figure 11.10(b) is the Vickers hardness of such a sample. It can be seen that the outer wear surface does not only have an increased wear resistance but also a greater hardness than the inner surface.

Future research on this topic will include the use of filler combinations such as aramid, to improve the wear resistance, and PTFE powders, to decrease the coefficient of friction by their dry lubricating property. One of the current projects is also the creation of a software tool to predict the filler distribution after centrifugation.

11.3 Artificial Neural Networks Approach for Wear Prediction

Simulation of material properties generally involves the development of a mathematical model derived from experimental data. For this reason, ANN were recently introduced into the field of polymer composites [67]. Neural networks are composed of simple elements operating in parallel. These elements are inspired by biological nervous systems. As in nature, the network function is determined largely by the connections between these elements, which are mostly nonlinear transfer functions in computer simulations. An ANN can be trained to perform a particular function by adjusting the values of the connections (weights) between the elements. For material research, a certain amount of research result is necessary to develop a well-performing neural network, including its architecture, training functions, training algorithms and other parameters. After the network has learned to solve the example problems, new data from the same knowledge domain can then be put into the trained neural network, in order to output realistic solutions. The greatest advantage of ANN is its ability to model complex nonlinear, multi-dimensional function relationships without any prior assumptions about the nature of the relationships, and the network is built directly from experimental data by its self-organizing capabilities.

As shown in Figure 11.11 [23], an ANN is conventionally constructed with three layers, i.e. input, output and hidden layers. Hidden layers can contain one or several layers for its practical application. In the present case, several measuring details, i.e. material compositions, mechanical properties and testing conditions, were selected as input parameters for the ANN, and wear characteristics such as coefficient of friction and specific wear rate were chosen as output results.

A total dataset of 103 independent wear measurements was used [21, 22] coming from fretting tests at different wear measuring conditions. These data were obtained for PA46 composites which were modified with different amounts of short glass or carbon fibres, PTFE and/or graphite. The database also contained the material composition (volume fraction of matrix, short glass fibres, pitch-based carbon fibres, PAN carbon fibres, PTFE and graphite fillers), mechanical properties of the composites studied (compression modulus and strength, impact strength, etc.) and testing conditions (temperature, normal force and sliding speed) as input parameters. To obtain an optimized neural network construction, the given dataset was divided into a training dataset and a test dataset. The training dataset was used to adjust the weights of all the connecting nodes until the desired error level was reached. Then, by the use of the test dataset, the network performance could be evaluated on the basis of the coefficient of determination. This coefficient describes the fit of the ANN's output variable

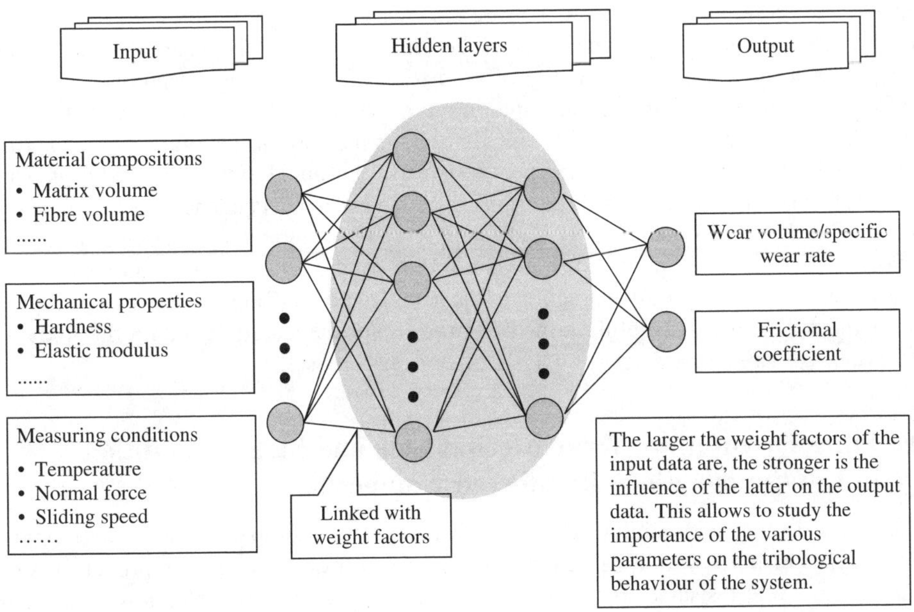

Figure 11.11 Input data, output data and schematic construction of an artificial neural network for correlating tribological properties with testing and material parameters (reprinted with kind permission from Elsevier) [23]

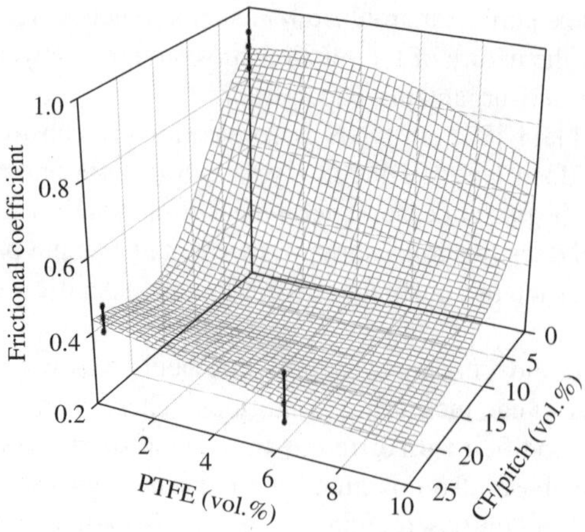

Figure 11.12 Frictional coefficient of a polyamide 4.6 (PA4.6) composite as a function of PTFE- and short CF-volume content. The experimental data points and their scatters are given by the black bars, whereas the rest of the 3D plane was calculated by an artificial neural network approach (fretting test configuration, measuring conditions: testing temperature $T = 20°C$; normal force $F = 10$ N; sliding speed $\nu = 0.04$ m/s) (reprinted with kind permission from Elsevier) [23]

approximation curve with the actual test data output variable curve. Higher coefficients indicate an ANN with better output approximation capabilities.

The predictive results of specific wear rate are shown in Figure 11.4, and Figure 11.12 gives a further prediction of the frictional coefficient. It became obvious that the material composition has a significant influence on the quality of prediction of the tribological properties, since this information was not yet used as an input in our previous work (in which the coefficient of determination was clearly lower) [68]. In addition, an increase of the number of data, 103 in Reference [23] instead of 72 in Reference [68], also contributed to this improvement. The major advantage of neural networks is their possibility to predict dependencies between many parameters (as shown in Figure 11.13), and to apply them to any given situation in a "black box" fashion. The quality of predictions based on ANN can be further improved by enlarging the datasets and optimizing the construction of the neural network (Table 11.2). A well-trained ANN is expected to be very helpful for the design of composite materials and for the study of the importance of various parameters on the predicted properties.

11.4 Fibre Orientation, Wear Mechanisms and Stress Conditions in Continuous Fibre Reinforced Composites

Many articles dealing with sliding wear of continuous fibre composites have in common that some types of fibres with a particular orientation can result in greater improvements of the composites' wear resistance than others which may be more favourable, however, when they are positioned under a different orientation relative to the wear direction. Bringing the favourable orientations and types of fibres together may even lead to further improvements or to the so-called "synergistic" effects [69].

In the following, sliding wear results of unidirectional continuous carbon fibre/polyetheretherketone (CF/PEEK) composites against 100Cr6 steel counterparts are

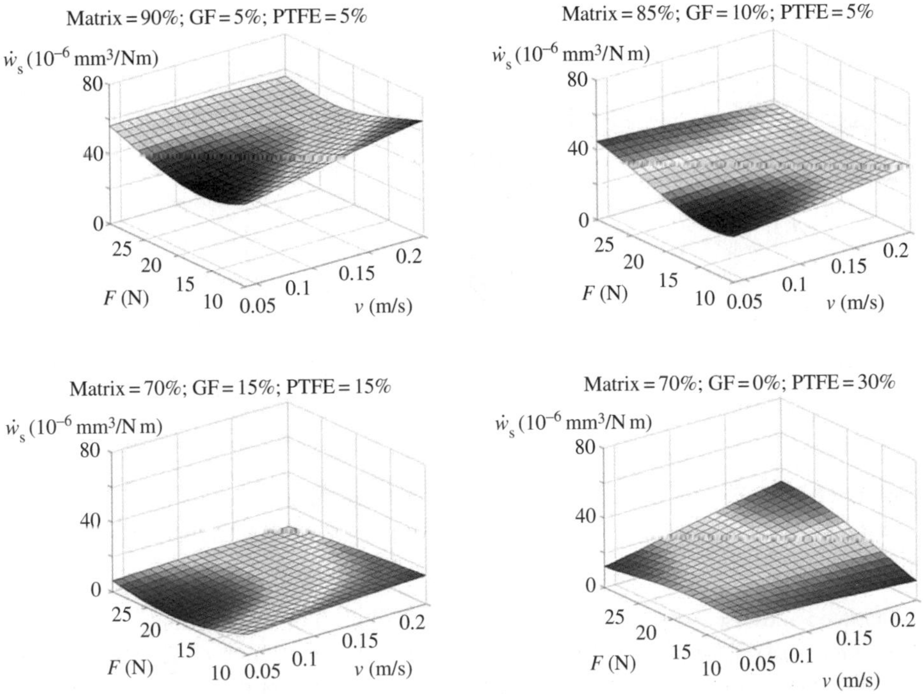

Figure 11.13 Results of a parameter study by the use of the neural network concept: specific wear rate as a function of normal load and sliding speed for various material compositions (reprinted with kind permission from Elsevier) [23]

Table 11.2 An overview of input data used for calculating specific wear rate and coefficient of friction by an artificial neural network approach

Input	Material Composition	Matrix Volume (70%~100%) Fibre Volume (0~30%) PTFE Volume (0%~10%) ......
	Mechanical Properties	Compression Modulus Compressive Strength Hardness Fracture Toughness ......
	Measuring Conditions	Temperatures (20~150°C) Normal Force (5~30 N) Sliding Speed (0.02~0.2 m/s) ......
Output	Wear Characteristics	Specific Wear Rate Frictional Coefficient ......

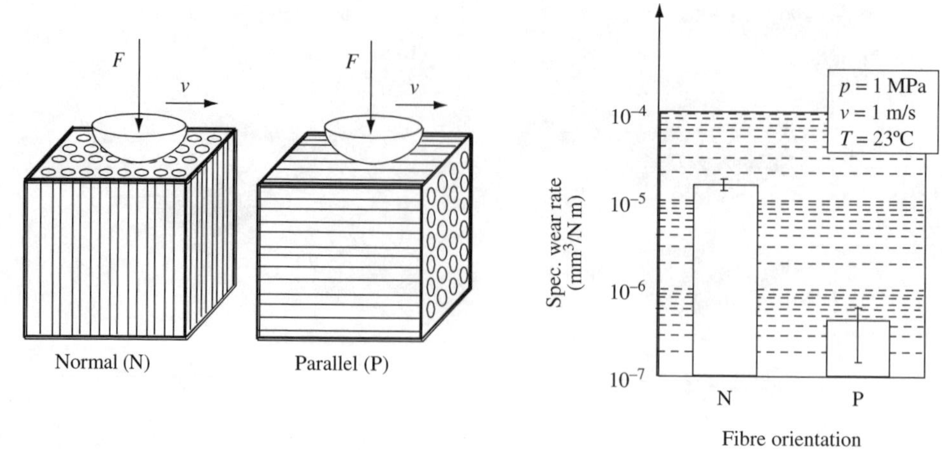

Figure 11.14 Specific wear rate of unidirectional continuous carbon fibre reinforced polyetheretherketone (PEEK) against 100Cr6 steel (German standard; ball bearing steel) as a function of fibre orientation [69]

demonstrated and discussed as a function of two principle fibre orientations, i.e. normal (N) and parallel (P) (Figure 11.14). The reason for the higher wear rate of the CF composites under N orientation relative to P orientation is based on the fact that the upper load limits (strength) of the fibres (under the same external load enhancement) are reached inside the composite earlier under N orientation. The results of the antiparallel (AP) orientation are found between the two others [70].

The experimental results could be verified by a finite element simulation of the stress conditions in a unidirectional composite, when being subjected to a moving asperity contact under different fibre orientations (Figure 11.15) [71, 72]. If these stresses locally exceed the

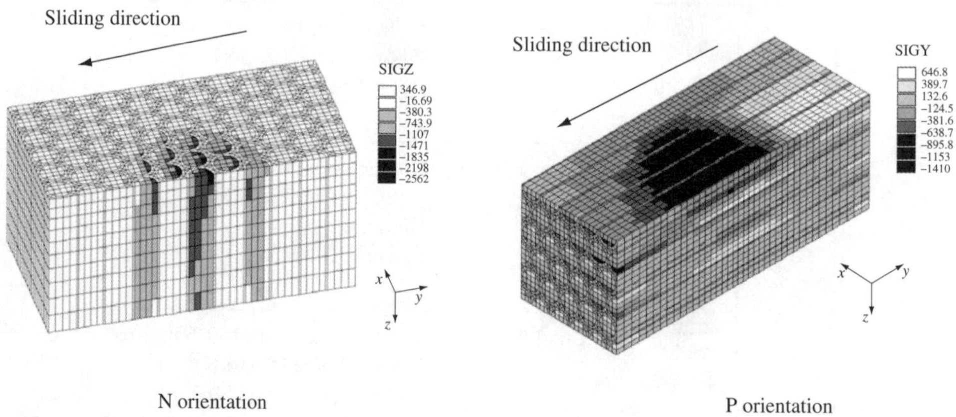

Figure 11.15 Example for the built-up of stresses (in MPa) in the fibre direction of the N- and P-oriented composites, achieved by FE modelling of a sliding asperity contact (purely elastic consideration) [70, 71]

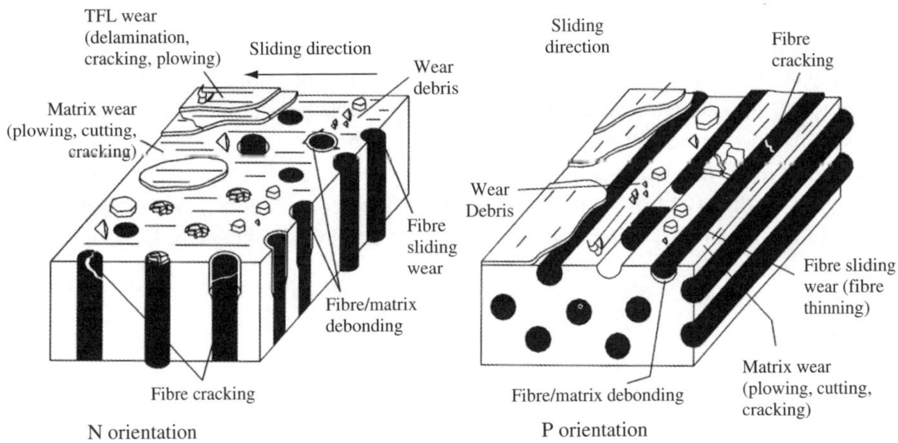

Figure 11.16 Schematics of wear mechanism in UD composites as a function of fibre orientation

strength of the fibre, fibre fracture according to the schematics in Figure 11.16 occurs, which, in turn, causes temporarily an enhanced contribution of abrasive wear by fibre fragments in the contact region. This statement holds also for the other mechanisms such as fibre/matrix delamination, fibre thinning and matrix shear failure, which could be proved by the worn surface micro-photographs in Figure 11.17.

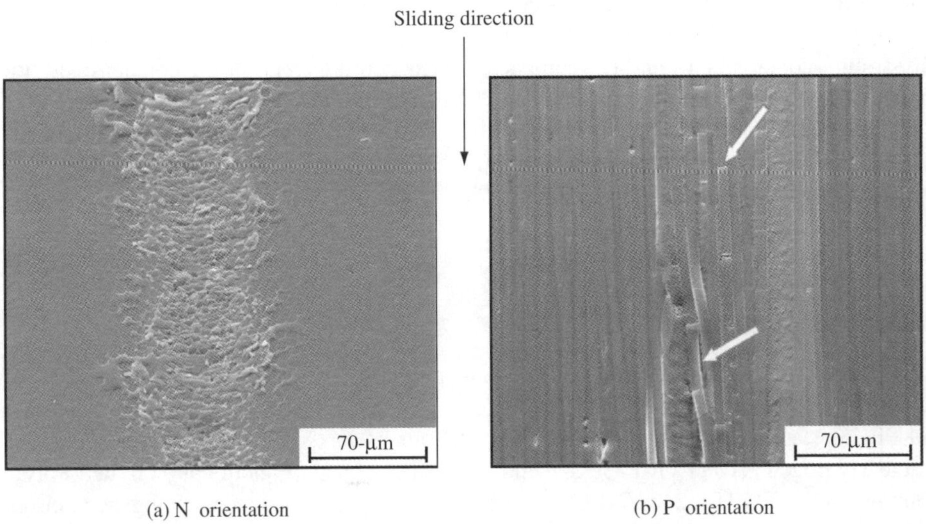

Figure 11.17 SEM photos of "single asperity" scratched surfaces of CF/PEEK showing the individual wear mechanisms expected from the FE predictions: (a) N orientation with fibre breakage at the rear edge of the fibre ends, delaminating phenomena and push-up, followed by shear failure of the matrix; (b) P orientation with fibre thinning, fibre fracture, fibre/matrix debonding and matrix shear failure

How these load limits, which at the end also determine the application limits of these materials as tribological components, change under the influence of temperature (due to the external effects or as a result of frictional heating) is presently under investigation [73, 74]. In these modelling studies, also the formation of a compacted debris layer for the reduction of friction and the protection of underlying material (both determining the wear behaviour of a wear couple in the steady-state range) is considered. In this respect, also other articles, e.g. dealing with the effect of such layers on the surface contact and wear of carbon graphite materials [75] and on the tribological surfaces of organic brake pads [76], are of great interest.

11.5 Conclusions

Over the past few years, interest in polymers and polymer-based composites for technical applications, in which low friction and low wear must be provided, has grown rapidly. For example, short fibre reinforced high temperature resistant thermoplastic materials (SFRP) are now being used as sliding elements which were formerly composed of metallic materials only. Their applications can reach temperatures up to 220°C, pressures of more than 10 MPa and sliding velocities of about 3 m/s [77]. A new tool in the form of ANN is now available for the prediction of the wear properties of these materials as a function of composition and testing conditions. It also allows systematic parameter studies on the computer for material optimization.

Continuous fibre reinforced high-performance composites are another class of tribo-materials which can endure even higher pressures than SFRP, depending on the fibre orientation relative to the sliding direction. The load limits, which finally determine the limits of their application, can be verified by finite element simulations. These studies on the micro-level give also information about the effects of (1) real asperity contact on the local temperature development, and (2) compacted wear debris layers on the contact conditions, contact stresses, temperature conditions and their final influence on the composite wear rate.

Quite recently, new fundamental studies have been concentrating on the development of FGM using different centrifugation techniques. It was the objective to demonstrate that a gradient in wear resistance and other properties as a result of a gradient in filler distribution can be created over the cross-section of samples or components produced.

Acknowledgements

Parts of these studies were carried out within the projects "Tribology of Composite Materials" (DFG FR 675/19-2; together with Professor K. Varadi, Hungary), "Properties of PTFE and PEEK Matrix Composites at Tribological Stresses in Cryogenic Media Particularly in Liquid Hydrogen" (DFG FR 675/32-2 and HU 791/2-2) and "Functionally Graded Polymer Composites" (DFG HA 2483/2-1), all supported by the Deutsche Forschungsgemeinschaft (DFG). Further thanks are due to the 'DFG' for support of the 'Artificial Neural Network' Project (Grant No. 'FR 675/45-1').

Z. Zhang is grateful to the Alexander von Humboldt Foundation for his Sofja Kovalevskaja Award, financed by the German Federal Ministry of Education and Research (BMBF) within the German Government's "ZIP" program for investment in the future.

Figures 4, 11, 12 and 13 are reprinted from Wear, Vol. 252, "Prediction of tribological properties of short fibre composites using artificial neural networks", pp. 668-675, Copyright (2002), with permission from Elsevier Science.

This material has been based upon an article that first appeared in the *Journal of Engineering Tribology* – Proceedings Part J, 2002, Vol. 216, No. J6, ISSN 1350–6501, published by Professional Engineering Publishing. Permission is granted by the Institution of Mechanical Engineers.

References

1. Verpoest, I., personal communication, 1998.
2. Czichos, H. and Habig, K.H., Tribologie Handbuch, Braunschweig, Vieweg, 1992.
3. Häger, A.M. and Davies, M., 'Short-Fibre Reinforced, High Temperature Resistant Polymers for a Wide Field of Tribological Applications', in *Advances in Composite Tribology* (ed K. Friedrich), Composite Materials Series (series editor R.B. Pipes), Vol. 8, Elsevier, Amsterdam, The Netherlands, 1993, pp. 107–157.
4. Lancaster, J.K., 'Composites for Aerospace Dry Bearing Application', in *Friction and Wear of Polymer Composites* (ed K. Friedrich), Composite Materials Series (series editor R.B. Pipes), Vol. 1, Elsevier, Amsterdam, The Netherlands, 1986, pp. 363–396.
5. Briscoe B.J., 'The Tribology of Composite Materials: A Preface', in *Advances in Composite Tribology* (ed K. Friedrich), Composite Materials Series (series editor R.B. Pipes), Vol. 8, Elsevier, Amsterdam, The Netherlands, 1993, pp. 3–15.
6. Friedrich, K., 'Wear Performance of High Temperature Polymers and Their Composites', in *Application of High Temperature Polymers* (ed R.R. Luise), CRC Press, Boca Raton, USA, 1997, pp. 221–246.
7. Stachowiak, G.W. and Batchelor, A.W, '*Engineering Tribology*', 2nd Edition, Butterworth-Heinemann, Jordan Hill, Oxford, UK, 2001.
8. Hübner, W., Gradt, T., Schneider, T. and Borner, H., 'Tribological Behaviour of Materials at Cryogenic Temperatures', *Wear*, **216**, 1998, 150–159.
9. Bahadur, S., 'The Development of Transfer Layers and Their Role in Polymer Tribology', *Wear*, **245**, 2000, 92–99.
10. Gao, J., 'Tribochemical Effects in Formation of Polymer Transfer Film', *Wear*, **245**, 2000, 100–106.
11. Kureha Chemical, Japan, personal communication, 1997.
12. Bijwe, J., Indumathi, J. and Ghosh, A.K., 'Evaluation of Engineering Polymeric Composites for Abrasive Wear Performance', *Journal of Reinforced Plastics and Composites*, **18**, 1999, 1573–1619.
13. Kim, S.J. and Jang, H., 'Friction and Wear of Friction Materials Containing Two Different Phenolic Resins Reinforced with Aramid Pulp', *Tribology International*, **33**, 2000, 477–484.
14. Kukureka, S.N., Hooke, C.J., Rao, M., Liao, P. and Chen, Y.K., 'The Effect of Fibre Reinforcement on the Friction and Wear of Polyamide 66 Under Dry Rolling–Sliding Contact', *Tribology International*, **32**, 1999, 107–116.
15. Marrs, H., Barton, D.C., Jones, R.A., Ward, I.M., Fisher, J. and Doyle, C., 'Comparative Wear Under Four Different Tribological Conditions of Acetylene Enhanced Cross-Linked Ultra High Molecular Weight Polyethylene', *Journal of Materials Science: Materials in Medicine*, **10**, 1999, 333–342.
16. Stuart, B.H., 'Tribological Studies of Poly(ether ether ketone) Blends', *Tribology International*, **31**, 1998, 647–651.
17. Kurokawa, M., Uchiyama, Y. and Nagai, S., 'Performance of Plastic Gear Made of Carbon Fibre Reinforced Polyetheretherketone', *Tribology International*, **32**, 1999, 491–497.
18. LNP Engineering Plastics Inc., Product Introduction, http://www.lnp.com/LNPSite.nsf.
19. RTP Company – Engineering Thermoplastic Compounds, Product Introduction, http://www.rtpcompany.com/.
20. Faigle Kunststoffe, Product Introduction, http://www.faigle.com/flashintro.html.
21. Reinicke, R., Haupert, F. and Friedrich, K., 'On the Tribological Behavior of Selected, Injection Molded Thermoplastic Composites', *Composites Part A*, **29**, 1998, 763–771.
22. Reinicke, R., Eigenschaftsprofil neuer Verbundwerkstoffe für tribologische Anwendungen im Automobilbereich, in IVW Schriftenreihe Bd. 21 (ed M. Neitzel), Institut für Verbundwerkstoffe GmbH, Kaiserslautern, Germany, 2001, ISBN-3-934930-17-4.

23. Zhang, Z., Friedrich, K. and Velten, K., 'Prediction of Tribological Properties of Short Fibre Composites Using Artificial Neural Networks', *Wear*, **252**(7–8), 2002, 668–675.
24. Hoffmann, C. and Müller, H.K., PTFE-Rotary Shaft Seals, MTZ Motortechnische Zeitschrift 56 (1995) 11, pp. 695–705.
25. Fusaro, R.L., 'Self-Lubricating Polymer Composites and Polymer Transfer Film Lubrication for Space Applications', *Tribology International*, **23**, 1990, 105–122.
26. Gardos, M.N., 'Self-Lubricating Composites for Extreme Environmental Conditions', in *Friction and Wear of Polymer Composites* (ed K. Friedrich), Composites Materials Series (series editor R.B. Pipes), Vol. 1, Elsevier, New York, 1986, pp. 397–447.
27. Theiler, G., Hübner, W., Gradt, T., Klein, P. and Friedrich, K., 'Friction and Wear of PTFE Composites at Cryogenic Temperatures', *Tribology International*, **35**, 2002, 449–458.
28. Drobny, J.G., '*Technology of Fluoropolymers*', CRC Press LLC, Boca Raton, 2001.
29. Pozzoli, M., Vita, G. and Arcella, V., in *Modern Fluoropolymers* (ed J. Scheirs), John Wiley and Sons, New York, 1997, p. 374.
30. Biswas, S.K., 'Friction and Wear of PTFE – A Review', *Wear*, **158**, 1992, 193–211.
31. Tanaka, K., 'Effects of Various Fillers on the Friction and Wear of PTFE Based Composites', in *Friction and Wear of Polymer Composites* (ed K. Friedrich), Composites Materials Series (series editor R.B. Pipes), Vol. 1, Elsevier, New York, 1986, pp. 137–174
32. Briscoe, B.J., Yao, L.H. and Stolarski, T.A., 'The Friction and Wear of Poly(tetrafluoroethylene)–Poly(etheretherketone) Composites: An Initial Appraisal of the Optimum Composition', *Wear*, **108**, 1986, 357–374.
33. Gong, D., Xue, Q. and Wang, H., 'Study of the Wear of Filled Polytetrafluoroethylene', *Wear*, **134**, 1989, 283–295.
34. Bijwe, J., Neje, S., Indumathi, J. and Fahim, M., 'Friction and Wear Performance Evaluation of Carbon Fibre Reinforced PTFE Composites', *Journal of Reinforced Plastics and Composites*, **21/13**, 2002, 1221–1240.
35. Khedkar, J., Meletis, E.I. and Negulescu, I., 'Sliding Wear Behavior of PTFE Composites', *Wear*, **9016**, 2002, 1–9.
36. Lu, Z. and Friedrich, K., 'On Sliding Friction and Wear of PEEK and Its Composites', *Wear*, **181–183**, 1995, 624–631.
37. Klein, P., Friedrich, K., Theiler, G. and Hübner, W., Mechanische und Tribologische Eigenschaften von PTFE/PEEK/CF-Verbundwerkstoffen, in Polymere Verbund-werkstoffe zwischen Raumtemperatur und 4,2K: Tribologische Eigenschaften und Methoden der Charakterisierung. BAM Tagungsbericht, (ed W. Hübner and K. Friedrich), BAM-Sonderheft 2003, Berlin: BAM, 2003, pp. 11–16.
38. Yang, E., 'Studies on the Transfer and Adhesion of Polytetrafluoroethylene (PTFE) and PTFE-Based Composites During Dry Contact', The Finnish Society of Sciences and Letters, Helsinki, 1992, p. 140.
39. Bahadur, S., 'The Development of Transfer Layers and Their Role in Polymer Tribology', *Wear*, **245**, 2000, 92–99.
40. Bahadur, S. and Gong, D., 'The Role of Copper Compounds as Fillers in the Transfer and Wear Behavior of Polyetheretherketone', *Wear*, **154**, 1992, 151–165.
41. Vande Voort, J. and Bahadur, S., 'The Growth and Bonding of Transfer Film and the Role of CuS and PTFE in the Tribological Behavior of PEEK', *Wear*, **181–183**, 1995, 212–221.
42. Bahadur, S., Gong, D. and Anderegg, J.W., 'Tribochemical Studies by XPS Analysis of Transfer Films of Nylon 11 and Its Composites Containing Copper Compounds', *Wear*, **165**, 1993, 205–212.
43. Bahadur, S. and Polineni, V.K., 'Tribological Studies of Glass Fabric-Reinforced Polyamide Composites Filled with CuO and PTFE', *Wear*, **200**, 1996, 95–104.
44. Yu, L. and Bahadur, S., 'An Investigation of the Transfer Film Characteristics and the Tribological Behaviors of Polyphenylene Sulfide Composites in Sliding Against Tool Steel', *Wear*, **214**, 1998, 245–251.
45. Zhao, Q. and Bahadur, S., 'The Mechanism of Filler Action and the Criterion of Filler Selection for Reducing Wear', *Wear*, **225–229**, 1999, 660–668.
46. Schwartz, C.J. and Bahadur, S., 'The Role of Filler Deformability, Filler-Polymer Bonding, and Counterface Material on the Tribological Behavior of Polyphenylene Sulfide (PPS)', *Wear*, **251**, 2001, 1532–1540.
47. Xue, Q. and Wang, Q., 'Wear Mechanisms of Polyetheretherketone Composites Filled with Various Kinds of SiC', *Wear*, **213**, 1997, 54–58.
48. Ng, C.B., Ash, B.J., Schadler, L.S. and Siegel, R.W., 'A Study of the Mechanical and Permeability Properties of Nano- and Micron-TiO_2 Filled Epoxy Composites', *Advanced Composites Letters*, **10**, 2001, 101–111.

49. Zhang, M.Q., Rong, M.Z. and Friedrich, K., 'Processing and Properties of Nonlayered Nanoparticle Reinforced Thermoplastic Composites', in *Handbook of Organic–Inorganic Hybrid Materials and Nanocomposites* (ed H.S. Nalwa), Vol. 2, Nanocomposites, American Scientific Publishers, Los Angeles, 2003, pp. 113–150.
50. Karger-Kocsis, J. and Zhang Z., 'Structure–Property Relationships in Nanoparticle/Semi-Crystalline Thermoplastic Composites', in *Mechanical Properties of Polymers Based on Nanostructure and Morphology* (eds J.F. Balta Calleja and G. Michler), Marcel Dekker Inc., New York, 2005, pp. 547–596.
51. Wang, Q., Xue, Q., Liu, H., Shen, W. and Xu, J., 'The Effect of Nanometer ZrO_2 on the Tribological Behavior of PEEK', *Wear*, **198**, 1996, 216–219.
52. Schwartz, C.J. and Bahadur, S., 'Studies on the Tribological Behavior and Transfer Film-Counterface Bond Strength for Polyphenylene Sulfide Filled with Nanoscale Alumina Particles', *Wear*, **237**, 2000, 261–273.
53. Wang, Q., Xue, Q., Liu, W. and Chen, J., 'The Friction and Wear Characteristics of Nanometer SiC and Polytetrafluoroethylene Filled Polyetheretherketone', *Wear*, **243**, 2000, 140–146.
54. Wetzel, B., Haupert, F., Friedrich, K., Zhang, M.Q. and Rong, M.Z., 'Impact and Wear Resistance of Polymer Nanocomposites at Low Filler Content', *Polymer Engineering and Science*, **42**(9), 2002, 1919–1927.
55. Rong, M., Zhang, M., Liu, H., Zeng, H., Wetzel, B. and Friedrich, K., 'Microstructure and Tribological Behavior of Polymeric Nanocomposites', *Industrial Lubrication and Tribology*, **53**, 2001, 72–77.
56. Friedrich, K., 'Wear of Reinforced Polymers by Different Abrasive Counterparts', in *Friction and Wear of Polymer Composites* (ed K. Friedrich), Composite Materials Series (series editor R.B. Pipes), Vol. 1, Elsevier, Amsterdam, The Netherlands, 1986, pp. 233–287.
57. Friedrich, K., 'Particulate Dental Composites Under Sliding Wear Conditions', *Journal of Materials Science: Materials in Medicine*, **4**, 1993, 266–272.
58. Zhang, Z., Haupert, F. and Friedrich, K., Enhancement on Wear Resistance of Polymer Composites by Nano-Fillers, German Patent DE 103 29 228.4, 2005.
59. Zhang, Z., Breidt, C., Chang, L., Haupert, F. and Friedrich, K., Enhancement of the Wear Resistance of Epoxy: Short Carbon Fibre, Graphite, PTFE and Nano-TiO_2, *Composites A*, **35**, 2004, 1385–1392.
60. Zhang, Z. and Friedrich K., 'Tribological Characteristics of Micro- and Nanoparticle Reinforced Polymer Composites', in *Polymer Composites – from Nano- to Macro-Scale* (eds K. Friedrich, S. Fakirov and Z. Zhang), Kluwer Academic Publishers, Dordrecht (Hingham, MA), 2004, pp. 169–185.
61. Chang, L., Zhang, Z., Breidt, C. and Friedrich, K., 'Tribological Properties of Epoxy Nanocomposites: I. Enhancement of the Wear Resistance by nano-TiO_2', *Wear*, **258**, 2005, 141–148.
62. Chang, L. and Zhang, Z., 'Tribological Properties of Epoxy Nanocomposites: II. A Combinative Effect of Short Carbon Fibre and Nano-TiO_2', *Wear*, 2005, in press.
63. Funabashi, M., 'Gradient Composites of Nickel Coated Carbon Fibre Filled Epoxy Resin Moulded Under Centrifugal Force', *Composites Part A*, **28A**, 1997, 731.
64. Vinarcik, E.J., 'Selectively Reinforcing Composites by Magnetic Processing', *SAMPE Journal*, **34**, 1998, 40.
65. Lee, N.J., Jang, J., Park, M. and Choe, C.R., 'Characterization of Functionally Gradient Epoxy Carbon Fibre Composite Prepared Under Centrifugal Force', *Journal of Materials Science*, **32**, 1997, 2013.
66. Klingshirn, C., Koizumi, M., Haupert, F., Giertzsch, H. and Friedrich, K., 'Structure and Wear of Centrifuged Epoxy Resin/Carbon Fibre Functionally Graded Materials', *Journal of Materials Science Letters*, **19**, 2000, 263–266.
67. Zhang, Z. and Friedrich, K., 'Artificial Neural Networks Applied to Polymer Composites: A Review', *Composites Science and Technology*, **63**, 2003, 2029–2044.
68. Velten, K., Reinicke, R. and Friedrich, K., 'Wear Volume Prediction with Artificial Neural Networks', *Tribology International*, **33**, 2000, 731–736.
69. Friedrich, K., 'Wear Models for Multiphase Materials and Synergistic Effects in Polymeric Hybrid Composites, in *Advances in Composite Tribology* (ed K. Friedrich), Composite Materials Series (series editor R.B. Pipes), Vol. 8, Elsevier, Amsterdam, The Netherlands, 1993, pp. 209–273.
70. Flöck, J., Beitrag zur experimentellen und modellhaften Beschreibung der Gleitverschleissmechanismen kohlenstoffaserverstärkter Polyetheretherketon (PEEK) Verbunde, in IVW Schriftenreihe Bd. 23 (ed M. Neitzel), Institut für Verbundwerkstoffe GmbH, Kaiserslautern, Germany, 2001, ISBN-3-934930-19-0.
71. Goda, T., Váradi, K., Friedrich, K. and Giertzsch, H., 'Finite Element Contact, Strain and Stress Analysis of Differently Oriented Fibre-Reinforced Composite Subjected to a Sliding Steel Asperity: Part 1: Normal Fibre Orientation', *Journal of Material Sciences*, **37**, 2002, 1575–1583.
72. Friedrich, K., Váradi, K., Goda, T. and Giertzsch, H., 'Finite Element Contact, Strain and Stress Analysis of Differently Oriented Fibre-Reinforced Composite Subjected to a Sliding Steel Asperity: Part 2: Parallel and Anti-Parallel Fibre Orientations', *Journal of Material Sciences*, **37**, 2002, 3497–3507.

73. Friedrich, K., Flöck, J., Váradi, K. and Neder, Z., 'Numerical and Finite Element Contact and Thermal Analysis of Real Composite-Steel Surfaces in Sliding Contact', *Wear*, **225–229**, 1999, 368–379.
74. Váradi, K., Neder, Z., Friedrich, K. and Flöck, J., 'Numerical and Finite Element Contact Temperature Analysis of Real Composite-Steel Surfaces in Sliding Contact', *Tribology International*, **31**, 1998, 669–686.
75. Williams, J.A., Morris, J.H. and Ball, A., 'The Effect of Transfer Layers on the Surface Contact and Wear of Carbon–Graphite Materials', *Tribology International*, **30**, 1997, 663–676.
76. Eriksson, M. and Jacobson, S., 'Tribological Surfaces of Organic Brake Pads', *Tribology International*, **33**, 2000, 817–827.
77. Häger, A.M., Polyaryletherketone für den Einsatz in Gleitlagern und Gleitelementen, Shaker Verlag, Aachen, Germany, 1997, ISBN 3-8265-2094-7.

12

Third-Body Reality – Consequences and Use of the Third-Body Concept to Solve Friction and Wear Problems

Y. Berthier

Abstract

This chapter presents an approach for providing a coherent and logical structure to contact mechanics models and linking them with tribological analyses in order to solve friction and wear problems.

This mechanical engineering approach deals with friction as a problem of third-body rheology, while wear is dealt with as a problem of third-body flows. The multi-scale effects are taken into account by the tribological triplet, which is composed of the mechanism that contains the contact, the two first bodies that form it, and the third body that attempts to separate them.

This approach using rheology and flows makes it possible to take into account the effects of the materials used in the contacts and the physicochemistry involved. By doing so, mechanical laws can be formulated with the aim of implementing solutions for each of the elements composing the tribological triplet.

The scientific argumentation has been kept to the strict minimum in order to keep the chapter brief.

Wear – Materials, Mechanisms and Practice Edited by G. Stachowiak

12.1 Introduction

Wear and friction are not expressed in terms of any specific values since they lack a unit of measurement. This apparent lack of unit in friction and wear description and lack of genuinely predictive models stem from the fact that various conditions can lead to the same friction or wear values [1]. As a result, in the models developed, these conditions need to be identified, together with the sequence of phenomena taking place. This would require placing the instruments inside the tribological contact. However, in most cases this is not possible without disturbing the contact conditions.

The first objective of this chapter is to demonstrate how the third body can be used as an indicator of contact conditions [2, 3]. The "third body" is the generic name used to describe the material imposed between the first bodies or generated as a result of interaction between the first bodies. This material layer separates the interacting first bodies and its thickness can vary from a monolayer to several micrometres [4].

The second objective is to show that the third-body concept can become a technological tool that can be used in tribological analysis, allowing for setting out guidelines for contact maintenance and benefiting from the feedback from the operation of contacts.

The third body can also be a scientific tool useful in understanding and modelling contact behaviour. It can be said that friction is a function of third-body rheology, while wear is a function of third-body flow (source flow, ejection flow, wear flow, etc.) [1, 5].

12.2 Relationship Between the Third Body and Friction

The concept of the third body can vary if the friction has a limit value in a structural mechanics problem or if it has to be adapted to a technological problem (e.g. low friction for a joint, high friction for a brake). This adaptation requires a friction analysis to be carried out first.

12.2.1 Boundary Conditions

When the friction value is only a boundary condition, it is not necessary to focus on the phenomena occurring in the contact, which is to say that in this case the notion of third body serves no purpose. The friction coefficient encompasses a large number of parameters and provides technological information that permits, for example, dimension selection in the design process without raising tribological questions. It can also be a parameter used to identify the vibratory behaviour of structures.

In such cases taking friction, in the form of Coulomb's or Tresca's models, into account is usually sufficient. The friction values deduced from more sophisticated models (molecular dynamics) can also be used [6]. When friction is used only as a boundary condition, it is often useful to parameterize its value in order to evaluate its effect on the behaviour of the structure. If this effect is too great, and depending on the risks involved, it may be necessary to carry out further friction analysis.

12.2.2 Friction Analysis

The contact phenomena must be controlled in order to know the friction coefficient value. Contrary to traction tests, which provide accurate data on the behaviour of materials, friction tests do not provide a sufficiently clear signature of the surface subjected to friction and there is no criterion that permits extrapolating the results from one contact condition to another. This requires the introduction of the third-body concept. For example, on the intuitive level friction can be linked to the "initial" surface roughness of bodies in contact. However, this surface roughness is often "erased" as soon as the contact is brought into operation, which results in (i) the production of the first third-body particles, and (ii) development of the new surface roughness. The particles produced are of submicrometre or micrometre size and they usually exhibit a large specific surface area that makes them reactive in the ambient environment. This reactivity is one of the factors leading to the adhesion of fine third-body particles that in turn, depending on the stresses to which the contact is subjected, results in a specific friction value.

Relationships exist between the initial surface roughness and particle adhesion, and thus with the friction coefficient. However, these relationships are too complex to be reliably formulated. This situation is quite similar to that of weather forecasts for which the fluttering of the wings of a butterfly in one hemisphere can affect the weather in the other hemisphere. Subsequently, the measurements need to be performed at the right scale in order to build models that permit understanding and control of the phenomena that, in turn, may permit the construction of increasingly accurate predictive models. Monitoring of third-body morphology and its distribution in a contact is a means of achieving a global and more uniform view as part of a mechanical approach to these different phenomena.

12.3 Relationship Between the Third Body and Wear

12.3.1 Wear Laws

Most of the wear formulae are based on three forms of Archard's equation [7]. The first formula

$$V_u = kF_nL$$

is based on geometric parameters where V_u is the worn volume, k the adjustment coefficient, F_n the normal force and L the distance travelled.

The second formula

$$V_u = KF_nS/\sigma_y$$

brings to light the influence of the softer material via its flow stress σ_y or its hardness. S is the area of the contact. The third formula

$$\frac{dh}{dt} = kPV$$

involves the energy dissipated in the contact via the product of PV, i.e. pressure $\times$ sliding speed, where dh/dt is the wear rate and h is the depth of the wear track.

These three wear laws or wear models have been modified to take into account various parameters characteristic of specific conditions, resulting in more than a hundred different formulae [8]. The higher the number of parameters in the formula, the easier it is to fit it to a complex experimental curve. This is the role of polynomial approximation. To achieve more general and more predictive models, Lim and Ashby developed wear maps based on "pin-on-disk" tests [9].

To allow extrapolation of wear map data to other types of contact, the maps providing information in terms of wear degradation (mild wear, severe wear, delamination, seizure, etc.) are presented as a function of non-dimensional velocity V and contact pressure P. These non-dimensional terms are obtained by using the combination of geometric and thermal parameters. However, the use of these non-dimensional graphs has not always proved accurate in transposing the results of one type of contact to another. It seems that taking into account the flow of the third body (tribological circuit) is one way of increasing the reliability of such transpositions [10].

12.3.2 Material Hardness and Wear

While friction is often intuitively associated with surface roughness, wear is often associated with the hardness of the materials in contact. The hardness of a material affects the contact area – since it is related to the material's Young modulus [11–13] – as well as the contact behaviour. Generally, the harder the material, the more brittle it is, and therefore more sensitive to the detachment of particles. Without going into much detail, it should be remembered that a material's hardness is a parameter that can have many effects, some of which are not desirable.

For example, although shot peening increases surface hardness, it also modifies its roughness. This surface roughness, comprising micro-pits, can trap the third-body particles produced by the degradation of asperities, while this degradation is being controlled by the compressive stress. Analysis of the effect of shot peening on the distribution and adhesion of the third body contributes towards distinguishing between the residual stresses and the stresses resulting from the geometry (roughness) of the interacting parts.

When measurements and criteria more suitable than hardness are found it would no longer be necessary to use the concept of a third body to facilitate understanding. The role of hardness and direct and indirect roles of materials in a wear problem will be considered in more detail in the following sections [1].

12.4 What Methods Exist for Studying Friction and Wear?

12.4.1 The Scientific Context Surrounding Tribology

All friction and wear models face the same problem that is dictated by a great complexity of mechanical and physicochemical contact phenomena and that affects the values of wear and friction used in these models. This complexity results in models having a large number of correlations between many parameters [8]. A mechanical and/or physicochemical interpretation exists in the literature for each of these correlations. Subsequently, these interpretations are transformed, step-by-step, into certitudes like on a Mobius band that on the scientific level hinders the development of a predictive model, while on the practical level ensures the sale of spare parts.

In this situation, solving a technical problem involving friction and wear requires the reduction of the number of parameters involved. This reduction begins with an in-depth mechanical analysis of the system, in order to identify the stresses transmitted across the contact. Studying the nature of the contacting materials becomes important only after this initial analysis. However, all too often these steps are carried out in reverse order.

12.4.2 Physical Difficulties Related to Studying Contacts

There are at least four different types of physical difficulties inherent to a scientific strategy.

12.4.2.1 The Contact's Geometry

The confinement of a contact, and often its small dimensions, means that it is difficult or even impossible to make the measurements without disturbing its operation. As a result, the contact operation is recreated on the basis of mechanical and physicochemical interactions with the outside world, i.e. via indirect measurements.

12.4.2.2 Effect of Tribological System

Three elements of the tribological system (machinery, first body and third body) act simultaneously, but it is often not clear which element is in control over the others. This is the eternal problem of the chicken and the egg.

(i) The machinery or mechanical device containing the contact imposes stresses and macroscopic geometries on the contact areas [1]. Thus this machinery has as much influence as the materials of the first bodies.
(ii) The contacting materials are subjected to a local compressive stress ranging, under specific conditions, from an instantaneous pressure reaching several GPa to an average pressure of only a few MPa, and also to considerable shearing stress gradients on the surface.
(iii) The third body, which separates the first bodies, is the medium that "produces" the friction and accommodates the velocity differences between the first bodies in relative motion. In the case of fretting, for example, this relative movement is manifested by local deformations.

12.4.2.3 The Response of Materials to Tribological Stresses

Apart from abrasion and adhesion [14, 15], which are relatively well controlled, materials respond locally to stresses transmitted by the mechanism, by cracking and especially plastic deformations. These deformations are flows of materials that can occur at a depth of several micrometres and distances parallel with the contact plane that can reach 500 μm. These flows are more considerable than classical plasticity and often occur with transformations of structure and phase. On a more general level, these behaviours are the scientific and natural response of materials subjected to tribological stresses, which is to say temperature, hydrostatic pressure and shearing gradients. This response is known as superficial tribological transformations (STT) (see also Section 12.7.2). The laws of behaviour of STT are in the process of being written [1, 16, 17].

12.4.2.4 Tribometry

The data supplied by tribometry are difficult to extrapolate since they do not correspond to specific local mechanical [1] and physicochemical contact conditions [18–20]. A tribometer is a mechanical device of a sufficient stiffness used to simulate stressed contacts in various physicochemical environments. During a test, the tribometer measures the integrated effect of these contact stresses on average tangential force, which greatly depends on contact conditions. At present, it is still difficult to obtain precise information on contact behaviour, and in particular on the phenomena that

(i) accommodate the difference in velocity between the first bodies (friction); and
(ii) cause the detachment of particles (STT, wear).

For example, differences in the value of tangential force can result from a different physical phenomenon activated, as well as from differences in the contact area. The physical phenomena activated can include the adhesion of the third body (humidity), the oxidation of the first and third bodies, the flow of the third body, etc. Regarding the contact area (domain of integration), observations during pin-on-disc tests with a sapphire pin have shown instantaneous contact area variations reaching 60% for the same normal load value and differences of the third-body thickness ranging from several nanometres to 60 μm (Figure 12.1) [3, 21]. Tribometry can almost be likened to a fishing net thrown around the

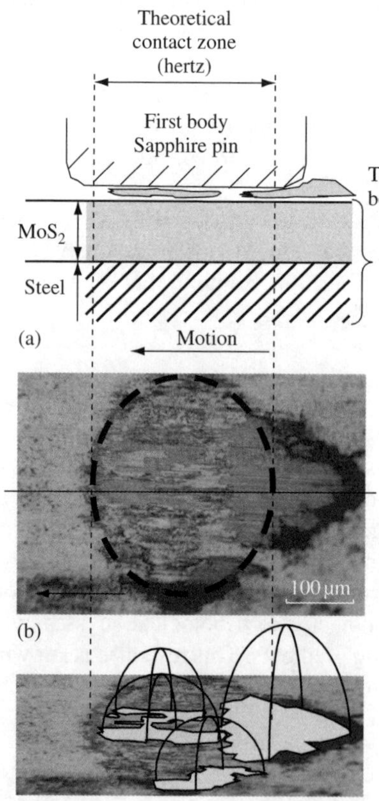

Figure 12.1 Instantaneous contact area with third-body particles of MoS_2 in pin-on-disc contact. (a) cross-section of a sapphire pin–disc contact. (b) Normal view of the contact. (c) Perspective view with the distribution of pressure zones [reproduced by permission of Dr. Yves BERTHIER]

contact, in which it is difficult to sort the catch; hence correlations between parameters are obtained that are neither in a general form nor predictive.

12.4.3 So Where to from Here?

To summarize, tribometric measurements are too global while *in situ* measurements are too delicate to carry out. This is why the essential parameters have not all been identified and their interactions (cause and effect chain) still remain to be determined. Thus greater progress in the area of test instrumentation is needed. One solution is (i) greater use of numerical simulations and (ii) increasing the validity of tribometer measurements by combining them with the results and knowledge of the third body, which can be used as a tracer of *in situ* contact conditions.

12.4.3.1 Numerical Simulation of Tribological Contacts by Calculation

These simulations can be performed for each individual element of the tribological system. The advantage is that it is possible to vary a single parameter at a time in a controlled way, which is practically impossible under experimental conditions. In addition, by calculating the resulting behaviour at a given point the simulations provide a form of instrumentation that does not disturb the contact, something that is not possible with physical instruments. At present, computational models exist for the machine design, the first body and, lastly, the solid third bodies (listed in order of decreasing efficiency).

(i) Numerical simulation models of mechanisms are developed using structural mechanics models. They are models of the mechanisms that contain the contact and do not pose any specific problems. They permit obtaining deformations of parts and the overall load patterns required to model the first bodies.
(ii) Models of first bodies range from the resistance of coatings and risks of cracking [22, 23] to those of the local dynamics of the contact. Local dynamics must be defined on a scale of a few micrometres, which is the relevant mechanical scale for current needs. For these dynamic models, special contact algorithms have been developed [24, 25]. These models permit, for example, highlighting the role of the Young's modulus, Poisson's ratio and relative velocities of the first bodies in the tribological condition of the local contact zones: adhesion, sliding and separation and impact (Figure 12.2) [26]. In fact, only the zones that adhere and slide in the contact effectively bear the normal load, which is to say that only a partial area of the contact is activated. In these models, friction is used as a genuine parameter in the same way as velocity and contact pressure. This means that friction is no longer a parameter used for "fitting" and adjustment, nor is it a kind of potentiometer used to fit models to match reality.
(iii) Discrete element models of solid third bodies are now being developed and used to simulate, for example, the effects of adhesion of third bodies on their flows [27]. These models have become a tool for differentiating between mechanical and physicochemical phenomena occurring in a tribological contact.

To summarize, the numerical models open up new possibilities for "numerical tribology" and also permit better targeting of the contact areas to be analysed. The detailed description of these models is beyond the scope of this work.

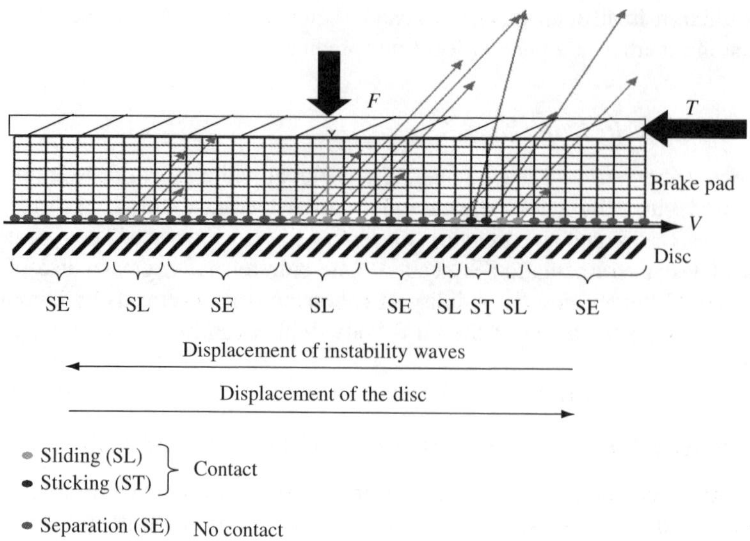

Figure 12.2 Visualization of instability waves at the brake pad–disc interface [reproduced by permission of Dr. Yves BERTHIER]

12.4.3.2 Experimental Simulations of Tribological Contacts

It is not necessary to discuss the experimental problems associated with both tribometers and industrial contacts as they are all well known. In the following sections, emphasis is placed on the scientific and technological use of flows and the rheology of solid third bodies.

12.5 The Third-Body Concept

12.5.1 Artificial and Natural Third Bodies

The third body is a concept introduced in 1974 by Mr Godet to (i) link friction and wear studies with those on lubrication, and (ii) take into account the influences and interactions of material and physicochemical parameters in a global mechanical approach. This concept is used in modelling as an artifice to provide a global view, otherwise known as the homogenization method in mathematics [28]. Physically, the third body is a continuous or discontinuous monolayer that separates the first bodies.

Thus the third body is a mechanical concept. This concept is known in lubrication as the lubricant film, the third body, which separates the two first bodies. In friction and wear, the existence of the third-body layer was first recognized in the following cases:

- three-body abrasion, in which the third body is composed of free abrasive particles;
- dead leaves, sand and other contaminants that come between wheel–rail contacts [29, 30]; and
- a layer of silver or MoS_2 placed in certain tribological contacts operating in vacuum (in satellites) to avoid “cold welds” and permit “solid lubrication” [31, 32].

In these cases, the third bodies are artificial, as the material composing them enters the contact from outside. It has taken much longer, however, to recognize the importance of natural third bodies produced in the contact and resulting from degradations of both the first bodies. This lack of knowledge stems from the fact that the first bodies were examined after stopping and dismantling the contact, therefore giving a static view of the contact that led to the generalized notion of transfer film.

The presence of the third-body layer is quite difficult to examine since its thickness is very low, often only a few micrometres, and the layer is often obscured by the edge effects of metallographic sections [2]. Since the 1980s, in industrial applications the notion of third body, known as "patina," has been used as a criterion for optimizing the composition of graphite brushes used for rotating electric collectors [33].

A natural third body can be composed of (i) oxide and contaminant layers (often called surface complexes), and (ii) particles detached directly from the first bodies. The lubricant leaking, for example, from the pores in porous bearings is considered as an artificial third body since it is initially introduced by impregnation [34]. Monolayers placed on first bodies are also artificial third bodies [35, 36]. The oil additives that only produce active third bodies if they react with surfaces can be classified as natural third bodies when the reactions involve local material degradation [37, 38].

12.5.2 Contact Without the Third Body

If a relative movement of first bodies eliminates the surface complexes (oxides, contaminants, etc.) and they do not re-form, the surfaces may become sufficiently reactive to stick together. Continuation of movement between the first bodies then leads to a surface degradation resulting in a production of a natural third body in the form of particles. In practice, a contact without a third body (surface or particle complex) does not exist. The production of a natural third body through the degradation of first bodies is the natural response of materials subjected to tribological stresses (see also Section 12.7). In other words, the surface is worn preferentially to protect the material beneath it.

12.5.3 Types of "Solid" Third Body from the Mechanical Viewpoint

From the viewpoint of mechanical models, a distinction must be made between fluid third bodies, which are modelled by fluid mechanics (Reynolds equation), and solid third bodies for which modelling tools are currently being developed. These models and the characteristic magnitudes at which solid third bodies act lead to separating these solid third bodies into two families, i.e. surface complexes and particles detached from the first bodies.

12.5.3.1 Surface Complexes

Surface complexes are physisorbed or chemisorbed layers of oxides and contaminants or monolayers deposited on surfaces. The thickness of these layers rarely exceeds 10 nm. These surface complexes significantly reduce adhesion strength, since they typically cause surface energies to fall from about J/m^2 to 40 mJ/m^2 in ambient atmospheres or less in the case of PTFE [39]. This reduction of adhesion strength by the presence of these layers between the first bodies has sometimes led to calling surface complexes "screens." This layer separating the first bodies can contribute towards accommodating their differences of velocity (see also Section 12.5.4).

12.5.3.2 Detached Particles

Particles directly detached from one or both first bodies are usually several nanometres in size (see also Section 12.7). They agglomerate together to form films of several micrometres in thickness. These agglomerates of "solid" material flow during the deformations and relative movements of the first bodies and therefore their morphology is fashioned by this flow. Subsequently, their morphology depends more on the flow conditions than on the initial formation conditions. This situation often leads to some confusion when interpreting test results with the aim of understanding the mechanisms underlying the detachment of the first particles. The agglomerates of the first particles are commonly called wear "particles." This terminology will be used in the following sections.

12.5.4 "Action Heights" of Third Bodies

The action heights of surface complexes and detached particles differ due to their sizes. Typically, the active height of a surface complex is several nanometres. For example, lipids bound by their hydrophilic heads to a first body form a carpet of molecules 2.3 nm thick. If this carpet is rubbed against an identical molecule carpet, this leads to a separation between the first bodies of $2 \times 2.3 = 4.6\,\text{nm}$. These two molecule carpets can shift in relation to each other over a distance where the surface roughness of the first bodies is less than the height of their separation [35, 39]. In industrial applications, it is not possible to obtain such low values of surface roughness by conventional machining, thus the surface complexes can only be active for contacts of a size less than 10 μm.

However, if the third body (detached particles) fills the surface pits during friction, it can form a new first body smooth enough for the surface complexes to act over large distances of displacement. For example, hard and brittle materials such as ceramics can be polished quite easily, and are capable of producing sufficiently flat surfaces to favour the action of surface complexes or very thin lubricant films (less than 0.1 μm).

Typically, solid third-body particles have "heights of action" (thicknesses) of several micrometres, and that allows them to act between the surfaces produced during normal machining conditions.

The activation of surface complexes can result in considerable differences in friction coefficient values. For chemically identical first bodies, the activation of surface complexes can lead to a coefficient of friction as low as 0.001, while the activation of particles can lead, for example, to a coefficient of friction of about 0.1.

12.6 Functions and Behaviour of the Third Body

12.6.1 Functions of the Third Body

A third body has four main functions, i.e.:

- To transmit normal load. The load-bearing capacity of solid third bodies (in e.g. bearings) exists statically, while that of fluids requires pressure (according to the Reynolds equation) that can be generated by the relative movement of the first bodies (hydrodynamics) or by external pressure (hydrostatics) [40].

- To separate the first bodies, i.e. avoid their direct interaction and contribute to reduce their deterioration.
- To accommodate the velocity difference between the first bodies, i.e. they can shear and then reform.
- To transform, usually by shearing, kinetic energy into heat and then to dissipate the heat generated via the first bodies or via a flow of the third body. In friction brakes, this energy conversion function is amplified, whereas it is reduced in "dry" bearings. Part of the energy dissipated during friction can be seen as the energy needed for keeping the third body in the contact.

12.6.2 Operation of Solid Third Bodies

Observation of contacts after operation gives a static view of the third body, which is usually known as a transfer film and in machining as a built-up edge. It is therefore necessary to distinguish whether this is a genuine material transfer from one first body to another, i.e. whether it remains bonded to it by adhesion, or whether it is a static view of a flowing film which accommodates the velocity. This occurs, for example, at the interface of the contact of a piezoelectric motor (Figure 12.3) and of a brake pad–disc contact (Figure 12.4) [2].

Examination of a contact through a sapphire first body revealed that a flow of "solid" third bodies occurs, thereby leading to the notion of third-body flows [3, 41]. As already mentioned, these flows of almost solid material can be explained in terms of (i) instantaneous local pressure in the contact reaching a tenth of a GPa for apparent pressures of only a few MPa, and (ii) very high shearing gradients in the layer of third body. It is known that hydrostatic pressure considerably reduces the value of the plasticity threshold and increases the deformation capacity of materials [16, 17, 42, 43]. It should be remembered that when the hydrostatic pressure is high enough liquids may behave like solids, while solids may behave like liquids.

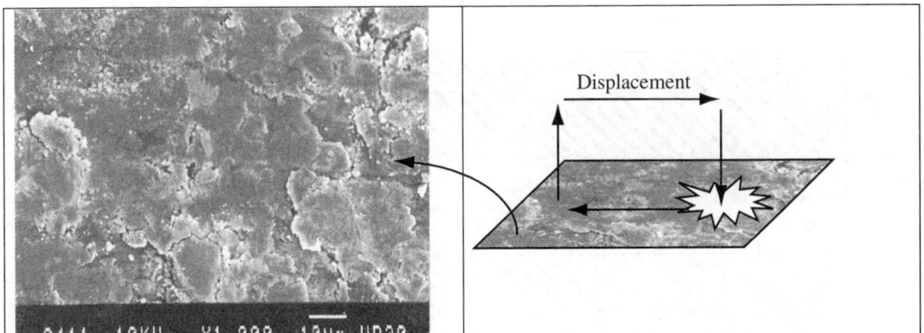

Figure 12.3 (a) Normal view of third body at the rotor–stator interface of a piezoelectric motor. (b) Kinematics – reciprocating and impact motion [reproduced by permission of Dr. Yves BERTHIER]

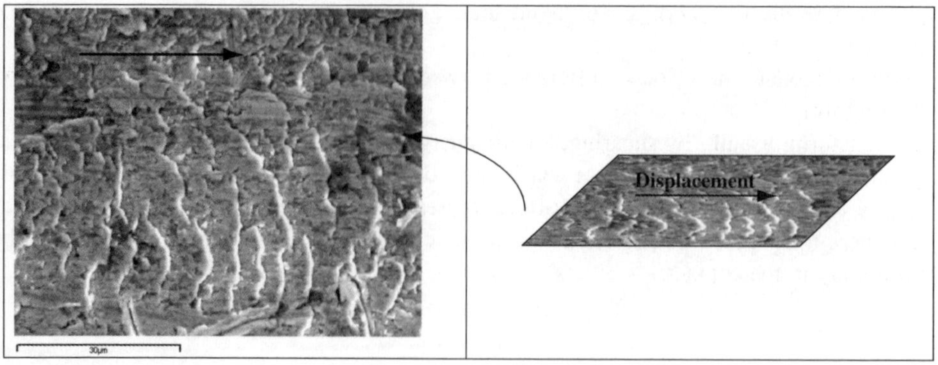

Figure 12.4 (a) Normal view of third-body interface in a brake pad–disc contact. (b) Kinematics – sliding motion [reproduced by permission of Dr. Yves BERTHIER]

To model the contact operation with a third body composed of solid particles, it is necessary to reconstitute and evaluate its flows and rheology. These flows make up the tribological circuit.

12.6.3 Tribological Circuit of Third-Body Flows

The tribological circuit, shown schematically in Figure 12.5, represents different flows of the third body which are likely to be activated in an elementary contact [1, 3].

As shown in Figure 12.5, flow Q, source flow Q_S and internal source flow Q_S^i correspond to the detachment of particles caused by STT, cracking, adhesion, etc. This leads to the formation of the natural third body. External source flow Q_S^e stems from the introduction of the artificial third body in the contact. Internal flow Q_I is the flow of the third body that circulates between the first bodies. External flow Q_E is the flow of

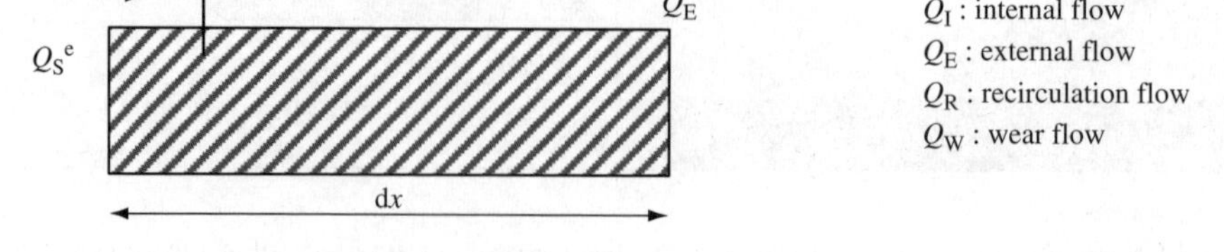

Figure 12.5 Tribological circuit [reproduced by permission of Dr. Yves BERTHIER]

the third body that escapes from the contact. It is divided into a recirculation flow Q_R and a wear flow Q_W. Q_R is composed of the third body that when reintroduced into the contact by, for example, dragging by one of the first bodies will contribute to velocity accommodation. However, flow Q_W is composed of the third body which, when finally ejected from the contact, no longer participates in velocity accommodation and load bearing.

In a real contact, the elementary tribological circuit is three-dimensional and must be extended to cover the entire contact by distinguishing the active flows that contribute towards separating the first bodies (load-bearing zones) and the inactive flows that, although transiting in the contact, do not contribute towards load bearing. These flows can, however, feed the load-bearing zones.

The tribological circuit is activated by mechanical (pressure, shearing, inertial effects, gravity, etc.) and physicochemical actions (oxidation, etc.) [18].

12.6.4 Rheology of the Third Body

Rheological studies of the third body and the formulation of corresponding flow models are in progress [16, 17, 27]. However, the main problem is associated with the measurement of the rheology of the solid third bodies over a wide range of stresses (shearing stress and its gradient). This is why characterization of the third body by nano-hardness measurements is inadequate. Indeed, stress fields under an indenter are very different from those found in real contacts. At present, the rheology of the third body is evaluated on the basis of observations of third-body particles whose morphology has been shaped by the flows. These observations allow for qualitative and relative evaluation of the ductility and adhesion of third bodies [1, 3]. The details of these works are not presented here.

12.6.5 Scientific and Technological Consequences of the Tribological Circuit

From the scientific viewpoint, friction in the mechanical model is a function of the rheology of the third body, whereas wear is a function of third-body flows. This model has been validated experimentally [1, 3]. However, discrete element models of third bodies still remain to be developed [27]. The research in this area is focused on the development of numerical models of contacts and the development of new rheological techniques for studies of solid third-body flows.

In technical terms, the tribological circuit serves as a background for the contact analysis and the formulation of a solution. The tribological circuit also permits the experimental identification of the parameters that activate the different flows and affect the adhesion of the third body. When formulating a solution, the tribological circuit and rheology of the third body, approximated by its morphology, are the two main criteria that can be used to transpose the results from a tribometer to the real contact. These two criteria are added to classical transposition criteria, i.e. contact pressure and the relative velocities of the first bodies. We shall now focus on their use after having specified the role of the materials in a tribological contact.

12.7 Roles of the Materials in a Tribological Contact

The effects of first body materials on wear and friction are often combined in order to simplify the problems. However, in practice this may produce over-simplifications leading to inaccuracies, which in turn cause erroneous interpretations of test results and erroneous formulations of solutions.

These errors are due to the following facts:

(i) the material of one of the first bodies transmits stresses to the other;
(ii) the first bodies respond to these stresses by undergoing complex and substantial local changes; and
(iii) the contacts (first bodies + third body) are a boundary condition of the mechanical device containing them.

Consequently, the mechanical system in which the contact is a part plays a role in the stresses to which the contact is subject. The problem is therefore both multi-scale (mechanism, first body, third body) and coupled (the stresses to which the contact is subjected depend on both the contact and the mechanism). This situation creates a kind of loop in which the mechanism and contact form a system analogous to a snake biting its own tail.

Thus it becomes necessary to specify the role played by the materials (i) at the scale of the actual mechanism (structural mechanics, elastic response), called the indirect role, and (ii) at the scale of the contact, i.e. the first bodies (contact mechanics, tribological response). The much more complex role, on the scale of the third body (discrete element mechanics, rheology of solid third bodies), will only be mentioned here.

12.7.1 Indirect Role of the Materials – Scale of the Actual Mechanism or Mechanical Device

The actual mechanism or device contains the contact that transmits the vector T_c of the forces to which the whole contact is subjected (first and third bodies). This vector depends, on the first approximation, on the forces applied to the device and the stiffness of its components. The components' stiffness depends on the material's mechanical properties and also on the geometry of the components. Subsequently, changing the material, which translates mechanically as a change of the Young's modulus, could be counterbalanced by a variation of geometry. In this case, the materials at the scale of the actual mechanism play a role via the geometry of its components; hence the term "indirect role" is used. At this scale (actual mechanism), the response of the materials to the stresses ranges from elastic deformations to volume fatigue. These responses are taken into account by structural mechanics and are determined by well-established tests (traction, fatigue, etc.).

12.7.2 Direct Role of the Materials – Scale of First Bodies

Recent advances in contact mechanics permit calculations of the stress fields and deformations, based on the vector T_c of the forces exerted on contact, at the scale of the first bodies, even though the presence of the third body is not taken into account. This stress field can be determined at a depth of several dozen micrometres with a resolution of 1 μm.

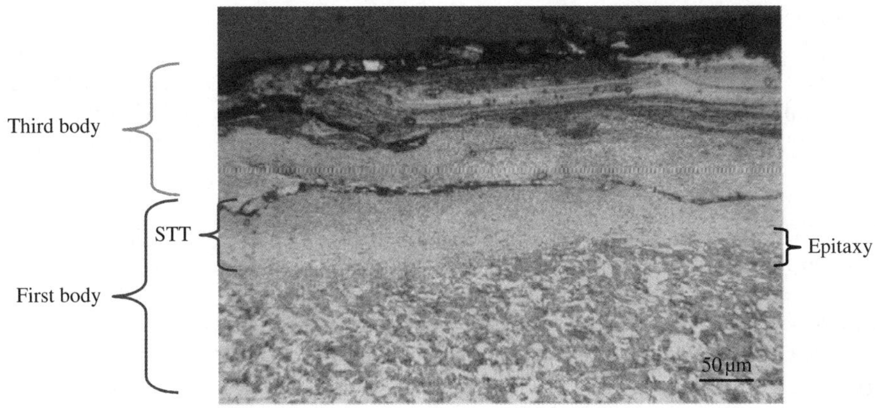

Figure 12.6 STT and third body obtained in a wheel–rail contact: longitudinal cross-section of the wheel [reproduced by permission of Dr. Yves BERTHIER]

These surface stress and deformation fields (first body) differ from the internal bulk body fields by (i) high stresses generated under hydrostatic pressure, and (ii) very high shearing gradients. These conditions are specific to tribological contacts.

The material's response to these surface fields is characteristic for a specific material. These responses are much more complex than the usual elastic response of the structural material (volume). They range from the initiation of cracks under tribological stresses to phase and structural changes known as STT. These STT in turn lead to the production of the natural third body (Figure 12.6) [21, 29]. As these responses bear the signature of the material, the term "direct role" is used.

It should be stated that the STT are the most common and natural response of materials to tribological stresses [2, 16, 17]. This response has been identified in polymers, metals and carbon materials. The STT formed by our skin are known as "stratum corneum." STT studies that use mechano-synthesis are not dealt with here. Only optimal direct response is briefly described.

12.7.3 Optimal Direct Response of Material to the Tribological Contact

The best responses that the first bodies can provide to reduce wear are:

- To ensure, without decohesion of the material, the transition between the bulk material stress field and deformations (structural mechanics, fatigue) and the surface stress fields (contact mechanics, fatigue under tribological stresses). This accommodation of "surface volume" is difficult to satisfy since it is the surface that transmits the stresses to the bulk material, although it is the bulk material that supports the surface. Physically, this problem of the first body is solved by the epitaxy between the STT, the volume of the first body and the nanostructure of the STT [2, 44] (epitaxy ensures the continuity of the atomic links of a crystallized structure when the latter is transformed into an almost amorphous structure).

- To produce natural third bodies (source flow) with minimum degradation of the first bodies. It should be remembered that if there are no artificial third body and surface complexes (screens), degradation of one of the two first bodies is required to produce the natural third body necessary for the contact to function. The STT are the responses permitting the production of the third body with the minimum amount of degradation.

12.7.4 Consequences on the Approach Used for Solving Technological Problems

To summarize, materials play an indirect role at the scale of the whole mechanism. This role is taken into account by structural mechanics. However, there is a direct role of the materials at the scale of the first bodies, a role relatively well taken into account by contact mechanics. The two functions of this direct role are (i) to accommodate the stress fields, and dilatational and surface deformations, and (ii) to ensure the source flow of the third body.

Studies of the first body material response as well as the studies into the role of the materials in the rheology of natural third bodies are in progress. The role of material is being progressively elucidated by discrete element mechanics (granular), while the study of solid third-body rheology is in its infancy.

In conclusion, the general trends of mechanical and predictive models of friction and wear are gradually being coordinated. However, an approach is used in which the pertinence of the models decreases as one goes from models of the whole mechanism to models of first and finally the third bodies. The solution to a friction and wear problem must offset these shortcomings in modelling by a phenomenological approach based on analysis of experimental data. In the following section the methodology used to offset the shortcomings of modelling at the level of each of the elements of the tribological system (first and third bodies) is described.

12.8 Taking into Account the Effects of the Mechanism

The finite element method (FEM) (structural mechanics) is used in mechanism modelling in order to determine the vector T_c of forces exerted on the contact. Iterations between structural and simplified contact mechanics make it possible to include the contact stiffness in the calculation of T_c using FEM [45].

12.8.1 Choosing the Conditions to be Modelled

During iterations between structural and contact mechanics, it is necessary to ensure that the operating conditions (forces, velocities, etc.) modelled at the scale of the actual mechanism are those that result in the most contact damage. It should to be mentioned that the maximum stresses at the mechanism or device level that generate the tribological stresses do not always result in most contact damage.

Accelerations or forces measured when wear has caused an excessive clearance in the machinery should not be included in the models. In such cases, the contact impact forces will certainly be greater than when the clearances were smaller, as they were at the beginning of the machine lifetime, i.e. when the degradations began. These forces when used in calculations will lead to the stress fields and the responses that greatly differ from those that would have resulted in wear, which must be studied in order to solve the problem. Thus the modelling conditions must be validated by an analysis of the mechanism studied.

12.8.2 Technological Consequences of the Effects of the Mechanism

From the technological viewpoint, the iterations between structural and contact mechanics permit solutions at the machine design stage, e.g. by focusing on the shapes of the machine components in order to minimize contact stress and deformations. For example, in the case of fretting, modifying the geometry of components can make them more deformable so that they can more easily accommodate relative movements across the contact. This approach permits contact simulations in different parts of the mechanism and thus reduces or avoids the risk that a solution found for one contact problem will produce a new problem elsewhere (this often occurs when dealing with fretting).

12.9 Taking into Account the Effect of the First Bodies

Iterations between structural mechanics and simplified contact mechanics provide the force vector T_c required in "fine" contact mechanics to calculate the local contact dynamics at the scale of several micrometres and the stress field together with the local deformations.

12.9.1 Local Contact Dynamics

Recent advances in contact mechanics (numerical tribology) permit calculations of the local dynamics of a contact with a resolution of a few micrometres. This resolution provides, for example, the tribological status of the finite element mesh nodes of two bodies in contact. During the macroscopic movement of first bodies at constant relative speed, the full cycle involves sticking followed by sliding, separation and impact when the nodes are back in contact [26].

In recent contact models the friction coefficient is no longer a setting parameter used to identify the behaviour observed experimentally. The friction coefficient is a real parameter whose effect upon the stress and deformation fields can be studied in detail. As mentioned previously, friction has become a parameter in its own right, with its own characteristics, rather than a parameter merely used to adjust or "fit" models to reality.

Even though the laws describing the material's behaviour, introduced in contact mechanics models, are not yet fully formulated (see also Section 12.6.4), the classical plasticity laws used permit reasonable approximations of local dynamics that can be adapted and validated. The local dynamics, which are quite reasonably approximated in these calculations, permit calculation of stress fields and deformations that in turn control the response of the materials (direct role). This material response controls the source flow of the third body.

12.9.2 Technological Consequences of the Effects of the First Bodies

Detailed studies of (i) material parameters such as Young's modulus, Poisson's ratio and thermal characteristics, and (ii) functional parameters such as residual stresses, friction and contact pressure permit highlighting the effects of these parameters on surface stress fields and deformations. Without even taking into account parameter–damage relationships, these calculations permit studying, for example, whether a substrate/coating pair can resist static thermomechanical stresses. The thickness of the coating could then be optimized.

More detailed analysis can even specify the relative values of the thermomechanical properties of the coating of different layers [12, 13]. Contact mechanics is gradually becoming capable of selecting coatings for given applications. Efforts must still be made, however, to measure the thermomechanical properties of coatings whose thickness is less than a few micrometres.

Whatever the case, contact mechanics at least helps to determine what levels of surface stresses are admissible. Subsequently, a number of friction tests with frictional pairs that would have no chance of surviving can be avoided, thereby saving much time and money.

Using local dynamics, contact mechanics also permits studying vibrations that can lead to a "tribological noise." This approach "at the source of the noise" is a new direction of research [46].

Lastly, to be really predictive, contact mechanics must provide laws of damage under tribological stress, i.e. STT formation and behaviour, crack initiation and the activation of source flows from STT. These damage laws refer to the material at a depth of the same magnitude as that affected by the procedures used to manufacture the machine parts (machining to remove material, the skin effect during polymer injection moulding, etc.). These laws are difficult to formulate. At the present time, this lack of laws is compensated by the expertise gained in first- and third-body studies.

12.10 "Solid" Natural Third-Body Modelling

Molecular dynamics is an efficient means of modelling the behaviour of surface complexes (screens), but these are rarely used in industry and so are not discussed here. The method, however, allows for the studies of natural third-body screens composed of particles detached from first bodies. These natural third bodies called "solid third bodies" are frequently present in industrial contacts, though they often go unnoticed. Controlling their flows (tribological circuit) and their rheology is the key to solving almost 80% of industrial friction and wear problems.

12.10.1 Reconstruction of the Tribological Circuit

Discrete element models of solid third bodies are progressively being used. They are the qualitative tools aiding the reconstruction of the tribological circuit and identification of stresses that activate different flows (contact life scenario). However, the reconstruction of the tribological circuit, which is activated in a real industrial contact, remains the most phenomenological and delicate part in studying a friction and wear problem.

In the simplest cases, this reconstruction can be performed on the basis of experiment. When this information is not available, then it is necessary to go back and forth between the expertise available on the specific industrial contact and test results obtained on a tribometer. In certain cases, it can be useful to use a special simulator that better approximates a particular industrial contact than a simple tribometer. These tests help to validate hypotheses made during the investigation of industrial contacts under known conditions. Furthermore, typical parameters such as velocity, contact pressure and the morphology of the third body serve as transposition criteria between the real contact and the tribometer. This phenomenological work consisting of formulating the contact life scenario (tribological circuit and mechanical and physicochemical actions that activate the flows) will be described in Section 12.13. Finally, when developing the practical solutions, it is the tribological circuit that can be used as the transposition criterion between the application and the tests on the tribometer.

12.10.2 Technological Consequences of the Third Body

To ensure the best control of the specific situation it is necessary, using the knowledge available, to make sure that (i) the natural third body is obtained from a single first body; (ii) STT is of the shallowest depth possible; and (iii) detachment of nano-particles is occurring. As a general rule the formation of STT is preferred to cracking, which should be avoided. On the other hand, in certain cases cracking can play a positive role by relieving stresses (e.g. stress cracking of disc brakes) [47].

The solid third body obtained must thus "reconstruct" itself [1] as easily as possible when sheared in the contact. This ease of reconstruction implies that the third body is almost physicochemically inert in relation to its environment, unless the formation of screens occurs, though this requires strong reactivity (see also Section 12.5.4).

12.10.2.1 Third-Body Flow and Wear

Analysis of the tribological circuit must determine whether the wear flow is high because the source flow is high or the ejection flow is high. If the source flow is high, then it generates a high internal flow downstream, which in turn generates a high ejection flow. In this case, to reduce wear it is necessary to focus on the source flow and thus on the local dynamics of the contact and/or on the formation of STT.

If the ejection flow is high, then it empties the contact of its third-body content downstream. In this case it is the lack of third body needed to separate the first bodies that reactivates the source flow. Here, to reduce wear, it is necessary to minimize the ejection flow, which is controlled by the following: the contact's geometry and/or the dynamics of the assembly and/or the adhesion of the third body. The geometry of the contact can be designed to increase retention of the third body. As for the dynamics of the assembly, this can lead to third-body ejection from the contact by gravity (mechanical action) or by oxidation. Oxidation can make the third body more powdery (physicochemical action), in which case it leaves the contact more easily in solid or in gaseous form [10, 18]. Lastly, the adhesion of the third body (rheology) can be modified by adding elements to the first body from which it stemmed.

12.10.2.2 Third Body and Friction

Friction caused by the solid third body depends mainly on its rheology. The rheology-activated flow depends on tribological stresses and especially on the sliding speed of the first bodies [29]. This problem of rheology is still being studied and depends on the coupling between mechanical and physicochemical actions. This coupling is especially important since natural third bodies often have specific surfaces that make them chemically reactive. Regarding discrete element models, these do not yet provide the resolution required to supply technologically viable results. Thus it is necessary to carry out laboratory tests.

12.10.2.3 Tests to Control the Source Flow of the Third Body

Even when using the tribological circuit as an aid, transposing the results obtained for a simple tribological state on a tribometer onto a complex tribological state in an industrial application is far from easy. This stems from the difference in stiffness between the two systems (tribometer and real contact) and also from the fact that any modification made to the composition of a material results in changes to its bulk properties and the properties of the third body produced. It is difficult to separate these two effects without using models. For example, the addition of special compounds to a polymer first body can significantly modify its thermomechanic properties, the local dynamics, chemical composition of the third body produced and also its rheology. It is even possible to control the magnitude and nature of the source flow by adding elements that "pre-fragment" the initial particles and use hard, i.e. brittle, particles (e.g. ceramics) that fragment on detachment into much smaller particles. If these particles are inert, they will mix together in the third body, thereby homogenizing the local properties of the latter (rheological properties), reduce its adhesion (by increased formation of powder) and most probably the friction. The advantage of fragmentation is to obtain particles that are much smaller than those that could have been dispersed during the material's manufacturing. Therefore, it is necessary to act on the rheology of the third body by modifying the microstructure of the materials used as the first bodies.

12.11 Correspondence of the Strategy Proposed to Reality

The strategy used in solving the problem described above merely applies the techniques used with fluid third bodies. However, the main differences are that little is known about the rheology of the solid third body as its production is difficult to control and it occurs *in situ*. The strategies described have been followed, often involuntarily, for the last 30 years and were used, for example, in car brake development, by modifying the clamping mechanism and the composition of the pads (first and third bodies) [48].

Modelling of the tribological circuit, i.e. allowing for the prediction of wear and friction, will become possible when

- structural mechanics, contact mechanics (first bodies) and the mechanics of the third body (discrete elements) are coupled, which still raises several mathematical problems; and
- the laws of behaviour under tribological stress of the first and third bodies are known.

Subsequently, during the next few years the design of a contact with a natural solid third body will require analyses of real parts and simulation tests to reconstruct the tribological circuit and estimate local stresses (contact life scenario). The information gained from the real parts influences the construction of a prototype.

12.12 Control of Input Conditions

12.12.1 Objectives

The objective is to formulate a contact operation scenario, i.e. to establish (i) the tribological circuit; (ii) its conditions of activation; and (iii) the rheology of the first and third bodies. As mentioned above, the study of the first and third bodies is one of the last solutions to offset the lack of models.

The studies conducted refer to each element in a tribological system by exploiting the possibilities of structural mechanics (machinery), contact mechanics (first bodies) and discrete element mechanics (third body) to the maximum.

Each study must be organized to capitalize on feedback from experience that may come from maintenance, diagnostic tools and by designing time-saving procedures in order to find solutions as far-reaching as "material design." These goals can be achieved by constructing a map of first- and third-body morphologies. These morphologies would be linked to the known behaviour of industrial contacts and reproduced on test rigs.

12.12.2 Procedure

The studies conducted can be likened to a difficult phenomenological task carried out by researchers in interactive fashion between real objects and tribometer tests. The goal of these tests is to establish the contact conditions that lead to certain third-body morphologies, which are the same in real and laboratory (tribometer) conditions. To achieve this the third body can be used as an instrument placed inside the contact, i.e. as a tracer of contact conditions (localization, local movements, etc.). Then it would be possible to establish whether the consequences of the controlled modification of a real mechanism correspond to that what was expected.

The study progresses from the exterior towards the interior of the contact, without any cleaning of the latter, since this would eliminate the third body and most of the information. When dealing with big parts, the study can be carried out using replicas or moulds. Only the major outlines of carrying out such studies are given here, beginning with the short description of the precautions to be taken.

12.12.3 Precautions

The tribological study is based on surface examinations (photographic, microscopic) and surface analysis; both are usually conducted after dismantling the contact. Surface analysis (mapping of species distribution) must then be interpreted dynamically to reconstruct the flows of the third body. These examinations are made on surfaces whose microstructures have been changed and which are usually covered with wear particles. Under these conditions, most examination techniques have reached the limits of their resolutions since they are calibrated for much smoother and bigger surfaces or much larger volumes of material than those available in tribological contacts. This situation therefore requires that great care is taken when interpreting the results obtained, followed by careful interpretation of these results in terms of mechanical consequences (flow, rheology).

It is also necessary to try to avoid conclusions, without confirmation, referring to apparently obvious features such as abrasion striations, traces of adhesion and thermal effects. Regarding the latter, one must know whether changes of phase may also have been caused by thermal effects or by pressure.

Comparison of the reconstituted first-body materials and particle behaviour to the behaviour of these same materials in their initial conditions should be avoided. For example, "hard" oxides in bulk state can have poor mechanical properties in powder state. The first bodies and particles of the third body often have specific surfaces and that makes them very reactive. Thus mechanical and physicochemical interpretations must be performed at the same time.

Even though it may appear trivial, it should be remembered that the study often focuses on visible evidence after operation, i.e. after plastic deformation has taken place. However, elastic deformations must also be considered and care must be taken during studies involving viscoelastic materials such as polymers.

Understanding of the contact behaviour is facilitated if the characterization of surfaces (normal view) is supplemented with characterization of metallographic cross-sections. At least it is necessary to visualize the phenomena taking place in section view, based on normal surface observations, i.e. as a function of the thickness of the third body and the first bodies.

Smooth transition between the scales of the different techniques used during observation and analysis must be ensured.

The contact condition at the time of examination is not necessarily representative of the conditions that initiated the problems to be solved. The contact can also manifest the superposition of several stress conditions, especially functional and parasite stresses. These stresses can be due, for example, to the operation of a particular mechanism assembled on another main mechanism, thereby adding to the global stresses affecting the whole assembly (vibration). Subsequently, it may prove to be useful to determine the contact conditions at the start of its life, i.e. under standstill conditions, in order to differentiate between the results obtained at different stresses.

12.13 Performing Experiments

12.13.1 Initial Conditions

It is necessary to know the initial mechanical and physicochemical condition of the first bodies (surface, skin) when assembling the contact, as the contact's initial condition will affect its subsequent performance. The reason for this is that the tribological stresses can affect the contact performance in a way similar to any disturbances introduced during the manufacturing process.

For example, when visualizing sections of machined material, different conformations can be seen such as hollows, "cavities," folded asperities and so forth (Figure 12.7). These conformations can lead to radically different mechanical behaviour (resistance) and physicochemical (oxidation) behaviour in the first body. The presence of fluid residues is also important (surface complexes).

In the case of polymer moulding, skin effects (a term specific to die cast moulding in metallurgy and plastics) must be taken into account together with the properties of the material's deformation gradients as a function of thickness, i.e. as a function of first-body wear. Periodic use of mould release agents in the form of surface complexes can also have a significant effect on friction by dispersing it.

In the case when coatings are used, the thinner they are, the more difficult they are to characterize, and the more sensitive they are to small variations of operating parameters.

To conclude, without knowledge of the operating parameters, it is necessary to establish protocols that would guarantee the reproducibility of the conditions used in the production of first bodies.

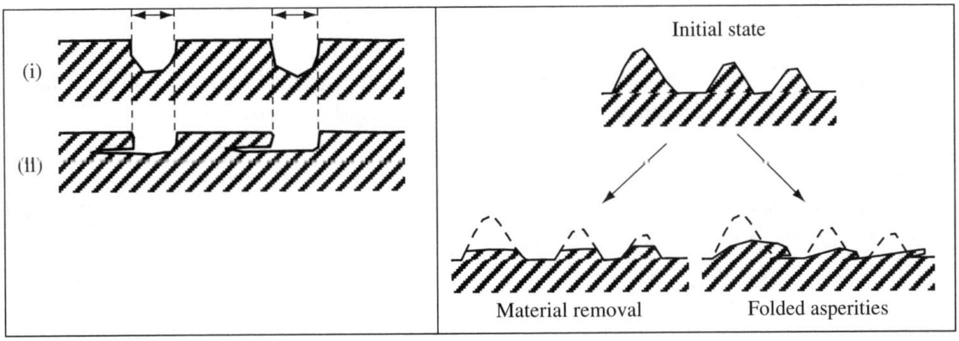

Figure 12.7 (a) (i) Hollows and (ii) "Cavities" (b) Different conformations of machined material [reproduced by permission of Dr. Yves BERTHIER]

12.13.2 Exterior of the Contact

It is necessary to localize the third-body ejection areas around the contact as a function of time. The third body can easily be ejected several centimetres from the contact. This ejection flow is characterized by quite violent local contact dynamics and is different from gravity flows that can occur when the contact is momentarily unloaded. It should be mentioned that violent local dynamics can occur with constant relative movements of the first bodies. These ejections can be observed when examining door hinges, which often bear the traces of considerable ejections of third body.

Examination of the morphology of ejected wear particles gives indications of their migration mechanisms inside the contact. For example, shapes that are more or less extruded, smoothed, drawn, etc. indicate different the migration mechanisms involved. In addition, the chemical composition of these particles indicates their origin (source flows).

The initial indication of the stresses in play must be correlated with the results obtained from structural mechanics models.

12.13.3 Interior of the Contact

This step of the experiment is focused on both the first and third bodies, as it is necessary to localize the preferential zones of contact. Usually it is quite rare for stresses to be transmitted uniformly from one of the first bodies to the other. This localization permits the evaluation of the contact apparent pressures. These preferential contact zones are generally degraded. A distinction must be made between the degradations resulting from a source flow of the third body and those that stem from considerable plastic deformations occurring in the first bodies. The problem of plastic flows of the first bodies could be solved, for example, by modifying their shape. Control of the activation of the source flow requires analysis of the third body. It is then necessary to understand when and how these preferential contact zones develop further, i.e. whether the degradation (source flow) takes place or whether it is the third-body movement that is occurring (internal flow), and localize new contact zones that in turn start to degrade. Action is required on the source flow or on the internal flow depending on the individual case.

Regarding preferential contact zones, it is necessary to identify the instantaneous contact zones and to estimate the instantaneous contact pressures if possible. As for the instantaneous contact zones, it is necessary to identify the orientations of third-body particles resulting from local stresses, i.e. evidence of impacts, sliding, etc. This evidence contains the clues for reconstructing the internal flows of the third body. Chemical analysis can be used to track the third-body flows. If the contact can be analysed at different instances of its life, known chemical species can be introduced to track the flow trajectories. More thorough observations often permit understanding the role of each of the constituents of the material producing the source flow.

12.14 Conclusions

Friction and wear problems are coupled and multi-scale problems. They are coupled in the mechanical meaning, since the stresses acting on the contact depend on both the contact itself and the actual mechanism. They are also coupled in the material meaning, since the materials of the first and third bodies respond to stresses by exhibiting complex behaviour. Lastly, they are multi-scale problems, since the influence of the mechanisms containing the contact is as great as the influence of the contact itself.

Subsequently, friction and wear problems must be treated using common sense and making sure that false evidence is eliminated. To achieve this, the notion of material changes needs to be complemented by realistic analysis of the mechanical stresses involved. The notion "it's the temperature that produces the flows of the first bodies" should be supported by taking into account the role played by pressure and deformation gradients. In reality, this entails a mechanical analysis of all three components of the tribological system, i.e. actual mechanism or device, and first and third bodies. This analysis uses numerical tribology that combines structural mechanics (mechanism), contact mechanics (first bodies) and the mechanics of solid third bodies. The current shortcomings of numerical tribology are offset by contact analysis used to reconstruct the tribological circuit and the contact life scenario.

The friction coefficient is a very practical parameter used as a boundary condition in structural models. However, to understand the model and solve a problem, use of the friction coefficient alone leads to errors, since it does not permit clear identification of the real cause–effect sequence. It is therefore necessary to combine friction with the contact life scenario.

Industrially, the judicious use of structural mechanics permits the control of the contact stresses starting from the design phase. This control significantly reduces risks of abnormal degradations that often lead to seeking remedies after the product has already been released. These remedies are usually in the form of coatings, surface treatments, etc., whose application outcomes are unpredictable and only increase uncertainty.

Acknowledgements

The work on wear and friction analysis presented in this chapter is a combined effort of all the members of the "Friction, Wear and Solid Lubrication" team at the Laboratoire de Mécanique des Contacts et des Solides de l'Institut National des Sciences Appliquées at Lyon, France. The members of this team also include industrial partners.

References

1. Berthier, Y. 'Background on Friction and Wear', *Handbook of Materials Behavior Models*, Chapter 8, Lemaître Academic Press, 2001, pp. 676–699, ISBN 0 12 443341 3.
2. Berthier, Y., Descartes, S., Busquet, M., Niccolini, E., Desrayaud, C., Baillet, L. and Baietto-dubourg, M.C., 'The Role and Effects of the Third Body in the Wheel–Rail Interaction', *Fatigue and Fracture of Engineering Materials and Structures*, **27**, 2004, 423–436.
3. Descartes, S. and Berthier, Y., 'Rheology and Flows of Solid Third Bodies: Background and Application to an MoS1.6 Coating', *Wear*, **252**, 2002, 546–556.
4. Godet, M., 'The Third Body Approach: A Mechanical View of Wear', *Wear*, **100**, 1984, 437–452.
5. Iordanoff, I., Seve, B. and Berthier, Y., 'Solid Third Body Analysis Using a Discreet Approach: Influence of Adhesion and Particle Size on Macroscopic Properties', *ASME Journal of Tribology*, **124**, 2002, 530–538.
6. Landa, A., Wynblatt, P., Hakkinen, H., Barnett, R.N. and Landman, U., 'Molecular Dynamics Study of Disordering, Roughening, and Premelting of the Pb(110) Surface', *Journal of Non-Crystalline Solids*, **205–207**(2), 1996, 767–771.
7. Archard, J.F., 'Contact and Rubbing of Flat Surfaces', *Journal of Applied Physics*, **24**, 1953, 981–988.
8. Ludema, K., 'Third Bodies: Perspectives on Modeling in Lubricated Contacts, in Close Fitting Contacts, etc: Following on the Concepts of Dr Maurice Godet', Proceedings of the 22nd Leeds–Lyon Symposium, The Third Body Concept: Interpretation of Tribological Phenomena, Elsevier Tribology Series 31, Lyon, France, 5–8 September 1995, pp. 3–19.
9. Lim, S.C., Ashby, M.F. and Brunton, J.H., 'Wear-Rate Transitions and Their Relationship to Wear Mechanisms', *Acta Metallurgica*, **35**, 1987, 1343–1348.
10. Rocchi, J., Berthier, Y. and Baillet, L., 'Tribological Expertise and Numerical Modelling to Determine the Conditions Acting on a Conforming Geometries Contact and to Reduce Wear', 3rd AIMETA International Tribology Conference AITC, Salerno, Italie, 18/9/2002–20/9/2002 CD.
11. Johnson, K.L., 'Contact Mechanics', Cambridge University Press, 1985, p. 452.
12. Plumet, S. and Baietto-Dubourg, M.C., 'A 3D Model for a Multilayered Body Loaded Normally and Tangentially Against a Rigid Body: Application to Specific Coatings', *ASME Journal of Tribology*, **120**, 1998, 4.
13. Baietto-Dubourg, M.C. and Berthier, Y., 'A 3D Coating Model for Predicting Tribological Behavior', 7th European Conference on Advanced Materials and Processes, Rimini, Italie, 10/6/2001–14/6/2001.
14. Jardret, V., Zahouani, H., Loubet, J-L. and Mathia, T.G., 'Understanding and Quantification of Elastic and Plastic Deformation During a Scratch Tests', *Wear*, **218**(1), 1998, 8–14.
15. Mathia, T.G. and Lamy, B. 'On the Applications of Sclero-Topometry for Measurement of Hardness, Toughness and Brittleness of Surfaces, Surface Coatings and the Interfaces', Proceedings of International Congress on 'Mechanics and Tribology of Transport Systems – 2003', Volume II, Rostov on Don, Russia, 10–13 September 2003, pp. 77–81.
16. Eleod, A., Baillet, L., Berthier, Y. and Torkoly, T., 'Deformability of the Near Surface Layer of the First Body', Proceedings of the 29th Leeds–Lyon Symposium on Tribology, Tribological Research and Design for Engineering Systems, Elsevier Tribology Series 41, Leeds, Royaume Uni Angleterre 3–6 September 2002, pp. 123–132.
17. Eleod, A., Balogh, T., Baillet, L. and Berthier, Y., 'Characterisation of the TTS and Numerical Modelling of the Particle-Detachment', EUROMAT 2000 'Advances in Mechanical Behaviour, Plasticity and Damage', Vol. 1, Elsevier, Tours, France, 7/11/2000–9/11/2000, pp. 555–556.
18. Gouider, M., Gouider, M., Berthier, Y., Jacquemard, P., Rousseau, B., Bonnamy, S. and Estrade-Szwarckopf, H., 'Mass Spectrometry During C/C Composite Friction: Carbon Oxidation Associated with High Friction Coefficient and High Wear Rate', *Wear*, in press.
19. Lancaster, J.K., 'A Review of the Influence of Environmental Humidity and Water on Friction, Lubrication and Wear', *Tribology International*, **23**(6), 1990, 371–389.
20. Lancaster, J.K. and Pritchard, J.R., 'The Influence of Environment and Pressure on the Transition to Dusting Wear of Graphite', *Journal of Physics D: Applied Physics*, **14**, 1981, 747–762.
21. Niccolini, E. and Berthier, Y., 'Progression of the Stick/Slip Zones in a Dry Wheel–Rail Contact: Updating Theories on the Basis of Tribological Reality', Proceedings of the 29th Leeds–Lyon Symposium on Tribology, Elsevier Tribology Series 41, Leeds, Royaume Uni Angleterre, 3–6 September 2002, pp. 845–854.
22. Baietto-Dubourg, M.C., Berthier, Y. and Vincent, L., 'Cracking Under Fretting Fatigue: Damage Prediction Under Multiaxial Fatigue', *Journal of Strain Analysis*, **37**(1), 2002, 1–15.

23. Lamacq, V. and Baietto-Dubourg, M.C., 'Modelling of Initial Crack Growth and Crack Branching Conditions Under Fretting Conditions', *Fatigue and Fracture of Engineering Materials and Structure, France*, **22**, 1999, 535–542.
24. Baillet, L. and Sassi, T., 'Numerical Implementation of Different Finite Element Methods for Contact Problems with Friction', *Comptes Rendus Mecanique*, **331**(11), 2003, 789–796.
25. Baillet, L. and Sassi, T., 'Finite Element Method with Lagrange Multipliers for Contact Problems with Friction', *Comptes Rendus Mathematique*, **334**(10), 2002, 917–922.
26. Linck, V., Baillet, L. and Berthier, Y., 'Modeling the Consequences of Local Kinematics of the First Body on Friction and on Third Body Sources in Wear', *Wear*, **255**, 2003, 299–308.
27. Iordanoff, I., Berthier, Y., Descartes, S. and Heshmat, H., 'A Review of Recent Approaches for Modelling Solid Third Bodies', *ASME Journal of Tribology*, **124**, 2002, 725–735.
28. Raous, M., Jean, M. and Moreau, J.J. (eds), '*Proceedings of the 2nd Contact Mechanics International Symposium, Held September 19–23, 1994, in Carry-le-rouet, France*', Plenum Press, New York, 1995, p. 470.
29. Niccolini, E. and Berthier, Y., 'Wheel–Rail Adhesion: Laboratory Study of the Role of the "Natural" 3rd Body on Wheels of Locomotives and Rails', 6th International Conference on Contact Mechanics and Wear of Wheel/Rail Systems, Vol. 2, Gothenburg, Suede, 10/6/2003–13/6/2003, pp. 495–501.
30. Kalousek, J., 'The Benefits of Friction Management: A Third Body Approach', Proceedings of the World Congress on Railway Research Conference, 1996, pp. 461–468.
31. Sliney, H.E., 'Dynamics of Solid Lubrication as Observed by Optical Microscopy', *ASLE Transactions*, **21**(2), 1977, 109–117.
32. Dellacorte, C., Zaldana, A.R. and Radil, K.C., 'A Systems Approach to the Solid Lubrication of Foil Air Bearings for Oil-Free Turbomachinery', NASA TM-2002-211482, ARL-TR-2867, October 2002, p. 15.
33. 'The Function of a Good Brush: What You Should Know', Technical note STA AE 101 GB – LE Carbone Lorraine.
34. Meurisse, M.H. and Giudicelli, B. 'A 3D Conservative Model for Self-Lubricated Porous Journal Bearings in a Hydrodynamic Steady State', *ASME Journal of Tribology, France*, **121**, 1999, 529–537.
35. Israelachvili, J., '*Intermolecular and Surface Forces*', 2nd edition, Academic Press, 1992, p. 450.
36. Bhushan, B., 'Principles and Applications of Tribology', John Wiley and Sons, 1999, p. 1020, ISBN 0 471 59407 5.
37. Kaleli, H. and Berthier, Y., 'The Mechanism of Layer Formation and the Function of Additives Used in Fully Formulated Engine Crankcase Oils', Proceedings of the 28th Leeds–Lyon Symposium on Tribology, Boundary and Mixed Lubrication: Science and Applications, Elsevier Tribology Series 40, Vienna, Austria, 4–7 September 2001, pp. 189–197.
38. Grossiord, C., Martin, J.M., Le Mogne, Th. and Palermo, Th., 'UHV Friction of Tribofilms Derived from Metal Dithiophosphates', *Tribology Letters*, **6**, 1999, 171–179.
39. 'Microscopic Aspects of Adhesion and Lubrication' (ed J.M. Georges), Elsevier Tribology Series 7, 1982, p. 812, ISBN 0 444 42071 1.
40. Frene, J., Nicolas, D., Degueurce, B., Berthe, D. and Godet, M., 'Hydrodynamic Lubrication: Bearings and Thrust Bearings' (ed D. Dowson), Elsevier Tribology Series 33, Amsterdam, 1997, p. 470.
41. Berthier, Y., 'Experimental Evidence for Friction and Wear Modelling', *Wear*, **139**(1), 1990, 77–92.
42. Karman, Th., 'Festigkeitsversuche unter allseitig', Druck, Zeitschrift VDI 55, 1911, pp. 1749–1757.
43. Bridgman, P.W., '*Studies in Large Plastic Flow and Fracture*', Harvard University Press, Cambridge, Massachusetts, 1964, p. 362.
44. Bertrand, J.P., Galtier, A. and Guelton, N., 'Phase blanche dans les rails: Caractéristiques métallurgiques et approche mécanique sur les conditions de formation', IRSID, Note No. MPM 97N 1126, 1997.
45. Baillet, L., Berthier, Y., Bontemps, O. and Brunet, M., 'Tribologie de l'interface fibre/matrice. Approche théorique et expérimentale', Revue des composites et des matériaux avancés, France, 1997, pp. 89–105.
46. Baillet, L., Berthier, Y. and Laulagnet, B., 'Modelling of the Origin of Tribological Noises. Application to a Braking System', International Congress and Exposition on Noise Control Engineering, Deaborn, Etats Unis d'Amérique, 19/8/2002–21/8/2002 CD.
47. Baietto-Dubourg, M.C., Vallet, F. and Floquet, A., 'Prediction of the Development and Evolution of Hot Spots During Braking', JEF 2002, GRTT, Lille, France, 13/3/2002–14/3/2002, pp. 147–154.
48. Akay, A., 'Acoustic of Friction', *The Journal of the Acoustic Society of America*, **111**(4), 2002, 1525–1548.

13

Basic Principles of Fretting

P. Kapsa, S. Fouvry and L. Vincent

Abstract

Fretting has to be considered as a complex phenomenon related to interaction between two sliding bodies separated by a third body. Various surface and subsurface processes involved in fretting have to be thoroughly investigated in order to understand its occurrence in real situations.

In this chapter the case of fretting is presented as an example to show how it is possible to understand and model the tribological damage. The industrial needs are important because fretting occurs frequently in assemblies, leading to damage. Fretting refers to a tribological loading of materials in the contact when a small-amplitude alternating motion is applied. Depending on the tribological parameters (sliding amplitude, normal force, frequency of motion, environmental factors, etc.), partial or gross slip can be observed. The consequent damage can be crack initiation and propagation or wear. The use of "fretting maps" showing the loading and the damage on a "normal force vs sliding amplitude" graph is very helpful in understanding and modelling of this phenomenon.

Modes of surface damage observed during fretting are described in order to model fretting wear. Crack initiation can be analysed using the classical tools of fatigue failure, while wear can be studied by considering the energy dissipated in the contact.

On the basis of the experimental data, the effects of typical tribological parameters on fretting are demonstrated.

13.1 Introduction

This chapter deals with some friction and wear issues from a general viewpoint with a particular reference to fretting. The aim of the chapter is to set up a methodology that can be of use to design engineers. The methodology is developed on the basis of the results obtained during dry friction and fretting studies.

Wear – Materials, Mechanisms and Practice Edited by G. Stachowiak

Wear analysis based only on the tests conducted under increasingly more severe conditions of contact speed or the normal load or by altering the nature of the contact is of limited use in solving practical problems. However, one could ask the question, how can reliable and accurate contact models be formulated? During the last few years, powerful tools have been developed to simulate the contact behaviour. When developing any wear model the following issues need to be considered:

(i) It is not convenient to describe wear by a simple law for a given mechanism. Test conducted under severe conditions should be avoided and a suitable test must be developed first to reproduce the same kind of damage as observed under the industrial conditions or the damage that is considered as the more detrimental. Only after that ranking of possible remedies or plotting of wear law can be undertaken. The test conditions must also consider environmental variables such as temperature or humidity.
(ii) Wear is a complex phenomenon and it is often useful to consider the initial damage and its consequences separately. For example, an initial wear particle can dramatically change the contact conditions as well as participate in rapid crack nucleation in the contact area. This is typical for fatigue experiments where at least a two-step process takes place, i.e. damage nucleation and damage propagation.
(iii) A contact problem occurs at a wide range of dimensions, i.e. from the nanometre scale, related to surface characteristics, up to a metre scale related to the wear of the entire component. This implies that a multiscale, multidisciplinary or even an interdisciplinary approach is needed to successfully analyse and solve an industrial problem. Specialists of tribochemistry are needed as well as the specialists in material science and mechanics.

The role of mechanics in tribology varies with the area studied. It dominates thick film lubrication but it took a back seat in friction and wear studies, at least until the early 1990s. Hydrodynamic and elastohydrodynamics theories are well developed and commonly used as a tool in machine design [1, 2]. The coefficient of friction between rubbing surfaces was proposed by Amontons [3] and later studied by Coulomb [4] who stressed its limitations. Contact theory, which was initiated by Hertz [5] and later developed by Johnson [6] and others, is limited to static case and to the determination of contact stress and displacement fields in mostly elastic solids. Stresses and displacements induced by normal loads are well described but tangential effects are often introduced using locally, and not just globally, the most criticized coefficient of friction known today as Coulomb's law.

Contact mechanics offers a wide area for investigation. Even with some specific applications, e.g. gears, ball bearings, wheel on rail contact, etc., the transfer of contact models to industrial problem is not always quick enough even with the use of modern numerical tools like the finite element method, and much fewer real applications than expected are considered. This is surprising as the subject is well described, its mathematical tools are well developed and efficient and are continuously being updated [6, 7]. The economic impact of contact mechanics studies of loaded contacts is considerable as it touches material performance and machine durability. However, third bodies (oil films, for example) are usually not included in the analysis while thermal aspects are often considered.

13.2 Wear

Wear is defined as material loss, often expressed in terms of mass loss per unit distance travelled. Durability is related to the machine function and is expressed in terms of duty or life. Wear modelling from the viewpoint of wear loss implies that each step in the process of wear particle formation and expulsion from the contact is identified and understood. Classical tribology often refers to wear mechanisms. Interface or third-body tribology distinguishes between particle detachment and wear. Third-body tribology [8, 9] divides the wear process into the following steps, which can occur simultaneously:

(i) **Particle detachment:** A wear particle is detached from first bodies, i.e. from the original surfaces in contact, by any one of the well-established mechanisms (adhesion, abrasion, corrosion, surface fatigue, etc.).
(ii) **Third-body life:** Once detached, the trapped particle is subjected to the strenuous conditions at the interface along with other debris. It becomes part of the 'third body' and it changes in both morphology and composition.
(iii) **Debris circuit:** Depending on many parameters related to the nature of the contact, such as vibrations or humidity, a particle can be trapped in the contact, re-circulated or expelled.
(iv) **Particle expulsion:** Finally, the wear particle is ejected from both the wear tract and the contact (re-circulation is avoided) and it only then becomes a "true" wear particle.

The third-body is a "medium" through which the load is transmitted from one first body to the other. It also accommodates, through the contact (flow), in a dissipative manner (friction) most of the velocity difference between first bodies. It is made out of bulk material from the two surfaces and its role in a given application can be investigated either theoretically, as in lubrication when third-body rheology and boundary conditions are known, or experimentally through visualization, as in dry friction, where under some conditions at least sites of motion can be identified. Third-body formation and elimination is thus a flow problem [10], with its sources and sinks, and a strong parallel between lubrication and wear can be drawn from this flow analogy.

Careful observations have also shown that

- different velocity accommodation mechanisms act simultaneously in the same contact;
- relays are taken between different mechanisms;
- many factors can activate or deactivate a wear mechanism, such as roughness, humidity or third-body rheology.

Various fundamental phenomena used in the description of surface damage need to be considered, e.g. cracks, plastic deformation, adhesion and physico-chemical transformations (as illustrated in Figure 13.1). Any modelling of wear must therefore take into account the following facts:

- adhesion, abrasion, delamination, etc. are just the particle detachment mechanisms, i.e. they are not wear mechanisms;
- the passage from "motion of matter in the interface" (internal flow) to "true wear" depends on the track shape;

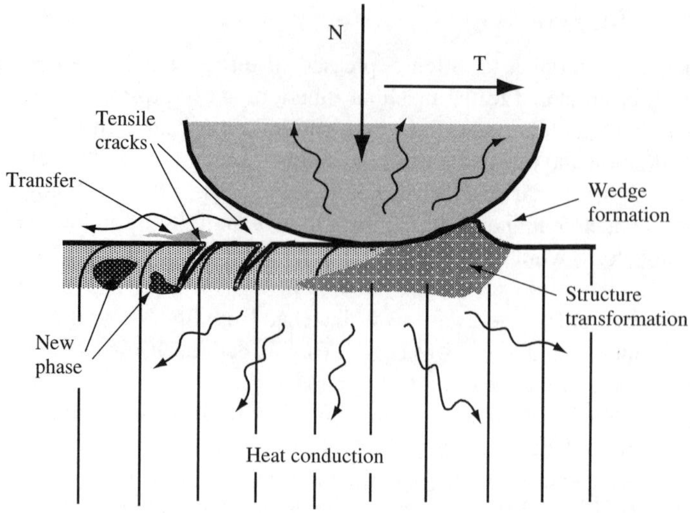

Figure 13.1 Various phenomena occurring during sliding process for a ball-on-flat sliding contact

- external flow (loss of matter from the interface) depends on the velocity field at the edge of the contact and therefore on the velocity accommodation mechanism;
- wear and friction are not intrinsic properties of materials or of a couple of heterogeneous materials.

13.3 Industrial Needs

In order to become a useful tool for industrial applications, tribology should allow for reliable prediction of contact areas, pressure and temperature distributions, strain hardening, toughness changes, flow stress modifications, crack formation and particle detachment. Contact mechanics provides accurate stress, deformation, temperature models, stress intensity factor, maps for homogeneous, coated, cracked, smooth or rough solids of different shapes, heat fluxes or temperatures or any combination of these factors. Contact mechanics contribution to engineering is thus considerable, but the gap between what is expected and what is provided is still great. Typically, four types of data seem to be necessary to bridge the gap between contact mechanics theories and first-body damage prediction, i.e.

(i) better description of the evolution of the contact geometry (considering surface roughness);
(ii) better material characteristics knowledge, for thin coatings for example (with the gradient of properties at the surface);
(iii) better understanding of crack propagation under conditions found in tribological contacts, with multicracking situations;
(iv) better formulation of the evolution of properties in materials operating under the contact conditions.

13.4 Fretting in Assemblies

Terms used to describe fretting can be quite confusing as fretting wear (FW), fretting fatigue (FF) and fretting corrosion (FC) are the expressions commonly used. They are usually derived from specific loading conditions or based on the observation of damage. According to the ASM glossary of terms [11], fretting

(i) comprises wear phenomena occurring between two surfaces having oscillatory relative motion of small amplitude. Note that fretting is a term frequently used to include fretting-corrosion and other forms of fretting wear. This usage is not recommended due to the ambiguity that may arise.
(ii) is caused by a small-amplitude oscillatory motion, usually tangential, between two solid surfaces in contact. Note that here the term fretting refers only to the nature of the motion without reference to wear, corrosion or other damage that may result.

In the second case it is then important to consider the relative displacement as compared to the contact size. This is illustrated in Figure 13.2 where the limit between "fretting": and "reciprocating" sliding wear is shown.

When considering a ball-on-flat contact, various modes of fretting can be distinguished depending on the type of the relative motion (Figure 13.3). All these cases can be encountered in practical situations but the occurrence of tangential fretting is more common.

While the word fretting describes agitation in the contact quite well, the reason for the displacement is usually not well identified. This drawback makes us think in terms of well-defined fretting sources. For example, it appears more convenient to consider fretting wear as fretting for which small displacements are a consequence of external vibrations while fretting

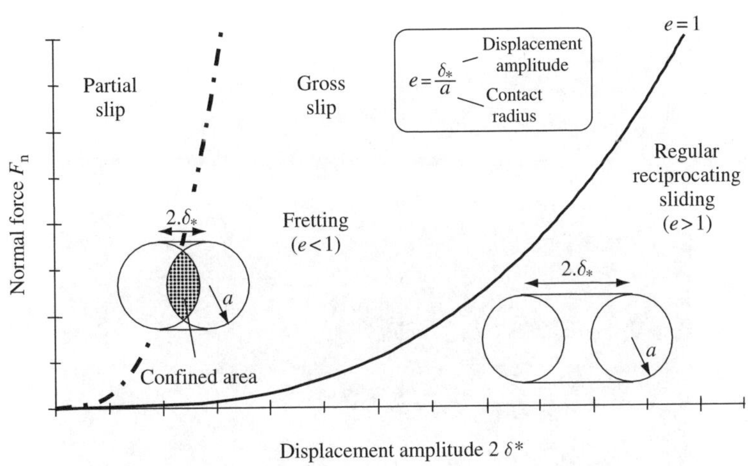

Figure 13.2 Various situation of sliding conditions defined in a graph normal force (vertical axis)–relative displacement (horizontal axis) for a ball-on-flat contact. The two cases of fretting, partial and gross slip (defined later), are situated in a domain with $e < 1$ corresponding to a situation where a part of the wear scar of the flat is always in the contact. Reciprocating sliding wear corresponds then to a situation where the entire wear scar is in the contact with the environment [12]

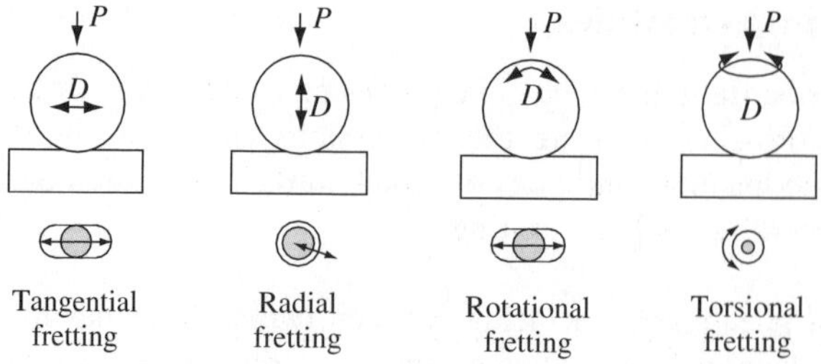

Figure 13.3 Schematic diagram of four basic fretting modes for a ball-on-flat contact

fatigue as a result of small displacements which are a consequence of the (cyclic) deformation of one of the two contacting parts. Fretting corrosion is usually used when a chemical reaction dominates during the tribological process also involving small-amplitude sliding.

Several syntheses have been proposed to justify the fretting-type degradations [13–16] and to propose means of combating this problem [17, 18].

13.5 Fretting Processes

The fretting damage can be described as

- wear induced by fretting (WIF) – this corresponds to classical material loss (Figure 13.4);
- cracking induced by fretting (CIF) – cracks initiated on the surface can propagate up to the final failure of the specimen (Figure 13.5).

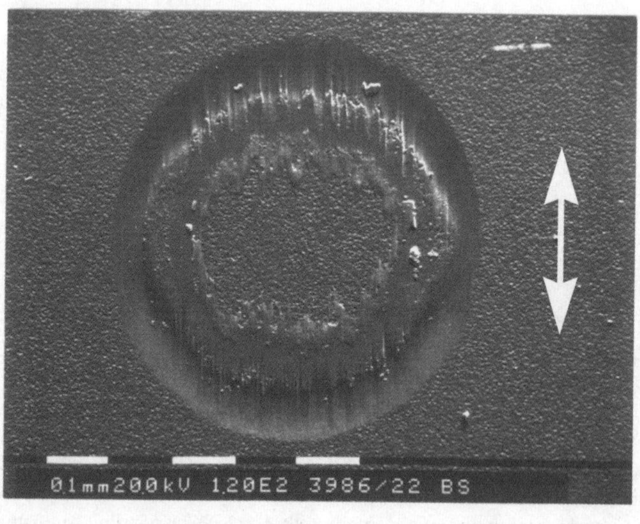

Figure 13.4 SEM image of the ball wear scar showing the loss of matter for a ball-on-flat test in a reciprocating motion [19]

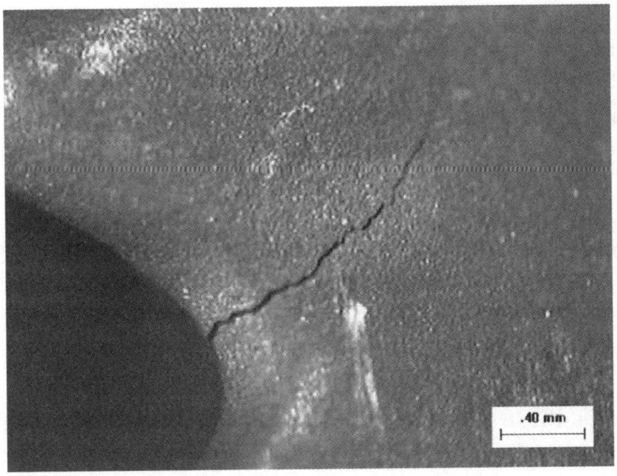

Figure 13.5 Fatigue crack propagating after an initiation due to the contact of the dovetail root in an aircraft engine, from Park *et al.* (reprinted with permission from Elsevier) [20]

Traditionally, WIF is related to the presence of debris powder (red powder for steels, black powder for aluminium or titanium alloys). For a wide range of metals, alloys and fretting conditions, WIF is described as follows:

- The removal of superficial oxide layers and the increased metal-to-metal contact.
- The formation of a new microstructure called the tribologically transformed structure (TTS) in the case of initially non-brittle materials. The TTS is a nanocrystalline structure which becomes too brittle to accommodate the imposed displacement other than by breaking [21, 22].
- The metallic debris are then trapped, crushed and oxidized in the contact which induces the formation of a powder bed (the red powder, for example, for steel).
- The establishment of the third body depends on the possibility of debris ejection (i.e. on the nature and the shape of the debris and the vibrating environment). If the powder bed is maintained in the contact, the bulk degradation can be stopped as the velocity is now entirely accommodated by the powder bed.

For CIF, several characteristic features have been observed for the cracks forming on the contact surface:

- In some cases, few cracks nucleated on each side (as related to the friction movement) of the contact limit. Then on each side, a main crack develops which can lead to spalling.
- Cracks can also initiate and be distributed anywhere on the contact surface. Often this induces the formation of coarse debris or spalls but usually only the most external cracks propagate because of the compressive stress field acting beneath the surface.

Material loss and cracking sometimes appear as competing processes as, for example, material loss can eliminate small superficial cracks or the opening of a deep crack which can accommodate the main part of the imposed displacement and thus strongly reduce the slip amplitude and subsequently the debris formation.

The terms FW, FF, WIF and CIF must not be confused. FW and FF are a loading of sliding materials due to sliding condition, while CIF and WIF are the consequences of the loading, i.e. a damage process. It therefore appears quite useful to develop a general approach to fretting with the intention of answering some of the questions discussed earlier.

(i) The first step is to describe the local loading. To do so, first the identification of the fretting condition is required. The fretting conditions can be classified as partial or gross slip (Figure 13.6). The partial slip condition is defined from closed cycles of tangential force Q versus displacement amplitude δ plots. Surfaces in the contact are partially "stuck" together and one can define a stick area (present in the centre) and a sliding area, as illustrated in Figure 13.6. Elastic deformation of the device and the sample accounts for the accommodation of the imposed displacement. Small amounts of partial slips can appear but no plastic deformation of the first bodies is noted. The gross slip condition is characterized by open or quasi-rectangular cycles: The entire contact is then sliding. In practice, the real displacement in the contact is smaller than the imposed displacement due to the limited rigidity of the devices and due to the elastic deformation of the sample. Once the condition is determined, stress and strain fields can be calculated in the case of elastic contact.

(ii) From these values, the properties of the two first bodies are considered. Risks of particle detachment, brittle cracking from the surface or the load concentration must be evaluated. The contact condition is modified just after the first few strokes. What happens during the very first cycles of component's life is of the greatest interest in the description of fretting behaviour. At this initial step of damage, the alteration of the surfaces must be taken into account. An increase in the contact area or a modification of the friction coefficient can occur. Humidity and the sensitivity of the surface to form new oxide layers are usually parameters that strongly modify the fretting condition.

(iii) The material's response affects the fretting condition; hence fretting needs to be considered from the dynamic viewpoint. On the basis of friction records, three fretting regimes have been identified (Figure 13.7). The partial slip regime is defined as the

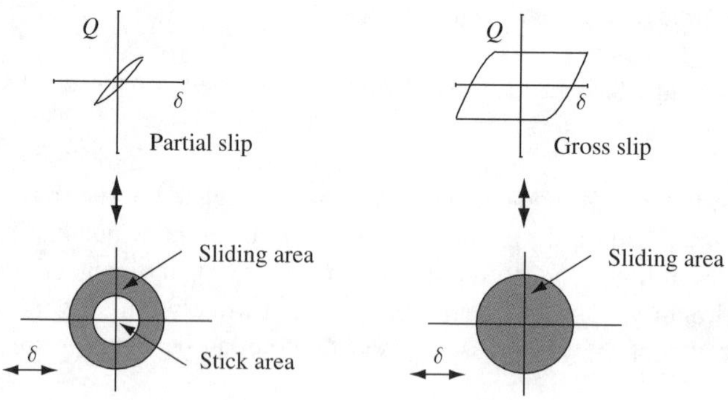

Figure 13.6 Partial and gross slip conditions in fretting for a ball-on-flat contact

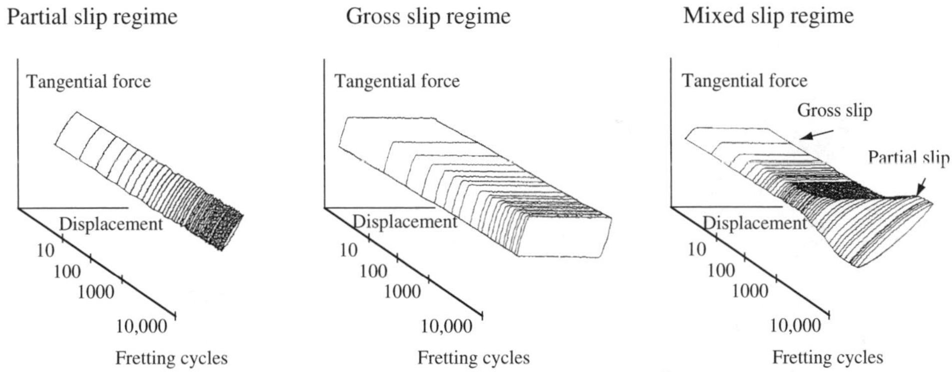

Figure 13.7 Fretting logs corresponding to the three fretting regimes

case where the partial slip conditions prevail during the whole test with Q–δ cycles remaining closed, up to a coarse damage initiation characterized by a decrease in the contact stiffness or in the Q–δ initial slope. The gross slip regime is defined as the case in which the gross slip condition prevails even if the coefficient of friction varies. A so-called mixed regime is a fretting regime characterized by a complex shape of the friction record [23]. Closed, quasi-rectangular and, very often, elliptic Q–δ cycles are successively observed during one test in a mixed regime. The elliptic cycle generally appears after some cycles. This shape indicates partial slip accompanied by plastic deformation in the upper layers of the stick area of the central contact region.

(iv) Depending on the first-body properties, these regimes have been associated with main damages. The gross slip regime always induces material loss, i.e. WIF, while the mixed regime is usually responsible for fatigue cracking.

(v) When the main damage is identified, the difficulty remains in controlling the evolution of the damage. This still appears to be a major difficulty, as the crack growth and the debris entrapment in the contact, which can establish a third body and accommodate the displacement, must be carefully studied. Tools already exist to describe cracking with identification of path and propagation law but the description of the third-body behaviour is not yet satisfactorily developed.

This step-by-step approach to fretting allows to propose a fretting chart from a set of the following fretting maps [24]:

(i) The running condition fretting maps (RCFMs) are plotted in a normal force–displacement diagram to locate the fretting regimes (Figure 13.8). RCFMs introduced initially displayed only two fretting conditions.

(ii) The material response fretting maps (MRFMs) were introduced to present, in a similar type of diagram, the various material damages that appear as a consequence of the fretting loading (Figure 13.9). Three main material responses have been identified during microscopy studies of both contact surfaces and cross-sections. These responses are either the absence of damage or the two types of damage described already as WIF, shown in Figure 13.9 as debris formation, while CIF is denoted as crack nucleation. It needs to be emphasized that these MRFMs are strongly time dependent.

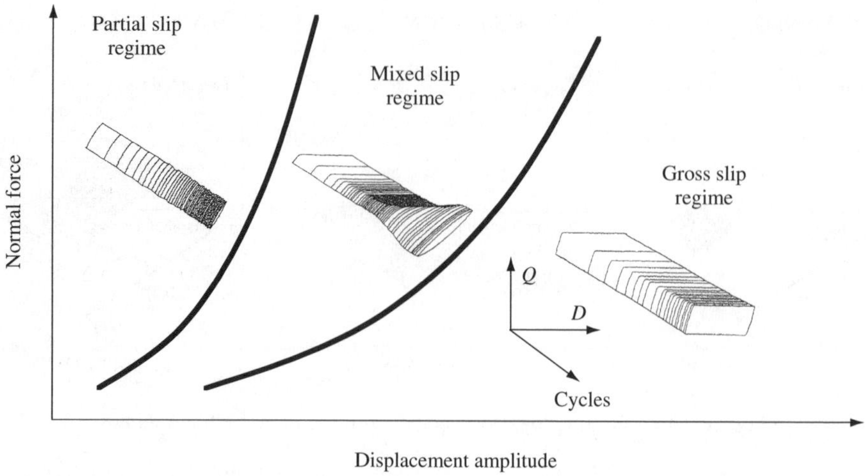

Figure 13.8 Example of running condition fretting map (RCFM) with typical corresponding fretting logs (force–displacement cycles with cycle number)

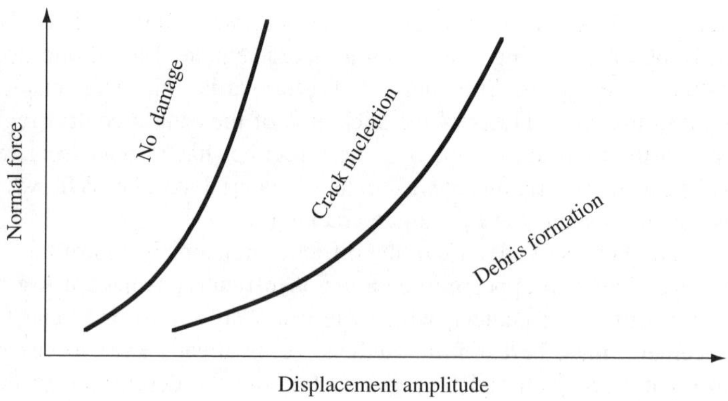

Figure 13.9 Example of material response fretting map

These maps give a dynamic "view" of wear surpassing the static instantaneous wear maps introduced by Lim and Ashby [25] and Vingsbo and Soderberg [26]. In order to propose guidelines for the understanding of the material behaviour and the material selection, this wear map approach must be further developed. The two maps can be theoretically generated by the following:

(i) Using Mindlin's theory [27], a ratio A was introduced to quantify and predict the boundary between partial slip and gross slip (Figure 13.10). This ratio refers to the energy dissipated in the contact over the total energy and was shown to be equal to 0.2 under elastic conditions. Its evolution during a test enables a precise definition of the three fretting regimes [28].

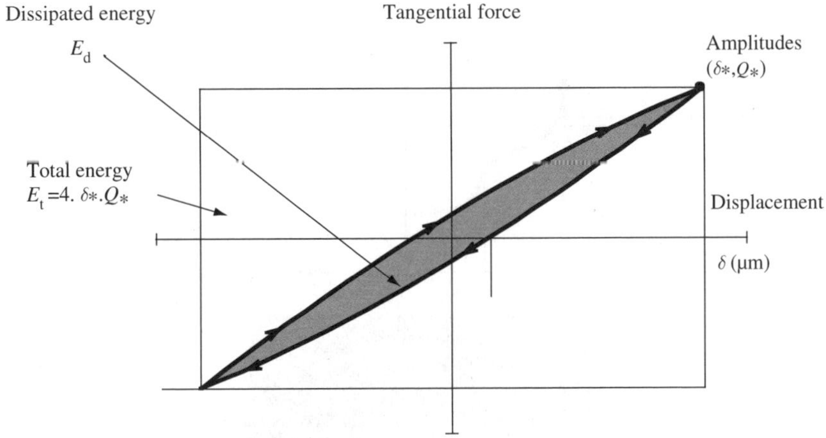

Figure 13.10 Definition of the ratio A separating partial and gross slip conditions. $A = E_d/E_t$ can be calculated from the tangential force–displacement cycles

(ii) Fatigue criteria can be used to predict the boundary between no damage and crack nucleation zones for a given number of cycles [29–31]. For instance, Dang Van's criterion well indicated the location of the crack nucleation at the limit of the contact zone [32] (Figure 13.11). However, to quantify the crack risk, the very sharp stress gradient imposed below the surface must be considered through a size effect analysis [33].

(iii) The growth of cracks and the interaction between several cracks beneath the contact area have been extensively studied and laws now exist for specific applications [34, 35].

(iv) An energy-based model has been recently introduced to relate the quantity of material loss due to the cumulative energy dissipated during the test [12, 36]. Various tribological processes which can be considered as energy dependent are shown in Figure 13.12.

In most cases, it has been observed that the wear loss is proportional to the energy dissipated in the contact. The slope found can then be considered as a wear characteristic of the sliding conditions. This slope depends on the material characteristics. For example, the application of hard protective coatings, such as titanium nitride, tends to decrease this slope (Figure 13.13).

When using this energy approach in fretting studies, it appears that at a fine scale, the initial period of sliding corresponds to a situation in which energy is dissipated without a material loss. It can be considered then that this energy is the energy necessary to drive the phenomena leading to the TTS formation (Figure 13.14) [21].

(v) Other aspects of fatigue cracking need to be considered also, for example:

- small surface cracks can be eliminated by wear which can prevent any catastrophic failure;
- small surface cracks can nucleate but they do not propagate. This can be a result of a crack arrest phenomenon, as it occurs in the case of polycarbonate or some surface treatments, where a local compressive residual stress field prevents any crack propagation.

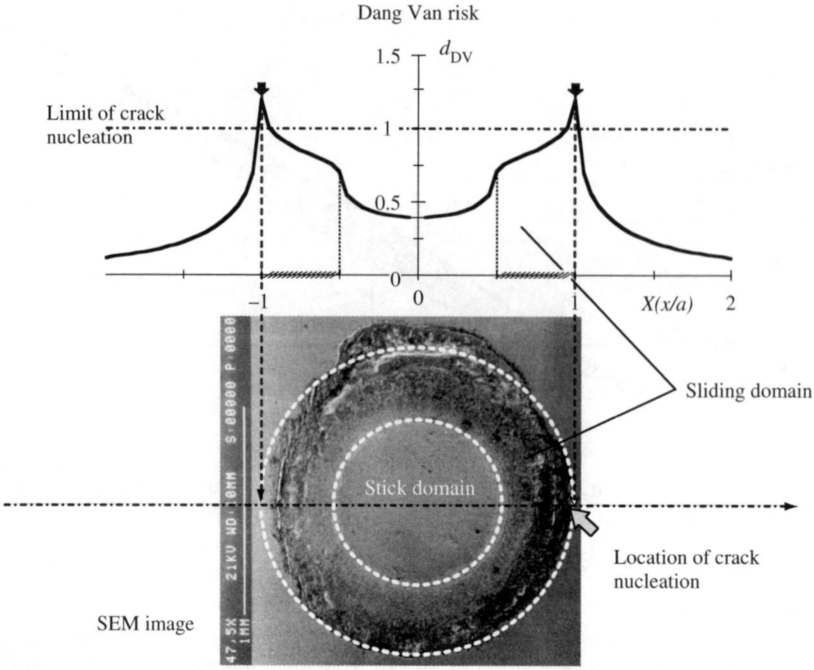

Figure 13.11 Example of crack initiation using the Dang Van criterion. The evolution of the cracking risk indicates the location of the first initiation of cracks as observed by SEM (bottom) for a ball-on-flat contact in partial slip regime fretting test

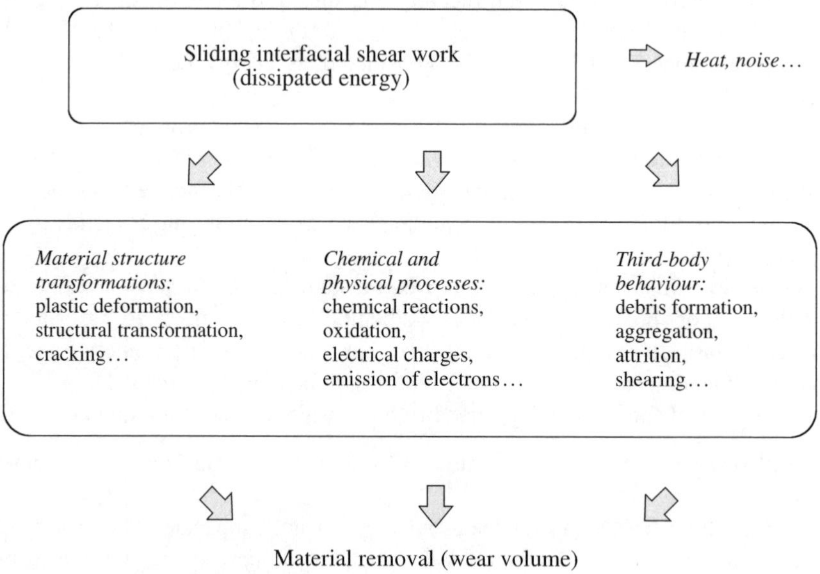

Figure 13.12 Energy approach to tribological phenomena [12]

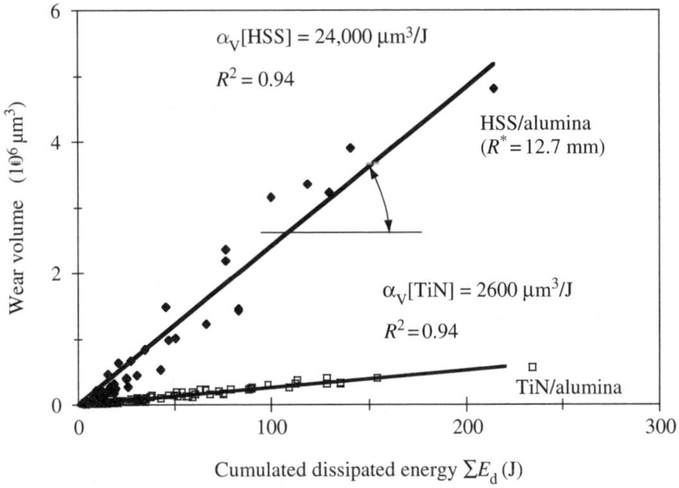

Figure 13.13 Evolution of the wear volume versus the cumulated energy dissipated in the contact. The wear/energy coefficient is dependent on the presence of a protective coating. Alumina ball sliding on HSS steel flat with and without a TiN coating [12]

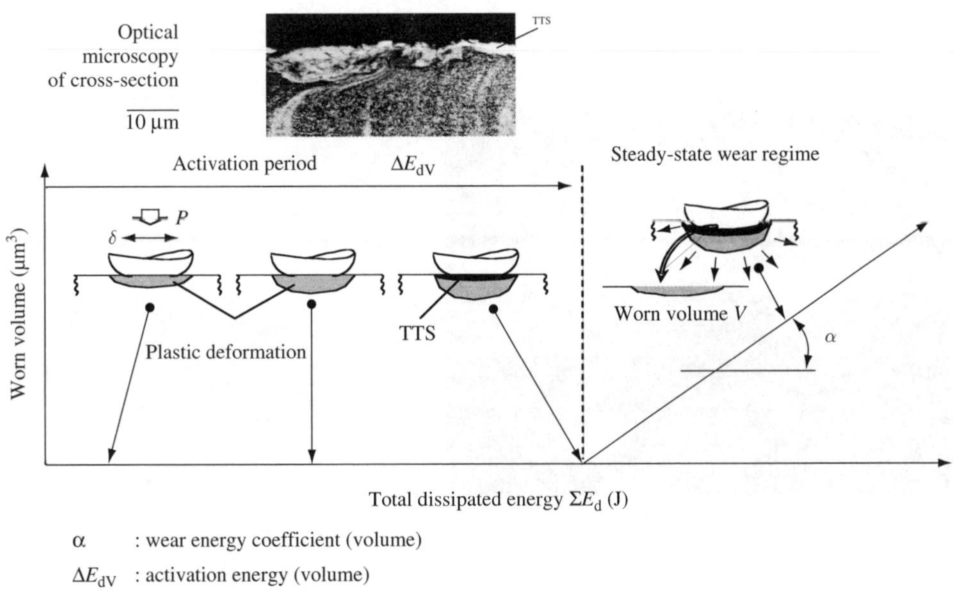

Figure 13.14 Worn volume versus the total dissipated energy. The first period in which energy dissipation is not accompanied by a material loss corresponds to surface phenomena leading to the formation of a new structure [21]

These phenomena mostly take place in the case of elastic contact behaviour. But it appears that also for elastoplastic and plastic contacts, the phenomena occurring are quite similar. Cracks were shown to nucleate first under the mixed regime. In this case, the amount of plastic deformation in the stick zone makes the identification and the calculation of the stress responsible for crack nucleation rather difficult.

13.6 Fretting Parameters

Fretting phenomena are complex and can be influenced by numerous parameters. Among these, the most important are considered in this section.

13.6.1 Nature of Loading

The load acting on contacting materials is one of the main parameters. It represents the "engine" of damage. The contact loading appears to be strongly dependent on the normal contact behaviour, in particular the friction coefficient. Its evolution with the typical tribological parameters, as normal force and sliding amplitude, needs to be considered.

The contact pressure distribution is also an important parameter. Typically, the tribologists consider a ball-on-flat contact leading to a parabolic pressure distribution. This is not always the case in many applied situations. It is then of prime interest to consider the real pressure distribution in the real contact which has to be reproduced during the laboratory tests. This is not always easy and adopted specific sample configuration has to be defined (Figure 13.15).

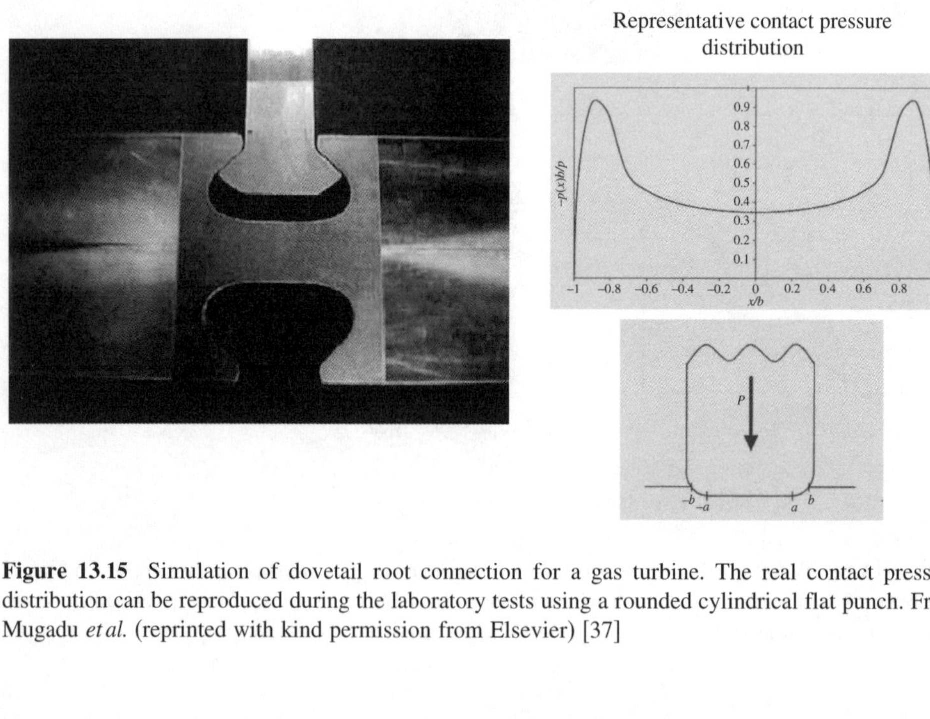

Figure 13.15 Simulation of dovetail root connection for a gas turbine. The real contact pressure distribution can be reproduced during the laboratory tests using a rounded cylindrical flat punch. From Mugadu *et al.* (reprinted with kind permission from Elsevier) [37]

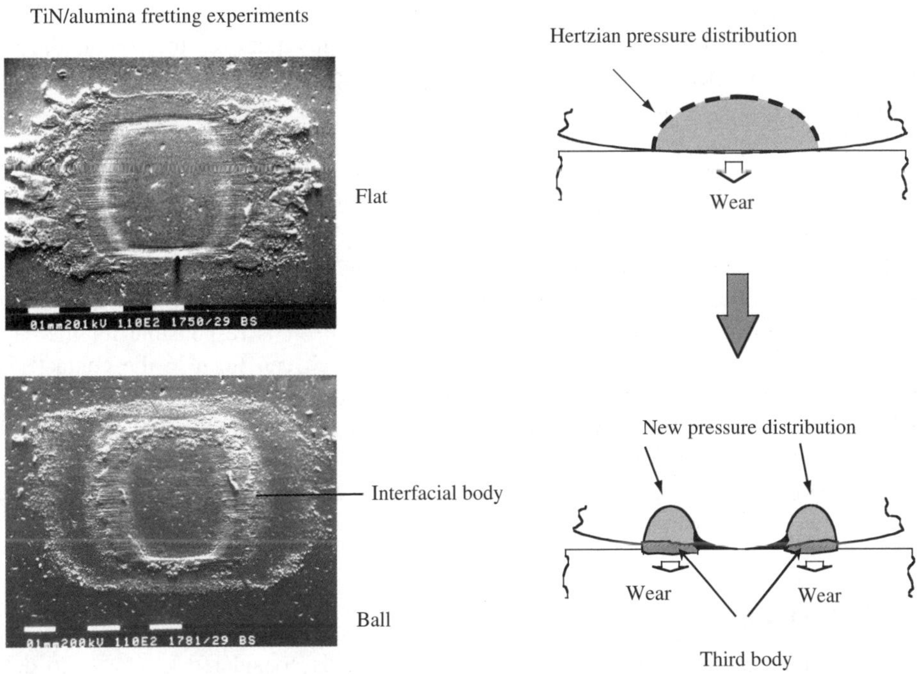

Figure 13.16 Evolution of pressure distribution with the formation of a third body and its effect on worn volume as observed in the SEM. From the initial parabolic Hertzian stress distribution, with a maximum pressure at the centre of the contact, the third-body presence creates high pressures near the external part of the contact. The high local wear rate is then displaced from the centre to the external part of the contact [12]

In most cases, if we consider that the local elemental wear volume is proportional to the local energy dissipated, the profile of the wear scar represents in fact a profile of the pressure distribution. As the sliding often generates a third body, the effect of this foreign body in the contact needs to be considered. The pressure distribution can be strongly affected and modified by tribological phenomena and maxima locations resulting in the modified local wear rate distribution (Figure 13.16) [12].

13.6.2 Nature of the First Bodies

In cases of homogeneous contacts (same material for the two sliding bodies), material properties have been used to describe CIF. Several aluminium alloys have been compared under the two, "partial slip" and "mixed", regimes and diametrically different behaviours were observed. These behaviours can be explained in terms of the resistance of these materials to crack nucleation and propagation [38]. In a similar way, for a Ti–6Al–4V alloy, the appearance of the cracking domain was clearly related to its fatigue limit.

Generally, in cases of WIF, particles detach well before the fatigue crack initiation and it seems reasonable to associate the particle detachment domain with the gross slip regime.

With the exception of brittle materials with low elongation ratio and low fracture toughness, particles are detached from the TTS. Thus, for the same alloy and even for several alloys of the same group, the particle detachment does not depend on the initial material's strength properties resulting from its thermo-mechanical history. The cyclic work hardening which induces the mechanical-like recrystallization, and then the TTS, changes the initial material's properties, such as yield stress, elongation and residual stresses. Consequently, it is quite difficult to rank the wear properties of several alloys belonging to the same group as the nature of the TTS is quite similar for all of them [21] and their wear properties depend mainly on the detached particle life, i.e. particle entrapment in the contact. The main factors contributing to WIF resistance are essentially mechanical, i.e. nature of the contact, vibrating environment, frequency of vibrations, etc. The material itself is responsible for the depth of the TTS and thus for the amount of wear debris which can form in the contact. The debris entrapment process can be affected by the chemical environment which can favour the formation of platelets which are harder to expel from the contact than free particles due to their different shape and size.

Determination of the fretting behaviour is much more complex in cases of heterogeneous contact. However, the same approach is recommended.

13.6.3 Coatings

Among other means to control fretting, surface coating is considered as a way to delay the fretting damage. Coating can be used to modify the interaction with the surrounding atmosphere and often to decrease the coefficient of friction. Any change in the coefficient of friction modifies the boundaries between the domains of the RCFMs (Figure 13.17). The value of the friction coefficient controls the maximum of the contact stresses applied and thus the cracking risk. It also controls the amount of the total energy dissipated in the contact and thus the wear rate.

The effect of coating can be analysed by considering the energy approach to wear. What is of importance is the first occurrence of a contact between the slider and the substrate material (at a point where the coating has been worn). Then, instead of a global energy

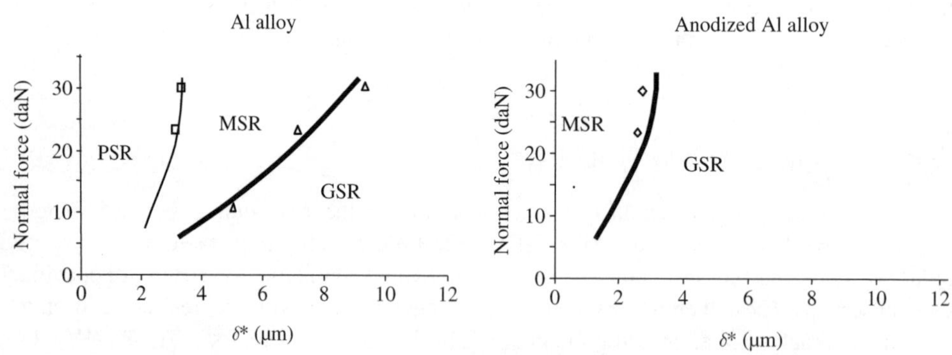

Figure 13.17 Effect of an anodized coating on the running condition fretting map, from Elleuch [39, 40]

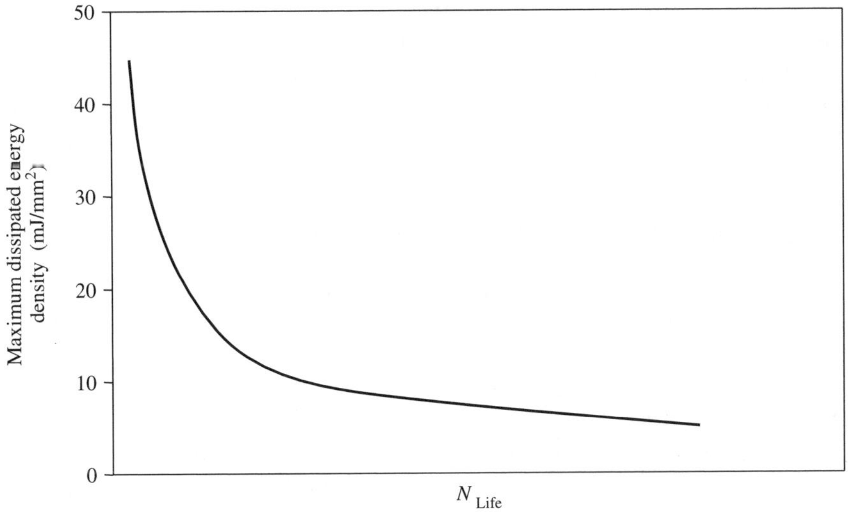

Figure 13.18 Illustration of a coating endurance chart showing the maximum dissipated energy density per fretting cycle versus the number of life cycles (N_{life}). The life cycles (N_{life}) correspond to the situation when the coating is locally eliminated, resulting in an increased friction coefficient [42–44]

approach where the total dissipated energy is related to the worn volume, it is preferable to consider a local approach where the local dissipated energy (related to the local pressure and local friction coefficient) is related to the local volume, i.e. the worn depth [41]. Then, the contact life related to wear of the coating could be linked to the maximum dissipated energy density (Figure 13.18) [39, 40].

The fretting behaviour is also modified as coatings change the mechanical properties of the material's upper layers (and then the contact area) and introduce residual stresses, the role of which in a crack propagation mechanism is very significant.

The use of coatings, however, requires a global approach that must begin before taking into account the contact dynamics. Leroy [45] analysed the effect of coating on unlubricated contacts and two main results were pointed out:

(i) Coating to substrate stiffness ratio can govern substrate failure. The substrate is not weakened but subject to higher and more localized stresses when stiff coatings are used.
(ii) Even modest temperature rises, commonly occurring in tribological applications, can generate high stresses in the coated systems if coating and substrate properties are mismatched.

Contact mechanics is needed to generalize the use of coatings by setting up the design rules. Specific substrate/coating combinations can thus be eliminated for particular applications after these analyses are conducted, allowing to avoid costly and wasteful experimental wear tests.

Typically, coatings can be classified on the basis of deposition methods, i.e. CVD, PVD, etc., composition, thickness, hardness (both indentation and scratch), crystallographic structure, residual stresses, coating to substrate adhesion, friction and wear data produced on a pin-on-disc machine [46]. However, Young's modulus, Poisson's ratio, coefficient of thermal expansion, thermal conductivity, density, specific heat, of both coating and substrate materials, along with temperature and stress limits for both materials, the coating thickness, information on residual stresses to assess the overall stress level exhibited by the working coated first body, are needed by the mechanical engineer to design coated elements.

Discrepancies between the results obtained on different test rigs can be reduced significantly by eliminating, by prior analysis, the contact mechanics aspects which can be addressed theoretically.

Regarding fatigue cracking, the mixed fretting regime always appeared as the most detrimental, whatever the materials used were (metallic or not). Thus, a first approach is to diminish the mixed regime of the RCFM, i.e. to favour the gross slip regime. Solid lubricants or varnishes are the potential industrial means of controlling fretting. If the RCFM cannot be changed, then the material property modifications are needed, for example, shot-peening reduces the cracking domain as it often limits cracks to very short lengths. Hard coatings quickly give rise to particle detachment which prevents the mixed regime by accommodating the displacement in the powder bed. The quick degradation of brittle materials modifies the RCFM by increasing the slip regime.

The main factor for the fatigue risk is the level of the maximum applied stress. Thus, the lower the coefficient of friction is, the lower the maximum tensile stress, and the lower the risk of cracking. Diminution of the maximum applied load can be obtained from a modification of the sites where the displacement is accommodated in the contact (the applied stress decreases strongly with the distance from the site). Guidelines to help select the appropriate surface treatments or coatings as a remedy against failure must include the use of the two fretting maps: RCFM and MRFM.

This approach to fretting using the fretting maps and the velocity accommodation mechanism concept can be extended successfully to fretting fatigue [47]. In a similar way, surface treatment or cladding can be chosen as possible means in fretting control due to their effects on the RCFMs (varnish, solid lubricant, thin film, etc.) or their effects on the MRFMs (increase in fatigue limit, introduction of compressive residual stresses, etc.). For example, the analysis of the shot-peening effect on fretting is an interesting case as

- the induced surface roughness can favour the stick domain of the RCFM;
- the compressive residual stresses can push up and reduce the cracking domain of the MRFM;
- the compressive stresses induced in the sublayers can arrest small cracks formed on the surface [44];
- the hardened, brittle peened layers can favour the debris formation in the slip domain, but, in this case, the residual stresses are eliminated after the first few cycles.

The above example shows how important it is to identify the fretting regime before attempting to explain the material behaviour and attempting to rank the material.

13.6.4 Environment

A chemically active environment can result in surface transformation and also it may accelerate the deterioration of contacting materials if fretting wear removes the protective surface layers. The damage induced on material surfaces is sensitive to the variables involved in the fretting process. These variables can be classified into three categories: the contact conditions (mechanical aspects), the environmental conditions (chemical aspects) and the contacting material properties (material's aspects). Fretting maps also appear to be a quite useful tool in the analysis of the environmental effects. The relative humidity can modify the tangential force required to impose the displacement and the transition between partial slip and gross slip conditions. Material property changes are taken into account by MRFM which can show the differences due, for example, to hydrogen embrittlement. In case of WIF, the formation of platelets can prevent debris from being ejected out of the contact and thus reduce wear.

13.6.5 Frequency

Frequency of motion is an important parameter for the dynamic behaviour of a mechanical system. The frequency can indeed change the sliding amplitude when some resonance or damping is activated. MRFMs are subsequently influenced by the stroke frequency. High stroke frequencies can induce heating and as a result modify the material properties and/or friction coefficient. The main effect of stroke frequency must be explained in terms of the third-body concept. Once formed in the contact, debris is trapped. Several cycles are necessary before its ejection and this mechanical phenomenon can be influenced by the speed of motion (related to the frequency). In cases of low alloyed and stainless steels, small changes in frequency (from 0.1 to 12.5 Hz) were shown to increase the crack nucleation risk; the higher frequency favours the ejection of debris and thus prevents the beneficial effect of the third body. When considering the contact geometry, the easier the debris is ejected from the contact, the easier the cracks tend to nucleate. This factor is of particular importance in industrial applications due to the large range of frequencies (0.01 Hz to several kilohertz) observed in real systems, and further work is needed to fully analyse its effects.

13.6.6 Temperature

Contact temperature, often wrongly interpreted as the temperature rise and one of the main factors affecting fretting, is estimated from metallographic observations only [15]. For instance, white etching areas observed in cross-sections were used as an indicator of very high temperature. Equilibrium diagrams showed this increase but did not take into account the pressure effect. Results based on monitoring the electrical potential of the interface [48] indicate that the average contact temperature increase is in the order of 20 K, for a medium steel carbon in given experimental conditions. However, temperature effect cannot be neglected as the temperature rise at the contact interface has a significant effect on both the mechanical and chemical aspects of fretting wear and fatigue processes. This effect can directly be related to the dependence of the oxidation rate and the mechanical properties of the fretting surfaces on temperature. It is quite difficult to precisely establish the level of temperature increase and its effects on friction.

A direct measurement of the contact temperature in fretting is difficult to perform because of the nature of the temperature field (both in the space and in time domains) and due to the limitations of measurement techniques. Thus, theoretical and numerical models are emerging as the most viable and practical approach. Recently, both analytical [49, 50] and numerical (using the finite element method) [51] analyses have been conducted in order to determine the temperature rise during fretting. A heat flow channel (containing a single micro-contact and extending some distance in the solid) model was used in both cases with an aim to calculate the microscopic thermal constriction resistance and to determine the 3D temperature field in the vicinity of a contact asperity. Results obtained from these two methods are consistent and indicate that the maximum temperature rise (flash temperature) can reach a great range of values (from a few kelvins to hundreds of kelvins) depending on the material properties (mechanical and thermal), the surface properties (roughness and coefficient of friction) and the fretting process operating parameters (frequency and slip amplitude of the stroke). The influence of these factors can explain the certain discrepancies in the results reported in the open literature on the temperature rise in fretting. Moreover, the calculations show that there is a significant temperature gradient (up to thousands of kelvins per millimetre) in a very thin layer (few micrometres to tens of micrometres), called the thermally disturbed zone.

13.7 Conclusions

The literature survey shows that widely varying tribological properties of materials result in high scatter of fretting results. This often is the result of the secondary wear stage that can be related to the effect of the initial damage on the contact loading and on the velocity accommodation mechanisms. The fretting approach presented in this chapter can be extended to other wear analyses. Uses of RCFMs and MRFMs can provide useful tools that could be used in predicting the work hardening effect and the crack nucleation risk.

The complexity of the material effect under wear is generally associated with many parameters. This, however, often results from attempts to analyse a phenomenon in which the two materials in contact have properties which vary during the loading time. This is generally performed using final observation of the contact degradations and/or some physical measurement (such as material loss and friction coefficient) which can be interpreted to know the contact condition.

Any fretting process needs to be considered as a two-stage phenomenon:

(i) an incubation period leading to the formation of the first damage; and
(ii) a complex period during which this damage modifies the local loading and participates in arresting, stabilizing or increasing the degradation process.

In the first stage, the material participates in the establishment of the wear regime, i.e. the contact dynamics, through the surface chemistry, its ability to form a thin layer according to the chemical environment, and its roughness. Once the regime is identified, the material is studied from the damage viewpoint, as it will suffer from overstraining and/or overstressing. The properties required to delay the damage will depend on this loading. Contact mechanics theories can provide tools to calculate these local loading conditions. Fatigue criteria already exist and can be used to predict crack initiation in the case of overstressing. After this stage of crack initiation, fracture mechanics is used to predict propagation paths, crack growth rates and lifetimes even in the case of short cracks, complicated crack networks and fretting fatigue experiments. Since there are too many unknown factors in contact behaviour, it will be some time before a modelling of a particle detached from TTSs can be conducted.

In a similar way, it remains doubtful that the competition existing between the two main damage modes can be reliably described. However, dynamic mapping will be extensively used, particularly to include the effects of parameters such as temperature and frequency. In this chapter, two fretting maps were suggested, i.e. the running condition map that permits to calculate overstrains and overstresses and the material response maps to rationalize the study of material behaviour in close connection with the mechanical approach. Wear mapping has been successfully used in many cases of fretting.

References

1. Dowson, D. and Higginson, G.R., '*Elastohydrodynamic Lubrication*', 2nd edition, Pergamon Press, Oxford, UK, 1966.
2. Frene, J., Nicolas, D., Berthe, D., Degueurce, M. and Godet, M., '*Lubrification Hydrodynamique: paliers et butées*', Eyrolles, Paris, France, 1988.
3. Amontons, G., 'Du frottement de diverses matières les unes contre les autres', Histoire de l'Académie des Sciences, Paris, France, 1699.
4. Coulomb, M., 'Essai sur une application des règles de Maximimis et Minimis à quelques problèmes de statique relatifs à l'architecture', Mémoires de Mathématiques et de Physique, Paris, France, 1773.
5. Hertz, H., '*Miscellaneous Papers by Hertz*' (ed Jones and Schott), English version, McMillan, London, UK, 1896.
6. Johnson, K.L., '*Contact Mechanics*', Cambridge University Press, Cambridge, UK, 1985.
7. Kalker, J.J., 'Three-Dimensional Elastic Bodies in Rolling Contact', Kluwer Academic Publishers, Dordrecht, The Netherlands, 1990.
8. Godet, M., 'The Third-Body Approach: A Mechanical View of Wear', *Wear*, **100**, 1984, 437–452.
9. Berthier, Y., Vincent, L. and Godet, M., 'Velocity Accommodation in Fretting', *Wear*, **125**, 1988, 25–38.
10. Godet, M. and Berthier, Y., 'Continuity and Dry Friction: An Osborne Reynolds Approach', Elsevier Tribology Series, Elsevier, Amsterdam, The Netherlands, 1987.
11. Blau, P., 'Glossary of Terms', *ASM Handbook*, Vol. 18, 1992, pp. 1–21.
12. Fouvry, S., Kapsa, Ph. and Vincent, L., 'Quantification of Fretting Damage', *Wear*, **200**, 1996, 186–205.
13. McDowell, O.J., 'Fretting Corrosion Tendencies of Several Combinations of Materials', Symposium on Fretting Corrosion, ASTM, 1953, pp. 40–53.
14. Waterhouse, R.B., '*Fretting Corrosion*', Pergamon Press, Oxford, UK, 1972.
15. Waterhouse, R.B., '*Fretting Fatigue*', Applied Science Publishers, UK, 1981.
16. Waterhouse, R.B., 'The Problem of Fretting Fatigue Testing', ASTM STP 1159, 1992, pp. 13–22.
17. Chivers, T.C. and Gordelier, S.C., 'Fretting Fatigue and Contact Conditions: A Rational Explanation of Palliative Behaviour', *Proceedings of the Institution of Mechanical Engineers*, **199**, 1985, 325–337.
18. Bill, R.C., 'Fretting of AISI 9310 Steel and Selected Fretting-Resistance Surface Treatments', *ASLE Transactions*, **21**(3), 1977, 236–242.
19. Fouvry, S., Kapsa, Ph. and Vincent, L., 'Fretting Behaviour of Hard Coating Under High Normal Load', *Surface and Coating Technology*, **68/69**, 1994, 494–499.
20. Park, M., Hwang, Y.H., Choi, Y.S. and Kim, T.G., 'Analysis of a J69-T-25 Engine Turbine Blade Fracture', *Engineering Failure Analysis*, **9**, 2002, 593–601.
21. Sauger, E., 'Contribution à l'étude de la transformation tribologique superficielle en fretting', Ph.D. thesis, Ecole Centrale de Lyon (no. 97-25), France, 1997.
22. Sauger, E., Fouvry, S., Ponsonnet, L., Martin, J.M., Kapsa, Ph. and Vincent, L., 'Tribologically Transformed Structure in Fretting', *Wear*, **245**, 2000, 39–52.
23. Zhou, Z.R. and Vincent, L., 'Mixed Fretting Regime', *Wear*, **181–183**, 1995, 531–536.
24. Vincent, L., Berthier, Y. and Godet, M., '*Testing Methods in Fretting Fatigue: A Critical Appraisal*' (eds M.H. Attia and R.B. Waterhouse), ASTM STP 1159, ASTM, Philadelphia, 1992, pp. 33–48.
25. Lim, C. and Ashby, M.F., 'Wear Mechanism Maps', *Acta Metallurgica*, **35**, 1987, 1–24.
26. Vingsbo, O. and Soderberg, D., 'On Fretting Maps', *Wear*, **126**, 1988, 131–147.
27. Mindlin, R.D. and Deresiewicz, H., 'Elastic Spheres in Contact Under Varying Oblique Forces', *Journal of Applied Mechanics, ASME Transactions*, **20**, 1953, 327–344.

28. Fouvry, S., Kapsa, Ph. and Vincent, L., 'Analysis of Sliding Behaviour for Fretting Loadings: Determination of Transition Criteria', *Wear*, **185**, 1995, 35–46.
29. Fouvry, S., Kapsa, Ph., Vincent, L. and Dang Van, K., 'Theoretical Analysis of Fatigue Cracking Under Dry Friction for Fretting Loading Conditions', *Wear*, **195**, 1996, 21–34.
30. Szolwinski, M.P. and Farris, T.N., 'Mechanics of Fretting Crack Formation', *Wear*, **198**, 1996, 93–107.
31. Neu, R.W., Pape, J.A. and Swalla, D.R., 'Methodologies for Linking Nucleation and Propagation Approaches for Predicting Life Under Fretting', ASTM STP 1367, 2000, pp. 369–388.
32. Dang Van, K., 'Sur la résistance à la fatigue des métaux', Science et Techniques de l'Armement, Mémorial de l'Artillerie Française, 3ème fascicule, 1973.
33. Fouvry, S., Elleuch, K. and Simeon, G., 'Prediction of Crack Initiation Under Partial Slip Fretting Conditions', *Journal of Strain Analysis*, **37**(6), 2002, 549–564.
34. Dubourg, M.C. and Lamacq, V., 'A Theoretical Model for the Prediction of Crack Field Evolution', ESIS 18, Mechanical Engineering Publication, London, UK, 1994, pp. 135–147.
35. Araujo J.A. and Nowell D., 'The Effect of Rapidly Varying Contact Stress Fields on Fretting Fatigue', *International Journal of Fatigue*, **24**, 2002, 763–775.
36. Mohrbacher, H., Blanpain, B., Celis, J.P. and Roos, J.R., 'Oxidational Wear of TiN Coating on Tool Steel and Nitrided Tool Steel in Unlubricated Fretting', *Wear*, **180**, 1995, 43–52.
37. Mugadu, A., Hills, D.A. and Nowell, D., 'Modifications to a Fretting-Fatigue Testing Apparatus Based upon an Analysis of Contact Stresses at Complete and Nearly Complete Contacts', *Wear*, **252**, 2002, 475–483.
38. Zhou, Z.R. and Vincent, L., 'Cracking Behaviour of Various Aluminium Alloys During Fretting Wear', *Wear*, **155**, 1992, 317–330.
39. Elleuch, K., Fouvry, S. and Kapsa, Ph., 'Fretting Maps for Anodised Aluminium Alloy', *Thin Solid Film*, **426**(1–2), 2003, 271–280.
40. Liskiewicz, T., Fouvry, S. and Wendler, B., 'Impact of Variable Loading Conditions on Fretting Wear', *Surface and Coating Technology*, **163–164**, 2003, 465–471.
41. Elleuch, K., 'Comportement en fretting des alliages d'aluminium – Effet de l'anodisation', Ph.D. thesis, Ecole Centrale de Lyon No. 2002-06, France, 2002.
42. Fouvry, S., Liskiewicz, T., Kapsa, Ph., Hannel, S. and Sauger, E., 'An Energy Description of Wear Mechanisms and Its Applications to Oscillating Sliding Contacts', *Wear*, **255**, 2003, 287–298.
43. Fridrici, V., 'Fretting d'un alliage de titane revêtu et lubrifié : application au contact aube / disque', Ph.D. thesis, Ecole Centrale de Lyon No. 2002-22, France, 2002.
44. Fridrici, V., Fouvry, S., Kapsa, Ph. and Perruchaut, Ph., 'Impact of Contact Size and Geometry on the Coating Life Time', *Wear*, **255**, 2003, 875–882.
45. Leroy, J.M., 'Modélisation thermoélastique de revêtements de surface utilisés dans les contacts non lubrifiés', Ph.D. thesis, INSA Lyon, 1989, France.
46. Godet, M., Berthier, Y., Leroy, J.M., Flamand, L. and Vincent, L. 'Coating Design Methodology', Tribology Series 17, Elsevier, Amsterdam, The Netherlands, 1990.
47. Petiot, C., 'Développement d'une méthodologie prévisionnelle du comportement en fretting-fatigue couplant les approches expérimentales et théoriques', Ph.D. thesis, Ecole Centrale de Lyon No. 94-43, France, 1994.
48. Sproles Jr, E.S. and Duquette, D.J., 'Interface Temperature Measurements in the Fretting of a Medium Carbon Steel', *Wear*, **47**, 1978, 387–396.
49. Attia, M.H. and Yovanovich, M.M., 'A Model for Predicting the Thermal Constriction Resistance in Fretting', ASME Symposium on Contact Problems and Surface Interactions in Manufacturing and Tribological Systems, 1993.
50. Attia, M.H. and Comacho, F., 'Temperature Field in the Vicinity of a Contact Asperity During Fretting', ASME Symposium on Contact Problems and Surface Interactions in Manufacturing and Tribological Systems, 1993.
51. Fridrici, V., Attia, M.H., Kapsa, Ph. and Vincent, L., 'A Finite Element Approach to Temperature Rise in Fretting with Consideration to the Effect of Random Distribution of the Micro-Contact Areas", Proceedings of the International Conference on Tribology ITC 2000, 29 October–2 November 2000, Nagasaki, Japan.

14

Characterization and Classification of Abrasive Particles and Surfaces

G.W. Stachowiak, D.V. De Pellegrin and P. Podsiadlo

Abstract

Abrasive particles or grits are an inherent feature of many tribological systems. Two major factors controlling the abrasivity of a particle are its size and sharpness. It is intuitively felt that the particles with higher sharpness should generate higher abrasion or erosion wear rates. While it is relatively easy to quantify particle size, the numerical description of particle sharpness (or angularity) is much more difficult. An overview of the numerical descriptors of particle shape is presented in this chapter. While the traditional parameters such as shape factors are mentioned, the emphasis is put on the numerical descriptors that include sharpness of particle protrusions (or asperities) in their derivation. Sharpness describes the shape of the particle or surface protrusions in terms of its potential to abrade or erode.

Particle sharpness is correlated with abrasive and erosive wear rates the grits generate, using examples from the authors' laboratory and also from open literature. It is often observed experimentally that the particle size also influences particle abrasivity. The effect of particle size on abrasive wear is discussed in terms of its correlation with particle shape, shape deterioration during abrasion, material property changes at small scales and clogging of abrasive surfaces by debris.

The concept of asperity sharpness can also be used in the description of abrasive surfaces. Examples of the numerical characterization of surface sharpness are shown, using both model and real abrasive surfaces. The ability to characterize surface sharpness is important, both in the case where high surface abrasivity is desirable (e.g. grinding papers or wheels) and in the case where it usually should be avoided (two mating machine surfaces).

In the final section of this chapter, advances in the computer-based classification of surfaces are described with an example of the classification of abrasive surfaces.

Wear – Materials, Mechanisms and Practice Edited by G. Stachowiak

14.1 Introduction

Abrasive wear, two- and three-body, contributes to about 50% of industrial wear problems [1] but it is useful in machining or other surface-finishing techniques such as polishing. For example, the estimated cost of grinding in industrialized countries is about 20–25% of total expenditure on machining operations [2].

It is intuitively felt that, in addition to hardness and size, particle shape plays an important role in abrasion. However, determining the particle shape effects on abrasive wear rate is not an easy task. This is because wear depends on many different variables and the particle shape effect is often masked by stronger effects of other system variables. Particle shape in relation to abrasive or erosive wear is described by particle angularity or sharpness. Laboratory tests have confirmed that with the increase in particle angularity there is a significant increase in abrasive or erosive wear rates [3–8]. This chapter presents an overview of particle shape descriptors, with the emphasis on particle angularity parameters. Work conducted on abrasive and erosive wear has demonstrated that any measure of particle abrasivity must include particle angularity [7, 9, 10]. The relationship between the angularity parameters of particles and the abrasive and erosive wear rates they generate is shown using experimental data.

Similar approaches are made to characterize the sharpness of abrasive surfaces and the methods used are described later in this chapter. Surfaces with sharp asperities are undesirable in tribological contacts, especially where the counterface is much softer. In grinding, on the other hand, sharp asperities facilitate high material removal rate from the workpiece. All abrasive surfaces gradually change their sharpness due to chipping, fracturing or rounding of the asperities. It is of great importance to both tribology and manufacturing to predict how these changes would affect the abrasive power of the surface.

The chapter finishes with a brief description of a computed-based surface classification method. Practical example shows the classification of surfaces of different abrasivity. The surfaces are first divided into three classes based on the degree and distribution of surface roughness measured by three-dimensional (3-D) stylus profilometry. How the computer program automatically classifies an unknown abrasive surface is then demonstrated.

14.2 General Descriptors of Particle Shape

Traditionally, qualitative descriptors of particle visual appearance such as ‘spherical’, ‘semi-rounded’, ‘semi-angular’ or ‘angular’ have been used to classify and differentiate among various groups of abrasive particles [5, 6, 11–13]. Such a description is, of course, subjective in nature, especially when the gradation of the degree of angularity is sought. It may be sufficient only when the particles used in testing differ significantly in their shape, e.g. spherical glass beads versus angular SiC [5] or quartz [6]. The introduction of the computer-based image analysis systems has facilitated the use of numerical parameters in the description of particle shape [7, 14]. Typical shape parameters, often called shape factors, usually included in the image analysis software are the aspect ratio (width/length or sometimes length/width), roundness, form factor, convexity, elongation, etc. [14, 15]. Shape factors have been developed for general particle description, without specific considerations relevant to the particle abrasivity. These shape measures are simple combinations of size measurements, and their definitions can be found in References [14, 15]. They describe the tendency of a particle to deviate from an ideal shape of a sphere (or a circle, as the particles are often analysed in two dimensions). However, these parameters do not provide satisfactory information about the particle angularity since they do not indicate how sharp the particle protrusions or asperities are (particle protrusions are also called corners or spikes in the references cited in this chapter).

Despite the limitations, shape parameters have been used to describe various abrasive particles. For example, roundness factor (perimeter^2/4πarea) and statistical parameter R_{ku} (kurtosis) describing edge detail have been employed to differentiate among various groups of abrasives [16]. Width-to-length aspect ratio (W/L) and perimeter squared-to-area ratio (P^2/A) have been used to describe SiC, Al_2O_3 and SiO_2 particles [7]. Attempts have been made to correlate shape parameters with abrasive or erosive wear rates, and in some cases a good correlation has been found. It was shown in one work that the erosion rate increased with increasing P^2/A and decreasing W/L for SiC, Al_2O_3 and SiO_2 particles tested [7]. However, other work has confirmed the limitations of shape parameters, especially the aspect ratio, when used in the description of abrasive particles [17–19].

More complex particle shape descriptors are based on either Fourier analysis [14, 20] or fractal analysis of a particle boundary [21–23]. Invariant Fourier descriptors have been used to distinguish between abrasive particles of different angularities [3, 20]. In this technique the coefficients of the Fourier expansion are combined into the parameters such as radance and roughness. It was found that there is a relationship between these parameters and volume loss in two-body abrasive wear tests conducted on a rubber wheel test apparatus. Empirical equation relating the angularity of particles to abrasion was proposed [3]. Fractal analysis can also be applied to characterize the complexity or irregularity of a particle boundary [21–23]. A resulting parameter is called a fractal dimension, D, the value of which ranges between 1 and 2 for a particle boundary. The higher the fractal dimension D is, the rougher or more irregular the boundary. Although fractal dimension has been successfully used to describe irregular shapes of powders [23] or wear particles [22], it has shown limitations when applied to the description of angular shapes [17, 18].

It has been quickly realized that abrasive particles require numerical descriptors that include the measure of sharpness (or angularity) of particle protrusions [7, 24]. The problem was how to quantify particle angularity, i.e. determine the angle of every protrusion and then combine these values to obtain one parameter.

14.3 Particle Angularity Parameters

In one of the early works, a parameter called the degree of roundness, R, was incorporated in an empirical model to predict the erosion of brittle materials by natural contaminants [24]. The model predicted that the erosion rate should decrease with increasing effective roundness of the eroding particles. The R parameter was originally developed for the description of rock particles [25], and may be regarded as the precursor of the angularity parameters that followed later. It is defined as $R = \sum c/I_R N$, where c is the radius of curvature of the individual particle protrusions, I_R is the radius of the maximum inscribed circle and N is the number of protrusions measured. The R value is equal to 1 for an ideal sphere, and is lower than 1 for irregular shapes (angular particles have R values between 0.3 and 0.5) [24]. It appears, however, that the subjectivity involved in the determination of protrusion radii, especially when the particle contains very sharp protrusions, limits the practical value of the R parameter [26].

Another parameter, called the mean particle angle, has been proposed as a measure of particle angularity [27]. The automated program determines the mean angle of abrasive particles from curvature plots of digitized particle outlines. The particle protrusions (or corners), deemed to be of importance in the particle abrasive action, are these peaks on the curvature plots that exhibit high curvature. The selected peak angles are determined using two tangents calculated at the 'begin' and 'end' points of the peak [27]. The weaknesses of this method, such as subjective selection of the corners, digitization noise, discrimination power of the algorithm, have been indicated by its users [27]. Also, the experimental data generated using this technique are too limited to properly assess its value [26].

In another work, a method based on the Radon transform was developed to characterize abrasive powder particles [28]. This technique involves the generation of a convex hull where only boundary features supporting the hull are deemed of interest. An angularity factor is calculated on the basis of the curvature of the hull at these locations. The applications of this technique are rather limited because the concept of penetration is ignored, i.e. an n-sided polygon will yield the same angularity as the corresponding star shape [26]. Intuitively, one would feel that the star-shape particle would cause more wear damage than the n-sided polygon. On the other hand, it could be argued that, as most real particles are generally convex, the hull technique might give better results than initially anticipated. Experimental verification of this technique using typical abrasives such as quartz and alumina would be needed to confirm its usefulness for angular particles.

Over the last 10 years, extensive research has been conducted at the authors' laboratory on the angularity description of abrasive particles and the following sections describe the results in more detail.

14.3.1 Angularity Parameters SP and SPQ and Their Relation to Abrasive and Erosive Wear

The first parameter developed that includes the sharpness of particle protrusions is called the spike parameter (SP) [17, 18, 29]. This descriptor is based on representing the particle boundary by a set of triangles constructed at different scales (a fractal approach is used here). At each scale, triangles with the greatest spike value, defined as $\text{sv} = \cos(\theta/2)h$, where h is the perpendicular height of the triangle and θ is the apex angle, are fitted to particle protrusions. In this approach, it has been assumed that the sharpness and size of these triangles are directly related to particle abrasivity, i.e. the sharper (smaller apex angle) and larger (perpendicular height) the triangles are, the more abrasive is the particle. The maximum spike values at each scale are normalized by their height, i.e. $\text{sv/height} = \cos(\theta/2)$, and then averaged. The final SP is the averaged spike value from all the scales used in the calculation and it varies between 0 and 1 [17, 29]. The higher the value of the SP is, the sharper is the particle.

The performance of the SP has been compared with more traditional shape parameters using artificially generated shapes, as shown in Table 14.1 [18]. The analysis of Table 14.1 shows that the SP provides the angularity ranking of the computer-generated shapes that agrees well with the visual assessment of their abrasivity (the abrasivity increases from top to bottom). It can also be seen that the SP outperforms the other shape parameters in this task.

Experimental results using real abrasive particles have verified that there is a good correlation between the SP of abrasives and the wear rates they generate [29]. However, it was

Table 14.1 Artificial shapes with increasing angularity and the corresponding spike parameters, boundary fractal dimensions, roundness factors (perimeter^2/4π area) and aspect ratios (length/width) (adapted from Reference [18])

Shape	Spike parameter	Boundary fractal dimension	Roundness	Aspect ratio
	0.1332	1.0046	1.1145	1.0000
	0.1633	1.0064	1.1323	1.0078
	0.1721	1.0063	1.2933	1.0556
	0.1951	1.0115	1.1426	1.0514
	0.2119	1.0095	1.6127	1.6800
	0.7243	1.0155	3.1458	1.0000

found that the SP has certain imperfections [19]. At small scales it is sensitive to digitization errors, and this leads to some 'smooth' shapes being assigned an artificially high SP. For boundaries with convex shape and rounded protrusions, the apex angles of the triangles constructed at large scales are much smaller (sharper) than the corresponding protrusions. Also, since the method takes into account every boundary irregularity, the computing time can be long.

These limitations have prompted the development of another particle angularity parameter, called the spike parameter quadratic (SPQ) fit [19]. In this method, particle protrusions which are most likely to cause abrasion are isolated, while other boundary features are ignored. Only portions of a particle boundary, called spikes, protruding outside the circle of mean diameter, are considered in the calculation of the SPQ, as illustrated in Figure 14.1 [18, 19]. The shape of the protrusions (spikes) is approximated by quadratic curves instead of triangles [18, 19].

For each spike, the local maximum radius is found and this point 'mp' is treated as the spike's apex [19]. The sides of the spike, which are between the points 'sp-mp' and 'ep-mp', are then represented by fitting quadratic curves. Differentiating the curves at the 'mp' point yields the apex angle θ and the spike value $\mathrm{sv} = \cos(\theta/2)$. Finally, the SPQ parameter is calculated as $\mathrm{SPQ} = \mathrm{sv}_{\mathrm{average}}$ [19].

Great advantage of the SPQ parameter over the older SP is that it considers only the boundary features, i.e. protrusions, which are likely to come in contact with the opposing surface. It was found that the SPQ parameter correlates better with abrasive wear rates than the SP [19]. Also, since the SPQ parameter is almost insensitive to image focus [19], an optical microscope can be used for collecting particle images for the SPQ calculation.

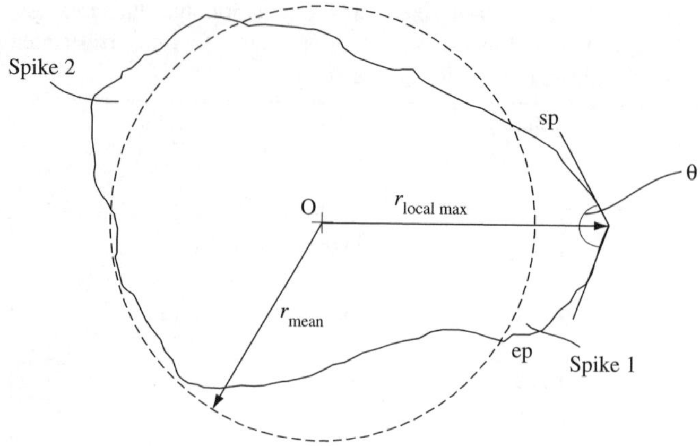

Figure 14.1 Method of the SPQ calculation by fitting quadratic segments to particle protrusions (adapted from [19])

Two-body abrasive wear experiments conducted on a pin-on-disc tribometer were used to assess the performance of the SPQ parameter. Various types of abrasive grits, such as glass beads, silica sand, garnet, natural industrial diamond, silicon carbide, quartz and crushed sintered alumina, were used to manufacture abrasive discs. The size of all the grits used in the experiments was 250–300 μm. Soft wearing material such as chalk was selected as a counterface to eliminate the extraneous effects of abrasive grit deterioration and clogging of the disc by wear debris. Low loads and velocities were used in the experiments in order to eliminate the fracture of the chalk and to ensure that 'pure' two-body abrasive wear took place. The data, plotted in Figure 14.2, show an almost perfect linear relationship between the two-body wear rates and particle angularity expressed in terms of the SPQ parameter [19].

The results of three-body wear experiments conducted on a ball-cratering tribometer and erosive wear experiments exhibited similar trends, as shown in Figures 14.3 and 14.4 [30].

The SPQ technique, as all the other methods, has its limitations. It does not consider the effect of load on the penetration depth (and therefore the wear rate) [31]. The important effect of particle orientation with respect to the wearing surface is measured only indirectly by averaging over all protrusions. Thus, the SPQ parameter is most reliable when the average penetration is small [31]. Similar criticism regarding the effect of load and orientation can also be directed at the other angularity parameters described so far [26, 31]. The technique presented in the following text tries to address these deficiencies.

14.3.2 Cone-Fit Analysis (CFA)

The ability of an abrasive particle to abrade depends strongly on its orientation to the wearing surface (or the angle of attack). For example, an elongated particle with sharp ends oriented along its longer axis to the wearing surface will not cause much damage. The situation will change when this longer axis is perpendicular to the wearing surface, as the particle ability to penetrate and cause wear damage will increase. Thus, a new technique involving angularity measurement at every orientation of the particle projection and over a large

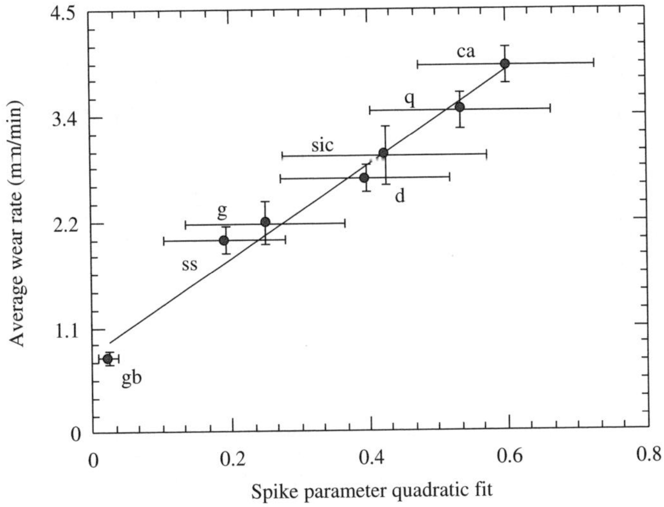

Figure 14.2 Relationship between the wear rates of chalk and the SPQ values for different abrasive grits (gb – glass beads, ss – silica sand, g – garnet, d – natural industrial diamonds, sic – silicon carbide, q – crushed quartz and ca – crushed sintered alumina) [18, 19]

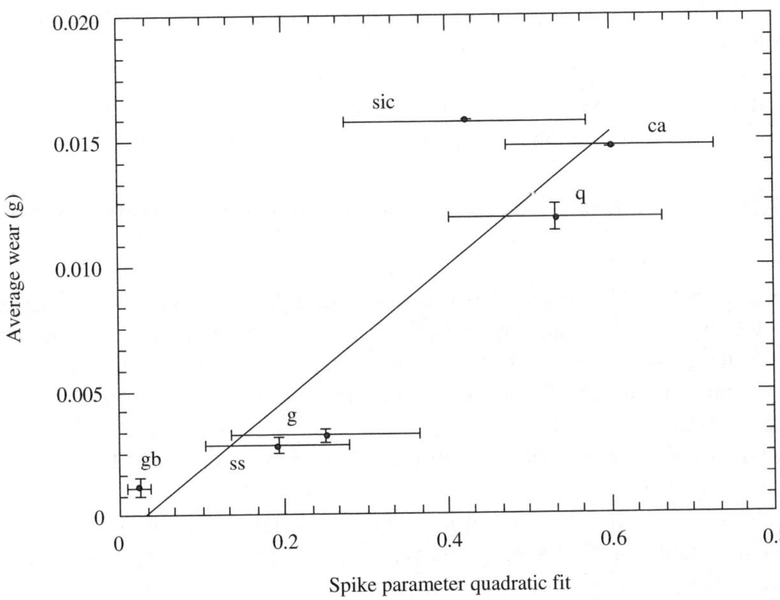

Figure 14.3 Relationship between three-body abrasive wear and the average SPQ parameter for different abrasive grits, error bars are ±1 standard deviation [30]

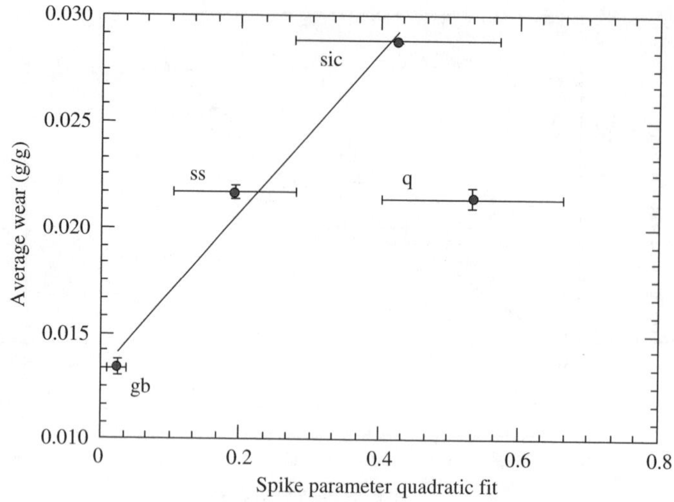

Figure 14.4 Relationship between erosive wear in air and the average SPQ parameter for different abrasive grits at 90° impingement angle (error bars are ±1 standard deviation). The erosion rates are normalized with respect to density and particle velocity [30]

range of penetration depth has been developed [31]. In this way, the statistical description of particle sharpness as a function of penetration depth is obtained [32]. The technique, called cone-fit analysis (CFA), has evolved from the classical abrasion model proposed by Rabinowicz [33], where the asperities of abrasive surfaces or particles are represented by cones. As the particle abrasiveness depends on the portion of the particle forced to penetrate and abrade the wearing surface, the severity of abrasion depends on particle orientation. Based on this notion, very abrasive particles might be represented by cones with a large angle of attack, while mildly abrasive particles may be represented by cones with a small angle of attack [31].

The classical abrasion model is schematically illustrated in Figure 14.5, where a single cone-shaped asperity with an angle of attack θ is pushed against and abrades a flat surface. Two areas shown in Figure 14.5 are of interest to CFA: the projected penetration area Ω and the groove area Λ. The projected penetration area Ω is defined as the intersection of the cone with the theoretically planar wear surface, while the groove area Λ is the orthogonal projection of the cone in the traversal direction. According to this model, the wear volume V is proportional to load P, sliding distance L and the tangent of the attack angle θ, and inversely proportional to hardness H.

The analysis of abrasive particles by CFA involves using a specially developed computer program to calculate Ω and Λ areas for cones fitted to digitized particle profiles. Detailed description of the technique is given in Reference [31]. The effect of particle orientation is included in the calculation. The average groove area Λ_{av} calculated for all orientations is then plotted against the penetration area (load) resulting in the CFA curve (also called the groove function). The gradient of the groove function (defined in CFA as an angularity ratio λ) is related to the abrasivity (or sharpness) of the particles tested. Linear character of the CFA curve indicates that particle protrusions or surface asperities behave like cones (the ratio of

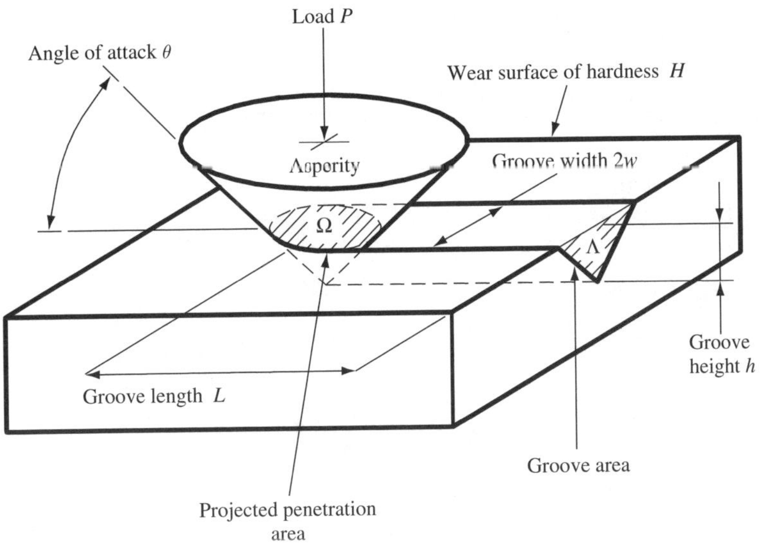

Figure 14.5 Schematic illustration of the projected penetration area Ω and groove area Λ concept (adapted from [31])

the groove area to the projected penetration area is constant and independent of penetration depth, $\lambda = \Lambda/\Omega = \text{constant}$). CFA curves for six types of real abrasives are shown in Figure 14.6. The standard deviation bars represent the variability of the data – in general the variability is greater for angular particles than rounded ones. The gradient of the curves indicates that glass beads are the least abrasive (the lowest gradient) and crushed alumina is the most abrasive (the highest gradient). However, the non-linearity of the curves shown in Figure 14.6 suggests that the real particle protrusions differ in shape from a perfect cone.

For most particle types, the gradient of the CFA curve increases with increasing penetration depth, suggesting that the wear rate should also exhibit a rise with increasing load or decreasing hardness [31]. This seems consistent with the observation made by Johnson [34] that a convex curved indenter causes an increase in strain intensity with increased penetration, augmenting the probability of chip separation from the wear surface. The increase in wear rate gradients, however, is not well documented in the literature. Rising wear rate gradients have been observed, for example, in manufacturing, in constant force grinding [33]. On the other hand, decreasing gradient curves are more frequently reported [35]. Decreasing gradients can be explained in terms of abrasive surface degradation or the ineffective mechanism of debris removal from a wearing surface as the load increases [31, 35].

Linear regression analysis was performed on the data points from Figure 14.6 and the average angularity ratio λ_{av} (resulting gradient) for each curve was determined. The average angularity ratio λ_{av} can then be used to find an average value of the asperity angle of attack $\theta_p = \tan^{-1}(\lambda_{av}\pi)$ [31]. The average angularity ratio for each particle type is plotted in Figure 14.7 against the wear rates of chalk abraded by similar particles (the wear rate data were taken from Reference [19] and are also shown in Figure 14.2). It can be seen from Figure 14.7 that the average angularity ratio calculated (and at the same time the average angle of attack) correlates well with the experimental two-body abrasion wear data.

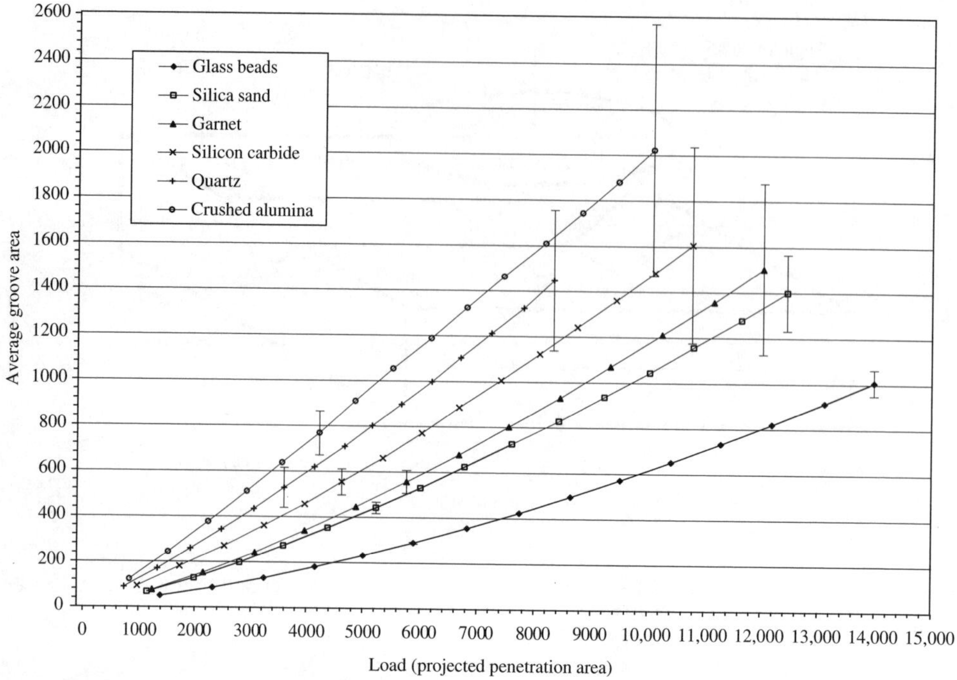

Figure 14.6 CFA curves for typical abrasive grits. The grits were sieved to 150–300 μm size range [31]

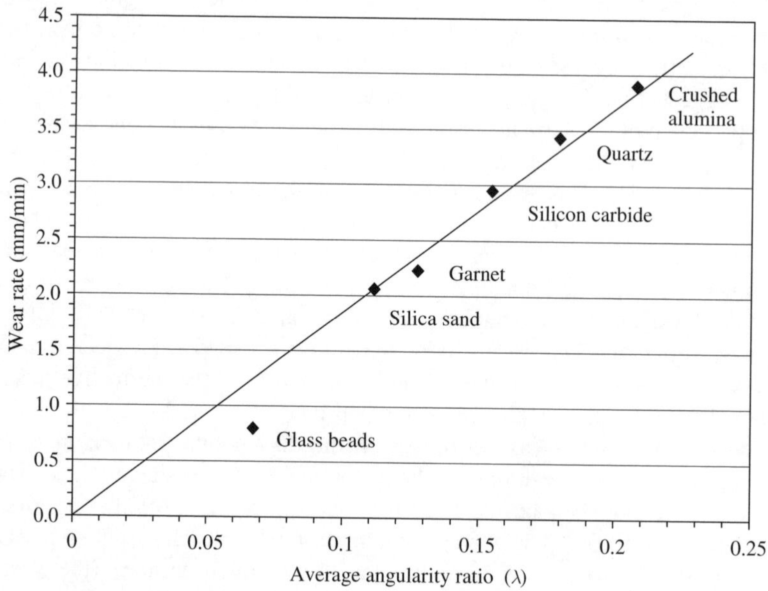

Figure 14.7 Relationship between two-body abrasive wear rates of chalk and the average angularity ratio λ_{av} of abrasives calculated by CFA [31]

14.3.3 Sharpness Analysis

Despite the apparent progress, it had been realized that the CFA must suffer some inaccuracy due to the inadequate approximation of asperity shapes by cones, as real particle asperities are generally not conical. A modified technique, called sharpness analysis (SA), was subsequently developed [26, 36]. The SA technique is more accurate as it uses the full integration of the particle boundary to determine the groove area and provides more detailed consideration of the averaging process and statistical variability of shape and size [26, 36]. The sharpness λ is defined again as the ratio of the groove area Λ to the projected penetration area Ω for a given particle at pre-determined values of penetration depth and sliding direction, as schematically illustrated in Figure 14.8. As natural particles may exhibit vastly different sharpness, depending on the penetration depth and orientation, the concept of average sharpness has been incorporated in the SA technique [26].

An example of the set of groove functions obtained for various orientations of an angular particle is shown in Figure 14.9 [36]. Each orientation of the particle can be treated as a single asperity. It can be seen from Figure 14.9 that groove functions for each individual orientation of the particle are rather irregular, reflecting the characteristic shape of the corresponding particle asperity [36]. Also, these curves tend to exhibit an increasing gradient trend, as expected from the bodies that are generally convex [36]. The data points from all the individual curves are then averaged and approximated by a high-order polynomial to obtain the average (smooth) groove function.

In practice, many thousands of particles constitute a typical abrasive surface in two-body abrasion. Although each particle orientation and shape is unique, the wear rates produced are repeatable between the experiments. This suggests that the statistical properties of the particles are stationary, i.e. they do not vary with the position, sample size or time [26].

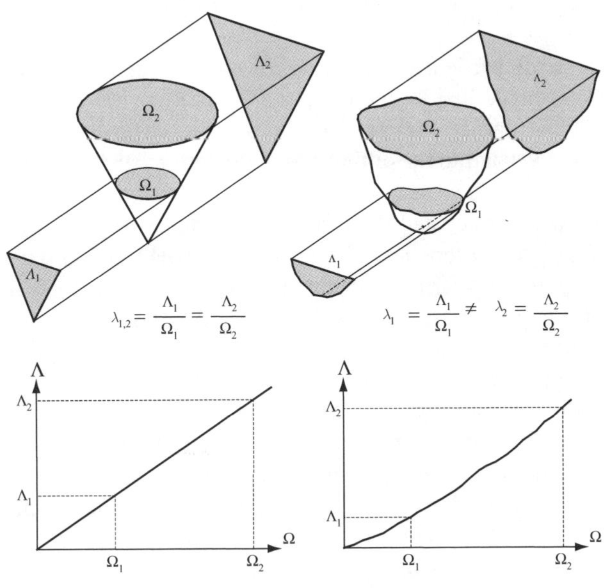

Figure 14.8 Schematic illustration of the sharpness λ concept for conical and realistic asperities [37]

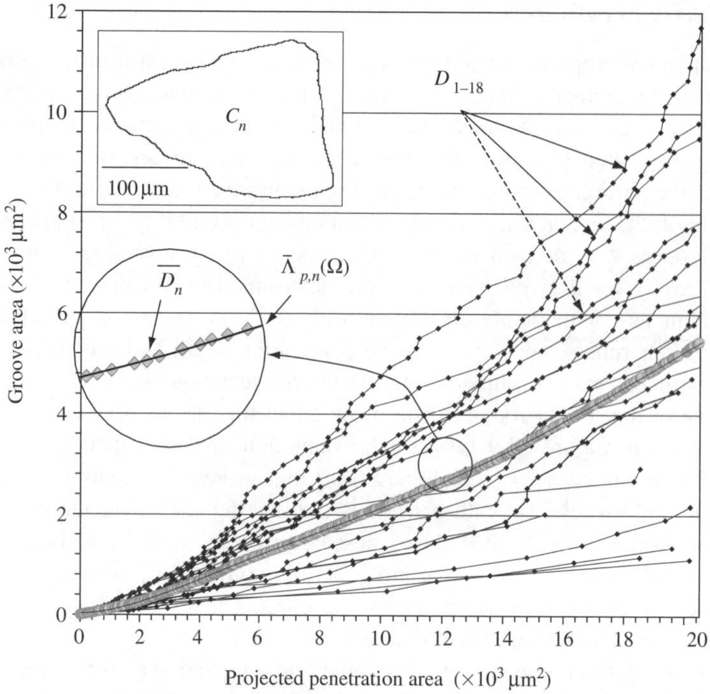

Figure 14.9 Groove functions for 18 orientations ($\Delta\varphi = 20°$) of a particle projection C_n together with the averaged groove function approximated by the polynomial fit. C_n is derived from a 212–250 μm silicon carbide particle, where D is the set of points representing the groove function for a unique particle orientation and $\overline{D}_n$ is the set of points representing the average groove function [36]

The correlation between particle sharpness, defined by SA, and the wear rates they generate has been verified experimentally [36]. Four types of particles of substantially different shapes, i.e. glass beads, alluvial garnet, alumina and quartz, sieved to the size range of 212–250 μm, were analysed by SA and the resulting groove functions are plotted in Figure 14.10(a). Two-body abrasive wear tests were conducted on specially manufactured abrasive surfaces with chalk used as a counterface and the results are plotted in Figure 14.10(b). It was found that the ranking of the groove functions for all the abrasives tested, Figure 14.10(a), is very similar to the ranking of the wear rate versus pin load curves, Figure 14.10(b).

It can be seen from Figure 14.10(a) and (b) that the groove functions for quartz and alumina particles are nearly identical and also the respective wear rates they generate are very similar. This implies that the groove function can be a good indicator of the particles' abrasive power (ability to cause wear). The groove functions for glass beads and alluvial garnet are significantly different but the correct correspondence between the respective groove functions and the wear rates is maintained. The average groove functions shown in Figure 14.10(a) were obtained for 100 particles and increasing the sample size from 100 to 1000 particles would unlikely produce a distinction between the groove functions for quartz and alumina [36]. A similar problem was described when differentiating particle size. For example, it has been suggested that as many as 10,000 images are necessary to accurately classify the size of ceramic powders [38].

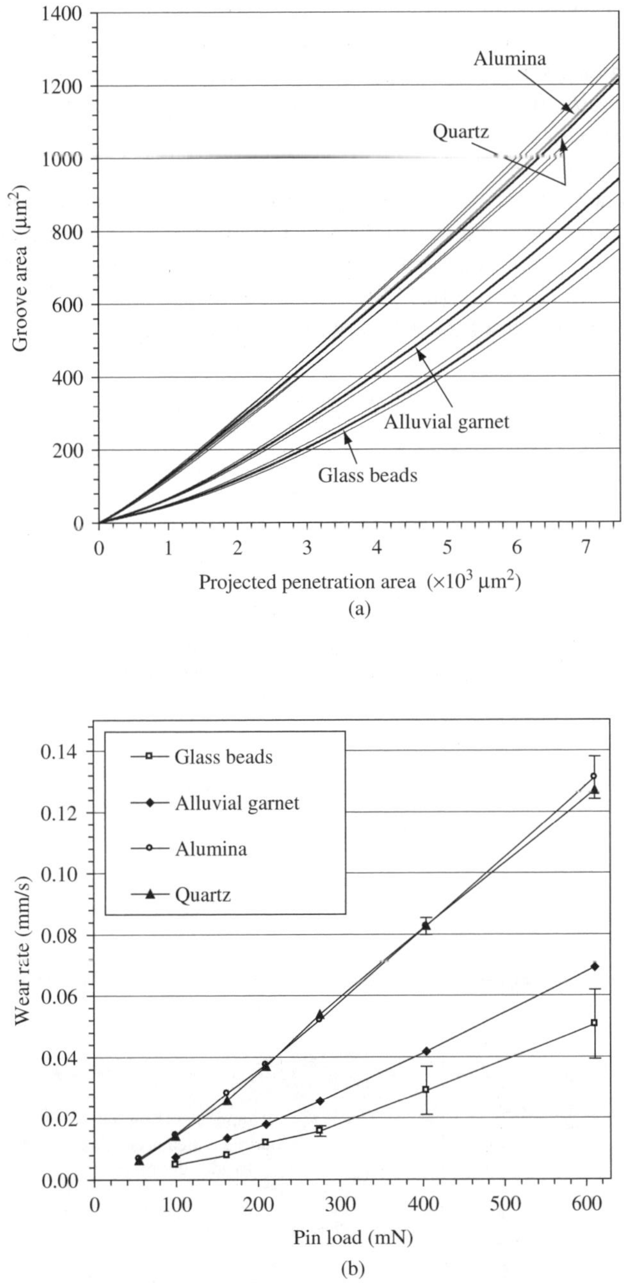

Figure 14.10 Plots of (a) average groove functions and 95% confidence intervals for different types of abrasive particles; 100 particles per sample, 212–250 µm; and (b) wear rate versus load for two-body pin-on-disc experiments with the same particles as in (a); average of two tests plotted, 200 mm/s sliding speed, ∅9 mm chalk pin. Bars are shown when error is greater than 0.001 mm/s [36]

While the sharpness parameter λ is a relatively complex and comprehensive shape property, in some cases it fails to completely describe the particle shape and provide a complete solution to a particle angularity characterization problem [36]. For example, preferential alignment of the particles, relative to the wearing surface, may change their behaviour and the wear rate induced [36].

As can be seen from Figures 14.9 and 14.10(a) SA produces groove functions with an increasing gradient for all particle types tested. This is in agreement with the findings of CFA (Figure 14.6) and the assumptions used when formulating SA [36]. Further investigations have revealed an interesting fact that groove functions can be approximated by the power-law, as illustrated in Figure 14.11(a) [37]. This indicates the fact that the power-law may be a convenient approximation of the abrasivity of many types of natural particles.

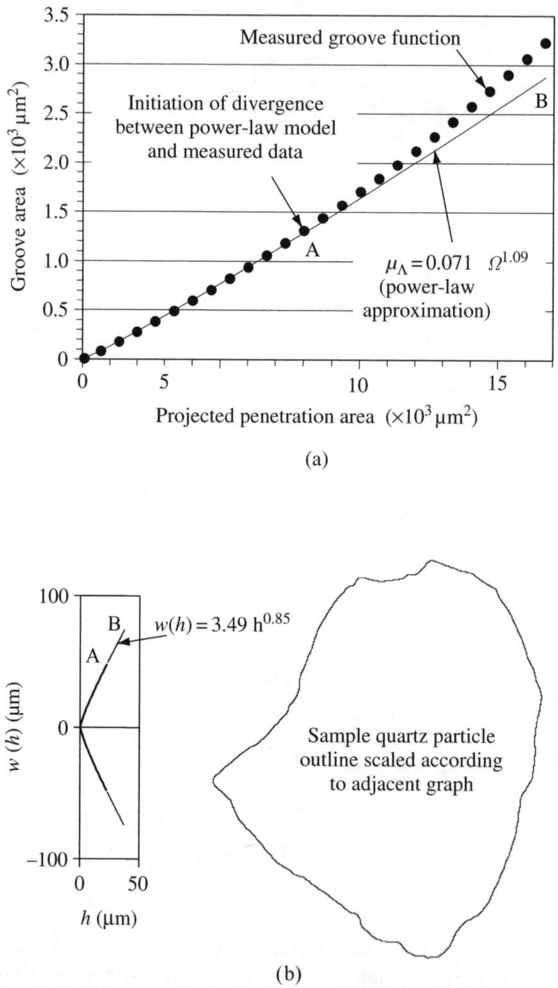

Figure 14.11 Illustration of (a) a groove function and power-law model for 100 quartz particles 212–250 μm and (b) equivalent asperity profile derived from the groove function [37]

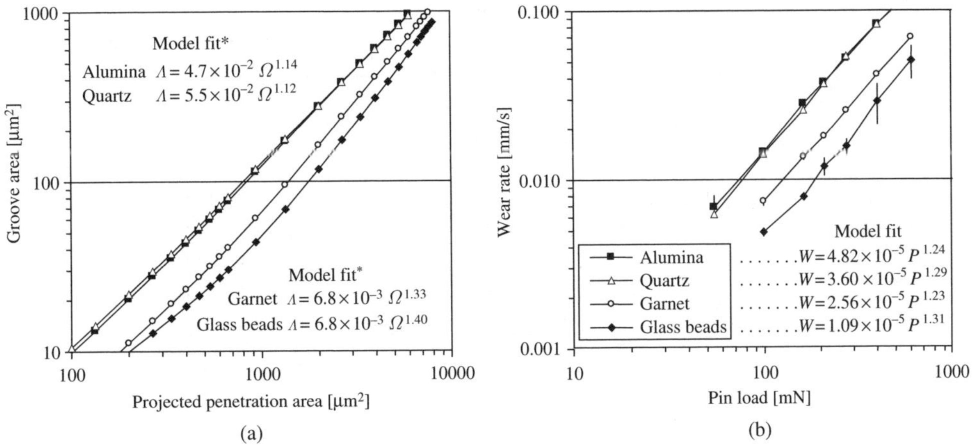

Figure 14.12 Power-law models for (a) groove functions of 100 typical abrasive particles of 212–250 μm size and (b) two-body abrasion wear results for the same abrasives [37]

The power-law approximation is consistent with fractal nature of particle boundaries and it can be incorporated in abrasive wear models [37]. However, it can be seen from Figure 14.11(a) that the power-law model eventually diverges from the experimental data. This can be explained by the fact that real particles have closed and mostly convex profiles that do not diverge indefinitely [36]. The accuracy of this model is thus dependent on the magnitude of penetration and, in principle, the model should work for low-stress abrasion where the penetration depth is moderate. From the average groove function, an average asperity profile can be derived as shown in Figure 14.11(b) [37]. This is a useful feature as these profiles could be used in the numerical simulation of abrasive wear [39]. It can be seen from Figure 14.11(b) that the asperity profile derived has an element of similarity to both hyperbolic and triangular profiles considered in other work [26, 33, 37, 40].

Experimental data shown in Figure 14.10(a) and (b) are re-plotted in Figure 14.12(a) and (b) using logarithmic axes to further assess the suitability of the power-law approximation. It can be seen from Figure 14.12(a) and (b) that the fractured-type particles such as quartz and alumina follow the power-law model over greater ranges of the penetration area Ω [37]. The exponents of the fitted model can be used to differentiate between various particle types. The exponent values tend to be lower for sharp particles than rounded ones [37]. Also, as shown in Figure 14.12(a) and (b), both the groove functions and the wear rates exhibit increasing gradient trends, as implied by the greater than unity exponents of the fitted models [37].

14.4 Particle Size Effect in Abrasive Wear

It is generally accepted that the particle size effect begins to manifest itself at particle sizes below 100 μm [26, 32, 41, 42]. This trend can also be observed in Figure 14.13 where the wear rates of chalk begin to decline when the quartz particle sizes drop below 150 μm for all three loads used in testing [36].

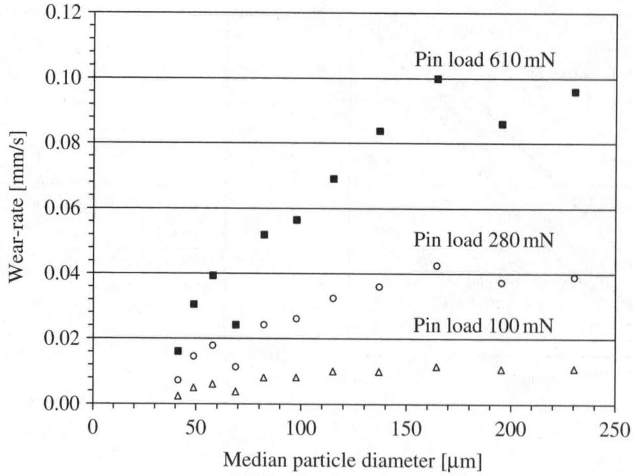

Figure 14.13 Effect of particle size on wear rate in two-body abrasion. Quartz particles, average of two tests plotted, 200 mm/s sliding speed, ∅9 mm chalk pin [36]

The interpretation of the apparent reduction in wear rate with the decreasing particle size is far from consistent. Some authors attribute this to the increased particle roundness with decreased particle size [43], while others suggest a material strengthening mechanism, often called the 'strain gradient effect', at reduced scales [44]. Similar strengthening is observed in micro- and nano-indentation tests [45–47]. This effect can be observed at indentation diameters below 50 μm [26]. With the reduction in indentation size down to 1 μm, the strength of the material, defined as the ratio of force to indentation area, can double or even triple [26, 46].

Recent computer simulations, where asperities are modelled as spherically tipped cones, have confirmed the apparent influence of the particle shape on the particle size effect [48, 49]. In these numerical simulations, the decreasing particle size was modelled by varying the spacing between cones, while the asperity tip radius and the cone angle remained unchanged. The results implied that the particle shape, represented by the relative asperity tip roundness, is a function of particle size and its roundness increases with decreasing size. However, the argument that the smaller particles are more rounded is often not supported by microscopic examination of real particles. It was found, for example, that the alumina particles about 1 μm in size do not appear much different in angularity than the particles 10 or even 100 times larger [26]. SEM images of quartz particles, ranging in size from 40 to 250 μm, are shown in Figure 14.14. It can be seen from Figure 14.14 that it is virtually impossible to notice any increase in roundness with decreasing size of quartz particles [26].

In real contacts during two-body abrasive wear only a small number of the most exposed asperity peaks support load. Large valleys remain between the contacting asperity points and, under low loads, they can act as a reservoir for debris accumulation. The wear debris clogging the valleys can play a substantial role in separating the surfaces, and mitigating wear [39]. This becomes more likely as the particle size decreases [39]. However, for the clogging theory to be plausible for the explanation of the particle size effect a stronger

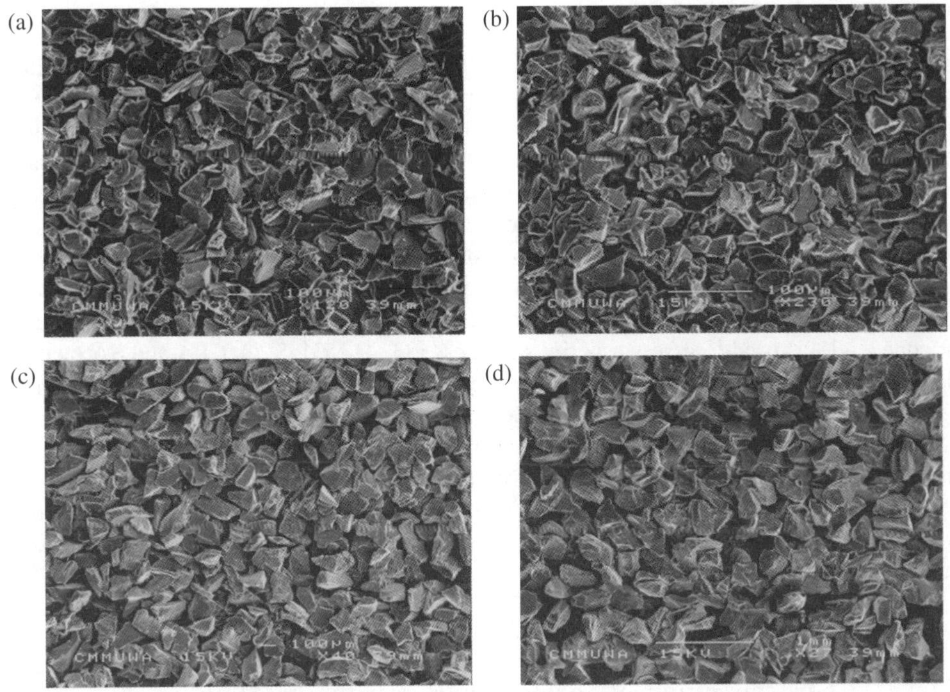

Figure 14.14 SEM images of quartz particles bonded to a flat surface: (a) 38–45 μm; (b) 63–75 μm; (c) 125–150 μm; (d) 212–250 μm [26]

relationship between the load, material hardness and wear rate would be expected. Why is this not the case? It is possible that with increased load and asperity penetration, debris are more effectively cleared into the surface valleys [39]. As it would be expected, increasing load produces more debris but at the same time it compresses them. The wear debris are then pushed out of the way during the abrasive contact. The combination of these effects may explain why the critical particle size effect is not more strongly pronounced with increasing load as illustrated in Figure 14.13 [36].

Deterioration of abrasive grits as the cause for wear rate decrease is proposed in another work [50]. It is argued that smaller particles endure abrasive contact for a relatively greater displacement and are therefore more prone to deterioration. It seems that the end effect of particle attrition is similar to the effect of clogging [39].

Considering all the evidence presented above, it appears that a combination of various mechanisms may contribute to the particle size effect, i.e. clogging/deterioration mechanisms [39, 50], and the scale-dependent strengthening of the abraded material [44, 51]. Changes to the mechanical properties such as hardness, stiffness and fracture toughness (as well as the adhesion forces) are due to the increased strain gradients at small scales [39, 52].

Particle size effect in abrasive wear is not only influenced by the shape and mechanical properties of individual asperities or grits that constitute the abrasive surface, but also by the way the asperities are distributed on the surface (surface topography or geometry). This aspect is discussed further in the next section.

14.5 Sharpness of Surfaces

It is generally recognized that the surfaces have capacity to abrade if their asperities induce stresses sufficient to cause plastic deformation and/or fracture of the counterface. As sharp and hard protrusions of particles can be agglomerated to form effective abrasive surfaces, these surfaces inherit properties directly attributable to these particles. Abrasive surfaces owe their properties to the shape, size and distribution of the constituent particles. Particle attrition by fracture or chipping causes great uncertainty when correlating the sharpness of individual particles with the sharpness of an abrasive surface formed by these particles [37].

Several parameters have been proposed to quantify the effect of particle shape on wear. For conical or pyramidal asperities the attack angle (and its variants such as included angle or cone angle) is used to assess asperity sharpness [33, 41, 53]. SPQ, CFA and SA methods described earlier utilize a similar concept to characterize asperity sharpness. For spherical asperities, the radius [41, 54], the ratio of depth to width [8] or the degree of penetration [55] can measure their effectiveness in the material removal processes. For abrasive surfaces, the average slope and curvature of grooves and asperities can be used to evaluate their abrasivity [56]. The parameters can be applied to individual particles and also can be used for describing abrasive surfaces. In the following sections, two methods of characterizing surface sharpness are presented.

14.5.1 Characterization of Surface Sharpness by the Modified SPQ Method

In an attempt to describe surface sharpness, the SPQ method has been adapted to characterize the shape of asperities (peaks) of surface profiles [57]. Surface profiles can be obtained by various techniques of surface roughness measurement and in this work the Talysurf profiles are used. Initially, a mean line running through the centre of the surface profile is fitted. It is assumed that the asperities (spikes) above this line would cause abrasion. The shape of each spike, formed by two segments from the left point (lp) to the apex and from the right point (rp) to the apex (Figure 14.15, [18, 57]), is approximated by quadratic functions. Differentiating these functions at the apex yields an apex angle 'θ'. SPQ_p value is calculated according to the formula [57]:

$$SPQ_p = \frac{1}{n}\sum_{n} \cos\frac{\theta_i}{2}$$

where SPQ_p is the spike parameter quadratic fit from a surface profile, θ is the apex angle and n is the number of spikes analysed.

The apex sharpness is the most important feature from the abrasion viewpoint. To obtain the true measure of the apex angularity, the small spikes present on the surface profiles are eliminated from the calculations as they do not contribute to wear, i.e. spikes whose height is less than 25% of the maximum spike height found in the profile are rejected [57].

The modified SPQ method has been applied to measure the sharpness of abrasive discs used in two-body abrasive wear tests, described in Section 14.3.1. Wear rate data shown in Figure 14.2 are re-plotted in Figure 14.16 against the SPQ_p values of abrasive disc surfaces.

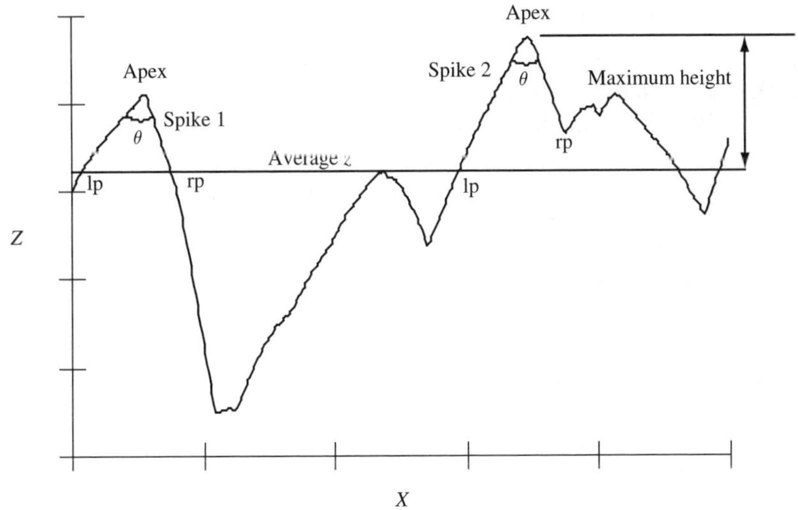

Figure 14.15 Method of evaluating the SPQ_p value from a surface profile [18, 57]

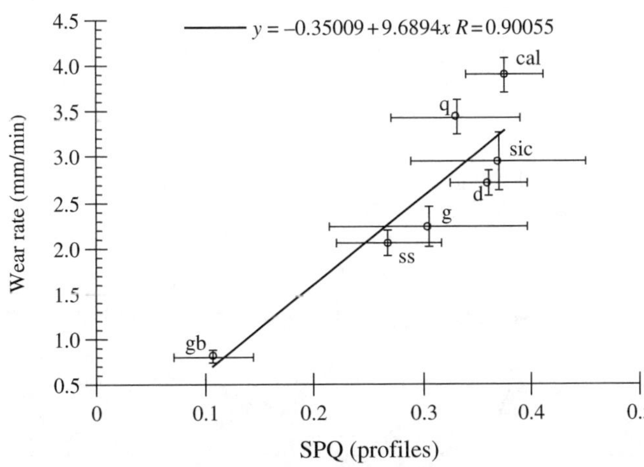

Figure 14.16 Relationship between the wear rates of chalk and the SPQ_p calculated for abrasive disc profiles [18, 57]

It can be seen from Figure 14.16 that there is a good correlation between the wear rates and the profile angularity described by the SPQ_p.

The SPQ_p values for angular abrasives such as quartz and crushed alumina were lower than expected [57]. This was caused by the Talysurf stylus with a finite radius that was unable to follow exactly the highly angular and re-entrant surfaces made of these abrasives [57].

14.5.2 Characterization of Surface Sharpness by SA

The concept of sharpness, SA, originally developed for particles and described in Section 14.3.3, has been recently extended to include the analysis of abrasive surfaces. It has been suspected that additional factors such as the density of asperities on the surface, their height distribution and orientation would complicate the surface sharpness estimation from particle attributes only [37]. The concept of surface sharpness is schematically illustrated in Figure 14.17 and detailed description can be found in Reference [37].

Horizontal sectioning technique was developed for the purpose of assessing geometrical similarity of abrasive surfaces at different scales [37]. Surfaces prepared with three sizes of SiC abrasive grits were nickel-plated and then successively ground to a specified depth to obtain serial sections. To make sure that the sample surface aligns itself according to the distribution of hard particles during grinding and polishing a 'floating' sample was used [37]. This arrangement eliminates the alignment problem that becomes increasingly significant with decreasing particle size and the distance between successive section planes [37]. Example of a typical section of the nickel-plated SiC surface is shown in Figure 14.18. The sections were examined by optical microscopy and two quantities, area fraction, a, and asperity frequency, f_N, were measured in relation to penetration depth.

It is theoretically possible, provided that there is enough data from the individual sections at sufficient resolution, to reconstruct a 3-D model of the abrasive surface. From this model, individual asperities can be isolated and their groove areas calculated. However, the technique has proved to be extremely time consuming as numerous samples are needed to obtain meaningful estimate of asperity density and area. From the preliminary investigation, it was found that both the area fraction, a, and the asperity frequency, f_N, increase exponentially as

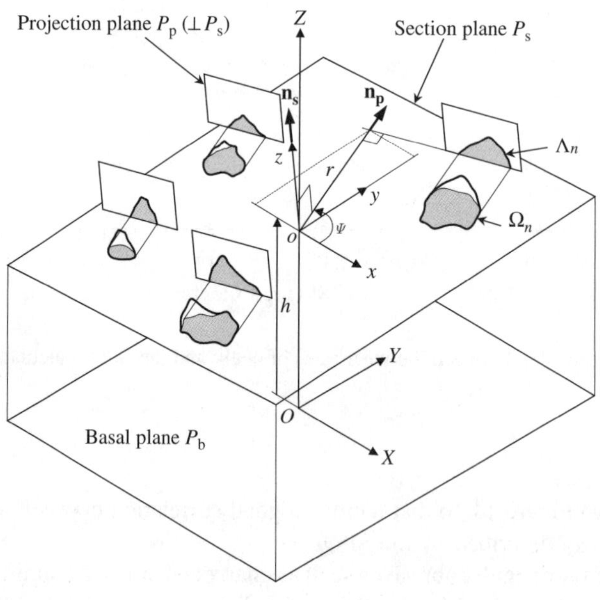

Figure 14.17 Development of sharpness for nominally planar abrasive surfaces [37]

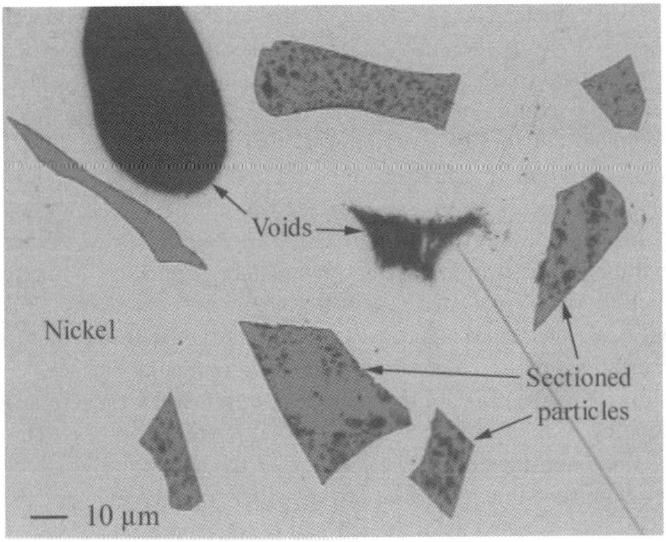

Figure 14.18 Typical section of the nickel-plated SiC surface [37]

the penetration depth increases into the surface. When the logarithms of f_N are plotted against a, a linear relationship is observed [37]. Similar results were reported when comparing scratch density as a function of load in two-body abrasion [58]. The data obtained from the sections, however, were insufficient for definite conclusions regarding the geometrical similarity of surfaces.

The preliminary data were subsequently used in statistical simulation of two-body abrasive wear [39]. A simple cutting model was employed. The simulation results showed that in pure cutting the wear rate increases with increasing asperity tip height variation and asperity sharpness [39]. It was also found that the tip height distribution could outweigh differences in sharpness exhibited by different particles. For example, sharper quartz particles are less abrasive than garnet if the height variation of their asperity tips is substantially lower than that of garnet [39].

14.6 Classification of Abrasive Surfaces

The application of computer-based technology and image-processing methods may soon lead to a fully automated classification system of tribological surfaces. Such a system would be very useful in classifying (i) surfaces during manufacturing processes (quality control), (ii) wear surfaces in the failure analysis of engineering components, or (iii) wear particles in machine condition monitoring.

From the practical viewpoint, the issue of prime importance when examining a surface is whether the given surface will do the job. Currently, various surface texture parameters of roughness and waviness help the engineer to make that decision. However, the limitations of the common surface roughness and waviness parameters are well recognized. In theory, surface description does not require parameters and methods used to classify the surface without numerical parameters are discussed in the following text. Numerical parameters are not necessary as long as the surfaces can be classified into certain groups based on some criteria, e.g. surface finish, surface topography, surface abrasivity, etc. Wear particles examined in machine condition monitoring can also be classified using this system.

One of the problems in surface description is that tribological surfaces are multi-scale objects (they show different length scales of surface features) and the surface topography data exhibit non-stationary characteristics (surface features are superimposed on each other and located at different positions on a surface). However, most of the methods used in surface topography characterization provide functions or parameters strongly depending on the scale at which they are calculated [59, 60]. Since these parameters are scale dependent they provide one-scale characterization, which is in conflict with the multi-scale character of tribological surfaces, and hence are not unique for a particular surface.

It has been shown that 3-D surface data can be represented by range images [61]. This is achieved by encoding a single elevation data point into pixel brightness. The horizontal and vertical image co-ordinates determine the location of the data point in an 'x–y' plane [61]. For a clearer visual presentation of surface texture patterns, range images can be transformed into shaded images and can also be displayed in a perspective view.

In the first approach, fractal methods were employed to address the problem of surface characterization in all directions [62–64]. To characterize the surface a modified Hurst orientation transform (HOT) method was employed to calculate Hurst coefficients (H), which are directly related to surface fractal dimensions (D), as $D = 3 - H$, in all possible directions across the surface [64]. The Hurst coefficients were then plotted as a function of orientation, in the form of a rose plot, to reveal surface anisotropy. Circular shape of the rose plot indicates that the surface exhibits the same fractal dimension in all the directions, i.e. the surface is isotropic. For the surface texture exhibiting some anisotropy, the shape of this plot is no longer circular and can be approximated by an ellipse [64, 65] as illustrated in Figure 14.19. It was also found that the Hurst coefficients, since they are related to the fractal dimension, can be used to assess the surface roughness, i.e. the rougher the surface, the lower are the Hurst coefficient values [62, 63].

Tribological surfaces exhibit a useful feature that allows converting parts of the image to fit approximately other parts located elsewhere in the image [65, 66]. Affine transformations which allow for this conversion contain the information on scale, translation, rotation, etc., and form the bases of the partition iterated function system (PIFS). These transformations encapsulate the complete information about the surface topography and boundary [65, 66]. The information stored in the PIFS can be retrieved by applying iteratively any initial image into the PIFS. The result is a sequence of images gradually converging to the original image of the surface [65, 66]. The limitation of this method is that it allows for surface characterization at all scales. As tribological surfaces are multi-scale and non-stationary objects their characterization based on the PIFS only is not accurate enough.

This problem has been rectified by applying a combination of fractals (the PIFS method) and wavelets in the hybrid fractal–wavelet method [67]. Wavelets are used to decompose surface topography into different scale components, i.e. roughness, waviness, etc. Fractals (PIFS) are then applied to describe these decomposed images in a scale-invariant manner. The hybrid fractal–wavelet method allows for multi-scale and non-stationary characterization of tribological surfaces. Its effectiveness in surface description is shown in Figure 14.20 where the outcome of the hybrid fractal–wavelet method is compared to that of the PIFS.

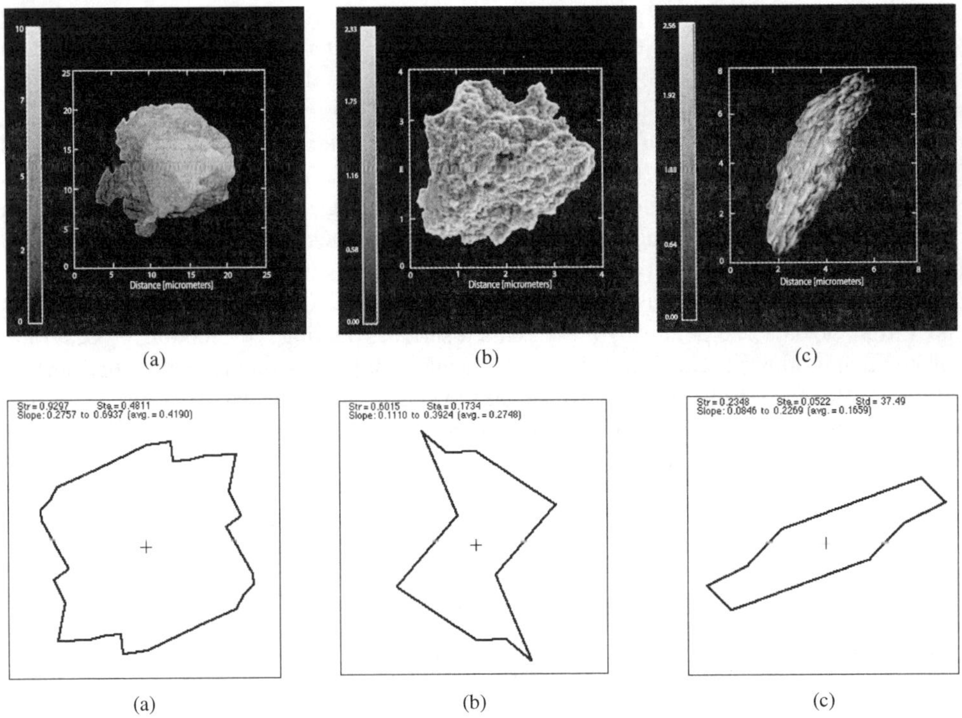

Figure 14.19 Examples of range images of cartilaginous wear particles exhibiting different surface topography with the corresponding rose plots of Hurst coefficients [65]

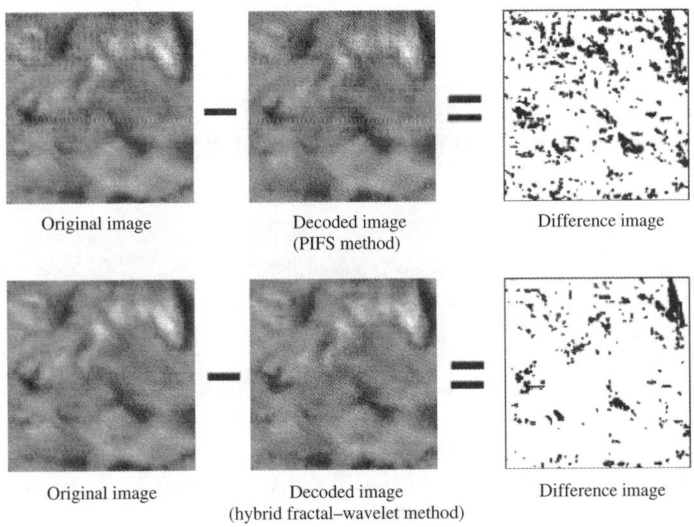

Figure 14.20 Differences between original and decoded images achieved by using PIFS and hybrid fractal–wavelet methods [67]

On the basis of the hybrid fractal–wavelet method, pattern recognition for tribological surfaces has been developed [68, 69]. The objective of pattern recognition is to assign an unclassified surface to a specific group/class of surfaces according to a pre-defined criterion, e.g. wear mechanism, surface texture, etc. The classification process works in the following manner. Initially, a number of surfaces are pre-classified into specific classes. This could be based on the traditional methods of surface topography assessment. The surface images are then decomposed into different scale components using wavelets, and then PIFSs are constructed for all the resulting decomposed images. An unclassified surface image is also first decomposed into different scale components by wavelets. Individual scale images of the unclassified surface are then used as the initial images in the decoding process using PIFSs taken from the database of already pre-classified surfaces. During the decoding process only one iteration is allowed. The decoded images are then compared with the unclassified surface image and Baddeley's distances (BDs) between the decoded and unclassified surface images are calculated, as schematically illustrated in Figure 14.21 [70].

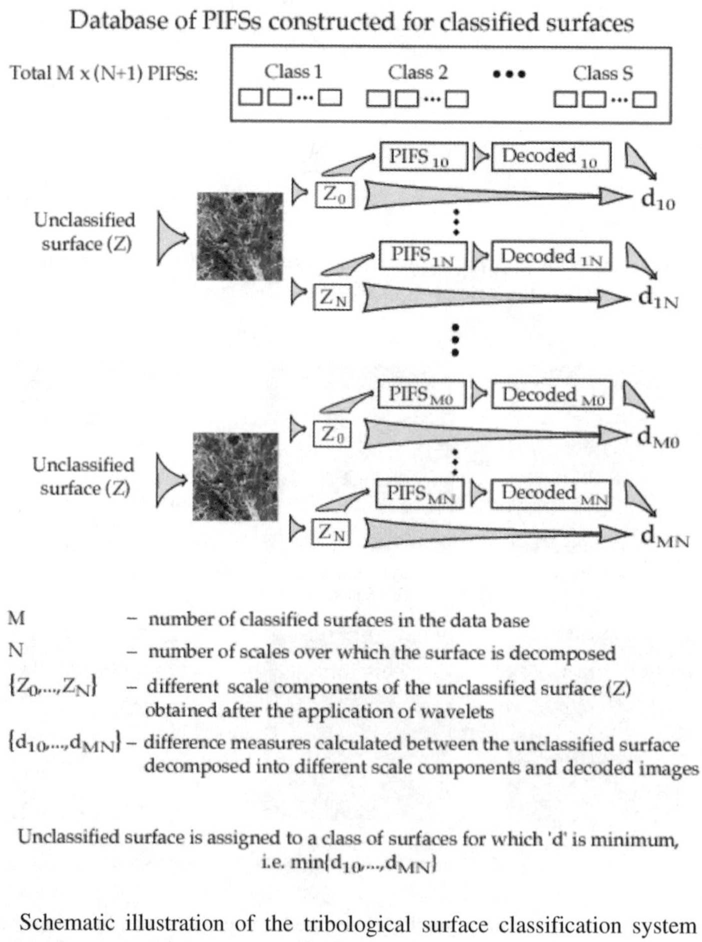

Figure 14.21 Schematic illustration of the tribological surface classification system based on the hybrid fractal–wavelet method (adapted from [70])

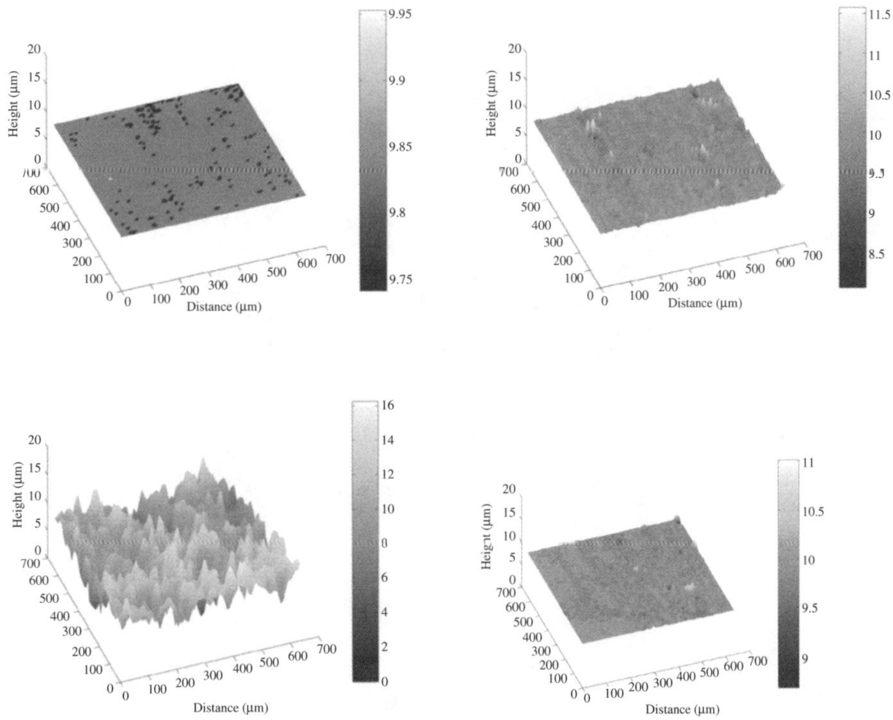

Figure 14.22 Examples of Talysurf shaded images of tribological surfaces: (a) non-abrasive; (b) moderately abrasive; (c) highly abrasive; and (d) unclassified

The hybrid fractal–wavelet method is used here to demonstrate how an unknown abrasive surface can be classified into one of three classes (groups) of abrasive surfaces. Three classes of surfaces of different surface roughness form a database: polished surfaces (non-abrasive), surfaces containing rough and polished areas randomly distributed (moderately abrasive) and rough surfaces (highly abrasive). Each group of already classified surfaces consists of 64 images. Examples of 3-D Talysurf shaded images of the surface from each class are shown in Figure 14.22. Visually one could guess that the unclassified surface belongs to a second group of moderately abrasive surfaces. The same result was obtained after the application of the surface classification system developed.

The performance of this system was evaluated on a database of 192 surface images (Figure 14.23), i.e. 64 images per class, using a leave-one-out cross-validation technique. It was found that in many cases the system correctly classified the unknown surface. The classification errors are shown in Table 14.2. However, despite apparent progress, there are still problems associated with the classification of surfaces, including the construction of a classifier that has learning abilities and can be accurately trained on a small dataset. Some of these problems have been addressed in the recent studies conducted on the classification of wear particle surfaces [71].

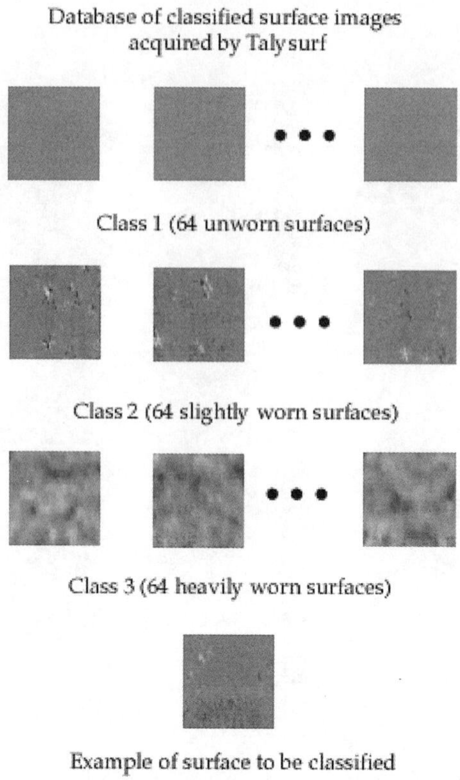

Figure 14.23 Schematic illustration of the database of classified surface images together with the example of unclassified surface

Table 14.2 Classification errors obtained for surface images shown in Figure 14.23. A leave-one-out cross-validation technique was used

Tribological surfaces	Classification errors (%)
Non-abrasive	3.48
Moderately abrasive	0
Heavily abrasive	3.48

14.7 Summary

Abrasive particle shape is an important parameter affecting the abrasive wear. It has been recognized that any measure of particle shape relevant to abrasive or erosive wear must include the sharpness of particle protrusions or asperities. But a sharpness parameter based solely on the particle geometrical shape is not a sufficient descriptor of particle abrasivity.

The amount of wear generated by a single protrusion depends also on the orientation of the protrusion to the wearing surface. Even a sharp protrusion, if oriented at a very small angle of attack, would not remove the material efficiently. Therefore, the effect of particle orientation on abrasive wear has to be included in the sharpness assessment.

A well-known experimental observation is the decrease in abrasive wear rates when the particle size falls below about 100 μm. Attempts to explain this size effect by arguing that small particles are less sharp are contradicted by microscopy images of small particles that do not look rounder. Several mechanisms that include particle deterioration, clogging of the abrading surfaces by wear debris and the effect of material strengthening at small scales should probably be taken into consideration when explaining this particle size effect.

Sharpness of abrasive particles affects the abrasive power of the surface made of these particles (e.g. grinding wheels or abrasive papers). However, the assessment of surface sharpness must also include additional influences from particle density and particle height distribution on the surface. Today, experimental testing of complex abrasive systems can be greatly augmented by computer modelling (statistical simulation). Simulation models should include parameters or functions that statistically describe particle shape and particle distribution on the abrasive surface (surface topography).

The new developments in computed-based image-processing techniques find the application in 3-D surface topography characterization and classification without the need of any numerical parameters. The methods developed allow for assigning an unclassified tribological surface to a specific group/class according to a pre-defined criterion such as wear mechanism, surface texture, etc. This would eventually lead to the development of an automated pattern recognition system which could be used in machine condition monitoring, failure analysis and manufacturing.

Acknowledgements

The authors would like to express their thanks to Elsevier for their kind permission to reprint Figures 14.3–14.19, 14.21 and Tribology Letters for their permission to reprint Figure 14.20.

This material has been based upon an article that first appeared in the *Journal of Engineering Tribology* – Proceedings Part J, 2002, Vol. 216, No. J6, ISSN 1350–6501, published by Professional Engineering Publishing. Permission is granted by the Institution of Mechanical Engineers.

References

1. Eyre, T.S, 'Wear Characteristics of Metals', *Tribology International*, **10**, 1976, 203–212.
2. Malkin, S., '*Grinding Technology: Theory and Applications of Machining with Abrasives*', Ellis Horwood, Chichester, 1989.
3. Swanson, P.A. and Vetter, A.F., 'The Measurement of Abrasive Particle Shape and Its Effect on Wear', *ASLE Transactions*, **28**, 1984, 225–230.
4. Swanson, P.A. and Klann, R.W., 'Abrasive Wear Studies Using the Wet Sand and Dry Sand Rubber Wheel Tests', Proceedings of the 1981 International Conference on Wear of Materials, ASME, San Francisco, CA, 1981, pp. 379–389.
5. Liebhard, M. and Levy, A., 'The Effect of Erodent Particle Characteristics on the Erosion of Metals', *Wear*, **151**, 1991, 381–390.
6. Raask, E., 'Tube Erosion by Ash Impaction', *Wear*, **13**, 1969, 301–315.

7. Bahadur, S. and Badruddin, R., 'Erodent Particle Characterization and the Effect of Particle Size and Shape on Erosion', *Wear*, **138**, 1990, 189–208.
8. Moore, M.A. and Swanson, P.A., 'The Effect of Particle Shape on Abrasive Wear: A Comparison of Theory and Experiment', Proceedings of the 1983 International Conference on Wear of Materials, ASME, NY, 1983, pp. 1–11.
9. Winter, R.E. and Hutchings, I.M., 'Solid Particle Erosion Studies Using Single Angular Particles', *Wear*, **29**, 1974, 181–194.
10. Torance, A.A., 'A New Approach to the Mechanisms of Abrasion', *Wear*, **67**, 1981, 233–257.
11. Glossary of Terms Relating to Powders, British Standard No. 2955, London, 1965.
12. Metals Handbook, 9th edition, Vol. 7, ASM 1984, pp. 233–236.
13. Prasad, S.V. and Kosel, T.H., 'A Study of Carbide Removal Mechanisms During Quartz Abrasion II: Effect of Abrasive Particle Shape', *Wear*, **95**, 1984, 87–102.
14. Russ, J.C., '*The Image Processing Handbook*', 2nd edition, CRC Press, Boca Raton, FL, 1994.
15. Robins, W.H.M., 'The Significance and Application of Shape Factors in Particle Size Analysis', in *The Physics of Particle Size Analysis* (ed H.R. Lang), The Institute of Physics, London, 1954.
16. Raadnui, S. and Roylance, B.J., 'The Classification of Wear Particle Shape', *Lubrication Engineering*, **51**, 1995, 432–437.
17. Hamblin, M.G. and Stachowiak, G.W., 'A Multi-Scale Measure of Particle Abrasivity', *Wear*, **185**, 1995, 225–233.
18. Stachowiak, G.W., 'Numerical Characterization of Wear Particle Morphology and Angularity of Particles and Surfaces', in *New Directions in Tribology* (ed I.M. Hutchings), Plenary and Invited Papers from the First World Tribology Congress, London, MEP Publications Ltd., 1997, pp. 371–389.
19. Hamblin, M.G. and Stachowiak, G.W., Description of Abrasive Particle Shape and Its Relation to Two-Body Abrasive Wear', *Tribology Transactions*, **39**(4), 1996, 803–810.
20. Luerkens, D.W., Beddow, J.K. and Vetter, A.F., Invariant Fourier Descriptors, *Powder Technology*, **31**, 1982, 209–215.
21. Mandelbrot, B.B., '*Fractals Form, Chance and Dimension*', W.H. Freeman and Company, San Francisco, 1977, pp. 25–35.
22. Podsiadlo, P. and Stachowiak, G.W., 'Evaluation of Boundary Fractal Methods for the Characterization of Wear Particles', *Wear*, **217**, 1998, 24–34.
23. Kaye, B.H., 'Fractal Description of Fine Particle Systems', in *Modern Methods in Fine Particle Characterization* (ed J.K. Beddow), CRC Press, Boca Raton, Florida, 1983.
24. Head, W.J. and Harr, M.E., 'The Development of a Model to Predict the Erosion of Materials by Natural Contaminants', *Wear*, **15**, 1970, 1–46.
25. Wadell, H., 'Sphericity and Roundness of Rock Particles', *The Journal of Geology*, **41**, 1933, 310–331.
26. De Pellegrin, D.V. and Stachowiak, G.W., 'Assessing the Role of Particle Shape and Scale in Abrasion Using "Sharpness Analysis", Part I – Technique Development', *Wear*, **253**, 2002, 1016–1025.
27. Verspui, M.A., van der Velden, P., de With, G. and Slikkerveer, P.J., 'Angularity Determination of Abrasive Powders', *Wear*, **199**, 1996, 122–126.
28. Leavers, V.F., 'Use of the Two-Dimensional Radon Transform to Generate a Taxonomy of Shape for the Characterization of Abrasive Powder Particles, *IEEE Transactions on Pattern Analysis and Machine Intelligence*, **22**(12), 2000, 1411–1423.
29. Hamblin, M.G. and Stachowiak, G.W., 'A Multi-Scale Measure of Particle Abrasivity and Its Relation to Two Body Abrasive Wear', *Wear*, **190**, 1995, 190–196.
30. Stachowiak, G.W., 'Particle Angularity and Its Relationship to Abrasive and Erosive Wear', *Wear*, **241**, 2000, 214–220.
31. De Pellegrin, D.V. and Stachowiak, G.W., 'New Technique of Grit-Profile Analysis by Cone-Fitting', *Wear*, **247**, 2001, 109–119.
32. De Pellegrin, D.V. and Stachowiak, G.W., 'Abrasiveness of Particles Measured by Cone-Fit Analysis (CFA), From Model Experiment to Industrial Problem' (eds D. Dowson *et al.*), Tribology Series 39, Elsevier, 2001, pp. 517–527.
33. Rabinowicz, E., '*Friction and Wear of Materials*', 2nd edition, Wiley, New York, 1995.
34. Johnson, K.L., '*Contact Mechanics*', Cambridge University Press, New York, 1985.
35. Larsen-Basse, J., 'Influence of Grit Diameter and Specimen Size on Wear During Sliding Abrasion', *Wear*, **12**, 1968, 35–53.

36. De Pellegrin, D.V. and Stachowiak, G.W., 'Assessing the Role of Particle Shape and Scale in Abrasion Using "Sharpness Analysis", Part II – Technique Evaluation', *Wear*, **253**, 2002, 1026–1034.
37. De Pellegrin, D. and Stachowiak, G.W., 'Sharpness of Abrasive Particles and Surfaces', *Wear*, **256**, 2004, 614–622.
38. Dragoo, A.L., Robbins, C.R. and Hsu, S.M., 'A Critical Assessment of Requirements for Ceramic Powder Characterisation', in *Advances in Ceramics, Ceramic Powder Science* (eds G.L. Messing, K.S. Mazdiyasni, J.W. McCauley and R.A. Haber), Vol. 21, American Ceramic Society, Westerville, OH, 1987.
39. De Pellegrin, D. and Stachowiak, G.W., 'Evaluating the Role of Particle Distribution and Shape in Two-Body Abrasion by Statistical Simulation', *Tribology International*, **37**(3), 2004, 255–270.
40. Torance, A.A. and Badger, J.A., 'The Relation Between the Traverse Dressing of Vitrified Grinding Wheels and Their Performance', *International Journal of Machine Tools and Manufacture*, **40**, 2000, 1787–1811.
41. Goddard, J. and Willman, H., 'A Theory of Friction and Wear During the Abrasion of Metals', *Wear*, **5**, 1962, 114–135.
42. Sin, H., Saka, N. and Suh, N.P., 'Abrasive Wear Mechanism and the Grit Size Effect', *Wear*, **55**, 1979, 163–170.
43. Xie, Y. and Bhushan, B., 'Effects of Particle Size, Polishing Pad and Contact Pressure in Free Abrasive Polishing', *Wear*, **200**, 1996, 281–295.
44. Clark, H.McI. and Hartwich, R.B., 'A Re-Examination of the "Particle Size Effect" in Slurry Erosion', *Wear*, **248**, 2001, 147–161.
45. Aifantes, E.C., 'Strain Gradient Interpretation of Size Effects', *International Journal of Fracture*, **95**, 1999, 299–314.
46. Begley, M.R. and Hutchinson, J.W., 'The Mechanics of Size-Dependent Indentation', *Journal of Mechanical Physics and Solids*, **46**(10), 1998, 2049–2068.
47. Shu, J.Y. and Fleck, N.A., 'The Prediction of a Size Effect in Micro-Indentation', *International Journal of Solids and Structures*, **35**(13), 1998, 1363–1383.
48. Jiang, J., Sheng, F. and Ren, F., 'Modelling of Two-Body Abrasive Wear Under Multiple Contact Conditions', *Wear*, **217**, 1998, 35–45.
49. Gåhlin, R. and Jacobson, S., 'The Particle Size Effect in Abrasion Studied by Controlled Abrasive Surfaces', *Wear*, **224**, 1999, 118–125.
50. Larsen-Basse, J., 'Some Effects of Specimen Size on Abrasive Wear', *Wear*, **19**, 1972, 27–35.
51. Misra, A. and Finnie, I., 'On the Size Effect in Abrasive and Erosive Wear', *Wear*, **65**, 1981, 359–373.
52. Gao, H., Huang, Y. and Nix, W.D., 'Modeling Plasticity at the Micrometer Scale', *Naturwissenschaften*, **86**, 1999, 507–515.
53. Mulhearn, T.O. and Samuels, L.E., 'The Abrasion of Metals: A Model of the Process', *Wear*, **5**, 1962, 478–498.
54. Hisakado, T. and Suda, H., 'Effects of Asperity Shape and Summit Height Distributions on Friction and Wear Characteristics', *Wear*, **225–229**, 1999, 450–457.
55. Hokkirigawa, K. and Kato, K., 'Ploughing, Cutting and Wedge Formation During Abrasive Wear', *Tribology International*, **21**(1), 1988, 51–57.
56. Torrance, A.A., 'The Correlation of Process Parameters in Grinding', *Wear*, **156**, 1992, 281–299.
57. Hamblin, M.G. and Stachowiak, G.W., Description of Surface Abrasivity and Its Relation to Two-Body Abrasive Wear', *Wear*, **206**, 1997, 69–75.
58. Wang, G. and Hutchings, I.M., 'The Number of Particle Contacts in Two-Body Abrasive Wear of Metals by Coated Abrasive Papers', *Wear*, **129**, 1989, 23–35.
59. Poon, C.Y. and Bhushan, B., 'Comparison of Surface Roughness Measurements by Stylus Profiler, AFM and Non-Contact Optical Profiler', *Wear*, **190**, 1995, 76–88.
60. Zahouani, H., Vargiolu, R., Kapsa, Ph., Loubat, J.L. and Mathia, T.G., 'Effect of Lateral Resolution on Topographical Images and Three-Dimensional Functional Parameters', *Wear*, **219**, 1998, 114–123.
61. Podsiadlo, P. and Stachowiak, G.W., '3-D Imaging of Wear Particles Found in Synovial Joints', *Wear*, **230**, 1999, 184–193.
62. Russ, J.C., 'Characterizing and Modeling Fractal Surfaces', *Journal of Computer-Assisted Microscopy*, **4**(1), 1992, 73–126.
63. Russ, J.C., 'Surface Characterization: Fractal Dimensions, Hurst Coefficients, and Frequency Transforms', *Journal of Computer-Assisted Microscopy*, **2**(3), 1990, 161–183.
64. Podsiadlo, P. and Stachowiak, G.W., 'The Development of Modified Hurst Orientation Transform for the Characterization of Surface Topography of Wear Particles', *Tribology Letters*, **4**, 1998, 215–229.
65. Stachowiak, G.W. and Podsiadlo, P., 'Surface Characterization of Wear Particles', *Wear*, **225–229**, 1999, 1171–1185.

66. Podsiadlo, P. and Stachowiak, G.W., 'Scale-Invariant Analysis of Wear Particle Morphology. I: Theoretical Background, Computer Implementation and Technique Testing', *Wear*, **242**, 2000, 160–179.
67. Podsiadlo, P. and Stachowiak, G.W., 'Hybrid Fractal–Wavelet Method for Characterization of Tribological Surfaces – a Preliminary Study', *Tribology Letters*, **13**(4), 2002, 241–250.
68. Podsiadlo, P. and Stachowiak, G.W., 'Scale-Invariant Analysis of Wear Particle Morphology. III: Pattern Recognition', *Wear*, **242**, 2000, 189–201.
69. Podsiadlo, P. and Stachowiak, G.W., 'Fractal–Wavelet Based Classification of Tribological Surfaces', *Wear*, **254**, 2003, 1189–1198.
70. Stachowiak, G.W. and Podsiadlo, P., 'Classification of Tribological Surfaces', *Tribology International*, **37**, 2004, 211–217.
71. Podsiadlo, P. and Stachowiak, G.W., 'Development of Advanced Quantitative Analysis Methods for Wear Particle Characterization and Classification to Aid Tribological System Diagnosis', *Tribology International*, **38**, 2005, 887–897.

15

Wear Mapping of Materials

S.M. Hsu and M.C. Shen

15.1 Introduction

Wear of material is a simple subject, at least on surface. If you rub two materials together for some time and measure the amount of material loss, you obtain wear. And if you rub a number of materials under the same conditions, then you can compare these materials in terms of wear resistance. On this basis, wear is hardly a subject worthy of scientific investigation.

The reality is that getting an accurate and meaningful wear result and comparing different materials are complex and difficult tasks. This chapter uses a wear mapping approach to resolve some of the difficulties.

Wear is not an intrinsic property of a material. Wear is a system-derived property. This means you can get any wear values if you change one of the many variables of the system. A system consists of the following: mechanical properties of the two materials in contact, material composition, microstructure, chemical phases, surface preparation process, surface roughness, mechanical design of the wear tester, contact geometry, motion and speed, relative velocity, load, alignment, vibration, temperature, pressure, lubrication method, lubricant chemistry, and environmental gases. For a given material pair, depending on how the system parameters are defined, wear could range from zero to seizure. Furthermore, different material pairs have different mechanical properties, so the contact conditions will need to be adjusted in order to interpret the conditions on an equivalent basis. The dependence of wear on so many variables underscores the complexity and subtlety of measuring wear in the context of a potential application. In other words, the meaning of wear result and the correct interpretation of the wear data are the critical issues.

Wear – Materials, Mechanisms and Practice Edited by G. Stachowiak

15.1.1 Wear – A System Perspective

If wear is a function of a complex system, then if we can define the system sufficiently, may be we can then define wear. So, let us examine what it means if we know the wear properties of a material. For a given system, if we know the wear property, it implies we know the following characteristics:

- different regimes of wear in terms of load and speed for a given contact geometry (mild wear, severe wear, and ultra-severe wear);
- occurrence of wear transitions and definition of the transition zones;
- positions in load and speed where transitions occur;
- the cause and mechanistic understanding of wear transitions;
- wear mechanisms in each regime;
- prediction of wear and wear transitions;
- effect of chemistry on wear and how it affects the wear, wear transitions;
- microstructural and compositional influence on wear.

This constitutes a comprehensive definition of wear characteristics of a material or a material pair. Given this set of knowledge and data, we are in a position to compare different materials for an application and select the best material combination for durable, predictable use.

Wear mapping is such an approach. In this chapter, we shall introduce the concept and show examples how this can be used.

15.1.2 Historical Material Selection Guide

Over the years, simulation studies and actual component testing have been the norm to select materials for a particular application. Since actual component testing is expensive and time-consuming, it is applied to very few selected materials that have gone through the gauntlet of bench screening tests, simulation tests, and material property analysis. Yet the result of this protocol is not foolproof; history is replete with unexpected premature wear failures costing millions of dollars and sometimes, even lives. Since the number of possible material combinations is huge, plus there is an ever-increasing number of new materials, proper material selection for wear resistance has become an art rather than a science.

The difficulty of evaluating wear of materials is not obvious. Since conducting a wear test is simple and a result is always available, literature is filled with wear data and comparison of materials under a particular set of conditions. One could assume that over time, sufficient data will emerge that will make material selection obvious. Unfortunately, wear is also very sensitive to material composition, defect population, history of fabrication, microstructure, and sometimes even grain boundary chemistry and minor chemical contaminations. Since we have insufficient knowledge on which material constituents are important to wear and the characteristics that govern mechanical properties, comprehensive material characterization is often lacking in published reports. Thus we have a complex system function on one hand and a complex material variation on the other. Is there any way we can select materials intelligently?

Common practice is to conduct laboratory bench simulation wear tests to rank materials under the same operating conditions. If the relationship between wear and the operating conditions (load and speed) is linear, relative ranking of materials is straightforward. Under the influence of chemistry or environment, unfortunately, the relationship between wear and the operating conditions is often not linear. The relative ranking of materials, therefore, will change when the operating conditions are changed. At the same time, material variations abound. So even though the material designation may be the same, there is no assurance that the chemical composition and microstructure are identical. Because of these factors, literature reports on material wear characteristics have wide ranges, as shown for ceramics in Table 15.1. For metals, similar ranges are observed for data from all sources. The metal wear data in Table 15.1 show the effect of lubricants on different metals on a single wear tester in a laboratory [1].

Currently, industrial practice is to make actual components and put them in actual field trials over a period of time. This way there can be no uncertainty of the suitability of the material for that application. However, cost and duration make such practice prohibitive, and a very low percentage of new materials are ever tested. To save costs, accelerated test cycles are often used to simulate long-term behaviors. Unfortunately, some environmental effects and long-term fatigue-induced interactions are unpredictable and premature failures occur.

This chapter discusses the use of the wear mapping concept in conjunction with laboratory bench wear tests to resolve some of these issues and provide a means to systematically compare and select materials on a common basis. While the discussion will focus on materials

Table 15.1 Wear data from the literature

Ceramics	Al_2O_3		Si_3N_4		SiC		PSZ	
	AVG	Range	AVG	Range	AVG	Range	AVG	Range
Fracture toughness ($MPa\,m^{0.5}$)	4	2–6	4.4	1.4–6	2.9	1.5–4	8	5–15
Hardness (GPa)	16	11–12	16.3	13–17	25	13–34	12	10–14
Elastic modulus (GPa)	370	340–410	313	290–333	397	390–402	206	196–216
Density (g/cm^3)	3.9	3.8 3.9	3.2	3.1–3.4	3.0	2.8–3.2	5.9	5.7–6.0
Wear coefficient ($K = HW/FD^a$)								
Dry air	10^{-5}	10^{-9}–10^{-3}	10^{-4}	10^{-5}–10^{-4}	10^{-3}	10^{-6}–10^{-2}	10^{-4}	10^{-7}–10^{-4}
Humid/H_2O	10^{-7}	10^{-7}–10^{-6}	10^{-5}	10^{-9}–10^{-4}	10^{-5}	–	10^{-6}	–
Lubricant	10^{-3}	10^{-7}–10^{-2}	10^{-6}	–	10^{-6}	–	10^{-7}	10^{-8}–10^{-6}

Wear surface	*Opposing surface*	*Atmosphere*	*Lubricant*	*Wear coefficient*
52100 steel	52100 steel	Dry air	None	1.0×10^{-3}
52100 steel	52100 steel	Air	None	1.0×10^{-3}
52100 steel	52100 steel	Air	Paraffinic oil	3.2×10^{-7}
52100 steel	52100 steel	Air	Paraffinic oil/TCP	3.3×10^{-9}
52100 steel	52100 steel	Air	Engine oil	$<2.0\times10^{-10}$
Mild steel	Mild steel	Air	None	2.3×10^{-3}
Carburized steel	Carburized steel	Air	Gear oil	1.6×10^{-9}
Aluminum bronze	Carburized steel	Air	Gear lubricant	2.5×10^{-8}

[a] K = wear coefficient; H = hardness; W = wear volume; F = load; D = distance slid.

in general, we will use advanced ceramics to illustrate the concepts and provide actual data for comparison. The applicability of the concept, however, is universal.

15.2 Basic Definition of Wear

15.2.1 Nature of Wear

Wear can be defined simply as material lost or detached from the rubbing surfaces. While this definition does not include subsurface deformation and surface damage which are an integral part of the wear sequence, it is a quantity easily measurable. Along this vein, a wearing system needs to have two surfaces in contact under relative motion with some mechanical loading acting on the surfaces. So the primary parameters will be the relative speed, the contact geometry, and the normal load. Given this set of information, material scientists often want to use contact pressure as a key parameter. Unfortunately, wear is related to load rather than contact pressure.

There are several reasons why this is the case. Surfaces are not flat and the rough spots (called asperities) receive the majority of the load. Asperity contacts typically represent a small fraction of the real area of contact, and for most contacting situations the real area of contact stays relatively constant. Therefore, direct machine loading translates into a proportional increase in asperity contact pressure which controls the asperity wear events.

As wear progresses, the apparent area of contact increases and the apparent contact pressure decreases, yet the asperity contact pressure may remain approximately the same. The change in apparent contact pressure could be very large, and if wear data are interpreted on the basis of the apparent contact pressure, erroneous conclusions may be drawn.

15.2.2 Wear Characterization

15.2.2.1 Wear Mechanisms of Materials

A large number of wear mechanisms have been shown in the literature. For metals, they include plastic deformation, adhesion, abrasion, third-body abrasion, delamination, fatigue, fracture, corrosion, stress corrosion, and oxidative wear. For brittle solids such as ceramics, they are asperity level fracture, abrasion, tribochemical wear, grain pullout, intergranular fracture, intragranular fracture, thermal shock induced fracture, transformation cracking, and corrosion. Broadly speaking, these can be classified into physical and chemical processes and their interactions. In most cases, several wear mechanisms occur simultaneously. Therefore, it is difficult to ascertain specific proportional contributions to wear from different mechanisms. Oftentimes, these wear mechanisms interact to produce the dominant wear outcome, e.g. fatigue leads to delamination which leads to third-body abrasion, and this results in observed abrasion grooves on worn surfaces.

Ying [2] conducted a series of two-ball collision experiments designed to understand the fundamental relationship between friction, material deformation, and wear at the single asperity basis. The results are summarized in Figures 15.1 and 15.2. For metals, wear is controlled by the accumulation of the shear strain underneath the contact. Lubrication tends to redistribute the stress over a much larger area, therefore delaying the onset of wear. For ceramics, wear is controlled by the stress intensity which induces crack propagation.

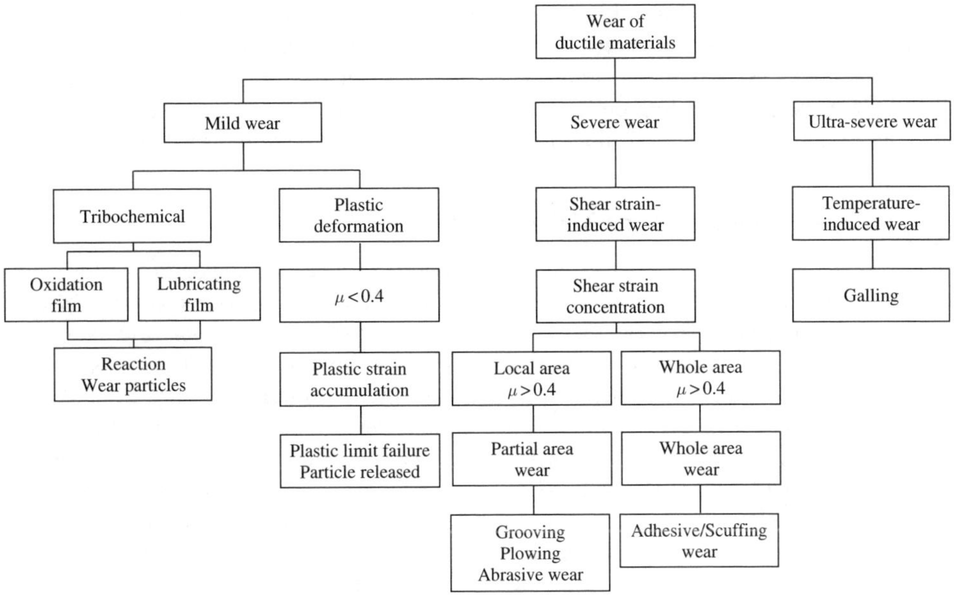

Figure 15.1 Wear mechanisms of metals

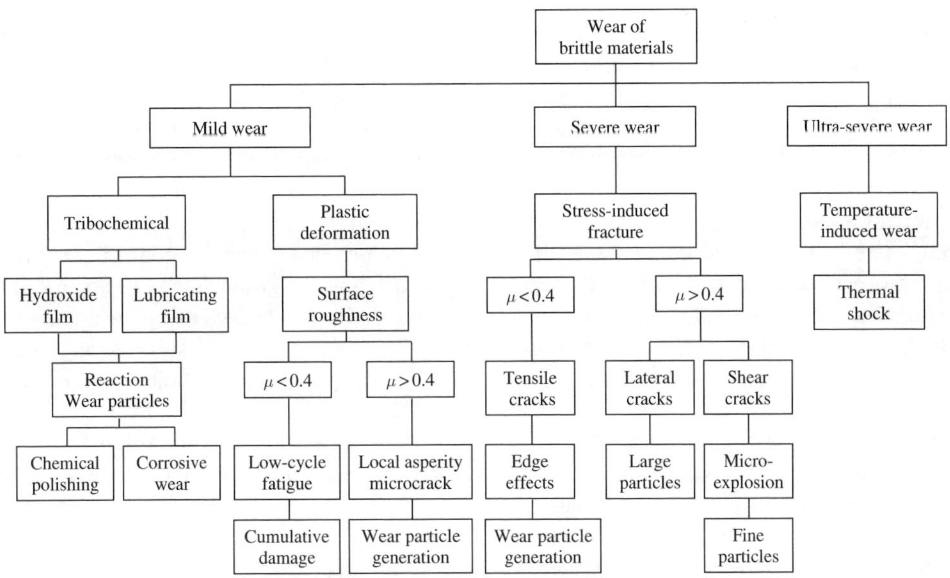

Figure 15.2 Wear mechanisms of ceramics

These observations, while useful, are based on single asperity contact experiments. In real contacts, there are multiple asperities in contact and the situation is considerably more complex. Contact asperity temperatures and wear debris interactions to form transfer layers all influence wear. Depending on the mechanical properties of the tribochemical reaction products forming the interfacial film, the wear processes can be diminished or enhanced.

15.2.2.2 Simulation Model and Material Selection

How well we understand the wear phenomena can be examined in the context of our ability to predict wear for a given situation. Modeling of wear in metals has been conducted by many researchers (see References [3, 4] for some approaches). Over the years, many models have been proposed for different materials and applications. Unfortunately, most models are correlational in nature and therefore system-specific, i.e. the model only works for the particular material pair, contact geometry, operating condition, and specific environment and lubricating conditions. The model is derived from the wear data of that system and therefore only applies to that data set. The inability of the models to transcend these restrictions results in a diverse collection of parameters and constants. The situation can be illustrated using erosion as an example [5]. A literature search on erosion models found 32 parameters being used by various researchers to describe their own erosion data. This results in a wide array of specific parameters describing a subset of the erosion phenomena. A lack of consensus on the most crucial parameters to describe erosion suggests different regimes, different contact geometries, different materials, and different operating wear mechanisms. Such confusion exists for most other wear phenomena, e.g. abrasion [6–9]. As a result, given the material properties and contact information, there is no model currently available that can predict metal wear *a priori*.

When it comes to ceramics, the situation is not any better. There are many proposed wear models and a clear consensus is lacking. Since the purpose of this chapter is not to provide a comprehensive list of models but rather suggest a particular line of attack to predict ceramic wear, so only certain selected articles are cited. Most ceramic wear models are material specific and under specific operating conditions [10–19]. These models are useful in elucidating the operating wear mechanism that occurred under those experimental conditions.

Evans first proposed a lateral crack model to predict ceramic machining rate, i.e. wear by machining process [11]. He developed an equation using the extension of lateral cracks to model removal of materials under high compressive stresses. Kato proposed the use of a contact severity index to classify the different wear modes under various contact severities to represent the propensity of brittle fracture by using applied stress intensity against the material's fracture toughness [12–14]. The severity index, S_c, was derived using the maximum Hertzian contact stresses as the dominating stress term [12] and an average surface roughness. The severity index was later modified to S_{cf} by including the contribution from friction; thus it represented a combined mode I and mode II stress intensity [13]. More recently, a further modified severity index, S_{cm}, was derived by taking the maximum tensile stress as the dominating stress term in the stress intensity modeling [14]. All the above severity indexes used the surface roughness to represent pre-existing cracks. In this approach, the material's microstructure was not included in the models. The tensile crack model proposed in Reference [10] presented yet another approach to predict wear of ceramics by using the fracture mechanics formalism. A concept based on energy balance was adopted to derive the wear equation. Microstructural parameter such as grain size was incorporated in the model.

Since ceramics are insulators, thermal diffusitivity is relatively low. Under high-speed dry sliding conditions, localized thermal heating can be substantial. Thermal shock accelerated wear was observed and modeled by Sibley in 1962 [15]. Winer proposed a thermomechanical model to include the friction-induced thermal stresses in the calculations of total stress [16]. A dimensionless parameter, Gt, was proposed to indicate the degree of thermal heating from asperity flash heating under atmospheric conditions [17] . Ashby proposed the use of both bulk and flash temperatures for the description of ceramic wear [19]. The effect of asperity flash heating under lubricated conditions was considered in these models. These models also suggest that asperity heating can dominate the wear processes even though the real area of contact is small.

Models described above are successful in explaining wear data and highlight the dominant wear mechanisms under different operating conditions. Lateral crack model assumes a severe contact stress under which the material's microstructure does not have dominant effect. In the development of the contact severity, rolling contact experiments were used. In formulating the contact severity index S_c, the use of the maximum Hertzian pressure to represent the dominating stress field therefore is reasonable. Under sliding condition, contributions from frictional traction to the estimate of stress intensity will need to be introduced, as in the S_{cf} and S_{cm}. But the fundamental assumption in this case is that no severe wear process is dominant. These hidden assumptions built into the various models need to be understood in order to apply these models.

Bayer [20] discussed at length the essential elements for design-oriented model-building and testing. The key is simulation – for all critical aspects of the application environment. The range of applicability of any model must be defined. The ruggedness of the model (whether it behaves properly when parameters are expanded somewhat out of the usual ranges) needs to be tested. Most importantly, the overall approach must be system-oriented, not merely focused on materials, lubricant, or mechanics.

Other discussions of this approach can be found in the literature. Godet [21] has identified the essential elements of simulation wear testing. He includes the need to measure the stiffness and damping characteristics of the test system as well as its wear behavior, all under a controlled mechanical environment. The aim is to ensure that the operating environment is adequately represented in the model and its validation. In the summary of a workshop on this subject [3], it was noted that design-oriented modeling also requires knowledge and identification of failure mechanisms, possibly in the form of maps that relate failure boundaries to operating parameters.

From this discussion on our current understanding on wear models, it is apparent that various models describe a subset of wear in a specific situation. Blind men touching an elephant may be an apt analogy. This leads to a conclusion that we need a more systematic, global approach to wear and wear modeling. Wear mapping presents a broad system view and wear map-based modeling can alleviate some of these issues.

15.3 Wear as a System Function

From the above discussion, wear should be recognized as a system function. Simply stated, wear is not a unique property of a material. The wear outcome depends heavily on the large number of parameters the wearing system consists of. Any changes in the system change the wear outcome.

The issue is how to define such a system? For a wear system, the primary parameters are load, speed, and contact stresses (surface roughness, contact geometry, elasticity, relative hardness, etc.). The secondary parameters are operating temperatures, duty cycles, lubricants used, lubricating conditions, environmental gases available, and circulation velocity. The tertiary parameters are alignment, vibration, and motion type (reciprocating, linear, fretting, etc.).

In a comprehensive way, we can divide wear into dry sliding, nonreactive fluid lubricated condition, reactive fluid lubricated condition, temperature, and time. If we can define wear of a system in all of these aspects, then the system is reasonably defined. Can wear mapping technique be used as a tool to define the system?

15.4 Wear Maps as a Classification Tool to Define the System

Wear mapping technique can be used to present wear data systematically according to a hierarchy of parameters to define the wear system. The dependent variable is wear. The independent variables can be divided into two types: continuous variables (speed, load, temperature, and time) and discrete variables (dry, nonreactive lubricant, reactive lubricant, environment, and contaminants). For a given material pair and a fixed discrete variable, there are five three-dimensional wear maps that can be used to describe wear systematically: wear versus speed and load; wear versus speed and temperature; wear versus speed and time; wear versus load and temperature; and wear versus load and time. Therefore, for a given material pair, a set of 20 wear maps will systematically define the wear behavior. These maps include five wear maps under dry sliding conditions; five wear maps under nonreactive fluid (to avoid chemical reactions but to remove the heat at the interface so that the true wear behavior can be observed); five wear maps under reactive lubricant conditions (formation of chemical films similar to industrial applications); and five wear maps under the same environment and contaminant conditions (e.g. engine blow-by gases, soot particles, and oxygen starvation conditions as in a typical diesel engine ring wear simulation). In many instances, a complete set of wear maps is not needed in order to define the wear behavior but a selected set of maps will serve to define the critical limits of the operational boundaries for the material pair in terms of acceptable wear behavior within those ranges.

Within each discrete parameter, the wear characteristics of the material pair will exhibit wear transitions, tribochemical reactions, oxide formation, plastic deformation, and fracture. The location of the speed and load (contact pressure, asperity temperature profile, and surface roughness evolution) at which such phenomena occur will differ as the discrete parameter changes from one to the other. Some phenomena will occur in one set of environments but will not occur in another set of environments.

These variations of wear behavior actually take place in many experiments; some are controlled and some occur accidentally. While these explain the wide variation in wear results and mechanisms reported in the literature, the resulting confusion about the definitive wear behavior inhibits theoretical development in effective wear modeling. When the wear behavior for a material pair is fully defined by these maps, it will become obvious that a single wear model will not be sufficient to describe or predict wear behavior for a material pair in general. Separate models will be required to describe the wear behavior under different operating conditions and environments.

15.5 Wear as an "Intrinsic" Material Property as Defined by Wear Maps

Lim and Ashby [22] proposed using normalized parameters such as the following to describe wear system of metals:

$$\widetilde{W} = \frac{W'}{A_n}$$
$$\widetilde{F} = \frac{F}{A_n H_0} \quad (1)$$
$$\widetilde{V} = \frac{V r_o}{a}$$

where W' = wear volume per unit distance slid, A_n = apparent contact area, r_o = radius of the apparent contact area, F = normal force, H_o = room temperature hardness, V = sliding velocity, and a = thermal diffusivity.

In their work on metals, Lim and Ashby suggested that these parameters were able to correlate data from different sources, using specimens of different shapes and sizes. $\widetilde{W}$ can be considered as a dimensionless wear coefficient, $\widetilde{F}$ is the nominal pressure divided by the surface hardness, and $\widetilde{V}$ is the sliding velocity divided by the velocity of heat flow.

Rabinowicz [6] defined an alternative wear coefficient, k, in the following form:

$$k = \frac{W H_o}{F D} \quad (2)$$

where W = wear volume, F = normal force, D = distance slid, and H_o = room temperature hardness. Thus, k could also be expressed as

$$k = \frac{\widetilde{W}}{\widetilde{F}}$$

Rabinowicz suggested that "wear coefficient represents the probability that, during the contact of the two surfaces at an asperity, a sizeable wear particle is produced." While these parameters have some intrinsic advantages in representing wear, they also carry with them some hidden assumptions about the wear behavior and also assume other parameters are not critical (such as grain size and microstructure in ceramics).

Which parameter to use in plotting the wear maps and the form of representation are important considerations in representing wear as an "intrinsic material property." To understand this, one needs to examine the actual wear processes in a wear test as a function of time. Figure 15.3 shows a typical relationship between wear volume, contact pressure, and the normal force in a ball-on-three-flats wear experiment. A step loading wear procedure was used. Two contact pressures are plotted as a function of the normal force. The Hertzian pressure is calculated for ideal elastic contacting surfaces. The mean contact pressure is the actual pressure calculated from the applied normal force divided by the actual wear scar area measured at the end of each wear step. Neither of these common indicators of contact stress in a wear junction accounts for the microscopic morphology. Both measures assume conforming surfaces, i.e. completely smooth flat surfaces in contact. Thus, the wear transition

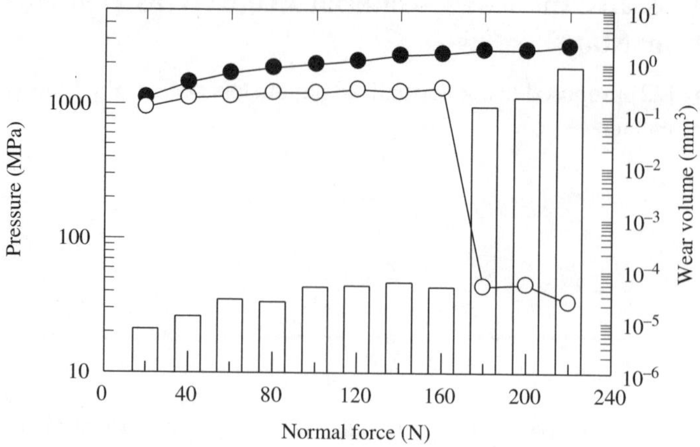

Figure 15.3 Relationship among wear volume, measures of pressure, and normal force illustrated with data of alumina

that occurred at 180-N normal force in Figure 15.3 is not reflected by the Hertzian pressure at all, and is seen as a sudden decrease in the mean pressure. In contrast, some results [23] seem to suggest that as the machine load is increased, the real contact area for ceramic-on-ceramic increases only fractionally, and hence the actual pressures at the tip of the microscopic asperities increase rapidly. This situation is quite different from ductile metals. For metals, surface conformity can be very high under certain speed and load regions, approaching 70–80% of the theoretical nominal contact area, and hence mean pressure may be a reasonable measure of the stress at the contact. For ceramics, the conformity typically is about 10–15%, and the resulting mean contact pressure, therefore, is not a good representation of the relevant contact stress. Consequently, it is preferable to use the normal force divided by the initial contact area to represent the load without any assumption about the microscopic contact morphology.

Wear volume, W, may be less susceptible to misinterpretation than other measures. Normalized measures of sliding speed have a similar difficulty in practice. The relevant thermal diffusivity is the value at the elevated contact temperature for which estimates are difficult to make. Again, it may be preferable to use linear velocity to represent the sliding speed.

15.6 Different Kinds of Wear Maps

Lim and Ashby [22] have demonstrated the use of a wear map to correlate the massive literature data on wear of steels. Their wear map is shown in Figure 15.4. Normalized parameters based on an assumed dominant variable were used to construct the map. The asperity temperature at the contact is assumed to dominate the dry sliding of steels on pin-on-disk wear testers. Wear regions are defined on the basis of the asperity temperatures they calculated. Various dominant wear mechanisms and models are also developed subsequent to the temperature analysis. These wear regions, however, describe fairly severe wear levels. In engineering applications, acceptable wear levels are orders of magnitude lower.

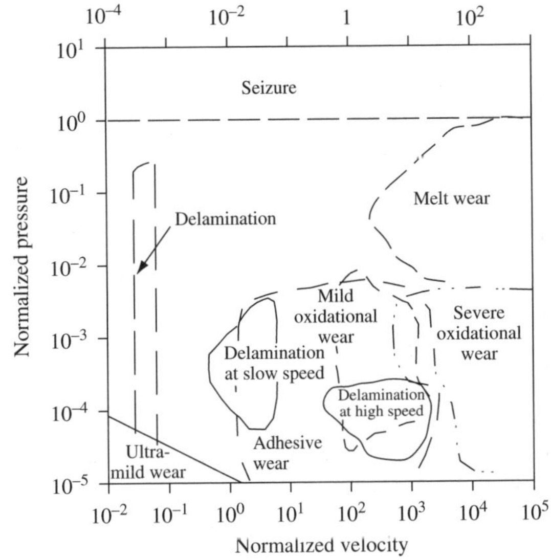

Figure 15.4 Wear map for steel under dry sliding conditions (from Reference [22])

Beerbower [24] proposed a conceptual wear mechanism diagram for steel under lubricated conditions as a function of the specific oil film thickness as shown in Figure 15.5. While the various mechanisms were reported in the literature, the diagram was constructed on the basis of inferences and isolated data. However, it illustrates the complex nature of lubricated wear of steel.

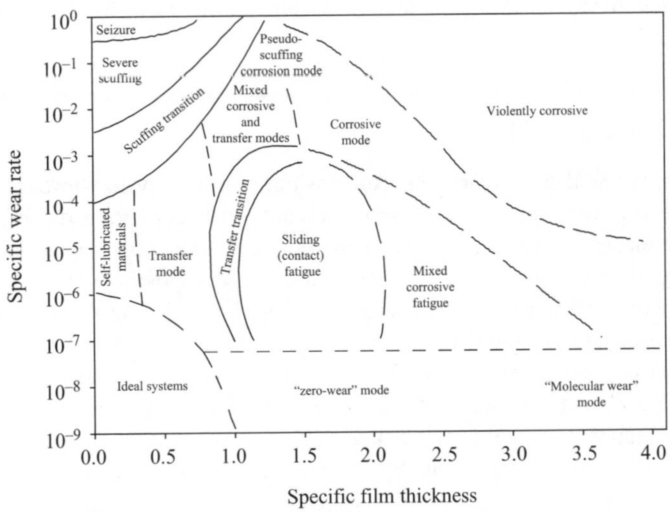

Figure 15.5 Wear map for steel under lubricated conditions (from Reference [24])

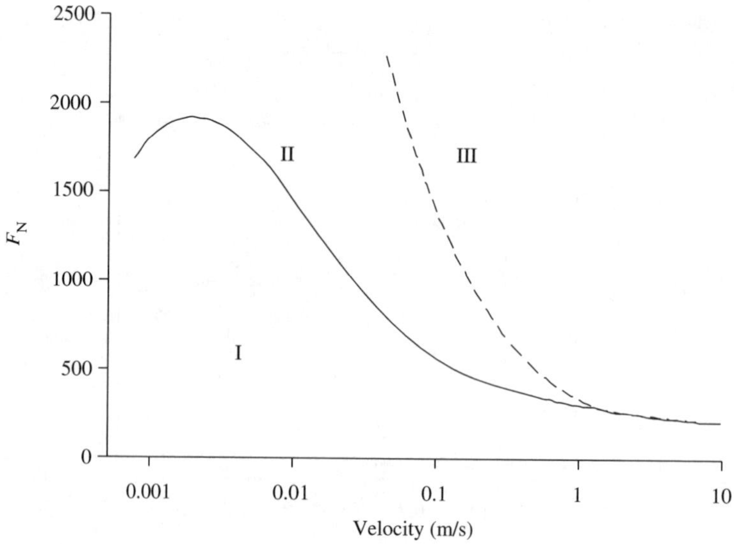

Figure 15.6 Wear transition diagram by Lim and Ashby [22]: (I) no wear; (II) mild wear; and (III) scuffing

deGee [25] proposed a simpler system on steel under well-lubricated conditions based on a large body of data generated under a set of standardized conditions. He referred to his maps as transition diagrams, Figure 15.6. He pointed out that as the severity of wear test increases, as reflected by speed and load, the wear of steel under well-lubricated conditions progresses from no wear to mild wear, and then to scuffing. Similar to Lim and Ashby's work, deGee found the concept of flash temperature useful in explaining the wear behavior.

These studies illustrate the complexity of examining wear in a comprehensive and systematic manner.

15.7 Application of Wear Maps

In this section, we will use wear data from a single laboratory to illustrate the wear map concepts. Four ceramic materials are used to create a self-consistent database to construct the maps. The materials are selected for their representative behavior within each material class. In ceramics, given a generic name and chemical composition such as silicon nitride, there are numerous variations in mechanical properties, microstructures, sintering aides, and surface properties. Tribologically, they may behave very differently for a given set of operating conditions and environments. Yet there are some general patterns of characteristics associated with each class of materials. This chapter focuses on these general patterns, but we caution the reader that each ceramic material should be treated as a unique sample.

Three-dimensional wear maps showing wear as a function of speed and load for alumina, zirconia, silicon carbide, and silicon nitride were constructed. Once the wear maps have been constructed for the materials, the maps can be used as effective tools to compare materials under the same environmental conditions.

Table 15.2 Material properties of the ceramics in wear map studies

Description	Al_2O_3	Y-TZP	SiC	Si_3N_4
Process	Sintered	Sintered	Sintered and post-HIP	HIP[a]
Sintering aid/Impurities	–	Al_2O_3, Y_2O_3	Al	Mg, Fe, Al, W
Density (g/cm^3)	3.9	6.05	>3.17	3.25
Phase	–	t-ZrO_2	α, β	α, β
Average grain size (μm)	2–15	~1.0	3–8	0.3–2
Elastic modulus (GPa)	372	**220**	430	310
Poisson's ratio	0.22	0.28	0.16	0.28
Hardness (GPa) (20 °C)	16±0.8	13±0.7	**31±1.6**	24±1.2
(500°C)	8±0.4	4±0.2	18±0.9	17±0.9
(1000°C)	4±0.2	3±0.2	10±0.5	13±0.7
Fracture toughness (MPa m$^{1/2}$)	4.5	**8.5**	3.2	5.4
Compressive strength (GPa)	2.6	1.9	2.5	3.0
Specific heat (J/g °C)	0.88	0.4	0.95	0.65
Thermal conductivity (W/m °C)	35.6	1.8	**110**	33.0
Thermal expansion (1/°C)	7.1×10^{-6}	10×10^{-6}	4.1×10^{-6}	3.5×10^{-6}

[a] HIP=hot isostatically pressed

(Hardness values are measured by Vicker's indentation with 1-kg load and a duration of 15s).

The material properties and the mechanical properties of the ceramics studied are listed in Table 15.2. The alumina used is a sintered α-alumina with a density close to the theoretical density. The grains are equiaxed and the average grain size is about 5 μm with the range from 2 to 15 μm. The zirconia ceramic used is a pressureless sintered polycrystalline zirconia doped with 4.7 wt.% yttria. The as-received material is 100% tetragonal. The grain size is about 1 μm in diameter. The density is 99% of theoretical, resulting in a porosity of about 1%. The silicon nitride used is a hot isostatically pressed (HIP) silicon nitride. The material is a mixture of α-Si_3N_4 and β-Si_3N_4. The α-phase is primarily in equiaxed grains of size $\approx$ 0.5 μm, while the β-phase forms elongated (rod-shaped) grains, up to 1 μm in diameter and 2–5 μm in length. Examination of the fractured surface by energy dispersive X-ray analysis indicates that trace amounts of magnesium, iron, and tungsten are present. The silicon carbide used is a sintered and post-HIP silicon carbide mixed with equiaxed grains and elongated grains. Elongated grains are up to 2 μm in diameter and about 10–20 μm in length.

15.7.1 Material Comparison Based on Wear Maps

15.7.1.1 Dry Sliding Conditions

Figure 15.7 presents the wear maps of the four ceramics together for the dry sliding case. The scale of the wear rate is the same for all four materials, from 10^{-10} to 10^{-2} mm^3/s. There are low-wear regions in the low-speed, low-pressure corners for all four materials. As the severity of the contact increases (an increase in speed and/or pressure), a rapid increase in wear takes place when the conditions approach the transition zones. The locations of the transition zones and the slopes of the increase depend on individual material. Wear

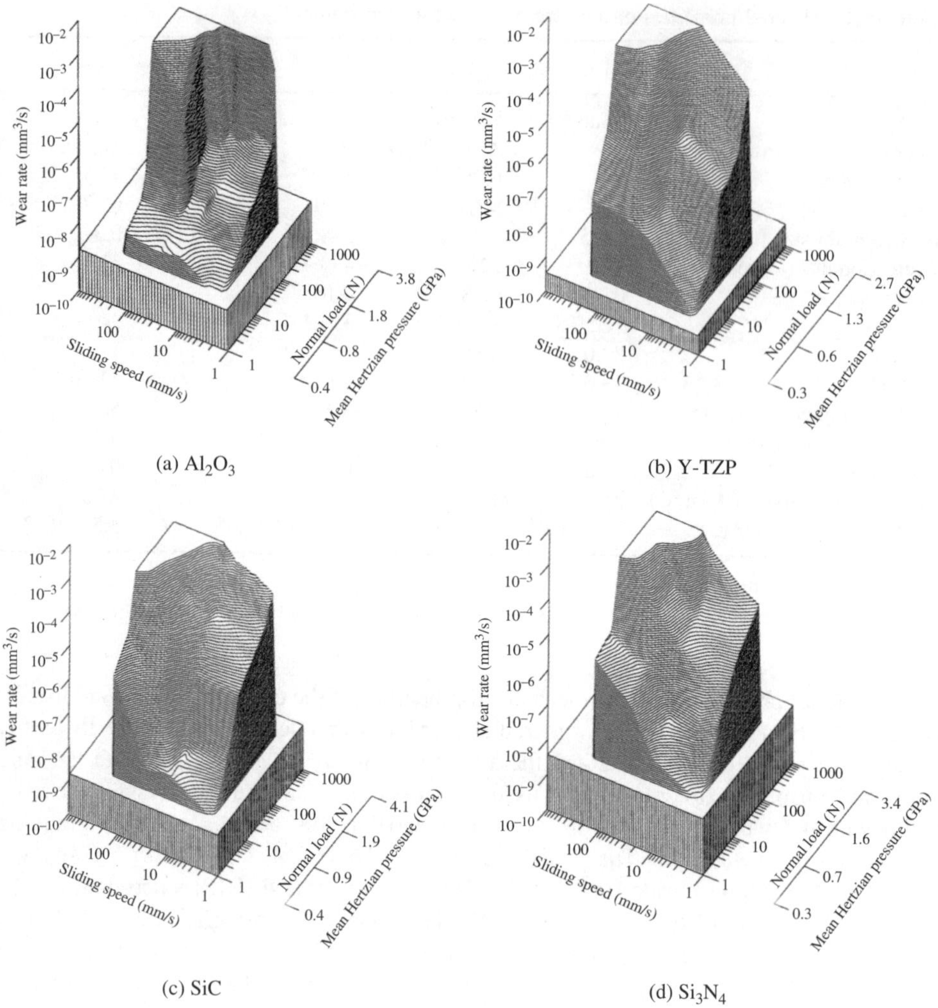

(a) Al_2O_3 (b) Y-TZP

(c) SiC (d) Si_3N_4

Figure 15.7 Wear maps under dry sliding conditions

accelerates as conditions become more severe. All four materials show both speed and load dependence and rapid transitions to severe wear. Alumina has a different speed dependence than the other three, probably because of the inadvertent reaction with water in the air. Zirconia shows more dependence on both speed and load than the other three. Therefore for zirconia, overdesign is necessary to protect the component/system. The silicon-based materials are relatively "tougher" in that the sensitivity to speed and/or load changes is less compared with either zirconia or alumina [26].

Under dry sliding conditions, material properties such as hardness, thermal conductivity, and density are important in determining wear resistance. Table 15.2 shows that in terms of hardness and thermal conductivity, silicon carbide has the highest value and zirconia has the lowest value. In terms of fracture toughness, zirconia has the highest value and silicon carbide has the lowest. This may explain the high sensitivity of zirconia with respect to change in speed and load, yet zirconia has some of the lowest wear rates at low-load and low-speed regions.

Microstructural features such as grain size [27], grain size distributions [28], and grain shapes [29–30] all have significant effects on wear of ceramics. For the four samples examined in this study, both silicon nitride and silicon carbide have duplex grains (elongated grains mixed with small- and medium-size equiaxed grains). This kind of grain design usually has a slightly higher wear rate under low load but lower wear rate under high load. Zirconia has the smallest equiaxed grains among the four materials; therefore, it has some of the lowest wear rates at the low-speed and low-load region. The equiaxed grain structure, however, once it begins to crack, is difficult to stop and wear accelerates rapidly [10, 31]. The picture that is emerging is that wear is a function of many parameters even for material properties and microstructures. There are no simple rules to predict wear behavior.

15.7.1.2 Paraffin Oil-Lubricated Conditions

Figure 15.8 shows the wear maps of the four materials under paraffin oil-lubricated condition. For these four ceramics, the purified paraffin oil itself does not react with the ceramics, nor does it contain impurities that will react with the ceramics [32–35]. Therefore, the effects of the presence of the paraffin oil are to remove the high interfacial temperatures and lower the flash temperatures in the contact. Therefore, the most dominant effect should be the moderation of the speed dependence under low or moderate loads. This indeed is the case. Figure 15.8 shows that the addition of paraffin oil clearly removes most of the speed dependence at low loads. For Si-based ceramics, the wear rates at low loads and speeds are much lower now than those of zirconia and alumina. In addition, at high loads and speeds, both alumina and zirconia exhibit rapid transitions to high wear. For alumina, the wear transition is basically load-induced with some speed effects beyond the critical speed. This probably reflects the fracture behavior of large grain-sized microstructure [31, 36]. For zirconia, both speed- and load-induced wear transitions are evident. The duplex microstructure of silicon nitride and carbide avoids the wear transitions [28].

15.7.1.3 Water-Lubricated Conditions

Water is a chemically reactive agent to alumina [37], zirconia [38], and SiC and Si_3N_4 [39–40] under certain tribological conditions. The wear maps in Figure 15.9 reveal different wear regimes directly contradictory to the wear characteristics observed for the same material under the dry sliding and/or the paraffin oil-lubricated cases. This can be illustrated using the silicon nitride case. At low loads, the wear increases with speed up to 15 mm/s; then the wear decreases as speed increases. Tribochemical reactions producing silicon hydroxides have been observed [40]. For the four materials, wear increases significantly in terms of the baseline data. This agrees with observations made by others [41]. Water has been observed to produce very low friction under certain operating conditions but accelerates the wear processes by corrosion and stress corrosion cracking.

In the alumina case, the tribochemical reactions change the primarily load-induced transition to more of a load- and speed-induced transition compared to the paraffin oil-lubricated case. Different aluminum hydrides can be produced and they have different

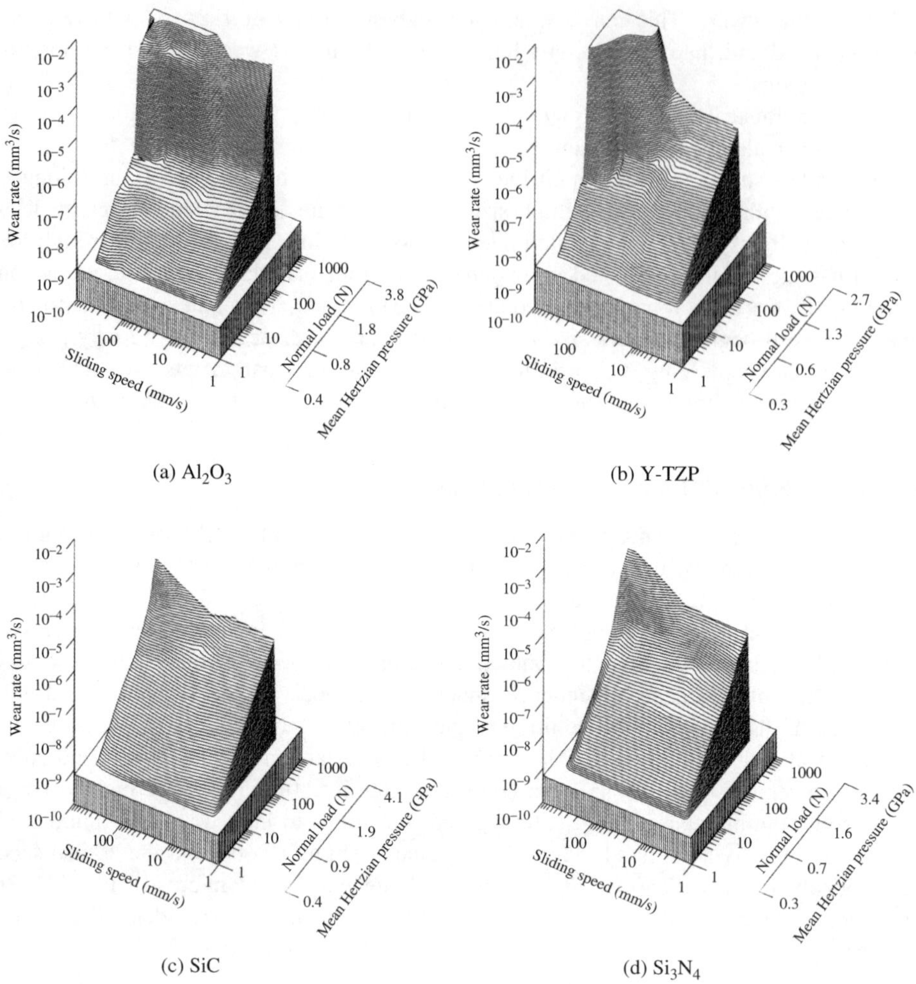

(a) Al_2O_3 (b) Y-TZP

(c) SiC (d) Si_3N_4

Figure 15.8 Wear maps under paraffin oil-lubricated conditions

"lubrication" characteristics [37]. Therefore, different wear zones are observed as a function of load and speed. In fact, for all four materials, speed dependence comes back as in the dry case. This suggests that speed activates the tribochemical reactions, and the reaction products change the wear behavior.

In the Si_3N_4 case, a low-wear zone occurs in the low-load, high-speed regime. In this regime, the wear is accompanied by an extremely low friction coefficient (<0.04). Silicon hydroxides have been found in this region in the form of slender rollers inside the contact zone. The reaction products are thought to provide some limited hydrodynamic lift to lower friction. However, the reaction in a water environment seems to be somewhat corrosive because the wear level is noticeably higher than the one in the same regime when paraffin oil is present. In the SiC case, an extremely low coefficient of friction is also measured in the high-speed, low-pressure regime. But the resulting wear does not show any noticeable

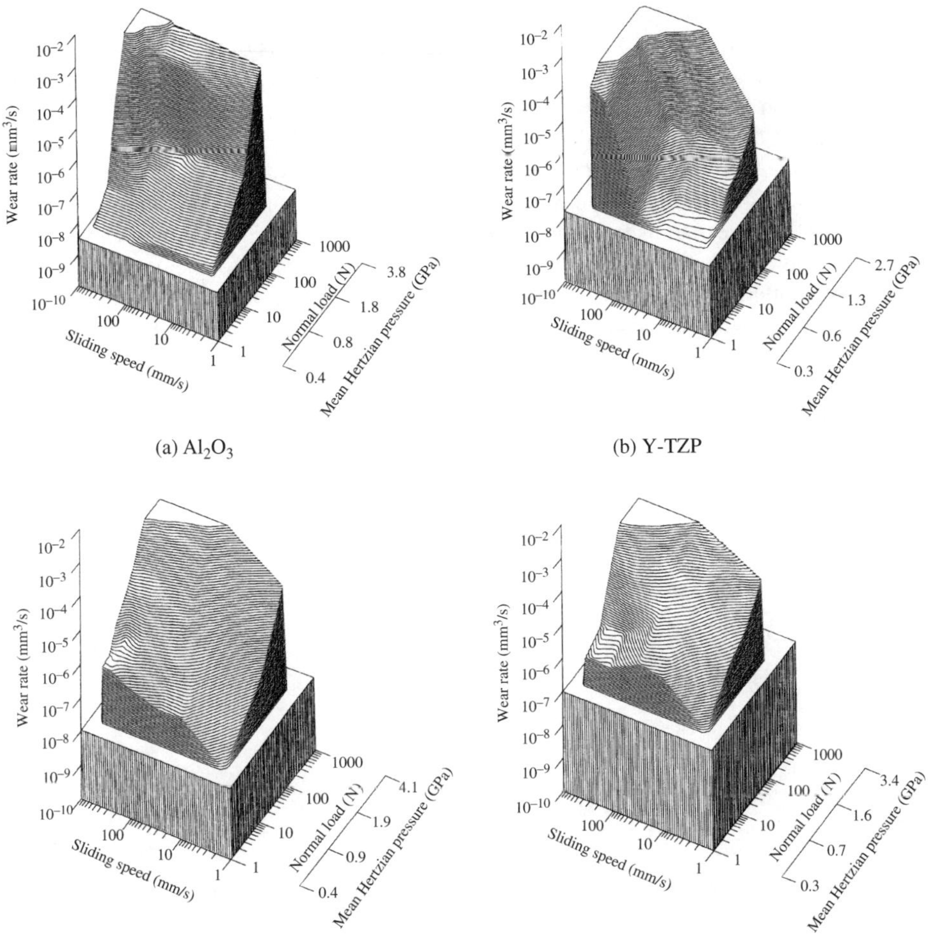

(a) Al_2O_3 (b) Y-TZP

(c) SiC (d) Si_3N_4

Figure 15.9 Wear maps under water-lubricated conditions

reduction as in the Si_3N_4 case. Based on the wear maps shown so far, wear maps of different lubrication environments are clearly required in order to describe fully the complete spectrum of wear characteristics for the same material pair under different environments.

15.7.2 Wear Transition Diagrams

Wear transitions can be defined as the sudden increase in wear over a small increment of the operating conditions (speed, load, temperature, and time). They usually signify a change in wear mode accompanied by a change in the dominant wear mechanism, e.g. onset of third-body wear induced by the generation of wear particles. Wear transitions have been observed in both metals and ceramics under unlubricated and lubricated conditions. Because the onset of transitions can often lead to rapid wear and eventual catastrophic failures, the transition phenomena have been extensively studied. In ceramics, the transition phenomena have been studied by many [10, 27, 31, 32, 36, 42]. The locations of the transition zone with respect to the operating parameters are important to design engineers to safeguard proper material selection and design.

Given a set of three-dimensional wear maps, the maps can be sliced at a plane to produce a set of contour maps of different wear rates as a function of speed, load, or any other parameters such as temperature and time under different environmental conditions. From the contour maps, the wear transition zone can be easily identified on the contour maps as the lines of constant wear rates bunch together to represent a steep ascend. These boundaries can then be plotted to show where the transitions occur. There may be one or more wear transitions in a given map, usually from mild to severe wear or a transition from severe to ultra-severe wear. The locations of these wear transition zones change with different lubricants.

Figures 15.10 through 15.12 show the wear transition diagrams of the four ceramics under different lubrication conditions. The effects of lubrication on the locations of the wear

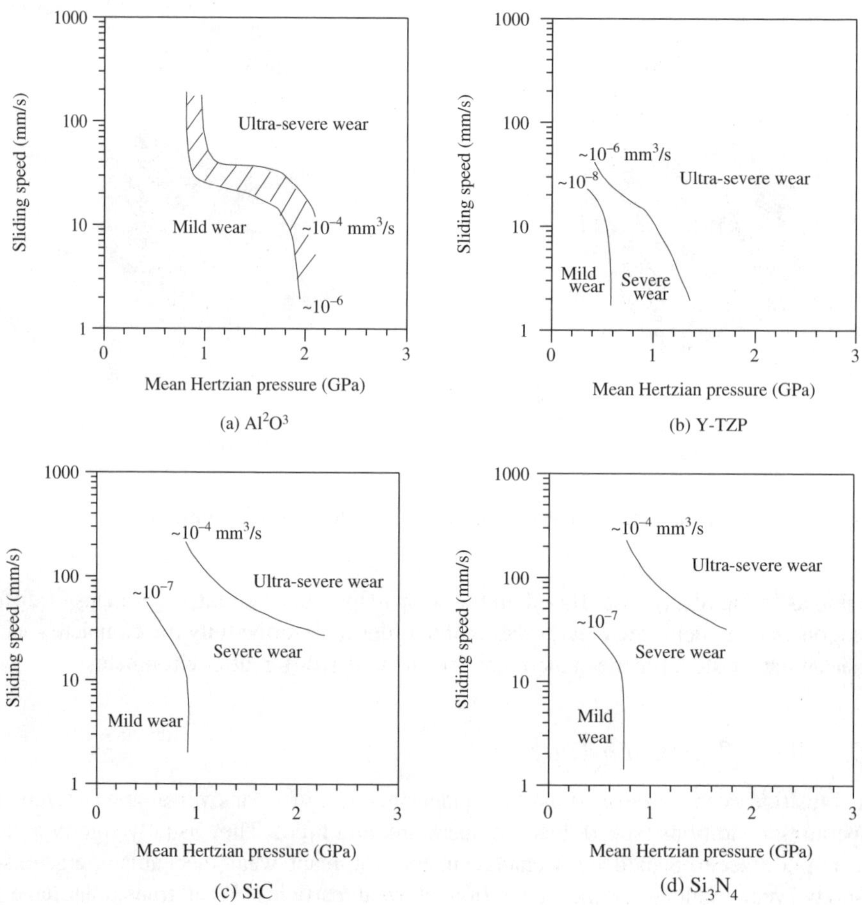

Figure 15.10 Wear transition diagrams under dry sliding conditions

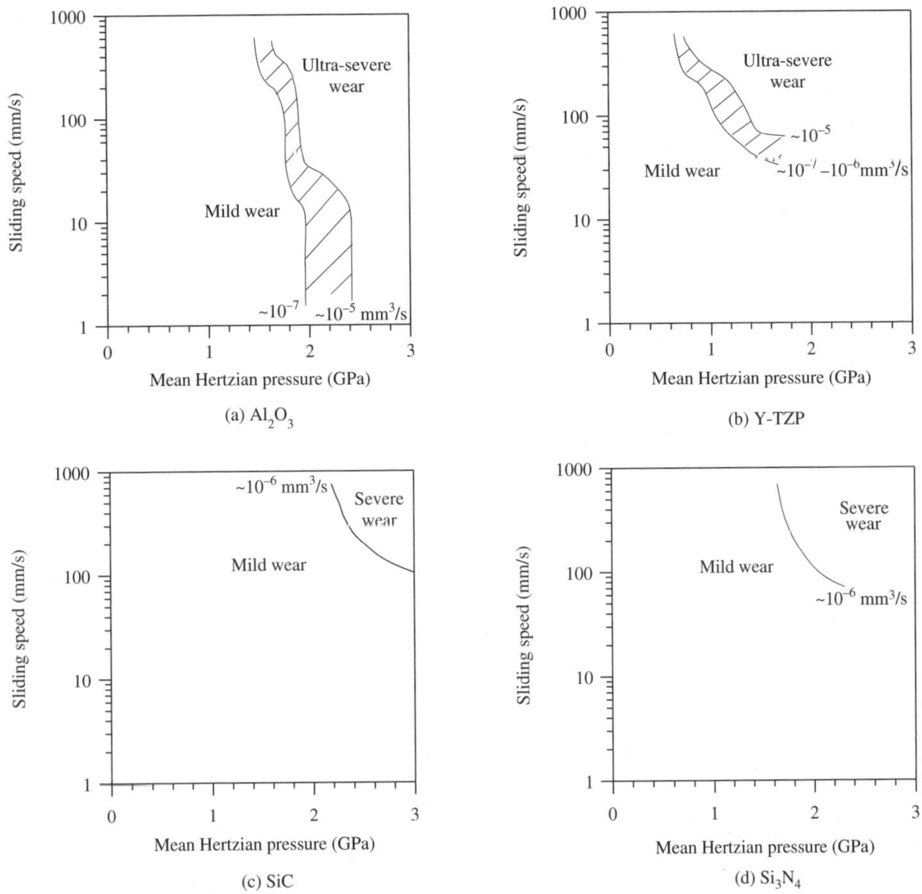

Figure 15.11 Wear transition diagrams under paraffin oil-lubricated conditions

transition zones can be clearly seen. In addition, the functional dependence of the wear transition lines on speed and load can also be observed.

Figure 15.10 shows such wear transition diagrams for alumina under dry sliding conditions; there is a critical speed and a critical load that will precipitate the transition from mild to ultra-severe wear. For the purified paraffinic oil (PPO)-lubricated case (Figure 15.11), it is basically a series of critical loads under different speeds. In the presence of water (Figure 15.12), the transition loads are much lower and another transition from severe to ultra-severe wear occurs at the high-speed and high-pressure region. The transition behaviors for zirconia under the three lubrication conditions are shown in these figures. The transitions depend on both speed and load due to the low thermal conductivity. Note the location of the mild to severe transition moves from the dry case to the PPO case. This measures the effects of the PPO on the transition behavior. For silicon nitride and silicon carbide, the behaviors are similar. The effects of PPO are quite dramatic. Water, in terms of wear, moderates the ultra-severe wear transition but does not change the mild to severe wear transition that much for silicon nitride.

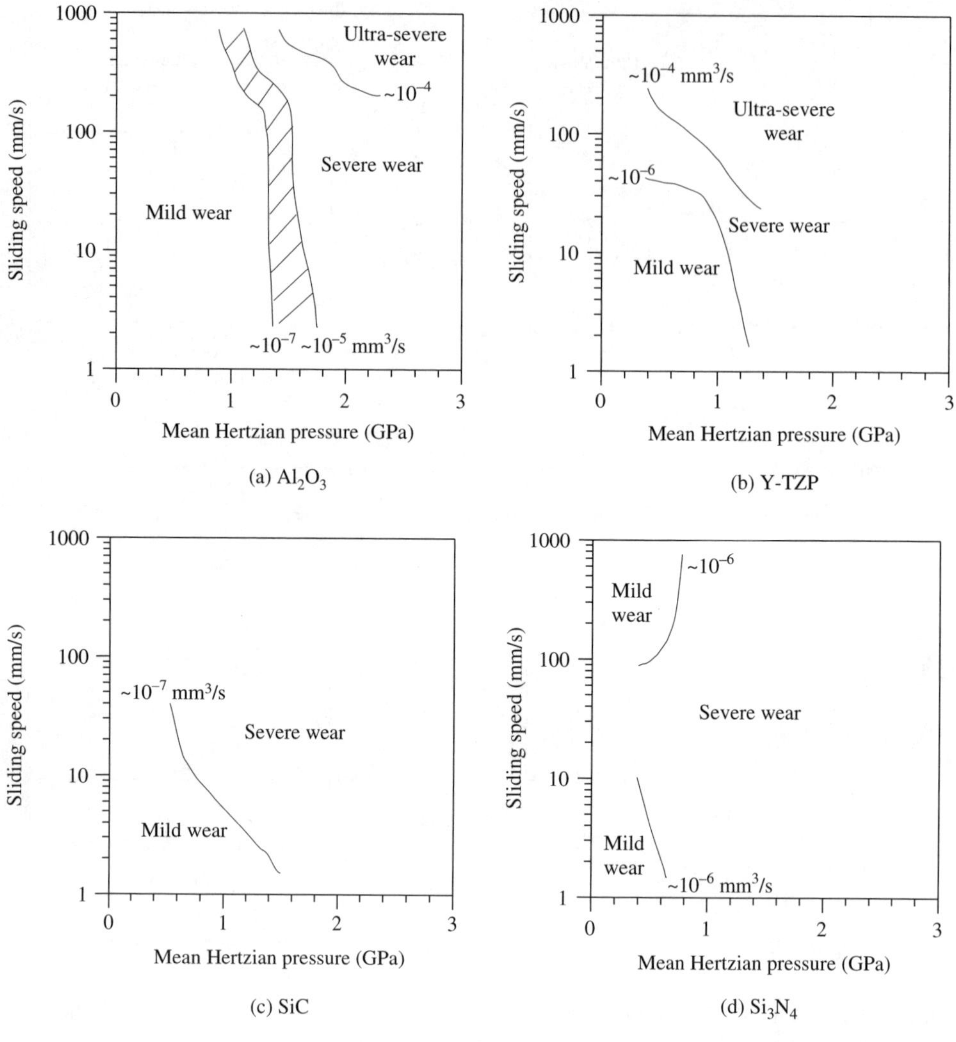

Figure 15.12 Wear transition diagrams under water-lubricated conditions

These figures also compare the transition diagrams of the four materials under the same lubrication condition. The locations of the transition zones as well as the functional dependence provide a powerful tool to compare different materials for a given application. Transient spikes in an application can be estimated and examined in the transition diagrams to assess the potential of premature failure.

The results shown above demonstrate the need to construct wear maps to better describe the wear characteristics of a material under a particular lubrication condition. Without such maps, it will be very difficult to grasp the complicated interactions of operating parameters, lubrication conditions, and material microstructures and properties. The results also reveal that significant differences are present in the wear characteristics of a material under different lubrication environments.

15.7.3 Material Selection Guided by Wear Maps

Once the wear maps for a given pair of materials are constructed, they can be used as material selection guides as well as design guides for different engineering applications. If we take the wear transition diagrams for different materials, we can superimpose them on a single diagram to compare the locations of the transition lines under different conditions. Figure 15.13 illustrates such diagrams for the four ceramics under dry sliding, PPO, and water-lubricated cases. The stress and the speed of the application can be calculated and examined in light of the transition diagrams. For a given safety margin, the material pairs that can safely operate in the region can be estimated.

Conversely, given the three-dimensional representations of the four materials, for a given contact pressure, these three-dimensional maps can be sliced along the constant contact

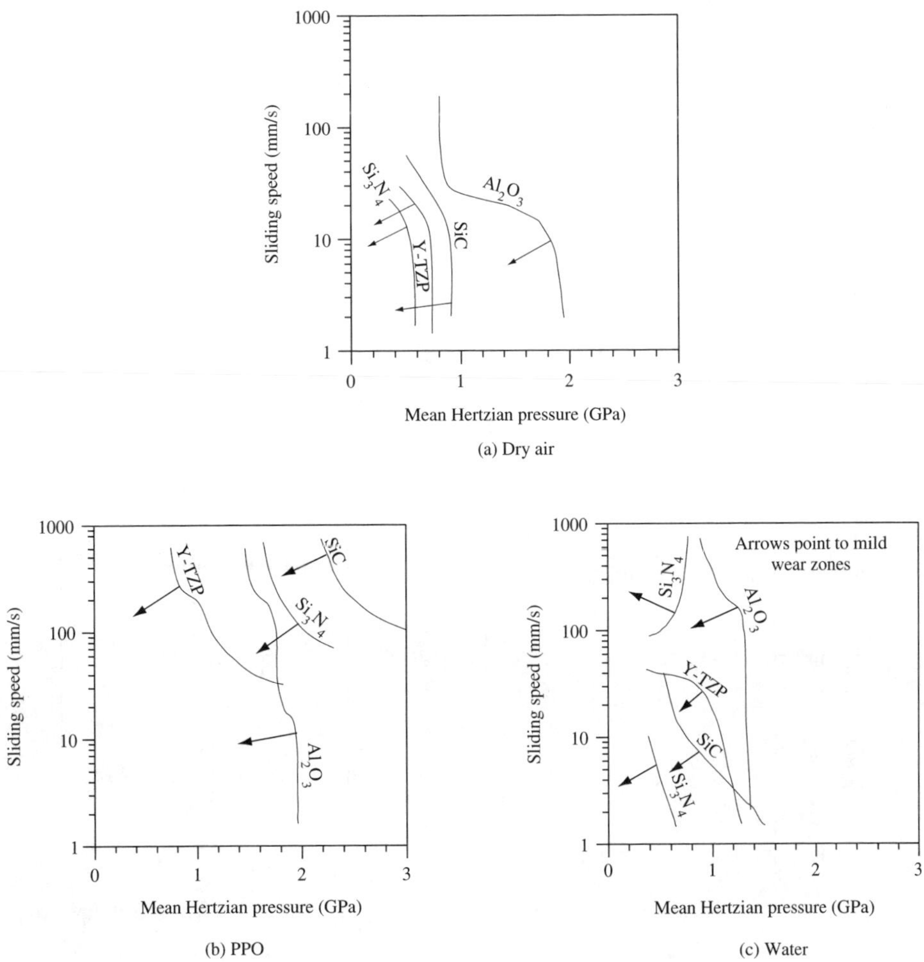

Figure 15.13 Material comparison based on mild-to-severe wear transitions under (a) dry air; (b) PPO; and (c) water-lubrication conditions

pressure lines and the four materials' wear levels superimposed in a single diagram for comparison. Figure 15.14(a) is such a diagram for 1-GPa contact pressure under dry sliding conditions. For this contact pressure, alumina has the lowest wear of the four materials. Similarly, the maps can be sliced along a single speed; then the wear rate as a function of load or contact pressure can be examined. Figure 15.14(b) shows the case for the four ceramics at 10 mm/s linear sliding speed under dry sliding conditions. In this case, alumina will work fine in the low-pressure region; SiC will be more suitable at higher pressures. Figure 15.14(c) illustrates the speed and contact pressure ranges that are available for use with these four materials under dry sliding conditions given a wear rate of 10^{-7} mm^3/s. Similarly, Figure 15.15 illustrates the case for these four materials under PPO-lubricated conditions. At 2-GPa contact pressure, SiC has the lowest wear (Figure 15.15(a)). Note the intersections of the constant wear lines among the four materials. Figure 15.15(b) shows the comparison of materials at a constant speed of 100 mm/s. For a constant wear rate of

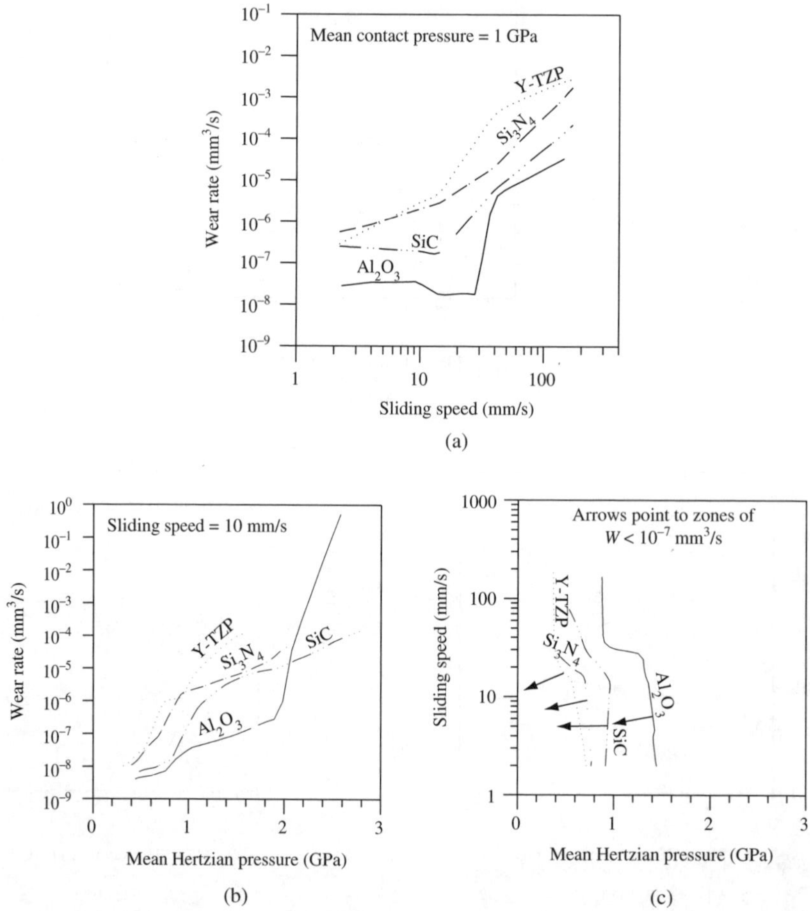

Figure 15.14 Material selection guides by using (a) fixed contact pressure; (b) fixed sliding speed; and (c) wear rate level (the condition is dry sliding)

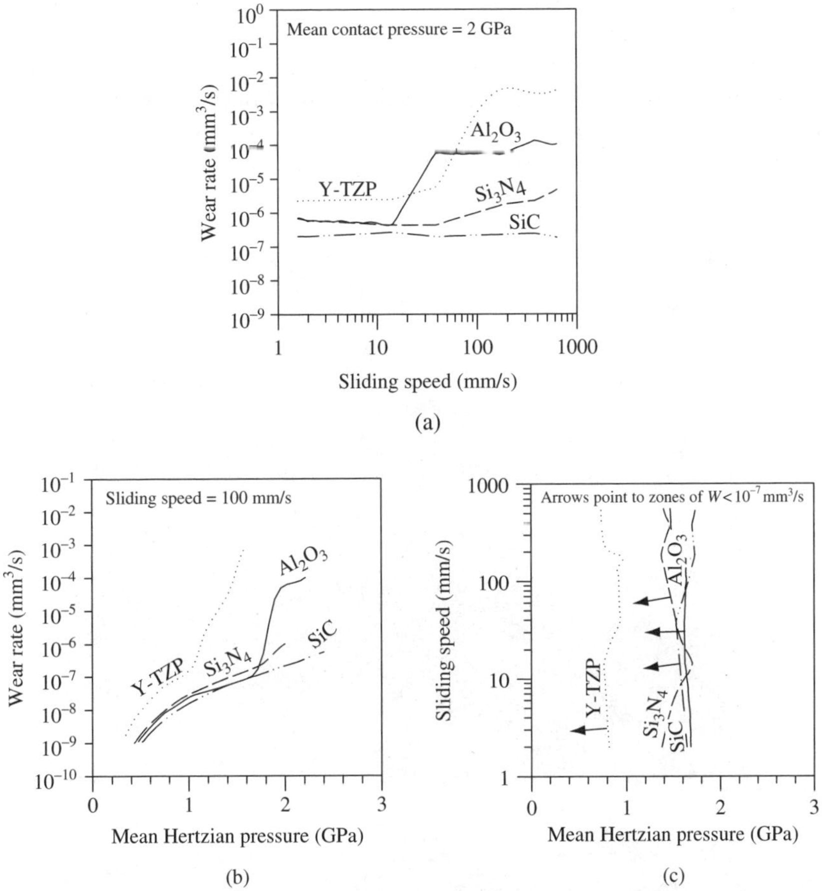

Figure 15.15 Material selection guides by using (a) fixed contact pressure; (b) fixed sliding speed; and (c) wear rate level (the condition is paraffin oil-lubricated sliding)

10^{-7} mm^3/s, Figure 15.15(c) shows the safe operating regions for the four materials. These derivative maps illustrate how one can select materials for a particular operating condition under a specific lubricating environment.

15.7.4 Wear Mechanism Identification

The availability of the wear transition diagrams provides the opportunity to study the different wear mechanisms in each region for a material under a lubrication environment. The transition diagram defines the minimum number of dominant wear mechanisms operating for a given system. By examining the contour maps (lines of constant wear rate), lines of equal spacing and lack of curvature usually indicate the same dominant wear mechanism. Valleys and plateaus usually suggest some change in the wear mode. In this way, regions with potentially different wear mechanism can be identified. Critical experiments can then be conducted within those regions to identify the dominant wear mechanism. In this way, wear mechanism maps can be constructed [43].

The experimental verification process is illustrated below. Based on the wear map and wear transition diagram of Y-TZP under paraffin oil-lubricated conditions, two sets of operating conditions can be chosen to examine the dominant wear mechanism in each region. Figure 15.16 shows two worn surfaces under high pressure but at different speeds – one is in

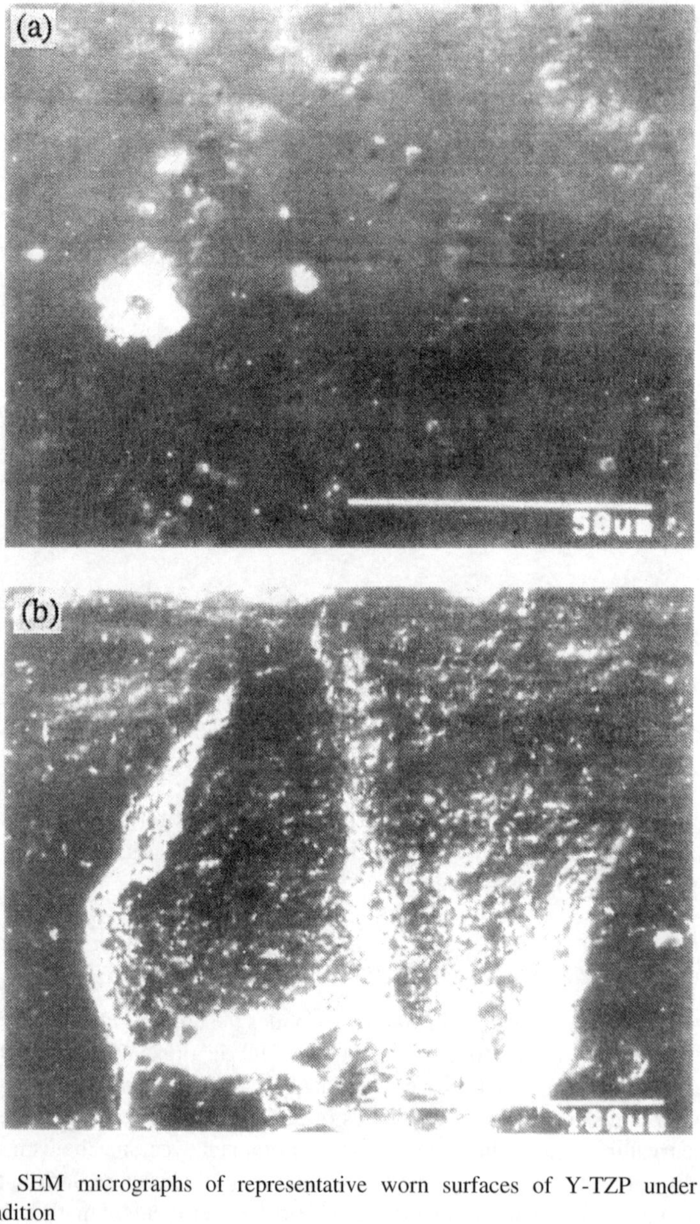

Figure 15.16 SEM micrographs of representative worn surfaces of Y-TZP under paraffin oil-lubricated condition

the mild wear regime and the other one is in the severe wear regime. The worn surface taken from the mild wear case shows mostly grooves, indicative of micro-abrasion and asperity scale fracture being the dominant wear mechanism. Meanwhile, the worn surface taken from the severe wear regime exhibits evidence of microfracture and brittle fracture. Brittle fracture has become the dominant wear mechanism. The onset of the brittle fracture is caused by intergranular cracks giving rise to wear particles in the interface. A series of separate experiments have been conducted to follow the onset of such transitions for alumina [36]. These results suggest that significant wear increase as well as change in wear mechanism is associated with this wear transition. In this manner, wear mechanism maps for the four ceramics are constructed and are shown in Figures 15.17 through 15.19.

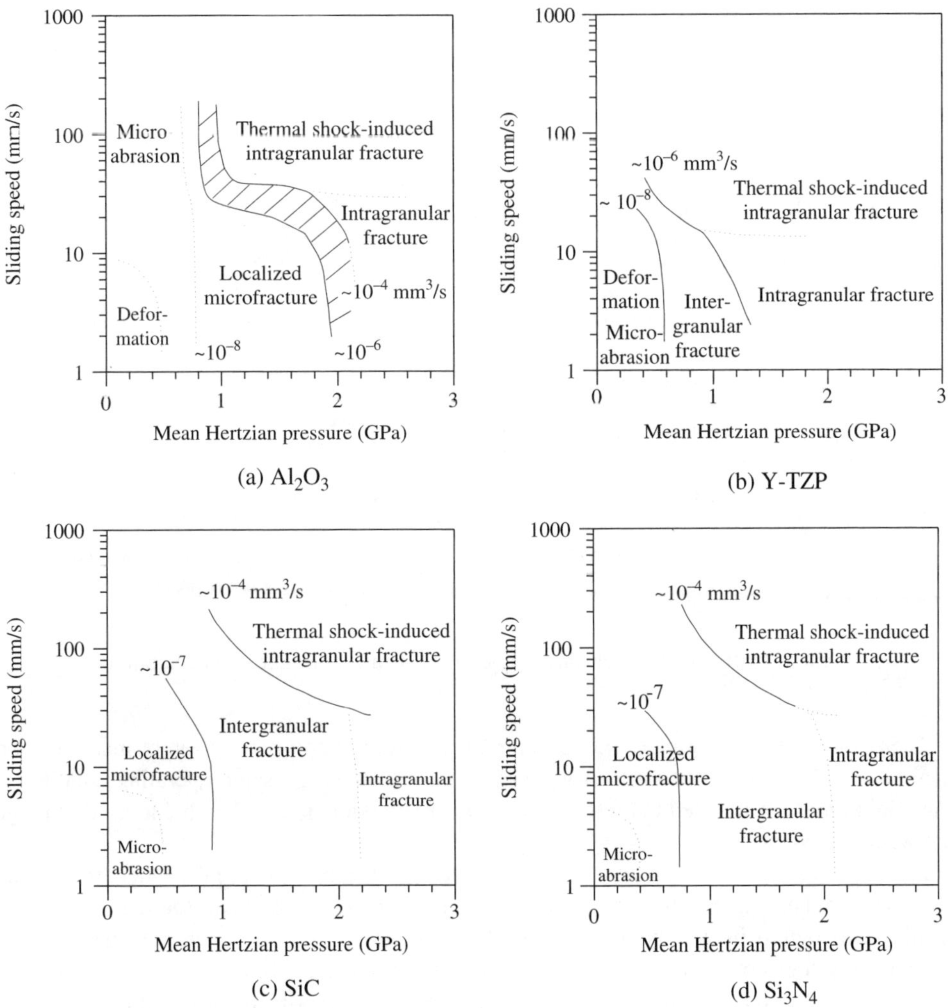

Figure 15.17 Wear mechanism maps under dry sliding condition

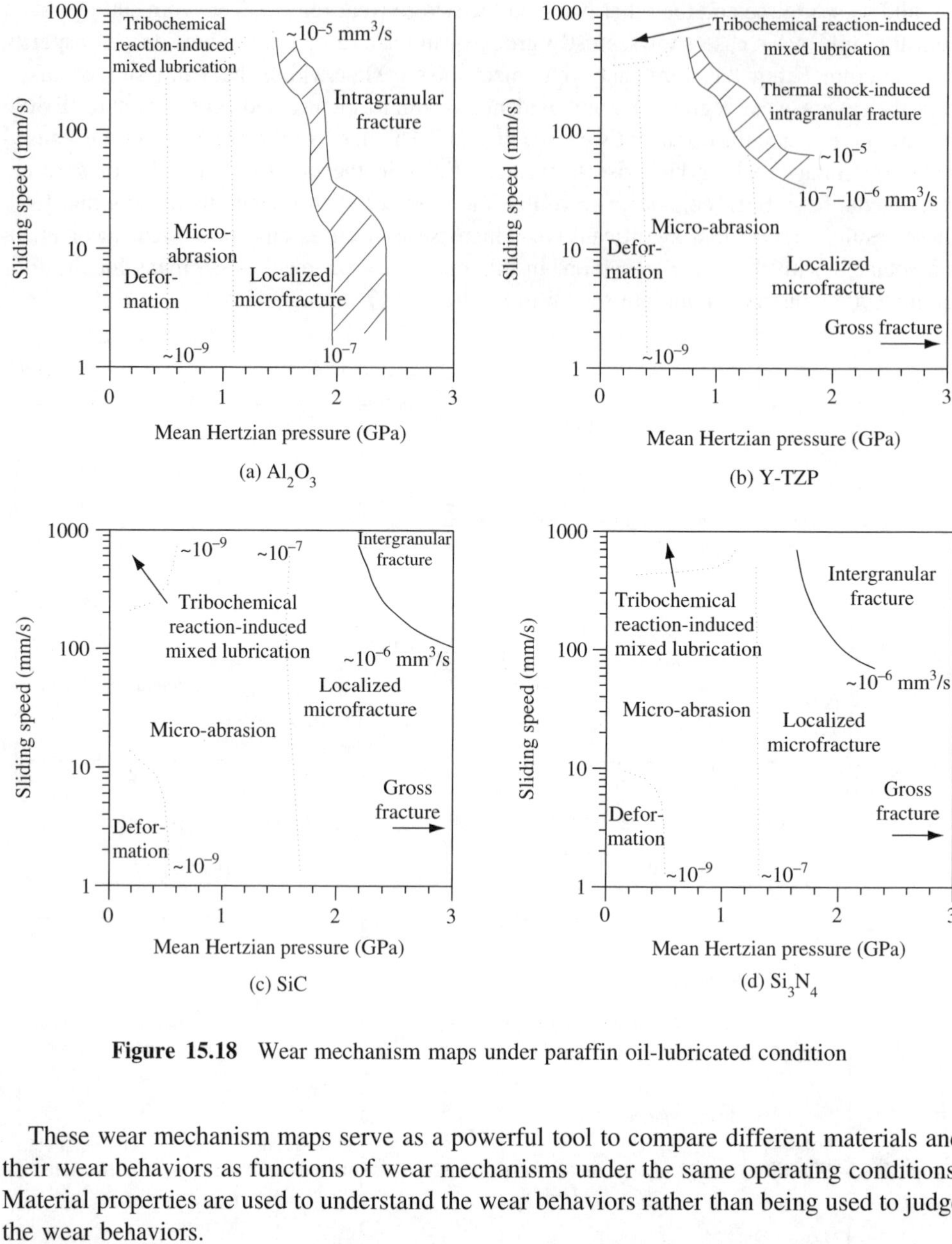

Figure 15.18 Wear mechanism maps under paraffin oil-lubricated condition

These wear mechanism maps serve as a powerful tool to compare different materials and their wear behaviors as functions of wear mechanisms under the same operating conditions. Material properties are used to understand the wear behaviors rather than being used to judge the wear behaviors.

Figure 15.17 shows the wear mechanism maps for the four materials under dry sliding conditions. When Figure 15.17 is compared with Figure 15.10, one can see there are zones within each region in which the wear mechanisms are different, but the change in wear mechanisms does not increase wear to the extent that wear transitions are formed. In a way, it can be pointed out that within a wear zone, there is a severity issue or a progression of severity until some events occur to trigger the wear transition. Figure 15.18 shows more delineation of the mild wear mechanisms of the materials when a nonreactive lubricant

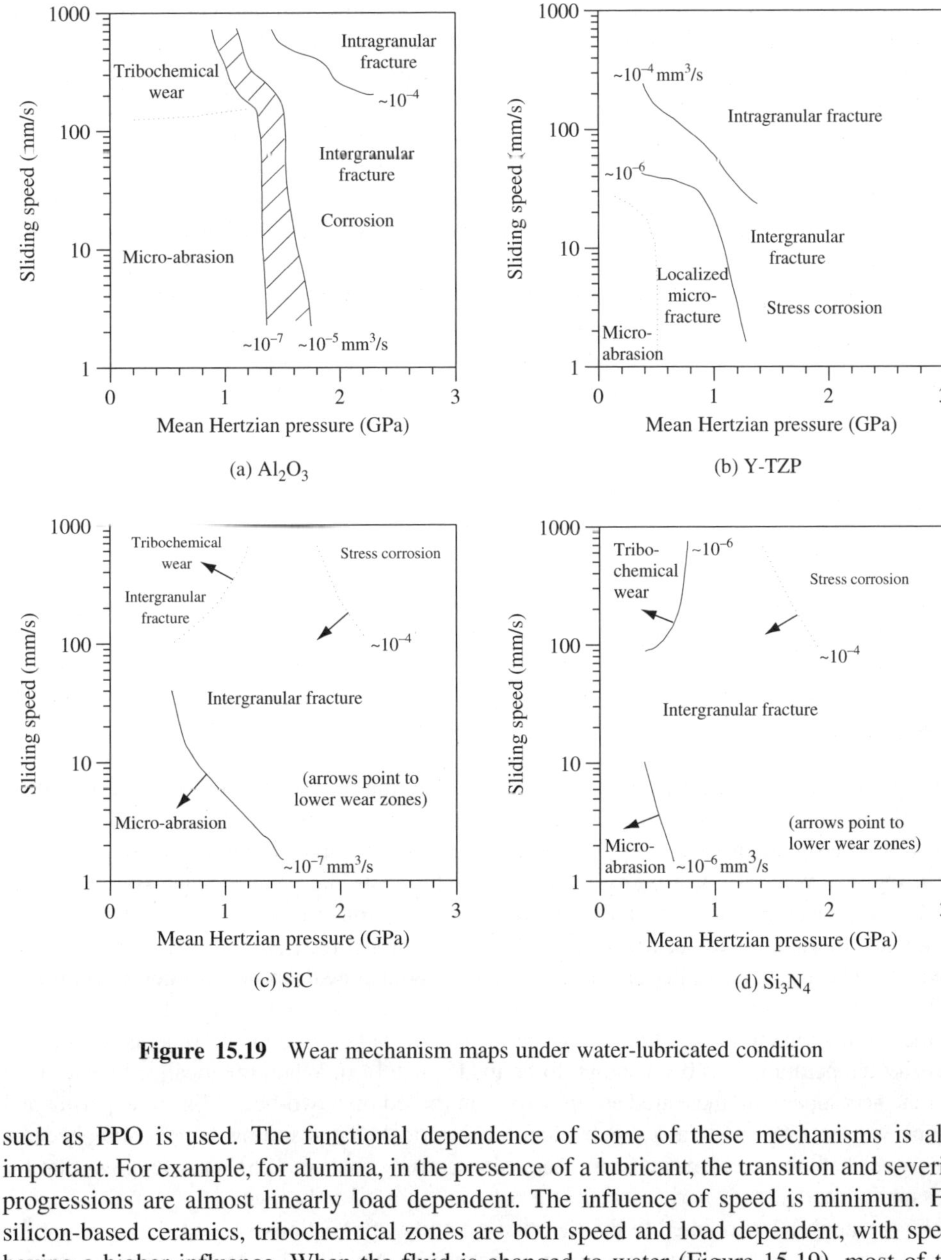

Figure 15.19 Wear mechanism maps under water-lubricated condition

such as PPO is used. The functional dependence of some of these mechanisms is also important. For example, for alumina, in the presence of a lubricant, the transition and severity progressions are almost linearly load dependent. The influence of speed is minimum. For silicon-based ceramics, tribochemical zones are both speed and load dependent, with speed having a higher influence. When the fluid is changed to water (Figure 15.19), most of the mechanisms are influenced by both speed and load, and the detailed mechanisms are difficult to determine because of the complexity of reactions taking place between water and the ceramics. This suggests that modeling tribochemical wear will be most difficult because of the lack of detailed mechanistic understanding of the processes.

Based on these mechanistic results, effective prediction of wear for a single material pair under different lubrication conditions will require a set of equations classified according to the wear regime and dominant mechanisms. This is similar in concept to Ashby's map; wear equations are developed on the basis of the different wear mechanisms. So, the first conclusion of wear mapping is that a family of equations is required to predict wear accurately.

At the same time, it must be pointed out that it is an issue of precision of prediction. As will be observed in Section 15.7.5, for one or two orders of magnitude prediction, all four ceramics can be treated as a single material pair under a particular lubricating condition. This is the global model based on fracture mechanics.

If one demands higher precision in describing wear of a material under a specific set of conditions, then a model can be developed on the basis of the dominant specific wear mechanism operating within that region. So according to wear maps, one could have a global mechanism governing all operating conditions and a wear model built on the basis of that assumption. Within each wear mechanism, a more refined model can be developed within that regime to describe the dominant wear mechanism more accurately and fully for that operating condition.

15.7.5 Wear Modeling Guide Based on Wear Maps

15.7.5.1 Metals

Lim and Ashby [22] used pin-on-disk (steel on steel) wear data under dry sliding conditions from the literature and constructed a wear mechanism map (Figure 15.4). The map uses normalized load and speed as parameters, and the data are successfully partitioned into different regions of dominant wear mechanisms. Asperity temperature is chosen as the key underlying parameter that caused all the wear phenomena, e.g. delamination wear, oxidation wear, melt wear, and seizure. Even though the asperity temperature calculation is based on the asperity scale, the coefficients of friction used are the average friction coefficient values from the macro-contact. Wear equations are developed for individual mechanisms and these equations are shown in Table 15.3. As discussed before, this approach is successful for severe wear regimes where high asperity temperatures dominate the wear mechanisms. For mild wear and lubricated cases, asperity temperatures are important parameters but not controlling (boundary lubricating films, tribochemical wear, elastohydrodynamic lubrication, etc.) the outcome. This limits the utility of this map but it does illustrate the power of such an approach.

The asperity level temperatures are also critical in analyzing chemical reactivity and surface film formation [44]. The use of measured "average" friction coefficients to estimate asperity temperatures has been shown to be inadequate [45]. When the localized wear event and the accompanying measured asperity friction (based on a two-ball collision experimental setup) was incorporated in a mechanical wear model, the resulting temperature became consistent with those estimated from chemical reactivity analysis [26]. This suggests that for an asperity model the asperity coefficients of friction need to be used.

Wear mapping has pointed to the fact that a family of equations is needed to account for wear for a given material pair in a given environment. No effective wear models for metals are available for the mild wear and well-lubricated conditions.

15.7.5.2 Wear Mechanism-Based Modeling of Ceramics

The wear mechanism maps described above suggest that the dominant wear mechanisms in different regimes are different. These mechanisms are summarized in Table 15.4. The stress

Table 15.3 Wear equations derived from wear mechanism maps in Reference [12]

Seizure	$\tilde{F} = \frac{1}{(1+\alpha_t \mu^2)^{1/2}} \left[1 - \frac{(T_b - T_o)}{20\, T_m} \, ln \left(\frac{10^6}{\beta \tilde{v}} \right) \right]$
Melt wear	$\tilde{W} = \left(\frac{T_m - T_o}{T^*} \right) \frac{H_o}{L} \frac{1}{\beta \tilde{v}} \left[\alpha \mu \tilde{F} \tilde{v} \frac{T^* \beta}{(T_m T_o)} - 1 \right]$
Mild oxidational wear	$\tilde{W} = \left(\frac{C^2 A_o r_o}{Z_c a} \right) \exp \left[- \frac{Q_o}{RT_f} \right] \frac{\tilde{F}}{\tilde{v}}$
Severe oxidational wear	$\tilde{W} = f_m \frac{K_{ox}(T_m^{ox} - T_b)}{L_{ox}\, a} \frac{(\tilde{F}N)^{-1/2}}{\beta \tilde{v}} \left[\alpha \mu \frac{a H_o \beta}{K_{ox}(T_m^{ox} - T_b)} \left(\frac{\tilde{F}}{N} \right)^{1/2} \tilde{v} - 1 \right]$
Plasticity-dominated wear	$\tilde{W} = \frac{2\, \gamma_o f_v}{f_A^*} \tilde{F}$

Definitions of the terms:

A_o	Arrhenius constant for oxidation	T_m	melting temperature
C	constant used in the model for mild oxidation wear	T_m^{ox}	melting temperature of oxide
		W	normalized wear rate
F	normalized pressure on sliding surface	Z_c	critical thickness of oxide film
H_o	room temperature hardness of metal	a	thermal diffusivity of metal
K_{ox}	thermal conductivity of oxide	f_A^*	critical area fraction of voids
L	latent heat of fusion per unit volume for metal	f_m	volume fraction of molten material during sliding
L_{ox}	latent heat of fusion per unit volume for oxide		
N	total number of contacting asperities	f_v	volume fraction of inclusions
Q_o	activation energy for oxidation	r_o	radius of pin
R	molar gas constant	v	normalized velocity
T^*	an equivalent temperature for metal	α	heat distribution coefficient
T_o	300 K	α_t	constant in Tabor's junction growth equation
T_b	bulk temperature		
T_f	flash temperature	β	dimensionless parameter for bulk heating
γ	cumulative plastic shear strain		
μ	coefficient of friction		

intensity in the contact is a critical parameter that underlies the formation of different wear regimes. In mild wear, the nominal contact produces stress intensity insufficient to cause macroscopic scale fracture. Wear occurs primarily at the asperity scale in the form of abrasion and microfracture. This micro-abrasion will generate subsurface damage/cracking at the micrometer scale. As the localized stress intensity exceeds K_{IC} of the material, microfracture

Table 15.4 Classification of the wear map data under dry sliding condition

Wear regime	Wear (mm^3)	Wear mechanism
Mild wear (stress intensity $< K_{1C}$)	10^{-7}–10^{-4}	(Asperity scale failure mode) – Abrasion – Intergranular cracks – Subsurface damage/cracks – Grain pullout – Tribochemical
Severe wear (stress intensity $> K_{1C}$)	10^{-5}–10^{-2}	(Nominal scale failure mode) – Fracture mechanics – Tensile cracks $\rightarrow$ edge effects – Third-bodies (linked to grain pullout)
Ultra-severe wear (stress intensity $>> K_{1C}$)	10^{-3}–10^{1}	(Nominal scale + few large debris) – Thermal shock (nominal scale)

occurs and generates intergranular cracks. This leads to subsequent grain pullouts. In severe wear, the nominal contact produces stress intensity that exceeds K_{1C} and causes macroscopic fracture such as tensile cracks [26]. The edge effect of those tensile cracks represents a source of wear particles [46]. These particles form third bodies in the contact zone and cause more localized fracture and grain pullouts. The combined wear particles and pulled-out grains are commonly observed in all four ceramics in the severe wear regime. In the ultra-severe wear regime, intragranular fracture is commonly observed. High loads, high sliding speed, and/or their combination cause the much increased stress intensity beyond K_{1C}. The end results are the presence of large quantities of large wear debris.

Once the wear mechanism is understood, specific models can be formulated to describe the specific wear process. These models will be tested against the specific wear data in the specific regime(s) where the model is expected to work. The lateral crack model [11] and the tensile crack model [10] were tested against the wear data of silicon nitride in the severe wear regime.

$$\text{Lateral crack model (LCM)} \quad \text{Wear volume} = C\frac{[E/H_V]^{4/5}}{K_{1C}^{1/2}H_V^{5/8}}N^{9/8}l \tag{3}$$

where C = constant, N = load, l = distance slid, E = Young's modulus, H_V = hardness, and K_{1C} = fracture toughness.

$$\text{Tensile crack model (TCM)} \quad \text{Wear volume} = C\frac{\sigma_{MAX}}{\sigma_D}\frac{Nl}{H_V(T)} \tag{4}$$

where σ_{MAX} = maximum tensile stress, σ_D = critical damage stress, H_V = temperature-dependent hardness, and T = temperature.

The presence of ultra-severe wear regime suggests that there is additional thermal stress to induce a higher rate of wear. Therefore, a thermal shock stress was modeled as follows [47]:

$$\text{Thermal shock stress} \quad \sigma_{\text{thermal}} = \frac{E\alpha}{(1-\nu)}\exp\left[\frac{-4}{\sqrt{\pi}}\left(\frac{T}{T*} - 5.3\right)\right] \tag{5}$$

where E = Young's modulus, α = linear expansion coefficient, ν = Poisson's ratio, T = flash temperature in Reference [48], T^* = bulk temperature from nominal contact = $\mu NV/4aK$, N = load, V = speed, a = nominal Hertzian contact radius, K = thermal conductivity, and μ = friction coefficient.

In this model, an exponential functional dependence of the thermal stress as a function of the ratio of flash temperatures to the surrounding surface temperatures was assumed. Subsequently, the maximum stress term in the TCM was assumed to be the sum of the mechanically induced tensile stress and the thermal shock stress, σ_{thermal}. The application of the thermal shock stress was limited only to the ultra-severe wear data.

Wear Transitions

Besides wear level predictions, the TCM was also tested for its applicability to predict the wear transitions. Owing to its assumptions, it was best suited for predicting the transition from mild to severe wear regimes. The onset of such wear transition was taken to be when the critical condition of $\sigma_{\text{MAX}}/\sigma_{\text{D}} = 1$ [10]. For the severe to ultra-severe wear transition, a thermal shock-related model was needed. The critical velocity model (CVM) proposed in Reference [18] was tested:

$$\text{Critical velocity model} \quad V_{\text{CR}} = \frac{4K^2}{(\mu\alpha E)^2\,\pi\kappa z} \tag{6}$$

where K = thermal conductivity, μ = coefficient of friction, α = coefficient of thermal expansion, E = Young's modulus, κ = thermal diffusivity = $K/\rho c$, z = width of slider, ρ = density, and c = specific heat. The onset of thermal shock damage was set to start when $V > V_{\text{CR}}$.

Figure 15.20 shows the predictions of LCM and TCM for silicon nitride data. The lateral crack model contained a correlational constant C that needed to be determined. Other parameters were determined by using the experimental conditions and material properties. The constant C was determined by fitting the data point at a sliding speed of 0.0019 m/s and a

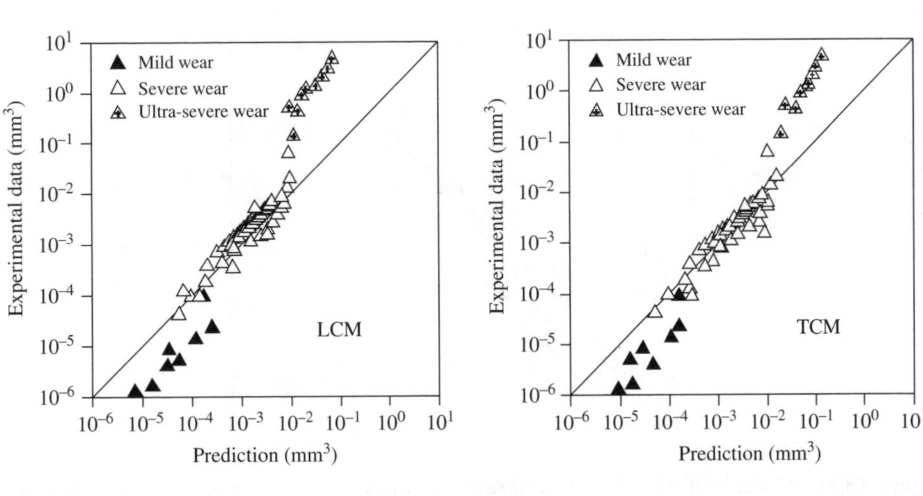

Figure 15.20 Model predictions of LCM and TCM for Si_3N_4 under dry sliding conditions

normal load of 20 N. For the TCM, the critical damage stress, σ_D, for alumina was previously estimated from various sources [49], but the value of σ_D for Si_3N_4 was not available. We assumed that the damage stress was a constant and therefore could be incorporated into the fitting constant C to form a new constant C^*, i.e. $C^* = C/\sigma_D$. The friction coefficients were taken from experimental measurements. The temperature-dependent hardness, $H_V(T)$, was determined experimentally and the data were fitted into exponential functions to give $H_V(T) = 24e^{-0.00064T}$ for Si_3N_4. The temperature for each experimental wear datum was estimated by using the flash temperature model in [48]. Finally, the new constant C^* was determined by fitting the same data point used in testing LCM, i.e. at a speed of 0.0019 m/s and a normal load of 20 N. In the TCM, the flash temperature calculations required the estimation of the real contact area. A sensitivity analysis was performed by changing the radius of real contact area from 5 to 20 μm. The resulting flash temperatures were from 300° to 600 °C for Si_3N_4. Within these temperature ranges, the C values changed only about 5%. Thus, the model was fairly insensitive to the temperatures and real contact areas. Overall, good agreements with the experimental data were exhibited in the severe wear regime by both LCM and TCM. However, the models overestimated the wear level in the mild wear regime and underestimated in the ultra-severe wear regime.

Figure 15.21 shows the prediction of TCM after the thermal shock stress was incorporated. Very good agreement between the model and the data in the ultra-severe wear regime was obtained. This result confirmed that the additional thermal shock stress was responsible for the extremely high wear in the ultra-severe regime. Figure 15.22 shows the predictions of wear transitions by using TCM and CVM. The TCM prediction had a slight deviation when the sliding speed was beyond 0.01 m/s. Meanwhile, the results from CVM matched well with experimentally determined wear transition, which was quite consistent with the wear level prediction by the modified TCM, i.e. additional stress by thermal shock could be accountable for the ultra-severe wear regime.

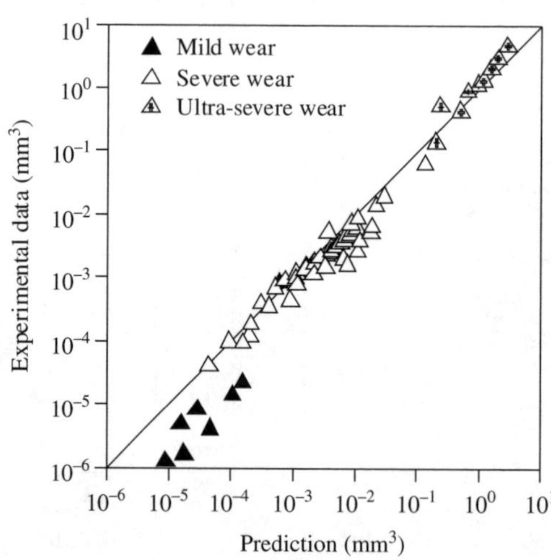

Figure 15.21 Model predictions by TCM in severe and modified TCM in ultra-severe wear regimes of Si_3N_4 under dry sliding conditions

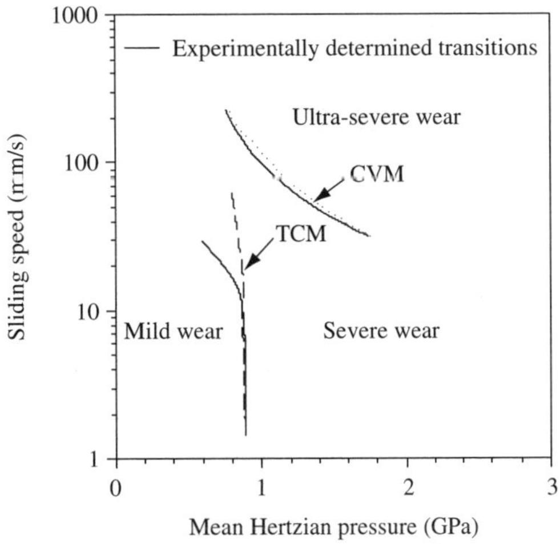

Figure 15.22 Wear transition predictions by TCM and CVM for Si_3N_4 under dry sliding conditions

The results obtained by the existing brittle fracture-based models were encouraging. Since wear of ceramics is dominated by brittle fracture, any fracture model with some reasonable fitting constants would describe the ceramic wear data somewhat satisfactorily. This may be the reason why the LCM and TCM perform similarly for the severe wear regime for silicon nitride. Speed was a factor, but only when the speed exceeded certain limits that it might induce thermal shock stresses that the fracture models began to underestimate. In the mild wear regime, the wear range was very large, from near-zero wear to 10^{-4}. This covers four orders of magnitude. Within this, tribochemical, plastic deformation, grooving, and single grain pullout are all dominant mechanisms. Fracture mechanics cannot account for these mechanisms, except maybe for grain pullouts. Therefore, it consistently overestimates the wear levels. The results of wear transition predictions also revealed some inadequacies in the existing models. The experimentally determined mild to severe wear transition could be predicted with some success by TCM. However, the model prediction overall was not quite satisfactory. This was probably due to the microstructural effects [29].

15.7.5.3 Correlational Models Based on Wear Maps of Ceramics

Given a large database of wear data, a new approach to examine wear modeling across materials over a wide range of operating conditions is to explore correlational models against the critical material and operational parameters.

Identification of Critical Parameters in Establishing New Correlational Models

Figure 15.23 shows the schematic of the different contact configurations under progressively more severe conditions. In the mild regime, wear can be treated as simple asperity scale abrasion (microfracture does not produce wear immediately and grain pullouts are only intermittent). In this case, wear by each asperity contact can be approximated by the product of wear depth, contact width, and length of sliding within each "sliding cycle," as illustrated in Figure 15.23(a). The "sliding cycle" can be described by the total distance slid l divided

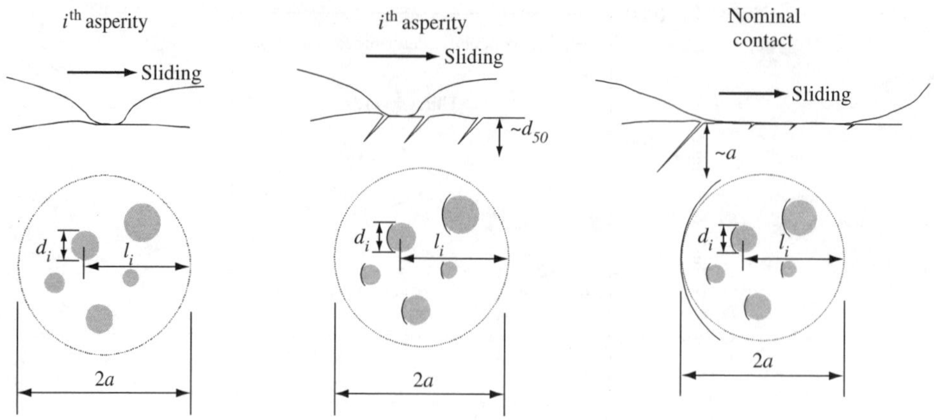

Figure 15.23 Illustrations of the contacts for modeling wear in different wear regimes: (a) asperity scale abrasion in mild wear regime; (b) tensile cracks inside the nominal contact in severe wear regime; and (c) gross fracture in the nominal contact in ultra-severe wear regime

by the Hertzian contact width $2a$, i.e. $l/2a$. Given this set of assumptions, wear can be expressed as

$$\text{Wear volume} \propto \left[\sum_i \left(\frac{P_m}{H_v} * C_i d_i\right) \times d_i \times 2a\right] \times \frac{l}{2a} \tag{7}$$

where P_m = mean Hertzian pressure [50], H_v = hardness, C_i = dimensionless geometric factor, and d_i = asperity contact width. The terms in the parenthesis represent the wear depth which is assumed to be proportional to the ratio of P_m and hardness and a fraction of the contact width d_i. The $2a$ term inside the bracket represents the upper bound of length l_i of such asperity scale abrasive wear within a single cycle. Since the ratio of P_m/H_v is constant throughout all the asperity contacts, the summation term will be proportional to the real area of contact which has been proposed to depend on [51, 52] the following relationship:

$$A_{real} \propto \frac{N}{E'} \times f \text{ (roughness parameters)}$$

where N = load and E' = composite Young's modulus. The roughness parameters included asperity radius [52], r.m.s. roughness [51, 52], and/or autocorrelation length [51]. By substituting the real contact area into equation. (7), the mild wear can be expressed by the following relationship:

$$\text{Wear volume} \propto \frac{P_m}{H_v} \times \frac{N}{E'} \times f \text{ (roughness parameters)} \times l \tag{8}$$

Since the surface roughness of all specimens in this data set was controlled to be nearly the same, the roughness term, f, can be treated as a constant in this analysis. Hence, the mild wear was directly proportional to $P_m \times N \times l$, related to the operating parameters, and inversely proportional to material properties, H_v and E'.

In severe wear regime, because of the presence of cracks, edge effects from those cracks and third-body wear particles, the asperity contacts can have very different characteristics, as illustrated in Figure 15.23(b). The total wear volume may be expressed as

$$\text{Wear volume} \propto \left[\sum_i \frac{\sigma_{\text{MAX}} \left(\frac{T^*}{T_{\text{O}}} \right) \sqrt{d_{50}}}{K_{1\text{C}}} \times b_i d_i \times d_i \times 2a \right] \times \frac{l}{2a} \tag{9}$$

where σ_{MAX} = maximum tensile stress [13], T^* = interfacial temperature of the nominal contact [53], T_o = ambient temperature at 20°C, $K_{1\text{C}}$ = fracture toughness, d_{50} = mean grain size, b_i = geometric factor, and d_i = contact width.

Here the wear depth is assumed to be proportional to the contact width d_i by using the ratio of a stress intensity from tensile stress divided by $K_{1\text{C}}$ and another geometric factor b_i. The crack length in the measure of the stress intensity is assumed to be equivalent to the mean grain size d_{50}. Because of the potential thermal effects from frictional heating, the tensile stress is further modified by multiplying a temperature ratio of T^*/T_o. The interfacial temperture T^* can be calculated by Archard's temperature equation [53]. In so doing, the nominal contact is treated as a single asperity and the interfacial temperature is applicable within a layer thickness of half of the Hertzian contact width, i.e. this temperature exists within a thickness of a. The multiple of the two d_i inside the bracket represents the real contact area, similar to the mild wear case. However, in the severe wear, a different interpretation may be needed to account for the edge effects due to cracking and third-body wear. This area is assumed to be proportional to the ratio of N/H_v [10]. Consequently, equation [9] could then be rewritten as follows:

$$\text{Wear volume} \propto \frac{\sigma_{\text{MAX}} \left(\frac{T^*}{T_{\text{O}}} \right) \sqrt{d_{50}}}{K_{1\text{C}}} \times \frac{N}{H_v(T^*)} \times l \tag{10}$$

In equation (10) the hardness value is further set to include the temperature effects. Equation (10) is similar to the tensile crack model proposed by Wang and Hsu [10] but with the added temperature effect. Also, the use of fracture toughness can eliminate the need for a special material property σ_{D}, the critical damage stress (10).

In ultra-severe wear regime, the thermal shock can be attributed to a temperature gradient between the heated spot and the surrounding low-temperature region. Since the interfacial temperature exists in a layer with a thickness of half Hertzian width, the wear equation in equation (10) can be extended by substituting the d_{50} term with a, namely, the crack length that determines the stress intensity is equivalent to a, as illustrated in Figure 15.23(c). This will give rise to an equation of the following form:

$$\text{Wear volume} \propto \frac{\sigma_{\text{MAX}} \left(\frac{T^*}{T_o} \right) \sqrt{a}}{K_{1\text{C}}} \times \frac{N}{H_v(T^*)} \times l \tag{11}$$

The increased T^*/T_o ratio and crack length a with a decrease in $H_v(T^*)$ should produce much higher wear level, as compared to the severe wear described in equation (10).

Modeling Approach

Armed with the existing models and the functional dependence of the fundamental wear processes, the wear data set is tested with different models. Many of the existing models cannot fit the large data set satisfactorily. Therefore, we embarked on an approach: the wear data set is plotted against a group of key parameters indicating the severity of the contact conditions. After the data are lined up according to the contact severity, the materials will be normalized using material property data. In principle, such material normalization should eliminate the differences in material properties among different materials (provided that we know all of the important material properties related to wear). If this approach is successful, then a generalized wear equation for all materials over the entire operating range will line up according to some functional dependence.

Contact Severity The severity of the contact has been modeled by many [12–14, 26, 31]. As a group parameter, it is successful in the wear map concept [14] illustrated by Adachi and Kato. However, we will show later on that the contact severity index alone has limited applicability when it is used against a large data set. In the present approach, the severity parameter is defined as a group parameter directly reflecting the operating conditions. Let us re-examine each of the wear equations. In mild wear, equation (8) has two sets of parameters. The $P_m \times N \times l$ term is related to the operating conditions. The $H_v \times E'$ is related to material properties. Similarly, the $\sigma_{MAX} \times (T^*/T_o) \times N \times l$ term in equation (10) and the $\sigma_{MAX} \times \sqrt{a} \times (T^*/T_o) \times N \times l$ term in equation (11) are related to the operating conditions. The $\sqrt{d_{50}}/(K_{1C} \times H_v(T^*))$ term in equation (10) and $K_{1C} \times H_v(T^*)$ in equation (11) are related to the material property. So at this point, one can plot the wear data in each wear regime against the corresponding severity parameter and observe the data scatter among different materials. The degree of data scatter should reflect the validity of the choice of the operating parameters and the associated material parameters. Therefore, a simple material normalization should be able to pack all the data of the same predominant wear mechanism into a single function with the severity parameter. This is the concept.

Clearly, the severity parameters determined from the three wear regimes are not identical. In order to test across the entire operating conditions (the three wear regimes), one needs to make some simplifications. Firstly, the common factor in all three wear regimes is $N \times l$. The uncommon term in mild wear is P_m, in severe wear are $\sigma_{MAX} \times (T^*/T_o)$, and in ultra-severe wear are $\sigma_{MAX} \times \sqrt{a} \times (T^*/T_o)$. However, the P_m and σ_{MAX} are related via the coefficient of friction [26]. Moreover, the T^*/T_o is relatively close to unity in the mild wear regime for the four ceramics, except for Y-TZP. Therefore, the severity parameter can be generalized to a form of $\sigma_{MAX} \times (T^*/T_o) \times N \times l$.

Materials Normalization Materials have been characterized historically by their material properties such as hardness, elastic modulus, toughness, thermal conductivity, thermal diffusivity, and the temperature influence on these properties. Yet at the same time, which of these parameters controls wear is not clear. Whether there are hidden parameters that are not obvious is also not clear (parameters such as grain boundary strength, population of defects, machining damage, grain geometry, residual stress, etc.). We know that brittle fracture behavior of ceramics is often influenced by the relative energy rate of the grain boundary governed by grain size, ratio of the grain boundary energy release rate versus the grain [54], and pre-existing defect population from sintering and machining. If the microstructure is a duplex structure, i.e. elongated grains intermixed with equiaxed grains, crack deflection by the elongated grains has to be taken into account, i.e. the aspect ratio of grain defined by grain length divided by grain diameter. Iteration between the severity index and the material normalization parameters will test the validity of this concept and provide insight into an issue that often eluded us in the past.

In the following section, historical severity indexes will be tested against the wear database, followed by the generalized severity parameters developed in the present study.

15.7.6 Wear Prediction Based on Wear Maps

15.7.6.1 Historical Severity Indices

There were several versions of the so-called severity index proposed [12–14]. They are described below.

$$S_c = \frac{P_o\sqrt{R_{MAX}}}{K_{1C}} \tag{12}$$

where P_o = maximum Hertzian pressure and R_{MAX} = maximum surface roughness. The S_c was originally derived from modeling stress intensity factor by using the maximum Hertzian pressure and a crack length equivalent to the maximum surface roughness [12]. Subsequently, the severity index was modified by incorporating stresses introduced by friction, and a slightly different form was derived [13], as shown below.

$$S_{CF} = \frac{P_o\sqrt{(1+\mu^2)R_{MAX}}}{K_{1C}} \tag{13}$$

where μ = friction coefficient and the rest of the parameters were the same as in S_C. More recently, a new severity index derived from using the maximum tensile stress in estimation of threshold stress intensity, S_{cm}, was employed in describing the mild to severe wear of ceramic materials [14]. The maximum tensile stress was simplified in S_{cm}, as shown below.

$$S_{cm} = \frac{(1+10\mu)P_o\sqrt{R_{MAX}}}{K_{1C}} \tag{14}$$

where $(1+10\mu)P_o$ was the simplified tensile stress.

The maximum surface roughness, R_{MAX}, was the same in all specimens. Because it represented the equivalent crack length in the estimate of stress intensity, one could employ an alternative interpretation of this term by using the mean grain size.

Figure 15.24 shows the applications of the various severity indexes to the wear map data from the four ceramics. The data of the mean grain size, d_{50}, of the four ceramics are listed in Table 15.5. The plots by using S_C and S_{CF} have quite similar features. Within each material, different wear levels could be present at the same severity index. Wear was mainly correlated by maximum Hertzian pressure in S_C, while friction coefficient had a slight contribution in S_{CF}. When all four materials were combined, they formed two groups. Alumina and SiC were in one group at higher severity. Silicon nitride and zirconia were in another group at lower severity. These were due to the combination of larger mean grain size and relatively lower fracture toughness in the alumina and SiC. On the other hand, Figure 15.24(c) shows that the wear data mingled more by using S_{cm}. Because all these severity indexes already contained material properties, no material normalization could be applied. Therefore, one may conclude that they did not quite correlate with the current wear data.

Figure 15.25 shows the application of a single parameter of mean Hertzian pressure to the database. Again, the results showed that it was inadequate to represent a unified severity parameter for this database.

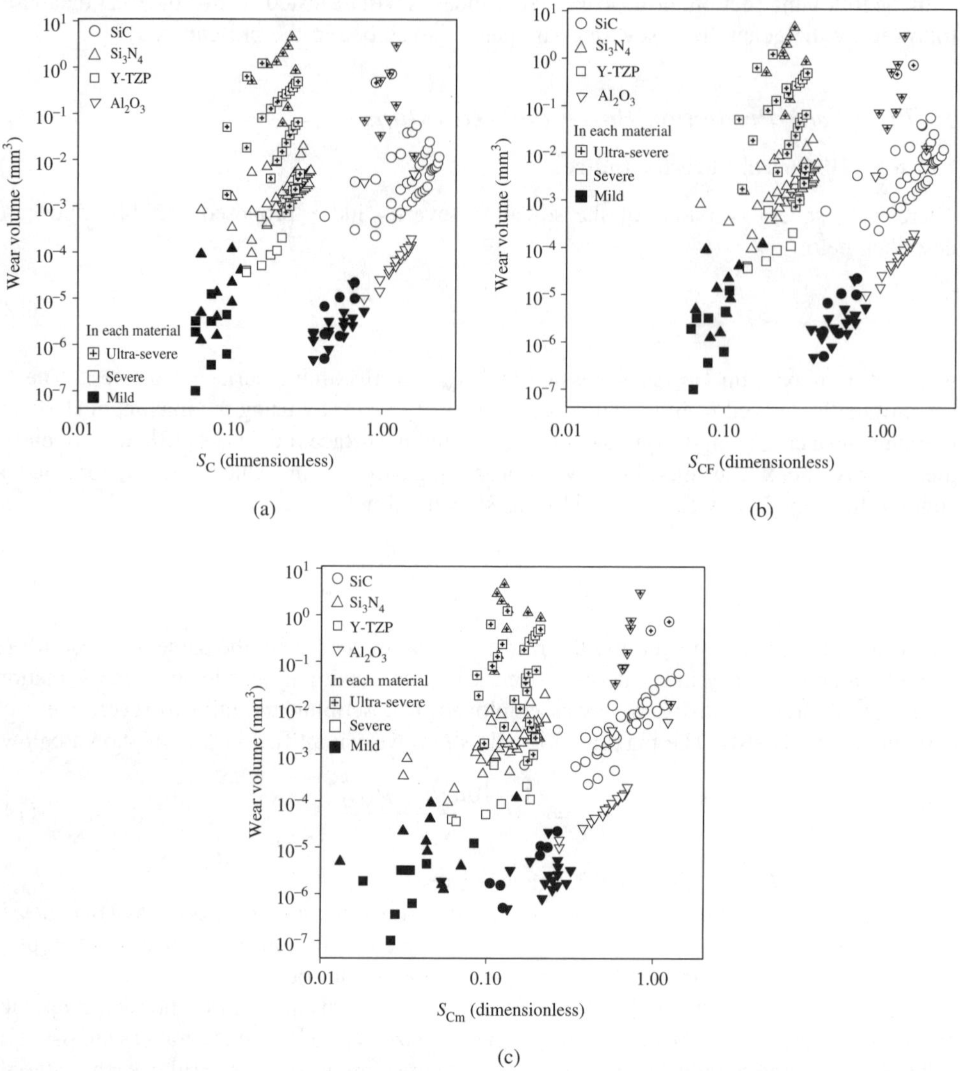

Figure 15.24 Correlational results form using historical severity indexes: (a) S_C; (b) S_{CF}; and (c) S_{Cm}. These indexes were defined in the text

Table 15.5 Microstructural parameters of the four ceramics

	Al_2O_3	Y-TZP	SiC	Si_3N_4
Mean grain size, d_{50} (μm)	5	1	3	0.5
Grain size ratio, d_{90}/d_{50}	3	2	2.6	2
Aspect ratio, A	1	1	5	4

Note: These are nominal dimensions based on the SEM micrographs with type B uncertainty of 50%.

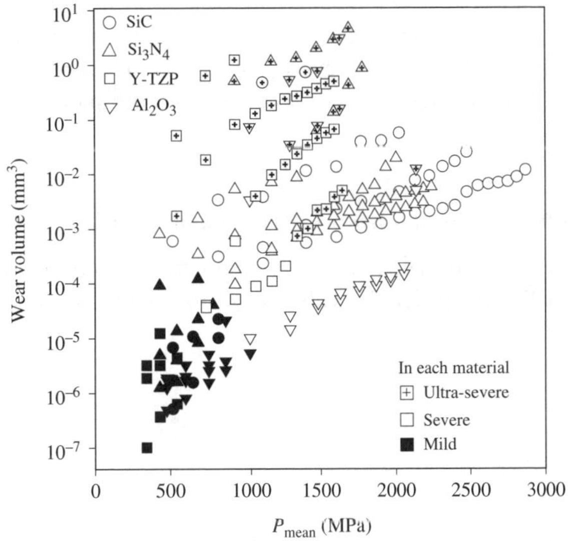

Figure 15.25 Correlational results from using mean Hertzian contact pressure

15.7.6.2 Severity Parameters from Correlational Models

Figure 15.26(a) shows the wear data plotted by just using the $\sigma_{MAX} \times (T^*/T_o)$. This parameter had about three orders of magnitude range. As compared to the total range of wear volume of nine orders of magnitude, this parameter was extremely sensitive to wear. On the other hand, Figure 15.26(b) shows the results from using severity parameter $\sigma_{MAX} \times (T^*/T_o) \times N \times l$.

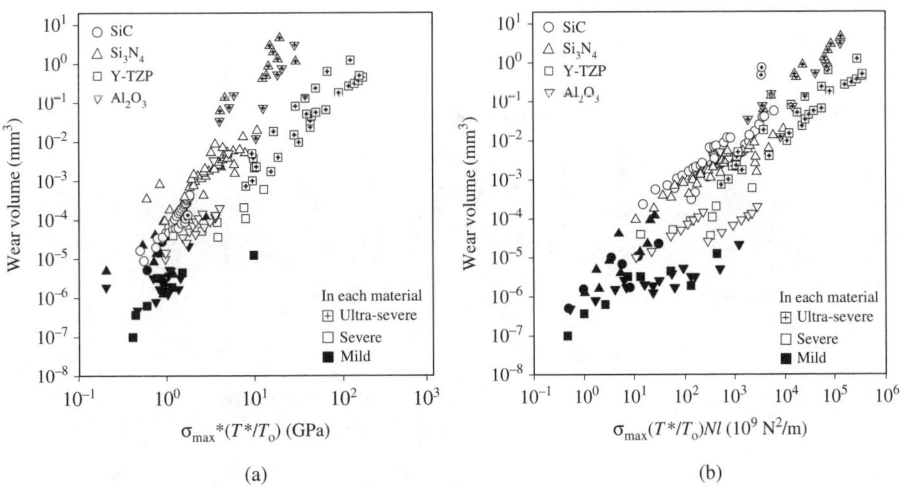

Figure 15.26 Correlational results from using (a) maximum tensile stress multiplied by a temperature ratio of T^*/T_o to take into account the thermal effects; and (b) a severity parameter derived from the current study

This parameter had a range of seven orders of magnitude; this was much more comparable to the wear data range. Also, the data in each wear regime appeared to form cluster, indicative of a rather reasonable selection of the severity parameter. Inclusion of $\sqrt{d_{50}}$ and $\sqrt{a}$ had also been tested. Only slight changes were obtained, so those plots were omitted from this presentation.

15.7.6.3 Material Normalization from Correlational Models

There are different ways to normalize the wear results for the purpose of direct comparison. One way is to normalize all materials with respect to a particular material. Another is to

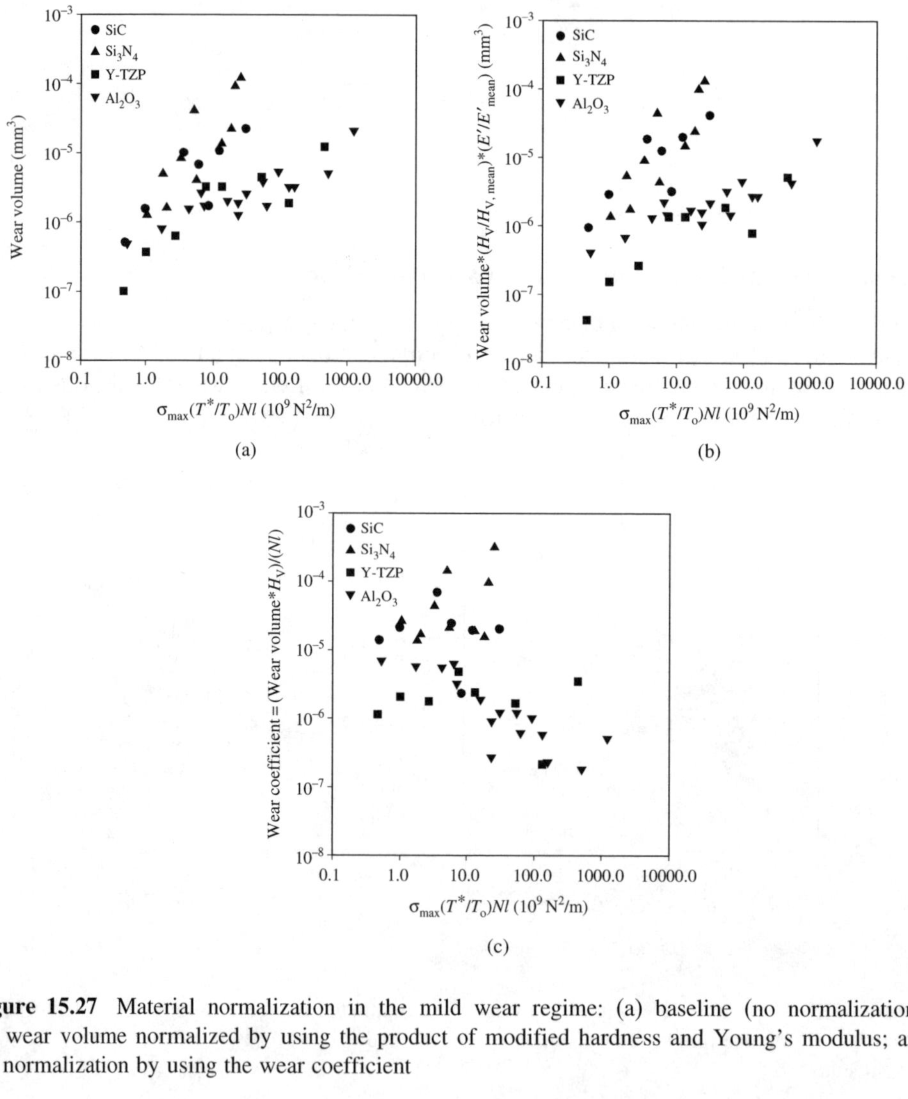

Figure 15.27 Material normalization in the mild wear regime: (a) baseline (no normalization); (b) wear volume normalized by using the product of modified hardness and Young's modulus; and (c) normalization by using the wear coefficient

determine the average value and then all materials will be normalized with respect to the average. The latter means was adopted in the current work.

Figure 15.27(a) shows the mild wear data separated from the whole group, and Figure 15.27(b) shows the data being normalized against average H_V and average E'. The average H_V for the four materials was 21 ± 8 GPa and the average E' was 176 ± 44 GPa. Because wear in this regime, equation (8), was inversely proportional to these material properties, material normalization was carried out by multiplication of the wear volume by the ratios E'/E'_{mean} and $H_V/H_{V,mean}$. For example, the wear volume data of SiC were multiplied by 1.48 and 1.26, since its hardness and Young's modulus were both higher than the mean values. Figure 15.27(b) displays that after material normalization, the Si-based ceramics were separated from the oxides. Because the functional dependence on the severity parameter was different between these two groups of data, different wear mechanism was controlling wear. One possible interpretation was that this could be caused by the tribochemical reactions, which were not included in the mechanism-based modeling. Figure 15.27(c) presents a conventional material normalization scheme using the wear coefficient, defined as wear volume $\times H_V/(N \times l)$. The results suggested that it was not applicable in this database.

Figure 15.28 shows the material normalization carried out for data in the severe wear regime. According to equation (10), wear volume was inversely proportional to the product of $H_V(T^*) \times K_{1C}/\sqrt{d_{50}}$. The average fracture toughness for the four materials was 5.4 ± 2.3 MPa$\sqrt{}$m and the average d_{50} was $2.4 \pm 2\,\mu$m. Figure 15.28(b) shows that the wear data among the two Si-based ceramics and Y-TZP appeared to be more packed. But the alumina data fell further away from the pack. There were two possible interpretations to such features. One was that the tribochemical reactions could have been extended into this regime among the fine wear particles. So the third-body wear has a significant effect on wear in alumina. The other could have been related to the data classification, i.e. the alumina data that fell away from the pack may need to be classified into mild wear regime. The material

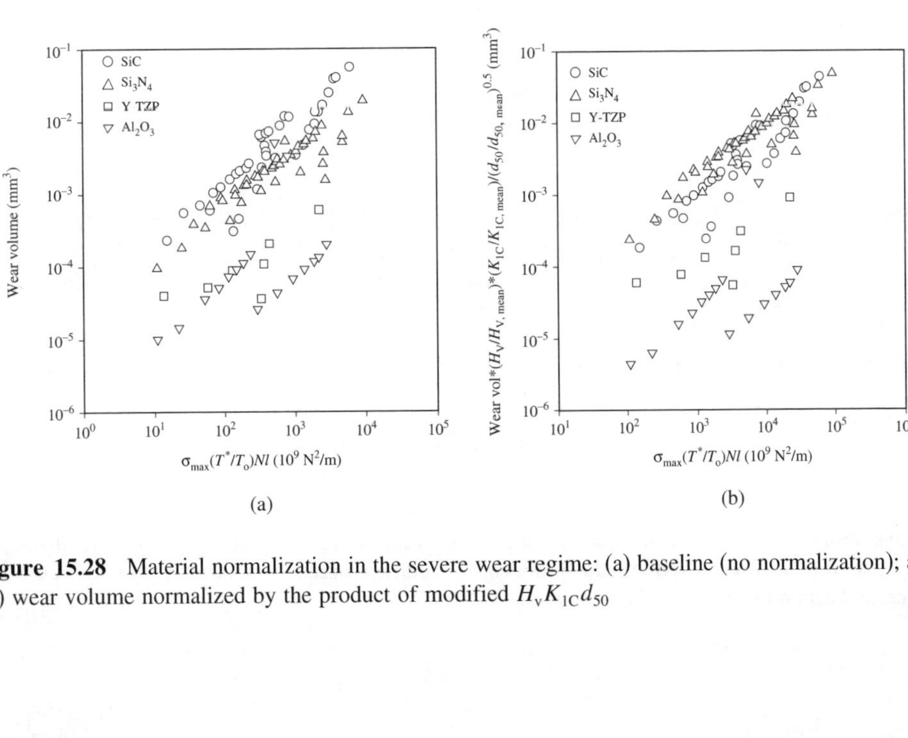

Figure 15.28 Material normalization in the severe wear regime: (a) baseline (no normalization); and (b) wear volume normalized by the product of modified $H_V K_{1C} d_{50}$

normalization used in Evans and Marshall [11] that correlated excellently with the grinding force data among different ceramics was also tested, i.e. $(K_{1C}^{0.625} \times H_v^{0.5})^{(8/9)}$. But only slight shifts were observed.

Figure 15.29 shows the material normalization carried out for the ultra-severe wear data. The material normalization included the product of hardness and fracture toughness. The improvement from the material normalization appeared to be only slight. Figure 15.29(c)

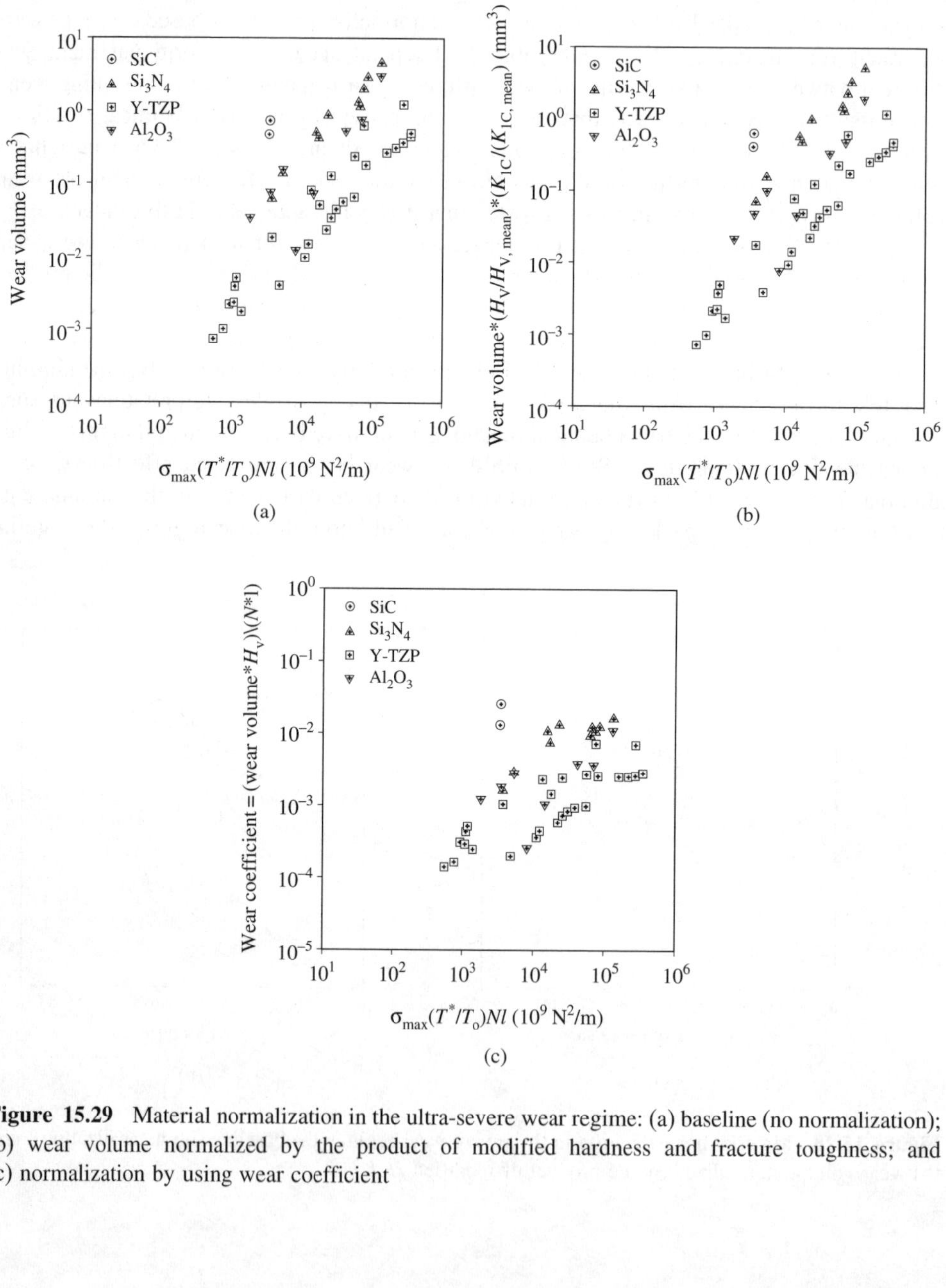

Figure 15.29 Material normalization in the ultra-severe wear regime: (a) baseline (no normalization); (b) wear volume normalized by the product of modified hardness and fracture toughness; and (c) normalization by using wear coefficient

shows the use of wear coefficient to normalize the materials. The results did not confirm that it was applicable in this fast wearing regime, either.

The material normalization schemes in the preceding text showed that different degrees of data shifts were obtained in the different wear regimes. Clearly, the process of selection of severity parameter and material normalization would be an iterative one, due to the simplification employed in the determination of severity parameters. Nevertheless, the process appeared to be in the correct direction. The overall data scatter after the material normalization was about ± one order of magnitude.

15.7.6.4 Summary of Correlational Modeling

The large wear database used to construct ceramic wear maps presents a unique opportunity to test correlational wear parameters. The results showed that grouping of data among different ceramics was achieved through using the first sets of severity parameter and material normalization parameters. This was quite promising, in light of the huge variations present in the experimental conditions and materials. Further refinements through iterations would be necessary to reveal or include other underlying factors that were not considered in the severity parameter as well as material normalization parameters. The contact severity concept proves to be more significant. The results from the material property normalization process suggest either there may be hidden material property parameters important to wear or there needs to be a new and novel normalization procedure.

15.8 Construction Techniques of Wear Maps

One of the common misconceptions about wear mapping is that a large number of wear experiments need to be conducted in order to construct such maps in a reasonably comprehensive way. The following sections describe the use of a step-loading technique to obtain wear data quickly over a range of loads. Properly used, a wear map can be constructed using 5–10 wear experiments. However, there are risks in using step-loading technique if the underlying assumptions are not obeyed.

15.8.1 Conducting Wear Experiments

Wear experiments are conducted on a four-ball wear tester using a ball-on-three-flats contact geometry. The ball and the flats used are made of the same ceramic material. The diameter of the ball is 12.7 mm. The flats are circular discs of 6.35 mm in diameter with a thickness of 1.59 mm. The final polish is done by using 1-μm diamond paste. Prior to testing, the ball and the flat specimens are cleaned in an ultrasonic bath by using successive solvents of hexane and acetone followed by a detergent wash in water. Afterwards, the specimens are rinsed with de-ionized water, then blown dry with dry nitrogen.

The test procedure used is a step-loading procedure at several fixed sliding speeds. At each sliding speed, the applied load increases in a stepwise manner. For loads lower than 20 N, the load steps follow a sequence of 2, 4, 8, 12, and 20 N. Beyond 20 N, the load increment of each step is 20 N. The maximum load is 360 N. The duration of each load step is 5 min. Wear tests sometimes are terminated before 360 N is reached due to extreme wear, such as in the dry sliding case. The sliding speeds are 1.9, 14.4, 38, 190, 380, and 570 mm/s. Additional speeds have also been used in some cases. In lubricated tests, an amount of 1.5-mL fresh lubricant is used in each load step. The worn surfaces are not washed before examination.

All tests are conducted at room temperature with the room air at a relative humidity of 50–55%. For dry air conditions, dry cylinder air is circulated through the wear tester at a rate of 0.76 L/min. The wear scar diameters on the three flat specimens are measured after each load step. For lubricated cases, PPO (4 CSt viscosity oil percolated through an activated alumina column prior to wear tests) is used as the nonreactive lubricant for ceramics. (Separate studies had been done to demonstrate that paraffinic oil did not chemically react with ceramics to form chemically active boundary lubricating films.) Water is used as the reactive fluid. Constant condition test had also been conducted to cross-check the step-loading procedure. Similar wear results were obtained.

The precision of the wear measurement is generally within ±10%. When the data are plotted in a three-dimensional plot, some data smoothing occurs. The data smoothing tends to reduce the uncertainty and throw out the outliers.

15.8.2 Wear Data

The raw data collected are the measurements of wear scar diameters from the three flat specimens. The diameters of the three wear scars are measured by an optical microscope. The measurements have an accuracy of 8 μm. Two diameters perpendicular to each other are measured in each wear scar. The average of the total of six measurements is taken to represent the wear scar diameter under the particular load. Figure 15.30 shows the wear

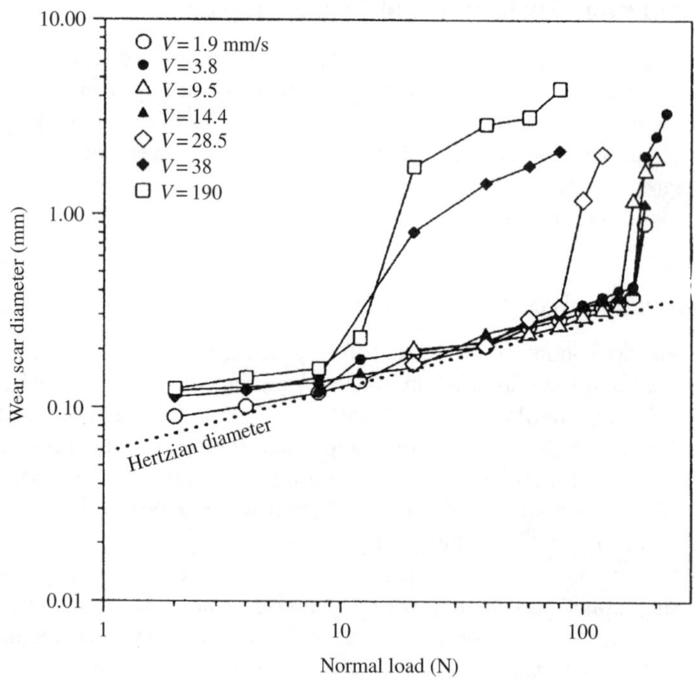

Figure 15.30 Measured wear diameter of alumina under dry sliding conditions

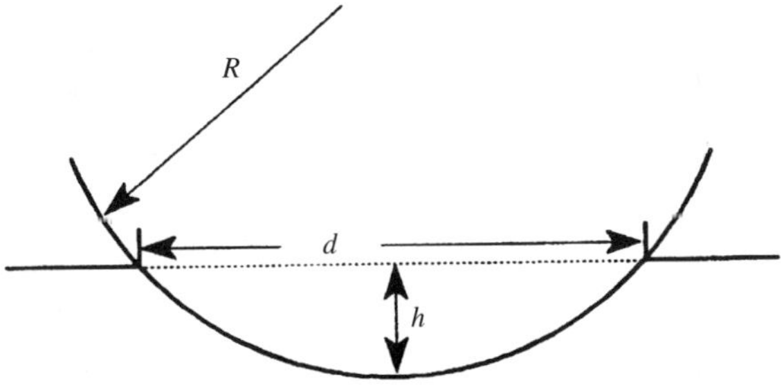

Figure 15.31 Terms used in calculation of wear volume from measured wear scar diameter (d); h, depth of wear scar; R, top ball radius

scar diameters measured in the case of alumina under dry sliding condition. The Hertzian contact diameters are also included for comparison, as shown by the dash line. The Hertzian diameters are calculated as follows [50]:

$$a = \left(\frac{3\,NR^*}{4\,E^*}\right)^{\frac{1}{3}}, \quad \text{with } \frac{1}{R^*} = \frac{1}{R_1} + \frac{1}{R_2} \text{ and } \frac{1}{E^*} = \frac{(1-\nu_1^2)}{E_1} + \frac{(1-\nu_2^2)}{E_2} \tag{15}$$

where N is the normal load and R_1, R_2 are the radii of curvature, E_1, E_2 are the Young's moduli, and υ_1, υ_2 are the Poisson's ratios of the two contacting bodies.

On the basis of the measured wear scar diameters, the wear volume is calculated by a geometrical consideration, as illustrated in Figure 15.31. The worn volume is assumed to be a spherical segment. The radius of the sphere equals to the radius of the top ball. Such assumption is reasonable when (a) the top ball has negligible wear such that no change in its radius takes place; and (b) elastic deformation does not change the shape of the wear scar or the curvature of the top ball. The wear volume can then be calculated by the following equation:

$$V = \frac{\pi h}{6}\left[\frac{3}{4}d^2 + h^2\right], \quad \text{with } h = \frac{D}{2} - \frac{1}{2}[D^2 - d^2]^{1/2} \tag{16}$$

where V is the wear volume, D is the diameter of the top ball, and d is the wear scar diameter.

15.8.3 Data Trend Analysis

As shown in Figure 15.30, the tests at relatively high speeds were terminated prematurely at normal loads lower than 100 N because of high wear. So the data set contains unevenly distributed data points. This makes the construction of three-dimensional map difficult. Therefore, some extrapolations are needed to obtain an evenly distributed data set. These are carried out by trend analysis of the entire data set. An example is shown in Figure 15.32 for the case of alumina dry sliding data. Extrapolation of data is done with extreme caution and at an absolute minimum. In case of doubt, higher wear values are used to provide a conservative estimate. It is only done to facilitate plotting up the operating limits where data under other

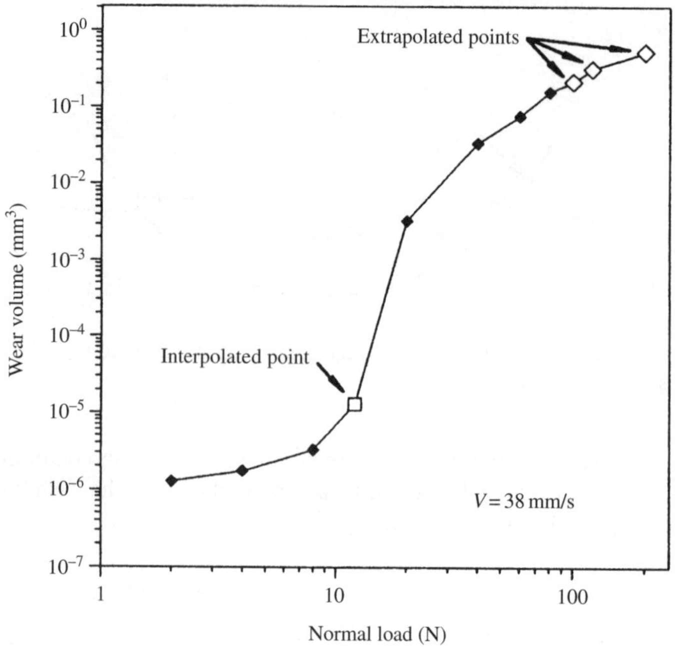

Figure 15.32 Illustration of adding data points through extrapolations and interpolations to prepare the data set for construction of a 3D map. The case shown is alumina under dry sliding

conditions exist. Interpolation of data is done to avoid automatic interpretation by the three-dimensional plotting software creating valleys and peaks unsubstantiated by data. This is illustrated in Figure 15.32. Subsequently, the adjusted data set will be plotted, as shown in Figure 15.33. Similarly, the wear data can be plotted by using wear volume and sliding speed, as shown in Figure 15.34. This plot is simplified to contain only the data of three loads.

15.8.4 Wear Mapping

15.8.4.1 Wear Transition Diagram

The locations of both load- and speed-dependent wear transitions are used to construct the wear transition diagram, as illustrated in Figure 15.35. First, all wear data points are plotted as functions of speed and load. Then the locations of load-dependent as well as speed-dependent wear transitions are identified by experimental observation and data trend analysis. Subsequently, a curve connecting all the transition points is traced, representing the wear transition boundary.

15.8.4.2 Three-Dimensional Wear Map

Because of the test procedure, the wear data set generally contains more normal load data than sliding speed data. In order to facilitate the construction of three-dimensional map by a computer software, additional interpolations, especially along sliding speed axis, are made.

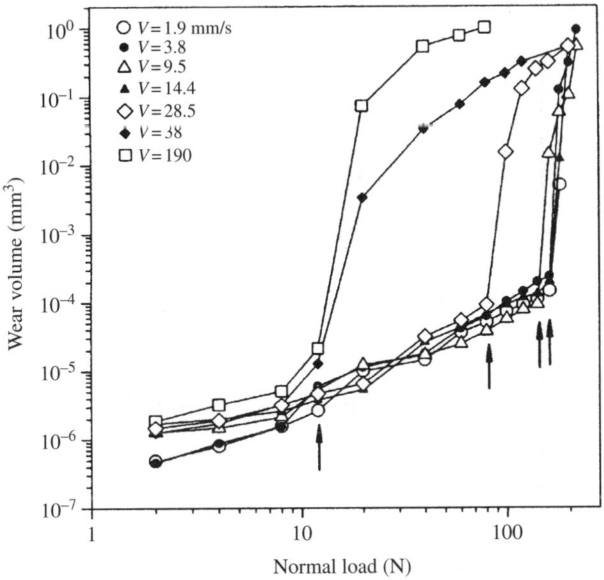

Figure 15.33 Wear volume data of alumina under dry sliding conditions plotted as a function of normal load (arrows indicate the on-set of load-dependent wear transitions)

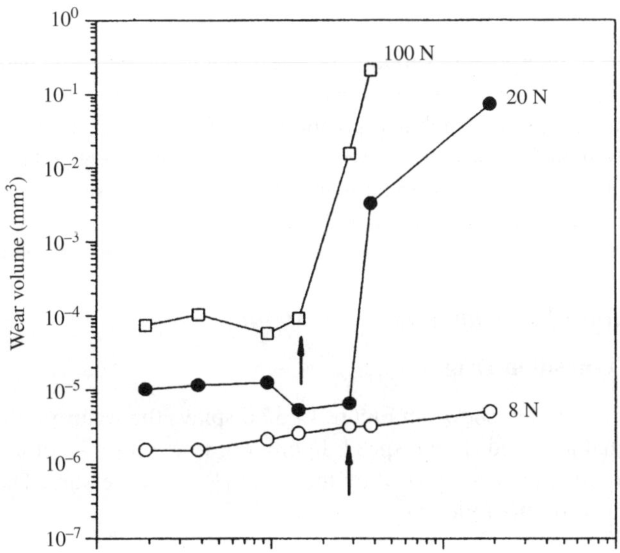

Figure 15.34 Wear volume data of selected normal loads taken from alumina under sliding conditions and plotted as a function of sliding speed (arrows indicate the on-set of speed-dependent wear transitions)

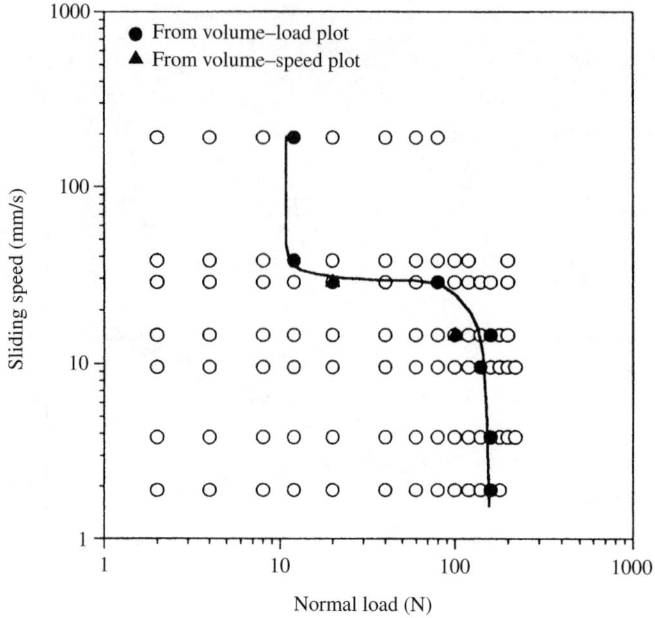

Figure 15.35 Mild to severe wear transition boundary of alumina under dry sliding conditions constructed using load- and speed-dependent wear transition points

This is done by linear interpolations in log–log scale. For example, more data points of 8 and 20 N shown in Figure 15.34 are added between speeds of 40 and 200 mm/s along the lines connecting the adjacent points. Such interpolations can also be made to the normal load data. Sometimes, a few iterations will be needed to determine the number of interpolated data required to construct a three-dimensional map. The wear maps shown in Figures 15.7–15.9 are all constructed with such interpolations.

15.8.5 Selection of Parameters for Mapping

15.8.5.1 Wear Transition Diagram

The wear transition diagram shown in Figure 15.35 displays the wear transition boundary as a function of normal load and sliding speed. In order to represent this information in a more usable form, the normal load is replaced by the mean Hertzian pressure. The mean Hertzian pressure is calculated by the following equation:

$$P_{\text{mean}}^{\text{Hz}} = \frac{N}{\pi a^2} \tag{17}$$

where N is normal load and a is the Hertzian diameter in equation (15). Once the mean Hertzian pressure is used, comparison among different materials can be easily made, since the elasticity constants are taken into account.

15.8.5.2 Three-Dimensional Wear Map

The wear maps shown in the main text are all constructed by using wear rate, a time rate at 5 min, normal load and sliding speed. Alternatively, wear coefficient can also be used. The wear coefficient is defined as follows:

$$\frac{V}{l} = k\frac{N}{H} \tag{18}$$

where k is wear coefficient, V is wear volume, l is distance slid, N is normal load, and H is hardness. Figure 15.36 shows the wear coefficient map of the alumina under dry sliding condition. This map shows different wear trends in terms of topographical features as compared to the wear rate map in Figure 7.15(a). The reason for this is the different normalization effects. The wear coefficient term itself contains the distance slid and the load term. So the wear volume is divided by the distance slid. Since in our experiments wear is measured at a fixed time of 5 min at different speeds, the net result of dividing the wear volume by distance slid is to skew the data with respect to speed.

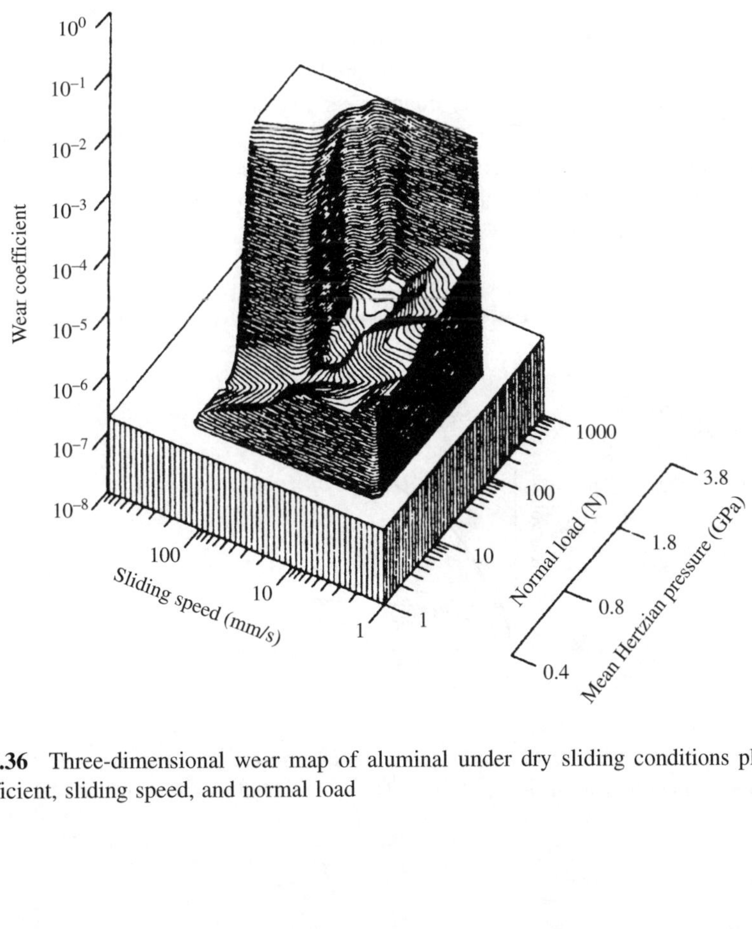

Figure 15.36 Three-dimensional wear map of aluminal under dry sliding conditions plotted using wear coefficient, sliding speed, and normal load

15.8.6 Assumptions in the Step-Loading Test Procedure

Before the assumptions in the step-loading test procedure are discussed, let us examine some characteristic wear behaviors obtained in the ball-on-three-flats contact geometry. Figure 15.37 shows typical wear results under paraffin oil-lubricated condition by constant loading tests. The tests are conducted under fixed loads, and wear measurements are made at different time intervals. At relatively low loads, e.g. L_1 and L_2, wear increases with time once the test starts, then it reaches a certain level and stays nearly unchanged. The asymptotic levels are a function of the normal load and the sliding speed. Such wear behaviors are presumably caused by two effects. One is that when wear occurs, the contact area increases and the contact pressure drops due to the ball-on-three-flats contact geometry. This would lead to a decrease in wear rate. The other is that the predominant wear mechanism in these cases is mostly micro-abrasion. As wear continues, a certain degree of conformity between the ball and the worn flat specimen can be reached. This will decrease the mean contact pressure further. So by prolonging the testing time from t_1 to t_2 and calculating the wear rate or wear coefficient, a smaller wear value will be obtained. When the steady-state wear level under the particular load and speed combination is sought, the test duration needs to be properly selected. If the test duration is too short, only the running-in wear will be measured. When the test extends well into the asymptotic region, the contact geometry effect will dominate the wear results. At higher load such as L_3 and L_4, the wear–time functions no longer have the asymptotic feature. Instead, wear continues to increase with sliding time. Because of fast wearing, the contact area will increase in a much faster rate than in the lower load cases. Thus contact pressure decreases more rapidly. But they do not seem to render an asymptotic wear level. This is because the predominant wear mechanisms are changed. Brittle fracture is generally the predominant wear mechanism in these cases. Large wear particles will be generated and the effects of third-body wear become more influential in the wearing events. Meanwhile, conformity between the two contacting bodies will be difficult to reach [55]. When a material wears in this fashion, the steady-state wear, if existing, will be very difficult to determine.

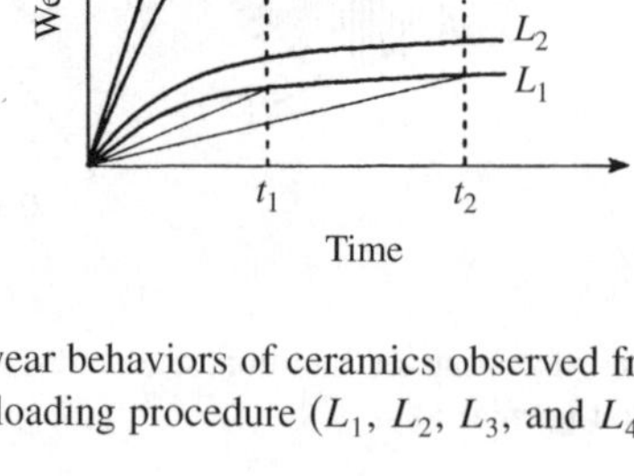

Figure 15.37 Characteristic wear behaviors of ceramics observed from tests using ball-on-three-flats contact geometry and constant loading procedure (L_1, L_2, L_3, and L_4 are normal loads)

Because of the wear characteristics discussed previously, different test durations have been explored to represent the steady-state wear of a particular load and speed combination in the development of the step-loading test procedure. A 5-min test duration appeared to be a reasonable approximation across the load and speed ranges employed. Another key factor in the step-loading test procedure is the increment of load in the load steps. Sufficiently large increment of load is necessary to minimize the wear history of the previous load steps. Figure 15.38 shows the initial mean contact pressures of the load steps in alumina under paraffin oil-lubricated condition. The initial mean contact pressure is calculated by the following equation:

$$P_{\text{mean}} = \frac{N}{\frac{1}{4}\pi\,\text{WSD}^2_{\text{prev}}} \tag{19}$$

where N is the normal load of the current load step and WSD_{prev} is the wear scar diameter of the previous load step. The results shown suggest that the load increment of 20 N is sufficient when in the mild wear regime because the initial mean pressures are comparable to the mean Hertzian pressures under the particular loads. Once wear transition occurs, the initial contact pressure decreases drastically. When this happens, wear history of the previous load steps will be included in the current load step, i.e. the step-loading results will become less reliable. This phenomenon generally occurs in the ultra-severe wear regimes.

Owing to the wear characteristics and the assumptions of step-loading test procedure, wear results in this work are represented by using the time rates at 5 min. This will avoid the decreasing trend of wear with increasing sliding speed in the mild wear regime. Once wear

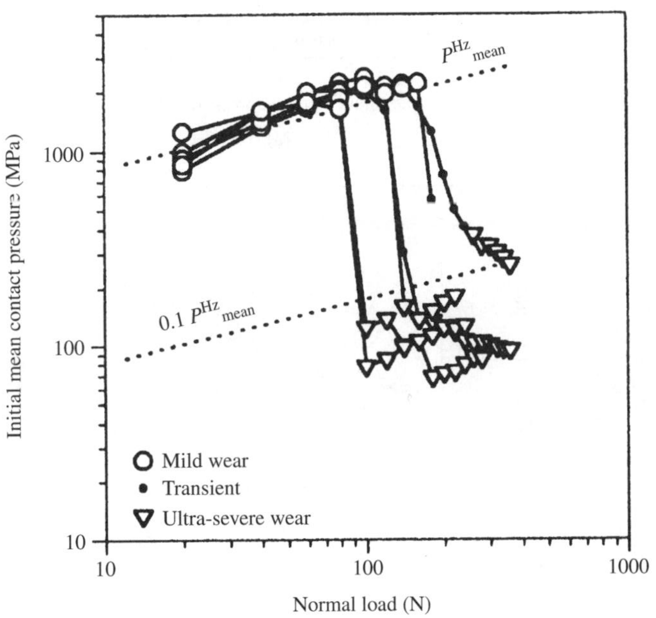

Figure 15.38 Initial mean contact pressures in the load steps of step-loading wear tests of alumina under paraffin oil-lubricated conditions

enters into the severe or ultra-severe wear regime, wear result representations by either time rate or wear coefficient do not seem to differ significantly.

15.9 Application Map Concept and Examples

Many applications of advanced materials require lubrication for lowering frictional loss and moderating material removal due to wear. The concepts of wear maps discussed previously can be readily extended to include the use of friction as an additional parameter in mapping the application system of interest. In the following text, we will use an example to illustrate the construction of application maps for the silicon nitride described previously.

Given the wear map of the silicon nitride under PPO-lubricated condition shown in Figure 15.8(b), one can readily convert the wear volume data to a commonly used parameter, such as wear coefficient, K, in equation (2). One can also establish a friction coefficient map from the measured steady-state friction data. Figure 15.39(a) displays the regimes where the

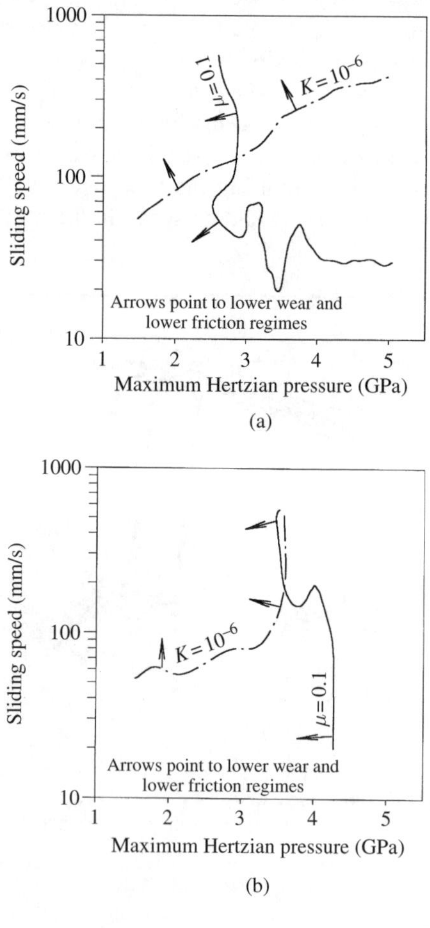

Figure 15.39 Application limits for silicon nitride, as defined in the text, under (a) pure paraffin oil; and (b) paraffin oil + 1% PPG lubricated conditions

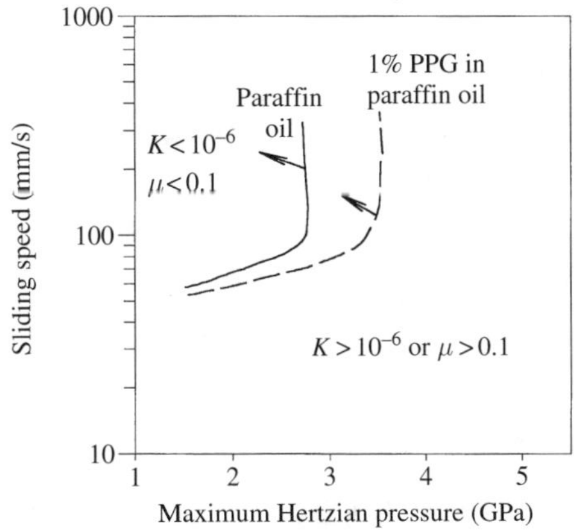

Figure 15.40 Application maps of silicon nitride under paraffin oil-lubricated conditions with and with out additive (1% PPG)

criteria of either $K < 10^{-6}$ or $\mu < 0.1$ can be met with this silicon nitride when it is lubricated by pure PPO. Clearly, when an application requires both criteria to be met simultaneously, the lubrication provided by pure PPO is acceptable only in the relatively low load (<2.5 GPa) and relatively high speed (>90 mm/s) regime. Similarly, the regimes where the same criteria can be met by the use of 1% PPG added to the PPO can be established as shown in Figure 15.39(b). The presence of 1% PPG gives rise to noticeable improvement in lowering the friction; however, the low-wear regime seems to dwindle a little bit. Figure 15.40 shows the combined applicable regimes, from both wear coefficient and friction coefficient, for this silicon nitride lubricated with PPO and PPO + 1% PPG. This application map displays clearly the improvement that can be expected from using the 1% PPG. When required, one can continue with other candidate additives or additive packages to further establish the desirable regimes that fit with specific applications.

15.10 Future Wear Map Research

Wear map research was originally developed in an attempt to provide a wear classification system for ceramics as well as to provide a sufficient database for the development of wear models. So far the first goal has been met. The progress in wear model development is ongoing. A single universal parameter to fit all the data has been demonstrated to be unrealistic, but one may ask if such parameter(s) exists in each wear mechanism zone. This question remains unanswered at this time. The current database is limited to a narrow speed and load range because of the need to simultaneously measure wear and lubricant chemistry effects. To extend the speed and load ranges will necessitate the use of several wear testing machines. How this will affect the internal consistency needs to be investigated.

In summary, wear mapping represents a new approach to examining the age old issue of wear and it presents an integrated view of wear that cannot be seen otherwise. With the understanding of wear mechanisms and development of wear models, both guided by the wear maps, the roadmap to establish wear as an intrinsic material property is clearer. There is much more work needed to fully develop this approach for practical applications.

References

1. Fein, R.S., 'AWN – A Proposed Quantitative Measure of Wear Protection', *Lubrication Engineering*, **31**, 1975, 581–582.
2. Ying, T.-N., Wear Mechanisms for Ductile and Brittle Materials in a Micro-Contact, Ph.D. dissertation, University of Maryland, College Park, MD, 1996.
3. Ling, F.F. and Pan, C.H.T., '*Approaches to Modeling of Friction and Wear*', Springer-Verlag, New York, NY, 1988.
4. Ludema, K.C. and Bayer, R.G., 'Cultural Impediments for Practical Modeling of Wear Rates', in *Tribological Modeling for Mechanical Designers*, ASTM STP 1105, American Society for Testing and Materials, Philadelphia, PA, 1991, p. 180.
5. Meng, H.C. and Ludema, K.C., 'Wear Models and Predictive Equations: Their Form and Content', *Wear*, **181–183**, 1995, 443–457.
6. Rabinowicz, E., '*Friction and Wear of Materials*', John Wiley and Sons, New York, 1965, p. 165.
7. Hokkirigawa, K. and Kato, K., 'Theoretical Estimation of Abrasive Wear Resistance Based on Microscopic Wear Mechanism', *ASME Wear of Materials*, **1**, 1989, 1–8.
8. Larsen-Basse, J., 'Success and Failure of Simple Models for Abrasive Wear', in *Tribological Modeling for Mechanical Designers* (eds K.C. Ludema and R.G. Bayer), ASTM STP 1105, American Society for Testing and Materials, Philadelphia, 1991, p. 51.
9. Komvopoulos, K., Suh, N.p. and Saka, N., 'Wear of Boundary-Lubricated Metal Surfaces', *Wear*, **107**, 1986, 107–119.
10. Wang, Y.S. and Hsu, S.M., 'Wear and Wear Transition Modeling of Ceramics', *Wear*, **195**, 1996, 35–46.
11. Evans, A.G. and Marshall, D.B., 'Wear Mechanisms in Ceramics', in *Fundamentals of Friction and Wear of Materials*, American Society of Metals, Metals Park, OH, 1980, p. 439.
12. Kim, S.-S., Kim, S.-W. and Hsu, S.M., 'A New Parameter for Assessment of Ceramic Wear', *Wear*, **179**, 1984, 69–73.
13. Kim, S.S., Kato, K., Hokkirigawa, K. and Abe, H., 'Wear Mechanism of Ceramic Materials in Dry Rolling Friction', *ASME, Journal of Tribology*, **108**, 1986, 522–526.
14. Adachi, K., Kato, K. and Chen, N., 'Wear Map of Ceramics', *Wear*, **203–204**, 1997, 291.
15. Sibley, L.B. and Allen, C.M., 'Friction and Wear Behavior of Refractory Materials at High Sliding Velocities and Temperatures', *Wear*, **5**, 1962, 312–316.
16. Ting, B.Y. and Winer, W.O., 'Friction-Induced Thermal Influences in Elastic Contact Between Spherical Asperities', *ASME, Journal of Tribology*, **111**, 1989, 315–320.
17. Yang, J. and Winer, W.O., 'A Comparison Between the Thermomechanical Wear Theory and Some Experimental Observations', *ASME, Journal of Tribology*, **113**, 1991, 262–267.
18. Dufrane, K.F., 'Sliding Performance of Ceramics for Advanced Heat Engines', in *Ceramic Engineering and Science Proceedings*, **7**, 1986, 1052–1058.
19. Kong, H.S. and Ashby, M.F., 'Wear Mechanisms in Brittle Solids', *Acta Metallurgica Materialia*, **40**, 1992, 2907–2920.
20. Bayer, R.G., 'Comments on Engineering Needs and Wear Models', in *Tribological Modeling for Mechanical Designers* (eds K.C. Ludema and R.G. Bayer), ASTM STP 1105, American Society for Testing and Materials, Philadelphia, PA, 1991, p. 3.
21. Godet, M., 'Modeling of Friction and Wear Phenomena', in *Approaches to Modeling of Friction and Wear* (eds F.F. Ling and C.H.T Pan), Springer-Verlag, New York, NY, 1988, p. 12.
22. Lim, S.C. and Ashby, M.F., 'Wear-Mechanism Maps, *Acta Metallurgica*, **35**(1), 1987, 1–15.
23. Hisakado, T., 'Wear Mechanism of Ceramics and Surface Topography', *Journal of Tribology*, **108**, 1986, 9–15.
24. Beerbower, A., Boundary Lubrication, US Army, Office of the Chief of Research and Development, Contract No., DAHC19-69-C-0033, 1972.
25. deGee, A.W.J., 'Wear Research for Industry – Examples of Application of the IRG Transition Diagram Technique', *ASME Wear of Materials*, 1989, 753–763.
26. Hsu, S.M., Shen, M.C., Ying, T.N., Wang, Y.S. and Lee, S.W., 'Tribology of Si-Based Ceramics', *Ceramics Transactions*, **42**, 1994, 189–205.
27. He, C., Wang, Y.S., Wallace, J.S. and Hsu, S.M., 'The Effect of Microstructure on the Wear Transition of Zirconia Toughened Alumina, *Wear*, **162–164**, 1993, 314–321.
28. Hsu, S.M., Nagarajan, V.S., Liu, H.Y. and He, C., 'Microstructural Design of Ceramics for Optimum Wear Resistance', *Proceedings of the International Symposium on Advanced Ceramics for Structural and*

Tribological Applications (eds H.M. Hawthorne and T. Troczynski), Canadian Institute of Mining, Metallurgy, and Petroleum, Canada, 1995.

29. He, C., Microstructural Effects on Wear of ZrO_2–Al_2O_3 Composites, Ph.D. dissertation, University of Maryland, College Park, MD, 1995.
30. Zutshi, A., Haber, R.A., Niesz, D.E., Adams, J.W., Wachtman, J.B., Ferber, M.K. and Hsu, S.M., 'Processing, Microstructure, and Wear Behavior of Silicon Nitride Hot-Pressed with Alumina and Yttria', *Journal of the American Ceramic Society*, **77**(4), 1994, 883–890.
31. Liu, H. and Hsu, S.M., 'Modeling of Micro-Fracture-Induced Wear and Wear Transition of Polycrystalline Alumina Under Sliding', *Wear*, **195**, 1996, 169–177.
32. Wang, Y.S. and Hsu, S.M., 'The Effects of Operating Parameters and Environment on the Wear and Wear Transition of Alumina', *Wear*, **195**, 1996, 90–99.
33. Deckman, D.E., Jahanmir, S. and Hsu, S.M., 'Wear Mechanisms of α-Alumina Lubricated with a Paraffin Oil', *Wear*, **149**, 1991, 155–168.
34. Gates, R.S. and Hsu, S.M., 'Effects of Selected Chemical Compounds on the Lubrication of Silicon Nitride', *Tribology Transactions*, **34**(3), 1991, 417–425.
35. Deckman, D.E., Chen, C.I. and Hsu, S.M., 'Effects of Selected Chemical Compounds on the Lubrication of Silicon Carbide', *Tribology Transactions*, **42**(3), 1999, 619–625.
36. Cho, S.J., Moon, H., Hockey, B.J. and Hsu, S.M., 'Wear Transition Mechanism in Alumina During Sliding', *Acta Metallurgica Materialia*, **40**(1), 1992, 185–192.
37. Gates, R.S., Klaus, E.E. and Hsu, S.M., 'Tribochemical Mechanism of Alumina with Water', *Tribology Transactions*, **32**(3), 1989, 357–363.
38. Michalske, T.A., Bunker, B.C. and Freiman, S.W., 'Stress Corrosion of Ionic and Mixed Ionic/Covalent Solids', *Journal of the American Ceramic Society*, **69**(10), 1986, 721–724.
39. Mizuhara, K. and Hsu, S.M., 'Tribochemical Reaction of Oxygen and Water on Silicon Surfaces', in *Wear Particles* (eds D. Dowson *et al.*), Elsevier Science Publishers B.V., Amsterdam, 1992, p. 323.
40. Tomizawa, H. and Fischer, T.E., 'Friction and Wear of Silicon Nitride and Silicon Carbide in Water', *ASLE Transactions*, **30**(1), 1987, 41–52.
41. Sasaki, S., 'The Effects of Surrounding Atmosphere on the Friction and Wear of Alumina, Zirconia, Silicon Carbide, and Silicon Nitride', *Wear of Materials*, **2**, 1989, 409–417.
42. Wang, Y.S., He, C., Hockey, B.J., Lacey, P.I. and Hsu, S.M., 'Wear Transitions in Monolithic Alumina and Zirconia–Alumina Composites', *Wear*, **181–183**, 1995, 156–164.
43. Hsu, S.M., Lacey, P.I., Wang, Y.S. and Lee, S.W., 'Wear Mechanism Maps of Ceramics', *Advances in Engineering Tribology* (eds Y.W. Chung and H.S. Cheng), STLE SP-31, Society of Tribologists and Lubrication Engineers, Chicago, 1991, pp. 123–132.
44. Hsu, S.M., 'Boundary Lubrication of Advanced Materials', *MRS Bulletin*, **16**(10), 1991, 54–58.
45. Hsu, S.M., Klaus, E.E. and Cheng, H.S., 'A Mechano-Chemical Descriptive Model for Wear Under Mixed Lubrication Conditions, *Wear*, **128**, 1988, 307–323.
46. Ying, T.-N., Shen, M.C., Wang, Y.S. and Hsu, S.M., 'Tribology of Si-Based Ceramics – Wear Mechanisms', *Tribology Transactions*, **40**, 1997, 685–693.
47. Shen, M.C. and Hsu, S.M., 'Wear Modeling of Si-Based Ceramics', in Proceedings of International Tribology Conference, Yokohama, Japan, 1996, pp. 403–408.
48. Kong, H.S. and Ashby, M.F., 'Friction-Heating Maps and Their Applications', *MRS Bulletin*, **16**(10), 1991, 41–48.
49. Cho, S.J., Hockey, B.J., Lawn, B.B. and Bennison, S.J., 'Grain Size and R-Curve Effects in the Abrasive Wear of Alumina', *Journal of the American Ceramic Society*, **72**, 1989, 1249–1252.
50. Hamilton, G.M., 'Explicit Equations for the Stresses Beneath a Sliding Spherical Contact', Proceedings of the Institution of Mechanical Engineers, 197C:53, 1983.
51. Lee, S.C. and Cheng, H.S., 'On the Relation of Load to Average Gap in the Contact Between Surfaces with Longitudinal Roughness', *Tribological Transactions*, **35**, 1992, 523–532.
52. Greenwood, J.A. and Williamson, J.B.P., 'Contact of Nominally Flat Surfaces', *Proceedings of the Royal Society A*, **295**, 1966, 300–310.
53. Archard, J.F., 'The Temperature of Rubbing Surfaces', *Wear*, **2**, 1958–1959, 438–446.
54. He, M.Y. and Hutchinson, J.W., 'Kinking of a Crack out of an Interface', *Journal of Applied Mechanics*, **56**, 1991, 270–278.
55. Wang, F.X., Lacey, P.I., Gates, R.S. and Hsu, S.M., 'Study of Relative Surface Conformity Between Two Surfaces in Sliding Contact', *ASME Journal of Tribology*, **113**, 1991, 755–762.

16

Machine Failure and Its Avoidance – Tribology's Contribution to Effective Maintenance of Critical Machinery

B.J. Roylance

Abstract

Following a brief historical overview of the principal maintenance strategies, the features that govern tribological phenomena are reviewed in terms of their relevance to the upkeep and care of operating machinery. Examination of the key factors involved leads to an assessment of what contribution tribology makes to the implementation of good maintenance practice. Some practical case studies are reported as a means of illustrating the various ways that application of tribological knowledge and expertise assist in the process of maintaining critical operating machinery.

16.1 Introduction

It has been stated that, 'The occurrence of a failure, without loss of life, is not so much a disaster, as the ultimate result of a compromise between perfection and economics' [1]. The operation of machinery that involves components transmitting power between sliding and/or rolling surfaces depends crucially on exploiting the fruits of tribological research and applying it to whole life design philosophy. This implies safe, efficient and reliable operation that is commensurate with diligent upkeep and care throughout a machine's working life.

Ideally, the goals that are sought to attain in the field of tribology are principally, first, to fully comprehend the principal modes of behaviour, and second, to define the underpinning

Wear – Materials, Mechanisms and Practice Edited by G. Stachowiak

mechanisms. By building upon these basic foundations, appropriate functional and predictive performance models ensue, leading to good design practice and application. But in a multi-disciplinary subject that draws heavily upon the combined knowledge and skills (including good communications) of the chemist, the material specialist and the mechanical engineer, it is always likely to fall short of ultimate perfection. Hence, we are forced to acknowledge the possibility that machines will experience tribological distress and failure. This is further exacerbated by the fact that it occurs within the demanding environment of a highly competitive industrialized and commercial world with the associated requirements to achieve better performance and profitable economic returns. It is an environment that also encompasses the fallibility of the people who work within it, the outcome being that the overall economic implications can ultimately result in 'failure' on a much grander and more costly scale.

The guiding principles governing the friction, lubrication and wear of critical machine components are examined here in the context of the need to try and ensure that critically operating machinery fulfil their intended function, and continue to do so for the duration of their expected lifetime. In essence, it requires a 'cradle-to-grave' philosophy in all its aspects.

Tribology's contribution to maintenance strategies and procedures must be examined in relation to the essential needs of detecting and diagnosing faults *before* catastrophic failure occurs. Consequently, it gives rise to the question as to what tribology can contribute to establishing appropriate prognostic techniques for determining the remaining useful life of failing machinery. Failure in this context can mean that which relates simply to deterioration not only in terms of a machine's functionality, but also in terms of a reduction in product quality to a level that is unacceptable to the customer.

16.2 Maintenance Practice and Tribological Principles

16.2.1 Maintenance Practice – A Brief Historical Overview

Prior to the Second World War (1939–1945), the practice of ensuring continued operation of machinery that depended largely on employing the human senses – eyes, ears, smell, touch, and even taste – to detect changes in performance that could mean imminent failure if nothing was done. In general though, maintenance was performed on the breakdown principle – 'If it isn't broken, don't fix it!' This approach to pursuing 'good' maintenance practice persisted into the latter part of the twentieth century. In the immediate post-war era, the practice of performing preventive maintenance began to evolve whereby equipment is subjected to regular inspection (conducted typically on an annual, or semi-annual, shutdown basis), in which machinery is stripped down, inspected and wearing or otherwise faulty parts are replaced. This is usually supplemented by additional periodic checks of ancillary equipment that involve, for instance, carrying out an oil change in an engine or gearbox system. This is a situation familiar to most people through the advent of the motor car. It is singularly instructive to note at this juncture that domestic vehicle users have experienced over the past 30 years a four- to five-fold increase in road miles between oil changes. The reason for this welcome improvement is due to the vast amount of research invested by the original engine manufacturers and oil companies in research – notably lubricants (and additives), bearings, gears and piston cylinder development. Over the same period, there have been corresponding advancements made in materials and manufacturing technology. Many research and development partnerships between industry and the universities (world-wide) have contributed fruitfully to these developments.

The next phase in the evolutionary process – condition-based maintenance – dates its origins back to the late 1940s for specific sectors of industry such as the railways and aerospace. More widely, recognition of the need for change in maintenance practices only really began to gather momentum from about 1970. More recently, a preference by some companies for adopting a proactive, root cause analysis approach to maintenance practice has attracted some interest. However, the cost beneficial aspects of these different strategies are not easily quantifiable [2]. Nevertheless, it is apparent that across the whole spectrum of industry all of the above types of maintenance activity are employed. This is often the case within a single company, depending on the kind of business involved, the type of machinery deployed, and the workforce availability – skill levels and turnover of staff. It also depends critically on the management's philosophy and approach regarding the importance of the maintenance activity in terms of the overall business strategy.

To respond to the enormous challenge posed by the increasing diversity and complexity of modern industrial machinery, a strategic framework for systematically evaluating the ways in which maintenance is carried out has been developed over the past 20 years. Known as reliability-centred maintenance (RCM), its principal purpose is to clarify the relationships between a company's management structure in terms of its physical assets, and those that operate and maintain those assets, so that a coherent strategy is developed that is of greatest benefit to the company [3]. The significance of adopting this approach is that it sets out from the very beginning to identify what failures could occur, i.e. by what means can an asset in question fail to fulfil its function, and what is likely to cause that loss of function?

16.2.2 Tribological Principles

Friction, lubrication and wear of interacting surfaces that experience relative motion trace their roots back to the very dawn of civilization. Scientific discovery of the laws governing friction behaviour took further root during the middle ages, and began to blossom forth following the birth pangs of the first industrial revolution in the eighteenth century [4]. In modern times, delivering the wherewithal for machines to provide more power places increased demands on the capability of the very small area available in the contact of interacting surfaces to transmit the requisite increase in load at higher speeds. On the basis of data published in 1968 [5], the range of bearing load and speed for three commonly used prime moving machines are presented in Table 16.1 for two different shaft sizes that encompass most applications.

Table 16.1 Typical load–speed characteristics for industrial machinery

Machine type	25-mm shaft diameter		250-mm shaft diameter	
	Load (kN)	Speed (rpm)	Load (kN)	Speed (rpm)
Electric motor	0.1–0.25	1000–7000	25–100	150–800
Piston engine	3.5–12.5	5000–30,000	240–800	200–1000
Turbine	0.025–0.075	20,000–80,000	10–50	2000–8000

The limiting load–speed characteristics for, respectively, dry rubbing, plain fluid film and rolling element bearings, in relation to shaft diameter, form the basis for selecting and designing bearings to function safely within their individual capabilities [6]. The issues limiting the life expectation of rolling element bearings are primarily load at low speed, and the increased likelihood of surface fatigue occurring as speed is increased.

Fluid film bearings, on the other hand, are governed at the lower speeds by excessive load, leading to a significant reduction in the operating film thickness and increased likelihood of higher temperatures occurring in the bearing contact. At higher speed, temperature and instability are the main threats to continued operation [7]. In the absence of concomitant improvements in the properties of materials (solids and lubricants), friction and consequential wear will increase, with a corresponding rise in temperature and removal of surface material, leading to either seizure or unacceptable wear, culminating in machine failure. The developments in lubrication theory – notably, hydrodynamic (late nineteenth century) and elasto-hydrodynamic (mid-twentieth century) – and their subsequent application to bearing technology were landmark achievements that restored the balance in favour of prolonging operating life in an era of unprecedented increase in power demand in the post-Second World War period. This is best illustrated by reference to the developments taking place in what were arguably the most significant scientific and engineering advances in the twentieth century – the motor car and the aeroplane (Table 16.2).

In relation to the operation of machinery, a vitally important aspect of oil-wetted systems is the regime of lubrication in the contact. Theoretically, a system that is designed to function in the fully hydrodynamic regime represents a zero wear condition, with infinite life expectation. Similarly, in the case of elasto-hydrodynamic lubrication of rolling elements, there is reasonable expectation of achieving enhanced life before pitting failure occurs.

Remarkable though these achievements are, the present-day performance capabilities of tribological components would not be attainable were it not for equally impressive improvements made in materials' physical and chemical properties – both solid- and liquid-based [8, 9]. Figure 16.1 provides some indication of the increase in the demand placed upon lubricants in terms of the thermal stresses to which they have been subjected over the past 60 years. At the same time there have been corresponding advances made in the development of

Table 16.2 Growth in machine performance in the twentieth century

Era	Car/aircraft	Automobiles		Military aircraft	
		Speed (mph)	Power (bhp)	Speed (mph)	Power/thrust (hp/lb)
1910–1930	Ford Model T	40	20		
	Sopwith Camel			118	120 hp
1930–1960	Ford Anglia	75	–		
	Spitfire			375	1300 hp
	Gloster Meteor			600	4000 lb
1960–1980	Ford Cortina	70	75		
	Jaguar (aircraft)			990	7300 lb
	Tornado			1452	15, 800 lb
1980–2000	Ford Mondeo	150	175		
	Eurofighter			1320	20, 250 lb

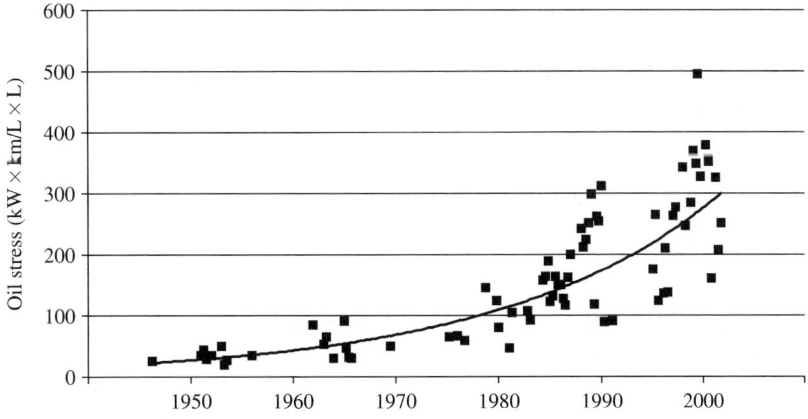

Figure 16.1 Increase in lubricant thermal stress since 1946

manufacturing techniques that are aimed specifically at enhancing solid surface properties. Table 16.3 shows the impressive life enhancement advances made by major rolling element bearing manufacturers over the past 50 years.

At the same time, parallel developments in sensor technology, assisted latterly by the huge expansion in information technology over the past 20 years or so, have led to technology transfer from the laboratory to field-based research and development, with spin-off application to everyday industrial operations. Among these initiatives are techniques that are employed routinely to monitor the maintenance status of machinery [10, 11].

One of the most significant landmarks in the history of tribological conquest in the UK was the publication in 1966 of the Jost Report [12]. The outcome of exhaustive enquiries

Table 16.3 Rolling element bearing life enhancement through material advances

Mid-decade	Process	Fatigue life enhancement		DmN rating
		High volume	Low volume	
1900	Air furnace (AF)	0.5		0.1
1920	AF	0.8		0.2
1930	AF	0.8		0.5
1940	AF	1.0		0.7
1950	AF	1.0		0.7
1960	Vacuum degassed furnace	3.0		
	Vacuum induction melted (VIM)		10	1.5
1970	Electro-flux remelted (EFR)		30	2.0
	VIM vacuum arc refined (VIM-VAR)		30	
1980	Ladle steel making	6.0		2.0
1990	Ladle steel making	10.0		2.5
	VIM-VAR		120	
2000				4.0

and discussions by leading tribologists of the day led to the identification of a number of areas where the failure to exploit existing tribological knowledge was costing the country an estimated £500 million at then current prices. Singled out for special attention in relation to increased savings potential arising from improvements in tribological practice were

- maintenance and replacement parts 45%
- losses resulting from breakdowns 22%
- investment through greater availability and higher efficiency 4%
- increased life of plant 20%

Other areas listed, where savings were anticipated included achieving reductions in energy consumption and manpower, and lubricant costs.

A later survey took into consideration the relative increase in the cost of energy arising from the world's energy crisis in 1973. It highlighted the need to conserve energy and raw materials through improved tribological design methods [13]. Consequently, and explicitly, the implication for industry in particular was that there was a clear mandate for focussing special attention on maintenance-related activities. On the evidence presented from a comprehensive survey of maintenance practice carried out in the mid-1970s [14], there was much that needed to be done. It turned out that many companies at the time were insufficiently aware of the possibilities for improvement in their practices afforded by adopting a condition-based approach to resolving their maintenance problems. Nor were they well placed to decide which machinery health-monitoring techniques were best suited to their purposes. However, at the very heart of the matter lay the fact that there were no clearly defined mechanisms available to guide senior management in determining the likely cost benefits to be expected from adopting such practices. In attempting to address these issues, Neales's report provided the much needed information and important 'first step' guidance on how to proceed against the background of the current maintenance procedures of the day. Industries that were already heavily committed to a condition-based maintenance strategy were mainly those in which issues such as safety or national security were a high priority. A number of high-tech companies facing strong international competition were also not slow to respond.

It is pertinent at this juncture to enquire:

- *To what extent have the technological advances in our knowledge and application of tribology over the past 40 years contributed to achieving these savings?*
  On the basis that 'Prevention is better than cure' considerable attention has been given to channelling the fruits of tribological research through improved design methods and manufacture of tribological components and systems that have found application in everyday practice – reference the example of the vehicle industry cited above. This has led to marked improvements in the performance and reliability of a whole range of industrial machinery. Nevertheless, it has proved difficult to rigorously quantify, globally, these technological advances in terms of actual cost savings.
- *Within the maintenance community itself, what has tribology contributed to achieving further cost savings?*
  When the Jost report was published, condition-based maintenance practice, in terms of the methods presently employed in industry, was in its infancy. It seems appropriate, therefore, to examine this aspect within the context of present-day practice.

16.2.3 Tribology and Maintenance

In summary form and in broad terms, Table 16.4 summarizes some of the major tribological developments during the twentieth century alongside the corresponding evolution in maintenance technology occurring over the same period.

Table 16.5 illustrates qualitatively the ways in which tribology contributes to the maintenance activity.

When, for example, a bearing component is correctly specified, designed to perform the task intended, manufactured and installed as per specification and operated within its design envelope, there is a reasonable expectation (probability) that it will function satisfactorily for its intended operating life. In the case of a rolling element bearing, the probability is conventionally stipulated to be 90%. This is in recognition that such bearings will ultimately fail as a direct consequence of pitting fatigue occurring due to the innumerable reverse stress

Table 16.4 Developments in maintenance and tribology during the twentieth century

Era	Maintenance strategy	Tribological developments
Pre-1950	Breakdown	Hydrodynamic lubrication theory – 1980s Tilting pad thrust bearings – early 1900s
1950–1970	Preventive	Elasto-hydrodynamic theory Jost Report – 1966 ESDU design guides
1970–present	Condition-based	Tribology Handbook – 1973 Wear Debris Analysis (Ferrography) – 1972 ASME deign guide for rolling element bearings ISO Standards
1980–present	Proactive	Wear-resistant material developments Rapid growth in computer-aided software systems

Table 16.5 Tribological input to the maintenance activity

Maintenance activity	Tribological input
Breakdown	Failure examination and diagnosis Designed out (maintenance) solution
Preventive	Detailed knowledge of: Lubricant/additive properties Material properties and treatments Surface finish and damage inspection (including boroscope examination)
Condition-based	Interpretation of: Oil and wear debris analysis results Vibration/acoustic emission data Performance measurement (displacement, temperature measurement)
Proactive	Root cause analysis Prediction of remaining wear life

cycles suffered in the rolling contact. For a plain fluid film bearing, on the other hand, infinite life is specified because, theoretically, there is no corresponding impediment to limit its life.

16.3 Failure Diagnoses

Provided the above criteria are adhered to, nothing should prevent a tribological component from functioning adequately for its expected lifetime. Nevertheless, surface distress-related failures continue to occur, and the causes are well known and documented [15–17].

The most commonly encountered problems that lead to wear-related failure are here listed in no particular order of importance or frequency of occurrence:

- lubricant starvation;
- contamination – hard solid, or liquid (water, fuel contamination);
- misalignment or looseness of parts;
- extraneous vibration;
- human error (supplying wrong materials or properties, operation outside the required performance envelope).

Detailed, post-failure analysis continues to provide the most important evidence of both cause and effect. When acted upon through appropriate designed out maintenance practice, accompanied where required by corresponding adjustments to operating procedures, the likelihood of further failures of the same kind occurring is often reduced or averted altogether. It is the unlikely combination of more than one seemingly unconnected event, leading to secondary consequences, which can lead to the more serious type of failure which may also prove to be the most difficult to diagnose and rectify.

16.3.1 Failure Morphology and Analysis

Engineering artefacts reach the end of their useful life by one of three ways: completion, obsolescence and failure (slow or sudden) [1]. Failure is defined as damage, fracture, breakage or rupture [18]. Damage is described generally in terms of, respectively, fatigue, wear and corrosion. Other forms of damage include impact and overload. Damage that occurs due to fatigue is macroscopically invisible. It often proceeds rapidly and is, therefore, inherently dangerous. Damage occurring as a result of wear and corrosion, on the other hand, is macroscopically visible and might be, therefore, expected to proceed more gradually.

The failure pattern that is probably most frequently described in the literature, known as the 'bathtub curve', commences with a high incidence of infant mortality during the 'running in' period ('A' in Table 16.6). This is followed by a constant or slowly increasing failure rate, and finally by a wear-out zone. In addition, Mowbray [3] identifies five other failure characteristics (B through F), and reports the breakdown in the incidence of failures for each mode in relation to statistics relating to failures occurring in civil aircraft.

Mowbray concluded that as equipment becomes more complex, there would be more instances of types E and F. He further observed that there was also no connection between reliability and operating age. The belief that the more an item is overhauled, the less likely it is to fail proves very often *not* to be the case.

Table 16.6 Failure patterns and their incidence in civil aircraft (Mowbray [3])

Failure pattern designation	Characteristic	Incidence (%)
A	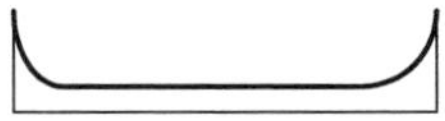	4
B	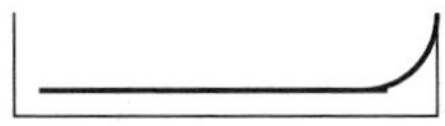	2
C	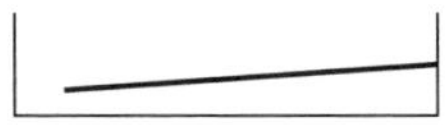	5
D	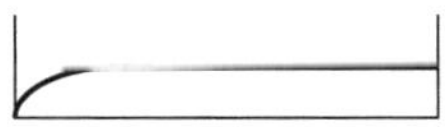	7
E	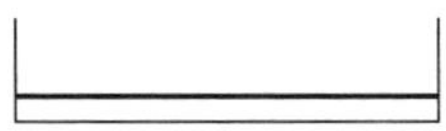	14
F	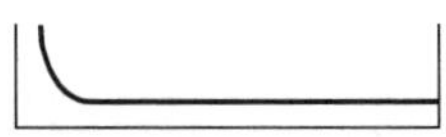	68

In relation to tribological failures, the specific techniques used to perform analysis depend in the first instance on whether a strip-down inspection of an individual failure is feasible and/or whether witness statements can be obtained.

To determine the most probable cause(s) of failure, it is important to take into account the following aspects:

- Carry out a detailed inspection and examination of the failed specimen, preferably *in situ* at the location.
- Continue the post-mortem in a suitably equipped laboratory in order to analyse physical and chemical changes in the surface and sub-surface regions of the failure.
- Collect relevant information from eye witnesses and collate it with data relating to operating conditions and procedures.
- Study the design and analyse performance within the whole design envelope. Postulate the conditions that would have had to prevail in order to cause the failure.

For situations involving tribological type failures, the combined resources and the knowledge and experience of the engineer, material specialist and lubricant chemist are more likely to lead to a satisfactory outcome of the investigation. This is particularly with regard to establishing the most likely cause of failure, its consequential effects and what needs to be done to put the matter right.

Alternatively, the failure history may relate to a number of similar occurrences taking place on more than one identical, or similar, machine over a period of time. In such cases, a Weibull analysis procedure may be usefully invoked to determine the failure characteristics [1] of the form

$$F(t) = 1 - \exp[\alpha(t - \gamma)^{\beta}]$$

$F(t)$ is the cumulative percentage failure and t is the time to failure of individual components. The three constants are, respectively, the scale parameter, α, the Weibull index, β, and the location parameter, γ.

To determine the failure characteristics, Weibull probability paper may be used and this effectively transforms the exponential form of the function to a linear plot of cumulative percent failure versus operating time, in which β is the gradient. This permits the failure to be specified in relation to, respectively, infant mortality, random or wear-out condition, in terms of the following criteria:

- infant mortality ($\beta < 1$)
- random ($\beta = 1$)
- wear out ($\beta > 1$).

16.3.2 Dealing with Failure – Two Short Case Studies

The first case highlights the solution to a severe breakdown failure situation. The second case is representative of a preventive maintenance activity in which a routine inspection revealed a serious flaw in the condition of a critical bearing operation. In each case, the application of known tribological knowledge brought about an immediate solution to the problem.

16.3.2.1 Sand Dredger Crane Bearing

In an investigation of a bearing failure in the pinion shaft of a sand dredger operating in the Bristol Channel, the diagram shown in Figure 16.2, referenced from Tribology Handbook, was utilized to show that the bearing was operating at the extreme limit of safe operation for grease lubrication. Initial attempts to resolve the issue through changing the bearing clearance and modifying the bearing groove geometry and dimensions proved unsuccessful. After conversion to a simple oil lubrication system, there were no further failures of this type reported.

16.3.2.2 Small-End Bush – Marine Diesel Engine

Another marine failure investigation concerned the connecting rod small-end bush of a ship's diesel engine. The problem was discovered during a routine inspection of the bearings while the ship was berthed at the dockside. Figure 16.3 shows the condition of one of a number of bearing shells similarly damaged in the same engine. Investigation revealed that replacement bearings had been previously installed incorrectly such that the entrance hole, seen in the figure at the top of the picture, had not been located directly in line with the hole running

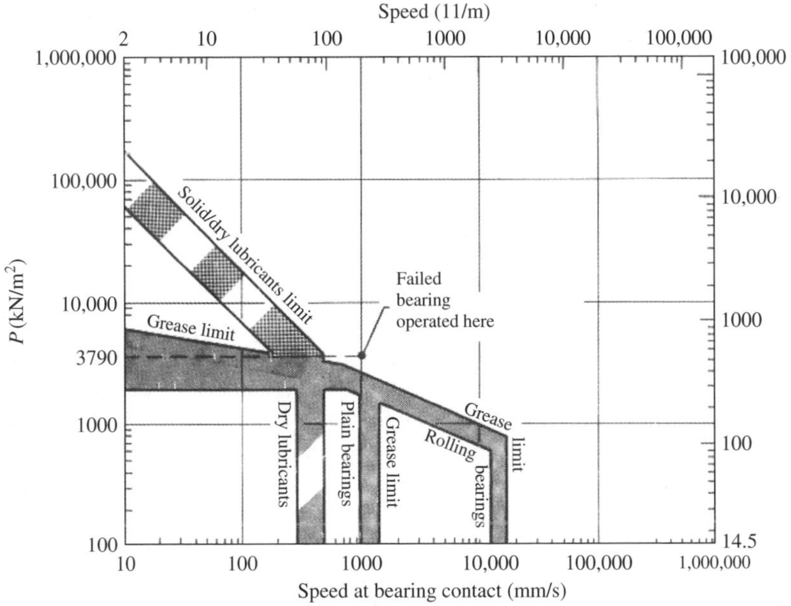

Figure 16.2 Speed–load characteristics for solid/dry lubricants and grease

Figure 16.3 Damaged marine diesel engine bearing

through the connecting rod from the big-end bearing. This was the sole means whereby the bearing received a continuous supply of pressurized oil lubrication. In addition, subsequent metallurgical analysis revealed that these particular bearings were defective. They were cast phosphor bronze but analysis confirmed them to have a higher than permissible level of porosity. Calculations, based on well-established squeeze film bearing theory, confirmed the hypothesis that during operation the cyclic squeeze film pressures generated were sufficiently large to have caused the type of damage witnessed in the photograph. Evidently, over a period of time, the bearing had adapted to the change in its circumstances such that a new 'entrance hole' was forced through to the inside of the bearing shell. However, as fragments of metal began to break off there was a strong possibility that the bearing would suffer catastrophic seizure. Thus, a timely routine inspection avoided a potentially serious situation that may well have developed while the ship was at sea.

16.3.3 Comment

So far, it has been demonstrated that when tribologically related failures occur there is a considerable wealth of knowledge, practical skills and experience available to assist the process of resolving cause and effect issues and how best to rectify them. It also confirms that understanding and knowledge of tribological phenomena can contribute positively to dealing with problems that arise from time to time in breakdown and preventive maintenance situations. However, it begs the question, 'What contribution can the tribologist make to assist in preventing failure in the first place?'

To begin with it is useful to describe what is involved in carrying out condition-based maintenance. The tribological aspects will then be identified in relation to the techniques employed for monitoring the condition of machinery. How to distinguish between what the sensing and analysing systems capture by way of data and the associated wear mechanisms that contribute ultimately to failure of critical components is a crucially important aspect of the whole process.

16.4 Condition-Based Maintenance

The key components of any successful monitoring activity comprise

- dependable *detection*
- effective *diagnoses*
- reliable *decisions*.

The first component – detection – is an essential element of any condition-based maintenance strategy. If the sensing device, or analysis procedure employed, fails to detect a fault as it develops, it is automatically rendered redundant. When deployed correctly, it is a vital front-line tool for monitoring, or 'tracking', the fault trend and the associated decline in component performance. Ultimately, it defines the reliability envelope. Effective diagnosis provides important information that assists engineers in separating 'cause' from 'effect'. The third component comprises reporting the results linked to a recommendation whether to cease operation in order to carry out an inspection or repair. This is the most significant and critical step.

It is pertinent here to note that not all faults that occur in machinery are initiated through tribological activity. However, the transition from what might be termed 'benign' to 'active' wear is something that must be monitored very closely, especially when it deteriorates further to a 'failure' condition. This is self-evident where safety issues are paramount, but equally so where there are significant consequential cost implications typified by production processes, such as the rolling of steel strip.

Figure 16.4 depicts, simplistically, the transition from benign to active wear in relation to time, when expressed in terms of wear debris production rate, and subsequent failure, in which the monitoring opportunities and requirements are also highlighted. During the initial period of a machine's life, it may often appear as though a severe wear condition is occurring. The detection system employed will soon confirm whether the rate of wear is increasing or decreasing. In addition to measuring wear debris concentration, morphological analysis of the wear products will reveal that where the running-in stage is transforming rapidly to a much lower wear rate, active wear particles (>100 μm) are replaced by much smaller (<50 μm) particles in which their outline shape and surface features will also change – see next section. In a machine that is functioning within its operating design envelope, the low wear rate will persist for much of the remainder of its operating life. During this time, the 'status quo' will be occasionally punctuated by a planned maintenance activity (oil change, etc.). Unintended changes in the situation may, however, occur with little warning (e.g. contaminant ingress), and this calls for constant vigilance by the monitoring team. Otherwise,

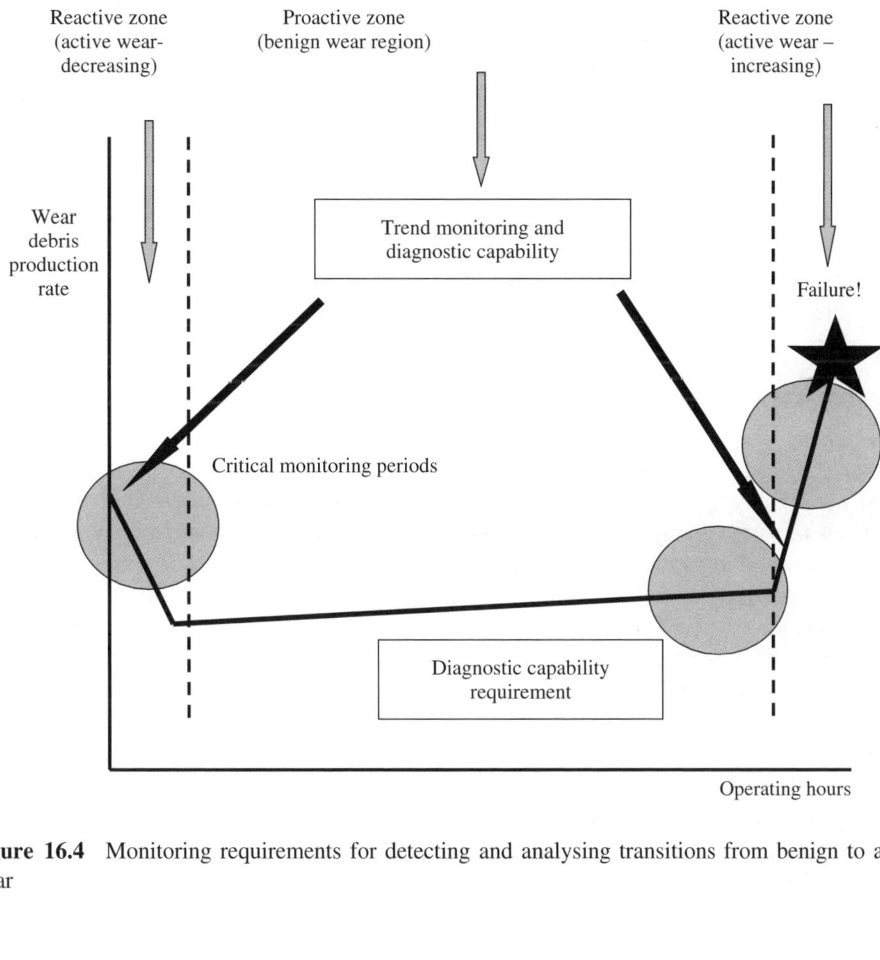

Figure 16.4 Monitoring requirements for detecting and analysing transitions from benign to active wear

the normal baseline 'signal' is established, thereby providing an important database against which any subsequent deterioration in the wear state can be measured and classified. As a transition from benign to active occurs the combined detection and diagnostic functions are harnessed to enable the severity and nature of the problem to be determined. At this juncture the main issue that has to be resolved is to decide how long to continue operations without interruption, i.e. what is the 'lead time' to failure? In some instances, it may be a matter of minutes or hours – for instance, helicopter operations. In other cases, it could be more like weeks or months before any failure avoidance action has to be taken; as, for instance, can occur in the case of steel finishing processes.

The most commonly utilized monitoring techniques deployed throughout general industry are

- vibration analysis (and latterly, acoustic emission);
- oil and wear debris analysis;
- performance analysis – temperature, electrical current, pressure, flow rate, etc.

Vibration analysis is, arguably, the most commonly used monitoring technique in general industry. It has the advantage that it yields relevant data in quantitative format and can be operated remotely in real-time mode. Pre-set alarm limits can be triggered automatically, based on signal velocity levels, or alternatively as a result of more intensive diagnostic interrogation of the data. In tribologically compromised conditions, it has proved to be very effective in sensing changes in rolling element bearing kinematics as a precursor to surface damage, and the subsequent detection of rolling pitting fatigue. This serves to provide a reliable indication of pending failure, notably in situations where the remaining time to a critical condition that requires maintenance action is of the order of days or even weeks. Its relevance to early wear detection lies in its capability to detect the incidence of loose fittings, or misalignment problems prior to the onset of a serious wear condition.

Acoustic emission technology has hitherto not been widely exploited in the field [19]. Nevertheless, the results of recent research demonstrate that it has the potential to respond unambiguously to changes in the sub-surface micro-structure that correspond directly to subsequent surface distress conditions in rolling mode (crack formation), and also in severe sliding mode (surface transformations and material transfer) – see Section 16.6.

Oil analysis is a heavily utilized facility within the condition-based maintenance community. From a single sample a number of specific measurements and analyses are carried out reliably, quickly and cheaply in a certified laboratory environment. Nevertheless, this means that processing and analysing the samples are nearly always carried out remotely from the scene of operations in which the turnaround time for results to be downloaded to the end user may be several hours or days. Interpretation of the results, and associated recommendations for further action, by the laboratory staff may sometimes be unhelpful to those charged with the responsibility for making the decisions. This can best be overcome by agreeing on strict guidelines to be followed on the basis of previous experience of the processes involved.

By its very nature, wear debris analysis reacts directly to wear occurring in the contact, usually as a consequence of failure of the lubricant film with consequential damage as the opposing surfaces come into contact with one another. Where a sudden transition to active wear occurs, as in the case of sliding motion, there is an increased risk of seizure. In this case, the use of off-line, remote analysis facilities to determine the type and severity of wear

Table 16.7 Monitoring capability in the early wear regime

Technique	Capability	Comment
Vibration analysis	Loose fittings, misalignment, shock loading. Changes in rolling element bearing kinematics. Later stages of severe surface distress – pitting fatigue, etc.	Primarily, an on-line monitor. Excellent for detecting dynamic instabilities, as a precursor to subsequent wear and failure
Acoustic emission	Sub-surface changes – fatigue cracks, surface changes as a consequence of high sliding contact temperature	On-line technique. Good prospects for proactive detection
Oil analysis	Monitor changes in lubricant properties and condition. Detects solid and fluid contaminants before wear occurs	Essentially off-line. Excellent for use in centralized data collection and analysis system. Low sample cost
Wear debris analysis	Wear severity and type of wear	Primarily off-line. Good prospects for on-line detection/diagnosis in the early wear regime

would be relatively ineffective. Where this does not occur, the fact that the type and severity of wear can be determined using this technique facilitates a more informed assessment of the level of criticality.

The relative merits of these techniques in terms of their capability for capturing data in the early wear regime are summarized in Table 16.7. Achieving a clear distinction between cause and effect needs to be addressed. It is heightened by the incidence of unforeseen events that subsequently lead to the onset of wear. One of the principal causes is that due to contamination – solid or fluid. Oil analysis techniques have proved to be very effective; one of which is spectrometric analysis. This method, in effect, detects the elemental composition of the small ($<5\ \mu m$), constituent particles of the solid contaminant collected in the oil sample.

Table 16.8 Reactive versus proactive monitoring opportunities

Wear status	Best-suited technique
Reactive – benign wear (low rate of wear)	Wear debris analysis
Reactive – active wear (run-in, wear out, random failure)	Wear debris analysis Vibration analysis Acoustic emission Oil analysis
Proactive (zero wear)	Oil analysis Vibration analysis Acoustic emission

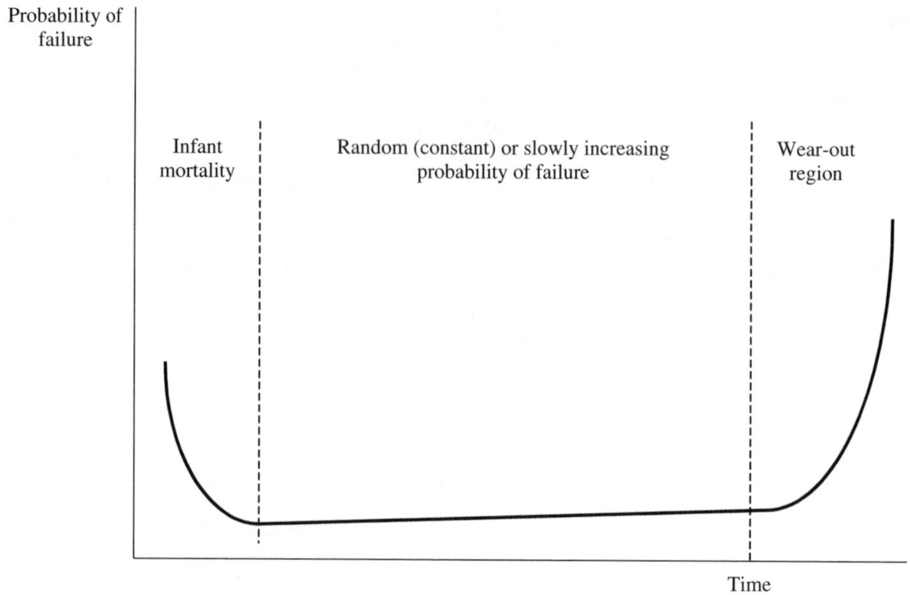

Figure 16.5 Principal modes of failure probability

Likewise, the presence of water in the oil can be determined. Thus, through application of such methods the means exist for avoiding the harmful by-products of contaminants in the form of aggressive abrasive wear. More generally, the suitability of each technique for distinguishing between proactive (pre-wear) and reactive (wear occurring) is presented in Table 16.8.

In respect of the three principal modes of failure probability, the potential suitability for the techniques to satisfy the requirements in each category is summarized in Figure 16.5.

A pictorial representation of how each monitoring technique can be used to detect and diagnose wear-related failure is shown in Figure 16.6.

16.5 Wear and Wear Debris Analysis

Wear debris analysis was first developed in the UK about 50 years ago as a viable machinery health monitoring technique through the use of magnetic debris plugs (MDPs) installed in the lubricating systems of aircraft of the Royal Air Force (also, British Commonwealth countries – Canadian and Australian Air Forces) [20]. Arguably, the most significant development was the introduction in the early 1970s of the ferrography technique [21]. The specific descriptors used at that time to identify and distinguish the morphological characteristics of the debris have since achieved worldwide acceptance within the condition monitoring community Table 16.9.

The links between the definitions in Table 16.9 and the descriptions of conventional wear mechanism and mode descriptors found in the literature are tenuous. Barwell [22] depicted the main wear modes and mechanisms in relation to the type of motion (sliding

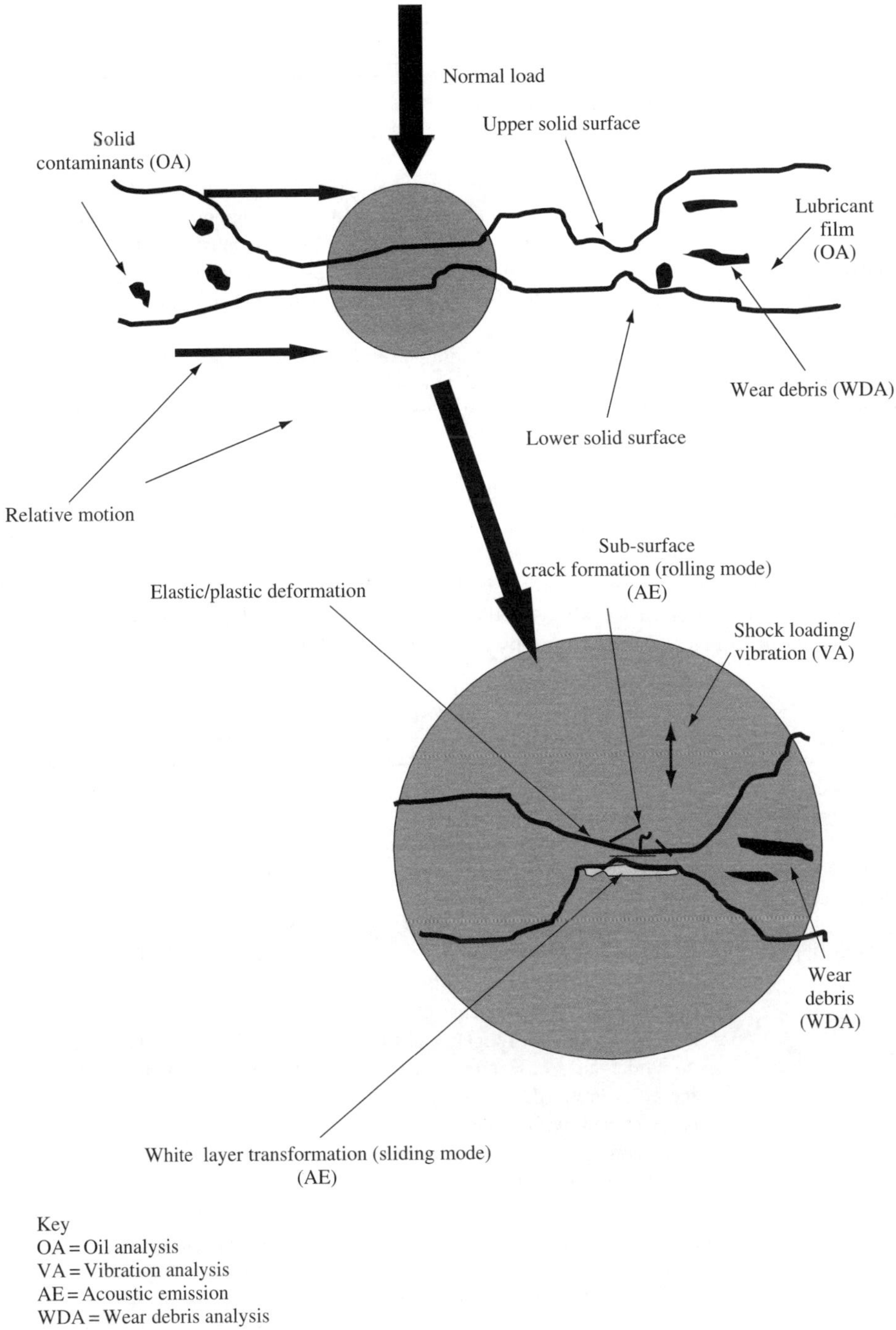

Figure 16.6 Deployment of monitoring techniques in the early wear regime

Table 16.9 Wear particle morphology – ferrography descriptors

Particle type	Description
Rubbing	Particles, 20-μm chord dimension and approximately 1 μm thick. Results from flaking of particles in the shear-mixed layer
Cutting	Swarf-like chips of fine wire coils caused by abrasive 'cutting tool' action
Laminar	Thin, bright, free metal particles, typically 1 μm thick, 20–50 μm chord width, with holes in the surface and uneven edge profile. Emanates from mainly gear/rolling element bearing wear associated with fatigue action
Fatigue	Chunky, several micrometres thick, 20–50 μm width.
Spheres	Typically ferrous, 1–50 μm diameter, generated from micro-cracks arising mainly from rolling contact conditions
Severe sliding	Large, 50 μm chord width, several micrometres thick. Surface heavily striated with long, straight edges. Typically occurs in gear wear

or rolling/sliding) and severity of the contact (dispersed or Hertzian). Table 16.10 is a summary of observations relating worn surface appearance to associated wear particle type; they do not appear to complement very well the wear particle descriptors listed in Table 16.9 [23]. The main differences may be due to the fact that conventionally, wear modes and mechanisms are based on observations and deductions relating to dry or boundary lubricated conditions in the contact. The ferrography descriptors, on the other hand, were based mainly

Table 16.10 Worn surface appearance and associated debris (from Reference [23])

Type of wear	Worn surface appearance	Wear debris
Mild wear	Fairly uniform loss of material; slight surface roughening	Fine, free metal particles <10 μm, unoxidized
Adhesive	Tearing and transfer of materials from one surface to another	Large irregular particles >10 μm, containing unoxidized metal
Two-body abrasive	Harder surface – little or no damage. Softer surface exhibits scores, grooves or scratches corresponding with rough asperities on harder surface. Harder surface scored or grooved, may appear to be machined, with grooves corresponding with hard embedded particles in counterface	Consists mainly of softer material (fine swarf), unoxidized material Contains swarf-like, unoxidized material from harder surface. May also contain softer material in lump form
Three-body abrasive	Surfaces have deep scratches or grooves	Fine, may contain some unoxidized metal, but mainly loose abrasive material
Fretting	Surfaces heavily pitted: pits may be small, or larger, producing roughened surface area; oxidized appearance	Fine, fully oxidized, if from ferrous metal. Will contain Fe_2O_3 (rust coloured). Spherical particles, sometimes

on samples obtained under fully lubricated contact, and comprise, essentially, a formal description of the appearance of the particles when viewed in an optical or scanning electron microscope.

Since the introduction of the ferrography method, many developments have been presented at international conferences and published in the literature, including the first ever international conference devoted entirely to the method, held in Swansea in 1982 [24]. The advent of computer technology over the past 20 years has contributed significantly to the development of enhanced image analysis techniques and the establishment of comprehensive databases [25]. Much of this technology can be deployed using laptop computers, thereby enhancing their use in the field, especially when accompanied with downloading facility via satellite from remote locations to a centralized base.

16.5.1 Wear Modes and Associated Debris Characteristics – Some Experimental Results and Their Application to RAF Early Failure Detection Centres

To establish a more fundamental and definitive relationship between different wear conditions and their associated wear debris, a programme of systematic testing was undertaken with the express aim of creating a database of wear particles generated under known conditions in a controlled laboratory-based environment. The programme was initiated as a consequence of the need to provide technical personnel at the RAF's Early Failure Detection Centres (EFDC), with enhanced techniques for diagnosing pending bearing and gear failures in engines and gearbox transmission systems [26]. Among the analyses performed, there is a requirement to assess the morphological features of debris deposited on a substrate that have been transferred from an MDP.

An optical microscope is used to view the particles, following which an assessment is made of a representative sample of the captured particle population, in particular, to identify and assess the type and severity of the condition. A manually operated analysis procedure, the morphological analysis methods employed have been in service for many years. It was decided to exploit the facilities afforded by using modern image analysis technology to establish a new set of computer-based analytical procedures that would also be more user-friendly and efficient to process. At the same time, it was recognized that there was a need also to provide a better defined and verifiable wear debris classification system.

The first step in the process was to create a database of wear particles generated under known operating conditions, viz., the type of contact and motion, load, speed and regime of lubrication. The type and severity of wear was determined by post-test analysis of surfaces. Friction measurements were performed during the tests in order to assist in specifying the regime of lubrication.

The test facilities used to carry out the tests comprised:

- four-ball machine (sliding and rolling contact)
- pin-on-disc machine
- gear test machine
- rolling element bearing fatigue test machine.

Table 16.11 Test machine characteristics

Test conditions	Test machine		
	Four-ball machine	Pin-on-disc machine	IAE gear machine
Motion			
Sliding	+	+	
Rolling	+		
Sliding/rolling			+
Geometry			
Point	+	+	+
Line		+	
Conformal		+	
Nominal pressure (MN/m^2)	150–600	700–2100	350–1400
Sliding velocity (m/s)	0.25–1	0.1–10	2–12
Material test			
534A99	+	+	+
665M17		+	
080M40		+	
Lubricants tested			
SAE10	+	+	
Mobile Jet II	+	+	+
ETO 25	+		+

The test machine characteristics relating to the first three machines are presented in Table 16.11.

A comparison between test machine test conditions and those representative of machines operating on-board ships and aircraft in the United States Navy is presented in Table 16.12, [27].

Table 16.12 Range of operating conditions for shipboard and airborne applications

Test machine	Load range (MN/m^2)	Speed range	Application
Four-ball machine	150–600	0.25–1.00 m/s	Ship main reduction and auxiliaries. Gearboxes, pump compressor bearings, helicopter transmission output module and blade bearing
Pin-on-disc machine	700–2100	0.1–10 m/s	Shaft support bearings, ship service turbine gearbox, engine accessory gearbox, helicopter mast bearing, transmission input module and engine nose box
Gear test machine	350–1400	1000–6000 rpm (2–12 m/s)	Ship propulsion, helicopter main gearbox, helicopter IGB and tail rotor gearbox

16.5.2 Summary of Laboratory Test Results

The sample set of results shown in Figure 16.7 represents (a) post-test analysis of the wear scars and (b) associated debris generated during a sliding test in the four-ball machine. The lubricants tested comprised a mineral-based SAE 10 lubricant (A) and two synthetic

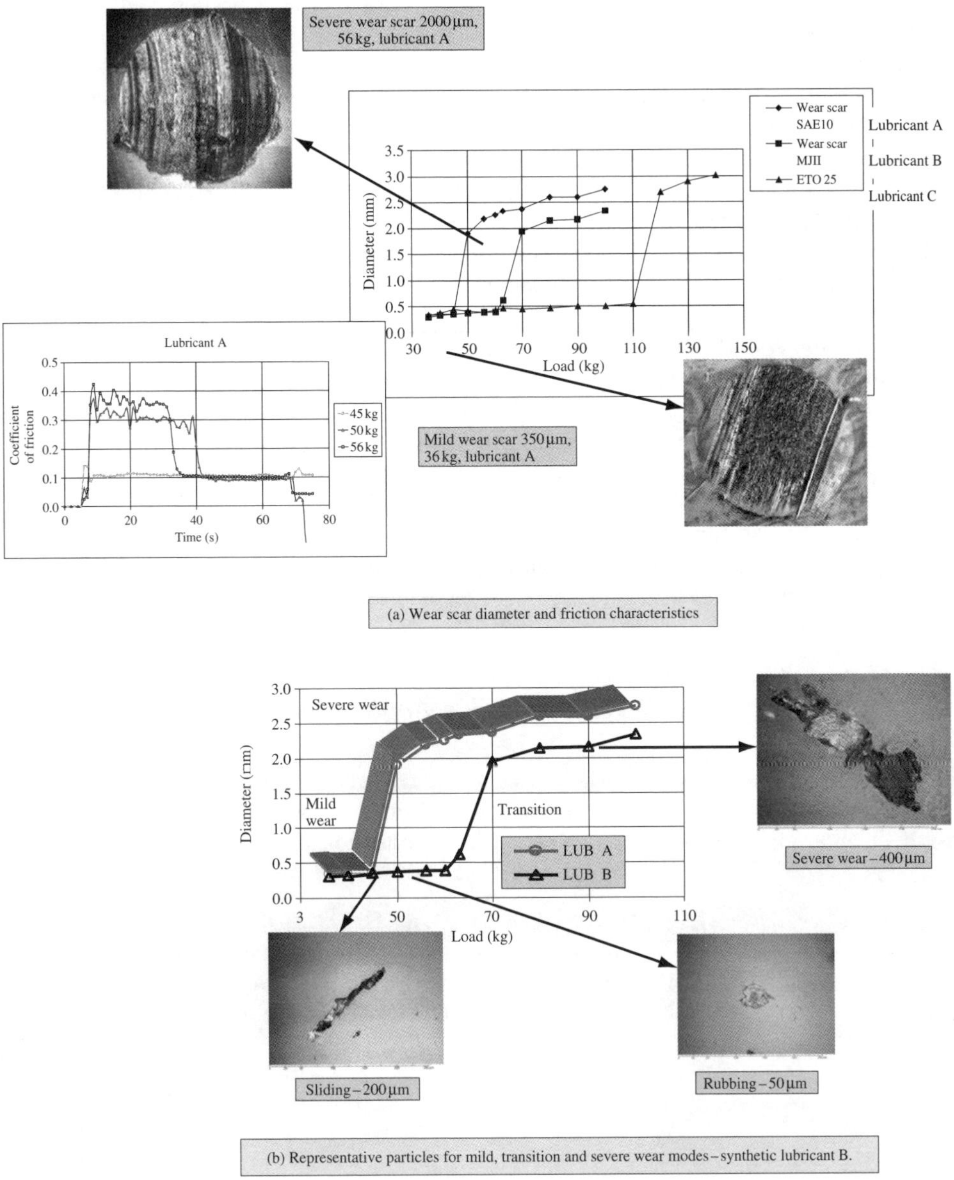

Figure 16.7 Results of analysis – four-ball sliding tests: (a) wear scar diameter versus applied load; (b) representative particles for mild, transition and severe wear modes – synthetic lubricant B

lubricants, respectively, Mobil Jet (B) and ETO25 (C). The results obtained when testing with the synthetic lubricants are plotted on the wear scar graph to enable comparison with the mineral oil. Continuous friction data, obtained during the test, confirmed that before seizure (45-kg load) there was no breakdown in lubrication with corresponding low-wear scar and only small (mild) rubbing wear debris generated. A transition from mild to severe wear occurred within the first 5s of running at 50 kg as a consequence of partial breakdown in the lubricant film, accompanied by a corresponding increase in friction coefficient from 0.1 to 0.3. The subsequent partial recovery in the film integrity is indicated by a reduction in friction coefficient after 40s. The characteristics at 56 kg are similar, though the seizure time is shorter and the wear scar and debris are larger and more severe in appearance. There were no observable differences in the particle shape or surface characteristics between the mineral and synthetic oil debris. There was, however, a distinct difference in the loads at which initial seizure occurred, and this was reflected in the higher debris concentration and size of particle examined at the higher loads.

It is pertinent to note that the chord dimensions measured for each specific type of particle obtained during these tests are generally larger than those commonly encountered when using the ferrography technique. This is because the particles captured in the laboratory tests were removed *in situ* with the small volume (7 mL) of oil remaining in the cup after completion of each test. Oil samples, typically taken from a large sump or directly from the lubrication flow system of an engine or gearbox, which are subsequently passed over a substrate (e.g. analytical ferrography technique) lead to deposition of smaller-size particles.

A large number of tests of this type were conducted by utilizing the machines available to evaluate the effect of different operating conditions and materials used. Thus, a comprehensive wear particle database was acquired and the main outcome is summarized in Table 16.13. At the completion of the test programme the database contained at least a hundred particles in each category. A more detailed account of the results programme is reported elsewhere [28].

16.5.3 Wear Particle Classification and Application

16.5.3.1 Off-Line Monitoring

To fully implement the morphological data derived from the laboratory tests (Section 16.5.1), as an integral part of a monitoring system, a classification procedure was established so

Table 16.13 Correlation between wear condition and debris characteristics

Debris descriptor	Four-ball machine	Pin-on-disc machine	Gear machine	Bearing fatigue rig	Wear condition
Mild rubbing	√	√	√	√	Mild abrasion
Sliding	√	√	√		Severe abrasion
Severe sliding	√	√	√		Scuffing (adhesion)
Cutting		√			Severe abrasion
Fatigue (chunky)	√ (Rolling)			√	Pitting fatigue
Fatigue (laminar)			√		Low cycle fatigue

that the analyst can compare 'live' particle features with those stored in the database. From the several methods available from artificial intelligence sources, a Bayesian belief network system approach was adopted because it assimilates the complex interactions between input and output using relatively simple mathematics [29].

The principal parameters used as inputs to the network are

- shape – regular, irregular, curved or elongated
- surface – rough or smooth
- edge – rough or smooth
- size – small (<100 μm), medium (100–300 μm), large (>300 μm)
- colour – bright, dark or heated
- cracks – yes or no
- striations – yes or no
- pits – yes or no.

Each particle stored in the database is classified according to the above criteria and also assigned to one of four specific particle types – mild, severe sliding, fatigue or cutting wear. The probabilities of all the possible combinations of inputs are calculated and yield results in terms of the likelihood of that specific type (expressed as a percentage). It is important to note that the particle sizes stated above differ from those characterized by the ferrography technique. This is because the latter particles are normally extracted from an oil sample taken from the lubrication system. MDPs capture the debris directly *in situ* and thus the size range of particle is much greater.

To convert the analytical procedures for application into condition monitoring programmes, a software programme was written resulting in a deliverable comprising a CD ROM disc (SYCLOPS) for use at RAF Early Failure Defence Centres. A graphical user interface (GUI) was developed which embodied the belief network as the classification system. Stylized images were adopted as representing the best format for establishing the input data. The analyst is first presented with eight different attribute sets displayed at the monitor screen to use as a basis for identifying the particle type, based on his own assessment of what he sees when he looks at the live image. On the basis of the eight selected attributes, a selection of the particles most closely resembling the image is presented, together with the probability of goodness of fit in each case. In addition to the classifier, the other 'push button' aids available are Tutorial (for novices and retraining); a Gallery containing images and notes of all the particles stored in the database; a 'Diagnoses' section; and also 'Glossary' and 'Help'. Customized versions of the generic version are provided for users of specific items of equipment, e.g. different aircraft types.

16.5.3.2 On-Line Monitoring

On-line monitoring of machine condition using vibration sensors has been common practice for many years. Devices for determining the wear condition based on monitoring the quantity and size of captured wear debris have also been in existence for quite some time, notably in military aircraft applications, but have not to date found wide application by industry at large [30]. In situations where deterioration through wear is well signposted over a period of time, the use of on-line, trending techniques as an aid to predicting the remaining useful life of failing equipment is likely to be a good investment. On-line determination of wear particle size and shape is achieved through the use of optically based monitoring technology [31].

Designated as 'LaserNet Fines' (LNF) it is presently under development by the United States Naval Research Laboratory at Washington DC, in conjunction with the Office of Naval Research and private enterprise, with the intention of making it also available for use by industry.

The US Navy is heavily committed to the implementation of a rigorous proactive shipboard maintenance programme as part of its intention to retain its global surface fleet supremacy. One of the primary goals is to permit automated morphological analysis of debris as it is discharged through the lubrication system of engine and gearbox systems. More specifically, it determines the type, severity and rate of progression of mechanical faults by measuring the size distribution, shape characteristics and rate of production of wear particles. The LNF instrument uses a pulsed laser diode illumination through a flowing liquid column. The transmitted image is magnified, collected and imaged onto a CCD (charge-coupled device) camera. The image is then processed in a computer, from which the size distribution of particles in the range of 5 to >100 μm is obtained. For particles >20 μm, shape information is obtained using artificial neural network (ANN) to specify the wear mode. The ANN expert system was trained using debris obtained from controlled laboratory tests of the type described in the previous section. The instrument is capable of being used either as an on-line system or, alternatively, as an off-line, bottle sampling technique that can be used at on-site locations. As the demands on machine performance and availability continue to increase, so also will the demand for automated monitoring systems – especially those that can be used as part of a proactive maintenance programme that is committed to predicting the remaining useful life of machinery. In this complex and uncertain environment, the tribologist engaged in the maintenance activity faces his greatest challenge. However, it also offers the rewarding prospect of contributing significantly to ensuring that a correct diagnosis of failing equipment is coupled to a reliable prognosis of the outcome and hence a realistic assessment of the prospects for continuing operations safely and cost-beneficially for as long as possible.

16.6 Predicting the Remaining Useful Life and Evaluating the Cost Benefits

16.6.1 Remaining Useful Life Predictions

Reliability-based projections of failure rates and their application to projections of the remaining useful life have been in use for some considerable time [32].

One such contribution to the research in this area was provided through investigating the condition-based maintenance activities in a steel works environment, in particular the finishing stages of a strip rolling mill [33]. The raw data from vibration analysis velocity measurements routinely collected were used in the development of a computer-based statistical (Weibull) analysis procedure. It effectively combined the historical data from previous failures on the mill with, additionally, provision to continuously update the prediction of remaining useful life based on current monitoring data. A useful way to apply the predictions was to assess the probability of failure occurring before the next downtime period (normally, every three weeks). On this basis, when the probability of failure was deemed to be less than a prescribed value, inspection of critical components, bearings, etc., would proceed with a view to carrying out repairs. Less critical repairs would be deferred to a major overhaul of the mill (conducted on an annual, or semi-annual, basis). A similar though entirely separate line of enquiry is being actively pursued at Salford University [34]. On the basis of the results of these researches, it is argued that automated wear debris analysis methods would be well placed to adopt the same approach. As in the case of basic velocity measurement obtained using vibration sensors, monitoring of debris production rates would provide the first indications of changes occurring in the wear state. In each instance, the raw data can be further processed to yield important diagnostic information about the type and location of the fault, from which realistic prognoses and predictions of remaining useful life would be established that are based partly on reliability data and partly on tribological knowledge. In principle, the knowledge and facilities required to achieve the main objectives already exist. It remains for the implementation of automated procedures to bring it to fruition. To achieve reliable methods for predicting the remaining useful life, combined, or integrated, monitoring methods are constantly being developed and evaluated. Maintenance personnel and other specialists have laboured for many years to try and exploit to best advantage the combined monitoring resources at their disposal. This has in turn spawned a number of research investigations, among them being work carried out in the UK funded by the United States Office of Naval Research (ONR) at, respectively, the Universities of Southampton and Swansea [35, 36]. The Southampton research has focussed attention on the role of electrostatic sensors, while Swansea's involvement is related to the further application of acoustic emission sensors. Correlating these respective methods with other techniques, such as vibration and wear debris analysis, has also been undertaken. The possibilities for detecting pending surface distress, *before* it occurs, have been examined with particular reference to contact fatigue behaviour. Based on results obtained from intensive laboratory-based friction and wear testing, there is evidence that it is possible to detect significant changes occurring in the sub-surface region of the contact [37]. In the light of further developments of the measuring and analysing techniques, coupled to rigorous substantiation of the phenomena, there are high expectations that it will have important implications for the future development of on-line sensor technology.

16.6.2 Evaluating the Cost Benefits

Within the broad spectrum of condition monitoring activities, there are a number of case studies cited that purport to demonstrate the cost benefits directly attributable to cost avoidance policies. Some of the instances cited relate to tribological issues, notably bearing applications [38].

However, the considerable difficulties experienced in establishing reliable cost benefits have been highlighted in a detailed investigation of maintenance practice as performed in a pharmaceutical process plant environment, in which the whole spectrum of maintenance strategies are utilized [39]. The problems encountered in providing adequate facilities and organization of inspection and monitoring are exacerbated by the multiplicity of machine types, operating duty requirements and product quality demands. Nevertheless, a viable model has been devised that takes all the numerous, sometimes competing, elements into account to yield an estimate of the benefits (or otherwise) of pursuing a condition-based maintenance programme. Not surprisingly, those functions in which the consequential costs incurred through lost production time and replacement of equipment are of paramount concern. There is no simple formula for determining the benefits in any one sector. Attempts to establish whether the same philosophy and methods could be exploited elsewhere, in this instance a specific sector of the steel industry, were only partially successful, and did not offer much prospect for a generic model that could be applied more widely across industry generally [40]. From a tribological standpoint, rolling element bearings functioning in critical applications is one area that would seem to benefit substantially from the deployment of monitoring technology. It is fortunate that it is also an area where there are very good prospects that reliable and timely data can be obtained to prevent the incidence of catastrophic failure. However, one of the most important remaining challenges is how to reliably reduce the incidence of premature cessation of operations to effect their replacement when it is subsequently shown that a bearing, for instance, could have been operated for much longer. This is an issue that continues to contribute significantly to inflating maintenance costs due to loss of production and consequential additional costs of parts and labour.

16.7 Closure

The tribologist's contribution to effective maintenance of highly stressed, critically operating machinery lies primarily in appropriately communicating his knowledge of the interactions that cause the breakdown of the lubrication system in the contacts of interacting surfaces, and the practical consequences in relation to the ensuing wear behaviour. Sound application of the fundamentals, supported by multi-disciplinary analysis techniques, for determining the cause and effect of surface failures during operation has been demonstrated over many years [41]. Similarly, advances in knowledge of how lubricants and solid materials function have led to the development of improved refining, processing and manufacturing procedures. This has culminated in impressive gains made in their properties, with concomitant gains in field operations through extending the oil change periods and by also achieving enhanced expectations of extended life.

Much of what has been gleaned from fundamental and applied research through the combined efforts of academia and industry has been profitably invested in achieving improved design methods, coupled to parallel improvements in the manufacture and assembly of tribological components, especially in bearing technology.

The advent of condition-based and proactive maintenance strategies has brought new challenges to dealing with the age old issue of component faults developing during operation that are due essentially to surface-related failures, notably in terms of wear phenomena. In responding to these challenges, there is a need for tribologists to get closer to the real-world issues that so plague the maintenance community. Among these, the matter of being able to predict the remaining useful life of failing equipment continues to be the most pressing need.

Acknowledgements

The funding of research at the author's laboratory relating to wear debris analysis, by the following bodies over the past several years, is gratefully acknowledged:

- Ministry of Defence (Procurement Executive); also, the RAF and QinetiQ.
- The Office of Naval Research, in conjunction with the Naval Research Laboratory, Washington, DC.

The private communication received from Paul Howard of Paul Howard Enterprises, New Hampshire, USA [27], was an important contribution to the research programme sponsored by the Office of Naval Research, and this is readily acknowledged.

References

1. Summer-Smith, D. and Neale, M.J., 'Failure patterns and failure analysis' *The Tribology Handbook* (ed M.J. Neale), 2nd edition, Section D1, Butterwoth Heinemann, London, 1995.
2. Rajan, B.S., Cost Benefit Analysis of Condition Monitoring in Batch Process plants, Ph. D. Thesis, University of Wales, 1998.
3. Mowbray, J., *'Reliability-Centred Maintenance'*, Butterworth Heinemann, London, 1991.
4. Dowson, D., *'History of Tribology'*, 2nd edition, Professional Engineering Publishing Ltd, London and Bury St Edmunds, UK, 1998.
5. Neale, M.J., 'Selection of Bearings' *Proceedings of the Institution of Mechanical Engineers*, **182** (Pt 3A), 1967–68, 547–556.
6. Engineering Sciences Data Unit Item No. 84031.
7. Martin, F.A. and Garner, L. 'Design of Plain Bearings. Use of Bearing Data Design charts'. *Tribology Handbook*, (ed M.J. Neale), 2nd edition, Section A9, Butterwoth Heinemann, London, 1995.
8. Landsdown, A.R., 'Selection of Lubricant Type', *The Tribology Handbook* (ed M.J. Neale), 2nd edition, Section C1, Butterwoth Heinemann, London, 1995.
9. Jones, D.B. and Hurricks, P.L., 'Rolling Bearing Materials', *The Tribology Handbook* (ed M.J. Neale), 2nd edition, Section A21, Butterwoth Heinemann, London, 1995.
10. Davies, A. (ed), *Handbook of Condition Monitoring*', Chapman and Hall, London, 1997.
11. Rao, B.K. N., *'Handbook of Condition Monitoring'*, Elsevier Advanced Technology 1996.
12. Jost, H.P., Lubrication (Tribology): Lubrication and Research, Department of Education and Science, HMSO, UK, 1966.
13. Jost, H.P. and Schofield, J., 'Methods by Which Savings of Energy Could be Made Through Improved Tribological Practice in UK industry', *Proceedings of the Insititution of Mechanical Engineers,* **195**, 1981, 151–195.
14. Neale, M.J., A Guide to the Condition Monitoring of Machinery, Department of Industry, Council for Terotechnology, HMSO, London, 1979.
15. Neale, M.J., *'Failures, The Tribology Handbook'*, 2nd edition, Section D, Butterwoth and Heinemann, 1995.
16. Cameron, A. *'Principles of Lubrication'*, Longmans, London, 1966.
17. Tallian, T.E., 'Rolling Contact Failure Control Through Lubrication', Proceedings of the Institution of Mechanical Engineers International Conference on Lubrication and Wear: Fundamentals and Application to Design, 1968, 205–236.
18. Nishida, S.I., *'Failure Analysis in Engineering Applications'*, Butterworth Heinemann, Oxford, 1986.
19. Holroyd, T.J., 'Using Acoustic Emission to Ease the Task of Condition Monitoring', *Proceedings of the International Conference on Condition Monitoring*, (ed. M.J. Jones), Coxmoor Publishing Company, Swansea, 1999, pp. 3–10.
20. Hunter, R.C., Engine Failure Prediction Techniques, Aircraft Engineering, **47**(3), 1975 , pp. 4–14.
21. Seiffert, W.W. and Westcott, V.A., 'A Methods for the study of Wear Particles in Lubricating Oil', *Wear*, **21**, 1972, 27–42.
22. Barwell, F.T., 'Wear of Machine Elements', *Proceedings of the International Conference on Fundamentals of Tribology,* MIT Press, Boston, Massachusetts, 1978, pp. 401–440.
23. Landsdown, A.R. and Price, A.L., *'Materials to Resist Wear'*, Pergammon Press, New York, 1986.
24. Jones, M.H. (ed), Proceedings of the International Conference on Ferrography, University of Wales Swansea, UK, 1982.
25. Ballard, D.H. and Brown, C.M., *'Computer Vision',* Prentice-Hall, Enylewood Cliffs, NJ, 1982.
26. Chapman, R.W., Hodges, D.J. and Nowell, T.J., 'Condition Monitoring – The Support of Aircraft Fleets', *Condition Monitoring – Engineering the Practice* (ed E.D. Yardley), Professional Engineering Publishing, London, 2002.
27. Howard, P.L., private communication.

28. Roylance, B.J., Sperring, T.P. and Barraclough, T.G., Bench Test Determination of Wear Modes to Classify Morphological Attributes of Wear Debris, Bench Testing of Industrial Lubrication and Wear Properties used in Machinery Applications, ASTM: STP1404, 2001, 235–257.
29. Viney, G.A., Sperring, T.P., Jones, M.H. and Roylance, B.J., The Role and Application of Artificial Intelligence Methods for Oil and Wear Debris Analysis in Condition-Based Maintenance Programmes, Proceedings of JOAP International Conference, Mobile, AL, USA, 2000, 237–246.
30. Crow, J. and Greenfield, S., 'The Application of On-Line Wear debris Monitoring', *Proceedings of the International Conference on Condition Monitoring, Swansea* (ed M.H. Jones), Coxmoor Publishing Company, London, 1999, pp. 477–486.
31. Reintjes, J.E. *et al.*, The Application of LaserNet Fines for the Detection of Mechanical wear and Hydraulic Contamination for CBM Systems, ASNE Fleet Maintenance Symposium, San Diego, 2001.
32. Davidson, J.F., 'Reliability of Mechanical Systems', Institution of Mechanical Engineers Guides for the Process Industry (Mechancal Engineering Publications, London), 1988.
33. Goode, K.B., Moore, J and Roylance, B.J., 'Plant Machinery Working Life Prediction Method Utilising Reliability and Condition Monitoring Data', *Proceedings of the Institution of Mechanical Engineers,* **214**, (Pt E), 2000, 109–122.
34. Wang, W. and Sharp, J., 'Modelling Condition-Based Maintenance Decision Support', *Condition Monitoring – Engineering the Practice*, (ed E.D. Yardley), Professional Engineering Publications, London, 2002, pp. 79–98.
35. Powrie, H.E.G., Tazbaz, O.D., Wood, R.J.K. and Fisher, C.E., 'Performance of an Electrostatic Oil Monitoring System During an FZG Gear Scuffing Test', *Proceedings of the International Conference on Condition Monitoring, Swansea* (ed M.H. Jones), Coxmoor Publishing Company, 1999, 155–190.
36. Roylance, B.J., Sperring,T.P., Price, E.D., Lees, A.R., Friswell, M.I., Reintjes, J. and Howard, P., 'Sensor Fusion in the Early Wear Regime for Condition-Based Maintenance' Symposium on Condition-Based Maintenance for Highly Engineered Systems, Pisa, 2000.
37. Price, E.D., High Frequency Techniques for Condition Monitoring Ph.D. Thesis, University of Wales, 2002.
38. Toms, L.A., *'Machinery Oil Analysis'*, 2nd edition, Coastal, 1998.
39. Rajan, B.S. and Roylance, B.J., 'Condition-Based Maintenance: A Systematic Method for Counting the Cost and Assessing the Benefits', *Proceedings of the Institution of Mechanical Engineer,* **214** (Pt. E), 2000, 97–108.
40. van Putten, L.S., 'A Condition Monitoring Justification Model for the Steel Industry', M.Sc. Thesis, University of Wales, 1997.
41. Roylance, B.J., Williams, J.A. and Dwyer-Joyce, R. 'Wear Debris and Associated Wear Phenomena – Fundamental research and practice', *Proceedings of the Institution of Mechanical Engineer*, **214** (Pt. J2000), 79–105.

Index